W9-AMT-618

ADVANCES IN CHEMICAL PHYSICS

VOLUME 109

Advances in
CHEMICAL PHYSICS

Edited by

I. PRIGOGINE

Center for Studies in Statistical Mechanics
and Complex Systems
The University of Texas
Austin, Texas
and
International Solvay Institutes,
Université Libre de Bruxelles
Brussels, Belgium

and

STUART A. RICE

Department of Chemistry
and
The James Franck Institute
The University of Chicago
Chicago, Illinois

VOLUME 109

AN INTERSCIENCE© PUBLICATION
JOHN WILEY & SONS, INC.
NEW YORK · CHICHESTER · WEINHEIM · BRISBANE · SINGAPORE · TORONTO

This book is printed on acid-free paper ♾

Published simultaneously in Canada.

Library of Congress Catalog Number: 58-9935

ISBN 0-471-32920-7

Printed in the United States of America

10 9 8 7 6 5 4 3 2 1

CONTRIBUTORS TO VOLUME 109

BIMAN BAGCHI, Solid State and Structural Chemistry Unit, Indian Institute of Science, Bangalore 560 012, India

RANJIT BISWAS, Solid State and Structural Chemistry Unit, Indian Institute of Science, Bangalore 560 012, India

ARIEL A. CHIALVO, Department of Chemical Engineering, University of Tennessee, Knoxville, TN 37996-2200

J. CRAIN, Department of Physics and Astronomy, The University of Edinburgh, Edinburgh EH9 3JZ, Scotland, United Kingdom

PETER T. CUMMINGS, Chemical Technology Division, Oak Ridge National Laboratory, Oak Ridge, TN 37831-6181

A. DE WIT, Service de Chemie Physique, Centre for Nonlinear Phenomena and Complex Systems, CP 231, Université Libre de Bruxelles, Campus Plaine, 1050 Brussels, Belgium

V. KOMOLKIN, Institute of Physics, Saint Petersburg State University, Saint Petersburg 198904, Russia

K. P. SCAIFE, School of Engineering, Department of Electronic and Electrical Engineering, Trinity College, Dublin 2, Ireland

INTRODUCTION

Few of us can any longer keep up with the flood of scientific literature, even in specialized subfields. Any attempt to do more and be broadly educated with respect to a large domain of science has the appearance of tilting at windmills. Yet the synthesis of ideas drawn from different subjects into new, powerful, general concepts is as valuable as ever, and the desire to remain educated persists in all scientists. This series, Advances in Chemical Physics, is devoted to helping the reader obtain general information about a wide variety of topics in chemical physics, a field that we interpret very broadly. Our intent is to have experts present comprehensive analyses of subjects of interest and to encourage the expression of individual points of view. We hope that this approach to the presentation of an overview of a subject will both stimulate new research and serve as a personalized learning text for beginners in a field.

I. Prigogine
Stuart A. Rice

CONTENTS

ON THE THEORY OF THE COMPLEX, FREQUENCY-DEPENDENT SUSCEPTIBILITY OF MAGNETIC FLUIDS

B. K. P. SCAIFE

School of Engineering, Department of Electronic and Electrical Engineering, Trinity College, Dublin 2, Ireland

CONTENTS

Advances in Chemical Physics, Volume 109, Edited by I. Prigogine and Stuart A. Rice
ISBN 0-471-32920-7

I. INTRODUCTION

Magnetic fluids are stable, colloidal suspensions of ferromagnetic materials [1]. The small size of the roughly spherical particles (diameter on the order of 10 nm) ensures that they each consist of a single magnetic domain. Extensive theoretical and experimental studies have been made of the magnetization of magnetic fluids in alternating magnetic fields of low intensity [2,3].

The behavior of magnetic fluids in such fields is characterized by a dimensionless, isotropic, complex, frequency-dependent susceptibility:

$$\chi_{\mathrm{mag}}(\omega) = \chi'_{\mathrm{mag}}(\omega) - \imath\chi''_{\mathrm{mag}}(\omega) \qquad \imath = \sqrt{-1} \tag{1.1}$$

Thus for a sufficiently small, alternating magnetizing force

$$\mathbf{h}(t) = \mathbb{R}\mathbf{H}(\omega)\exp(\imath\omega t) \tag{1.2}$$

($\mathbb{R}$ denotes "real part of") of frequency $f = \omega/2\pi$ will induce, along the axis of a long, thin, needle-shaped sample, a magnetization

$$\mathbf{p}_{\mathrm{mag}}(t) = \mathbb{R}\mathbf{P}_{\mathrm{mag}}(\omega)\exp(\imath\omega t) \tag{1.3}$$

such that

$$\mathbf{P}_{\mathrm{mag}}(\omega) = \mu_0\chi_{\mathrm{mag}}(\omega)\mathbf{H}(\omega) \tag{1.4}$$

$\mu_0(= 4\pi \times 10^{-7}$ H/m) being the absolute permeability. For simplicity it will be assumed that the colloid particles do not conduct electricity.

The purpose of this review is to derive, in as direct a manner as possible, and discuss an equation to describe the dependence of $\chi_{\mathrm{mag}}(\omega)$ on the angular frequency ω.

In this review we shall assume that all the particles of a magnetic fluid have the same size and are spherical in shape. Therefore, the magnitude of the magnetic moment of a particle of radius a_{p} is

$$m = v_{\mathrm{d}}M_{\mathrm{S}} = \frac{4}{3}\pi a_{\mathrm{p}}^3 M_{\mathrm{S}} \tag{1.5}$$

where M_{S} (Wb m^{-2}) is the saturation magnetization of the material and v_{d} is the magnetic-domain volume, which we shall take to be the same as the actual volume of the particle.

In a uniformly magnetized, crystalline ferromagnetic material the energy associated with the magnetization depends on the direction of the magnetic moment vector **m** with respect to the crystallographic axes. This phenomenon is known as magnetic anisotropy [4, §40]. In a single-domain particle there will be preferred orientations of **m** corresponding to minima of the anisotropy energy. In this article attention is confined to materials with uniaxial magnetic anisotropy; this means that at low temperatures the magnetic moment vector will lie along a particular axis through the particle. Consequently **m** can have two antiparallel equilibrium orientations. At finite temperatures in any particular particle, we must expect that thermal fluctuations will cause **m** to reverse direction in an abrupt and random manner.

If the energy of magnetic anisotropy is large, the direction of **m** with respect to the particle will not change for long periods of time, and we can assume that **m** is fixed rigidly in the particle. In this situation, we speak of the magnetic moment being "blocked," and all changes in orientation of **m** are determined by the rotational motion of the particle.

The calculation of $\chi_{\mathrm{mag}}(\omega)$ for a magnetic colloid in which the magnetic moments are blocked is a simple matter: We simply take over the Debye [5] theory of dielectric polarization in an assembly of weakly interacting, rigid electric dipoles. According to this theory, in which each spherical molecule carries a permanent electric dipole moment μ, the dielectric susceptibility

$$\chi_{\mathrm{elec}}(\omega) = \frac{1}{3}\beta\frac{N\mu^2}{\varepsilon_0 V}\frac{1}{(1 + \imath\omega\tau_{\mathrm{D}})} = \frac{1}{3}\beta\frac{N\mu^2}{\varepsilon_0 V}\Psi_{\mathrm{D}}(\omega, \tau_{\mathrm{D}}) \tag{1.6}$$

where $\beta^{-1} = k_{\mathrm{B}}T$ (k_{B} is Boltzmann's constant, and T is the absolute temperature), N/V is the molecule number density, ε_0 ($\approx 8.854 \times 10^{-12}$ F/m) is the absolute permittivity of free space, and τ_{D} is the relaxation time. Debye, in adapting Einstein's (1926) theory of translational Brownian motion [6] to rotational Brownian motion [7], treated each dipolar molecule as a small sphere, of radius a_μ, performing rotational Brownian motion. The molecular interactions, which give rise to the random changes in orientation, exert a damping effect on the rotational motion of a molecule; this damping may be accounted for in terms of the dynamic viscosity η. Debye showed that

$$\tau_{\mathrm{D}} = 4\pi a_\mu^3 \eta\beta \tag{1.7}$$

Applying the same analysis to a magnetic fluid, with each particle carrying a blocked magnetic moment m, and a particle number density N/V, we find that

$$\chi_{\mathrm{mag}}(\omega) = \frac{1}{3}\beta\frac{Nm^2}{\mu_0 V}\frac{1}{(1 + \imath\omega\tau_{\mathrm{B}})} = \frac{1}{3}\beta\frac{Nm^2}{\mu_0 V}\Psi_{\mathrm{D}}(\omega, \tau_{\mathrm{B}}) \tag{1.8}$$

with

$$\tau_{\mathrm{B}} = 4\pi a_{\mathrm{p}}^3 \eta\beta \tag{1.9}$$

In Eq. (1.9), the thickness of the surfactant coating of the colloidal particles is ignored and the effective hydrodynamic radius is assumed to coincide with the particle radius.

Eq. (1.8) can account quite well for the frequency dependence of $\chi_{\mathrm{mag}}(\omega)$ at low frequencies; however, such a simple theory neglects entirely the internal dynamics of the magnetic moment. When the magnetic moment vector **m** deviates from the magnetic axis of the particle, it is subjected to a mechanical torque that tends to return **m** to its equilibrium orientation; the result of this torque is that **m** undergoes Larmor precession [8, p. 234] about the magnetic axis. In a study of the frequency dependence of the magnetic polarizability of a fixed, single ferromagnetic domain, Landau and Lifshitz [9] proposed the following equation of motion for the magnetic moment **m**:

$$\frac{\mathrm{d}\mathbf{m}(t)}{\mathrm{d}t} = -\mu_0\gamma\left\{\Gamma(t) + \frac{\lambda}{m^2}[\mathbf{m}(t) \times \Gamma(t)]\right\} \tag{1.10}$$

where γ denotes the gyromagnetic ratio, λ is a damping constant, and $\Gamma(t)$ is the total torque exerted on **m** both by magnetic anisotropy and by any external magnetic field. Pending a discussion of Eq. (1.10) in Section III, we remark that for a uniaxial domain, with its axis parallel to the ζ-axis (Fig. 1), and a vanishingly small, external, perpendicular magnetizing force $h_\xi(t)$, we obtain the following equation for $m_\xi(t)$:

$$\tau_{\mathrm{pr}}^2\frac{\mathrm{d}^2 m_\xi(t)}{\mathrm{d}t^2} + 2\tau_{\mathrm{pr}}\frac{\mathrm{d}m_\xi(t)}{\mathrm{d}t} + (1+\omega_{\mathrm{pr}}^2\tau_{\mathrm{pr}}^2)m_\xi(t) = \frac{1}{\kappa}(1+\omega_{\mathrm{pr}}^2\tau_{\mathrm{pr}}^2)h_\xi(t) + \frac{\tau_{\mathrm{pr}}}{\kappa}\frac{\mathrm{d}h_\xi(t)}{\mathrm{d}t} \tag{1.11}$$

The constant κ takes account of the magnetic anisotropy, the characteristic angular frequency

$$\omega_{\mathrm{pr}} = \mu_0\gamma\kappa m \tag{1.12}$$

and the time constant for the damping of precessional motion is

$$\tau_{\mathrm{pr}} = (\mu_0\gamma\lambda\kappa)^{-1} \tag{1.13}$$

For

$$h_\xi(t) = \mathbb{R}H_\xi(\omega)\exp(\imath\omega t) \tag{1.14a}$$

and

$$m_\xi(t) = \mathbb{R}M_\xi(\omega)\exp(\imath\omega t) \tag{1.14b}$$

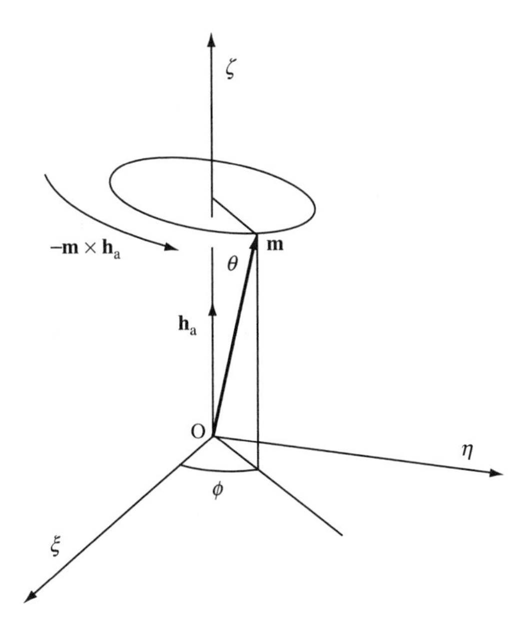

Figure 1. The geometry of the magnetic moment **m** of a spherical, single-domain, uniaxial ferromagnetic particle undergoing Larmor precession about the magnetic axis $O\zeta$. The coordinate system $O\xi\eta\zeta$ is fixed in the particle; the polar coordinates of **m** are θ and ϕ. The effective magnetizing force owing to magnetic anisotropy is $\mathbf{h}_a$ [Eq. (3.7)].

we find, from Eq. (1.11), that the magnetic polarizability

$$\alpha_{\xi,\xi}(\omega) \equiv \frac{M_\xi(\omega)}{H_\xi(\omega)} = \frac{1}{\kappa} \frac{(1 + \omega_{pr}^2 \tau_{pr}^2 + \imath\omega\tau_{pr})}{(1 + \omega_{pr}^2 \tau_{pr}^2 - \omega^2 \tau_{pr}^2 + 2\imath\omega\tau_{pr})} \tag{1.15}$$

Eq. (1.15), first derived by Landau and Lifshitz, describes what is sometimes referred to as ferromagnetic resonance, which we shall refer to as Type II resonance absorption. Although this type of resonance has many similarities with the resonance of a damped harmonic oscillator (which we shall call Type I), there are important differences, which will be discussed in Section II.C.

Unrelated to the work of Landau and Lifshitz were the studies by Van Vleck and Weisskopf [10] and Fröhlich [11,12], which were concerned with attempts to improve on the efforts of Lorentz [13] to take proper account of the effect of collisions on the shape of spectral lines. Our interest in these matters is twofold: first, the equations derived by Van Vleck and Weisskopf and by Fröhlich for the electrical susceptibility of a damped harmonic oscillator coincide with the Landau–Lifshitz relation, Eq. (1.15); second, one of the methods used by Fröhlich has a direct relevance in our derivation of an equation for $\chi_{\text{mag}}(\omega)$.

Our task is to combine the results of Landau and Lifshitz with those of Debye on rotational Brownian motion and, with the minimum of mathematical complexity, to obtain an expression for $\chi_{\text{mag}}(\omega)$. We briefly review and discuss the essential results of Lorentz, of Van Vleck and Weisskopf, and of Fröhlich in Section II and those of Landau and Lifshitz in Section III. In Section IV a short account is given of some important results from the theory of linear systems. Section V lists some relevant topics in the theory of fluctuations, the calculation of the transverse and longitudinal frequency-dependent polarizabilities of a spherical particle is described in Section VI, and the derivation of a formula for $\chi_{\text{mag}}(\omega)$ is given in Section VII.

II. RESONANCE ABSORPTION

A. Lorentz (Type I) Absorption

The study of collision-broadened spectra [14] has been a subject of continuing interest since the pioneering work of Lorentz [13, Note 57] who studied the effect of random collisions on the resonant absorption, in an alternating electric field, of an electric dipole capable, when isolated, of continuous harmonic oscillations with a particular frequency $(\omega_{\text{L}}/2\pi)$. Between collisions the equation of motion of the vibrating charge q of mass m_{q} is

$$m_{\text{q}}\frac{\mathrm{d}^2x(t)}{\mathrm{d}t^2} + kx(t) = qe(t) \tag{2.1}$$

where $x(t)$ is the coordinate of q, $kx(t)$ is the restoring force, and $e(t)$ is the external electric field. If we introduce the electric dipole moment

$$\mu(t) = qx(t) \tag{2.2}$$

we have

$$m_{\text{q}}\frac{\mathrm{d}^2\mu(t)}{\mathrm{d}t^2} + k\mu(t) = q^2e(t) \tag{2.3}$$

Lorentz envisaged a system made up of a large number N of such oscillators (all with their axes parallel to one another) in thermal equilibrium with a heat bath. The thermal agitation interrupts the oscillatory motion of the oscillators; electrical coupling between the oscillators is neglected. The collisions are assumed to be distributed in time according to a

Poisson distribution [15], with time constant τ_{coll}. By statistical arguments, Lorentz showed that the total electric dipole-moment $N\mu(t)$ induced by an external electric field

$$e(t) = \mathbb{R}E(\omega)\exp(\imath\omega t) \tag{2.4}$$

could be expressed by the equation

$$N\mu(t) = N\mathbb{R}\alpha_{\text{L}}(\omega)E(\omega)\exp(\imath\omega t) \tag{2.5}$$

in which the complex electric polarizability

$$\alpha_{\text{L}}(\omega) = \alpha_{\text{L}}(0)\frac{(1+\omega_{\text{L}}^2\tau_{\text{coll}}^2)}{[1+\omega_{\text{L}}^2\tau_{\text{coll}}^2 - \omega^2/\tau_{\text{coll}}^2 + 2\imath\omega\tau_{\text{coll}}]} \tag{2.6a}$$

$$= \alpha_{\text{L}}(0)\Psi_{\text{I}}(\omega,\omega_{\text{L}},\tau_{\text{coll}}) = \alpha_{\text{L}}(0)(\Psi_{\text{I}}' - \imath\Psi_{\text{I}}'') \tag{2.6b}$$

with

$$\alpha_{\text{L}}(0) = \frac{q^2}{k}\frac{(\omega_{\text{L}}\tau_{\text{coll}})^2}{[1+(\omega_{\text{L}}\tau_{\text{coll}})^2]} \tag{2.7}$$

and

$$\omega_{\text{L}} = \sqrt{\frac{k}{m_{\text{q}}}} \tag{2.8}$$

It is most important to appreciate that $\alpha_{\text{L}}(\omega)$ describes the average behavior of a large group of oscillators. An individual oscillator, as described above, does not have a polarizability.

In the limit of very light damping $\omega_{\text{L}}\tau_{\text{coll}} \gg 1$, Eq. (2.6a) reduces to

$$\alpha_{\text{L}}(\omega) \approx \frac{q^2}{k}\frac{\omega_{\text{L}}^2}{\omega_{\text{L}}^2 - \omega^2 + \imath g\omega} \tag{2.9}$$

where $g = (2k/\omega_{\text{L}}\tau_{\text{coll}})$. Lorentz pointed out that one could also derive Eq. (2.9) by assuming that a moving charged particle in the oscillator was continuously subject to viscous damping, so that in place of Eq. (2.3), which holds between collisions, one has the following equation, valid for all times:

$$m_{\text{q}}\frac{\text{d}^2\mu(t)}{\text{d}t^2} + g\frac{\text{d}\mu(t)}{\text{d}t} + k\mu(t) = q^2e(t) \tag{2.10}$$

For convenience we shall describe the resonance described by the complex function $\Psi_{\text{I}}(\omega,\omega_{\text{L}},\tau_{\text{coll}})$ [Eq. (2.6a)], as Type I resonance.

B. Van Vleck–Weisskopf–Fröhlich (Type II) Absorption

Van Vleck and Weisskopf [10], in an attempt to remove from the Lorentz treatment of collision broadening features which they regarded as deficiencies, proposed, in effect, that Eq. (2.6) should be replaced by

$$\alpha_{\rm vvw}(\omega) = \frac{q^2}{k}\frac{(1+\omega_{\rm L}^2\tau_{\rm coll}^2 + \imath\omega\tau_{\rm coll})}{(1+\omega_{\rm L}^2\tau_{\rm coll}^2 - \omega^2\tau_{\rm coll}^2 + 2\imath\omega\tau_{\rm coll})} \tag{2.11a}$$

$$= \alpha_{\rm vvw}(0)\Psi_{\rm II}(\omega, \omega_{\rm L}, \tau_{\rm coll}) = \alpha_{\rm vvw}(0)(\Psi'_{\rm II} - \imath\Psi''_{\rm II}) \tag{2.11b}$$

where

$$\alpha_{\rm vvw}(0) = \frac{q^2}{k} \tag{2.12}$$

The various forms that Eq. (2.11) can take are found in Appendix B.

The difference between the analyses of collision broadening by Lorentz and by Van Vleck and Weisskopf lies in their treatment of the mechanics of the collisions. As we have already mentioned, Fröhlich [11] obtained the same equation; the analysis for the harmonic oscillator given in his book [12, p. 172] is based on the Boltzmann equation. In his first publication [11], he points out that the method was originally developed for the case of a rigid dipole performing angular oscillations about an equilibrium position.

Fröhlich [12, p. 99] assumes the form of a time-dependent decay function and by linear-system theory obtains Eq. (2.1) for $\Psi_{\rm II}(\omega, \omega_{\rm L}, \tau_{\rm coll})$ [see Eq. (4.11)].

C. Comparison of the Two Types of Absorption

From the discussion and figures in Appendix A, it is clear that, near the frequency of maximum absorption, Type I absorption is similar to Type II. At low values of ω it follows from Eqs. (2.6) and (2.11) that

$$\Psi_{\rm I}(\omega, \omega_{\rm L}, \tau_{\rm coll}) \approx \left[1 + \omega^2\tau_{\rm coll}^2\frac{(\omega_{\rm L}^2\tau_{\rm coll}^2 - 3)}{(1+\omega_{\rm L}^2\tau_{\rm coll}^2)^2} + \cdots\right] - \imath\left[2\omega\tau_{\rm coll}\frac{1}{(1+\omega_{\rm L}^2\tau_{\rm coll}^2)} + \cdots\right] \qquad \omega\tau_{\rm coll} \ll 1 \tag{2.13a}$$

$$\Psi_{\mathrm{II}}(\omega, \omega_{\mathrm{L}}, \tau_{\mathrm{coll}}) \approx \left[1 + \omega^2\tau_{\mathrm{coll}}^2 \frac{(\omega_{\mathrm{L}}^2\tau_{\mathrm{coll}}^2 - 1)}{(1 + \omega_{\mathrm{L}}^2\tau_{\mathrm{coll}}^2)^2} + \cdots\right] - \imath\left[\omega\tau_{\mathrm{coll}} \frac{1}{(1 + \omega_{\mathrm{L}}^2\tau_{\mathrm{coll}}^2)} + \cdots\right] \qquad \omega\tau_{\mathrm{coll}} \ll 1 \tag{2.13b}$$

$$\Psi_{\mathrm{D}}(\omega, \tau_{\mathrm{D}}) \approx (1 - \omega^2\tau_{\mathrm{D}}^2 + \omega^4\tau_{\mathrm{D}}^4 - \cdots) - \imath(\omega\tau_{\mathrm{D}} - \omega^3\tau_{\mathrm{D}}^3 + \omega^5\tau_{\mathrm{D}}^5 - \cdots) \qquad \omega\tau_{\mathrm{D}} \ll 1 \tag{2.13c}$$

where, for completeness, we have included the expansion for $\Psi_{\mathrm{D}}(\omega, \tau_{\mathrm{D}})$ from Eq. (1.6). In the limit of large values of ω we find that

$$\Psi_{\mathrm{I}}(\omega, \omega_{\mathrm{L}}, \tau_{\mathrm{coll}}) \cong (1 + \omega_{\mathrm{L}}^2\tau_{\mathrm{coll}}^2)\left\{\left[-\frac{1}{\omega^2\tau_{\mathrm{coll}}^2} - \cdots\right] - \imath\left[\frac{2}{\omega^3\tau_{\mathrm{coll}}^3} + \cdots\right]\right\} \qquad \omega\tau_{\mathrm{coll}} \gg 1 \tag{2.14a}$$

$$\Psi_{\mathrm{II}}(\omega, \omega_{\mathrm{L}}, \tau_{\mathrm{coll}}) \cong \left[\frac{1}{\omega^2\tau_{\mathrm{coll}}^2}(1 - \omega_{\mathrm{L}}^2\tau_{\mathrm{coll}}^2) + \cdots\right] - \imath\left[\frac{1}{\omega\tau_{\mathrm{coll}}} + \cdots\right] \qquad \omega\tau_{\mathrm{coll}} \gg 1 \tag{2.14b}$$

$$\Psi_{\mathrm{D}}(\omega) \cong \left(\frac{1}{\omega^2\tau_{\mathrm{D}}^2} - \frac{1}{\omega^4\tau_{\mathrm{D}}^4} + \frac{1}{\omega^6\tau_{\mathrm{D}}^6} - \cdots\right) - \imath\left(\frac{1}{\omega\tau_{\mathrm{D}}} - \frac{1}{\omega^3\tau_{\mathrm{D}}^3} + \frac{1}{\omega^5\tau_{\mathrm{D}}^5} - \cdots\right) \qquad \omega\tau_{\mathrm{D}} \gg 1 \tag{2.14c}$$

It is clear from these results that it is at very high frequencies that the two types of resonance differ. For example, when ω is large, the real part of Ψ_{I} is proportional to ω^{-2} and is negative; whereas although the real part of Ψ_{II} is also proportional to ω^{-2}, it is negative only if $\omega_{\mathrm{L}}\tau_{\mathrm{coll}} > 1$. The imaginary parts of Ψ_{I} and Ψ_{II} have the same sign, but while Ψ_{I}'' is proportional to ω^{-3}, Ψ_{II}'' is proportional to ω^{-1}. It is this last difference that is at the heart of the physical difference between Type I and Type II resonance absorptions. As will be discussed in more detail when we deal with the spectra of polarization fluctuations in Section V, the fact that the frequency dependence of $\Psi_{\mathrm{II}}''(\propto\omega^{-1})$ falls off so much more slowly than that of $\Psi_{\mathrm{I}}''(\propto\omega^{-3})$ stems from the different treatments of the collision processes by Lorentz (Type I), on the one hand, and by Van Vleck, Weisskopf, and Fröhlich (Type II), on the other. In Type II collision broadening, the particle position can change abruptly; in Type I collision broadening, only the velocity changes abruptly. Since abrupt changes in position can occur only in the absence of inertia, Type I collision broadening is preferred as being more in harmony with basic principles.

Last, we draw attention to the fact that when the characteristic frequency $\omega_{\rm L} = 0$, Type II absorption reduces to Debye nonresonant absorption, i.e.,

$$\Psi_{\rm II}(\omega, 0, \tau_{\rm coll}) = \Psi_{\rm D}(\omega, \tau_{\rm coll}) \tag{2.15}$$

Since it is well known that the Debye theory of dielectric absorption does not take full account of the effects of inertia [16], it is not surprising that Type II absorption should reduce to Debye-type absorption. Notice that in Type I absorption, when $\omega_{\rm L} = 0$, there is still resonance—formally at least—because Eq. (2.6a) leads to the relation:

$$\Psi_{\rm I}(\omega, 0, \tau_{\rm coll}) = \frac{1}{1 - \omega^2\tau_{\rm coll}^2 + \imath 2\omega\tau_{\rm coll}} \tag{2.16}$$

which still has the same frequency behavior as the full Lorentz equation, except that now the characteristic angular frequency $1/\tau_{\rm coll}$ is determined by the frequency of the collisions.

III. FERROMAGNETIC RESONANCE: THEORY OF LANDAU AND LIFSHITZ

The equation of undamped motion of the magnetic moment $\mathbf{m}(t)$ of a spherical, single-domain ferromagnet is [17; 4, p. 270]

$$\dot{\mathbf{m}}(t) = \frac{\mathrm{d}\mathbf{m}(t)}{\mathrm{d}t} = -\mu_0\gamma\mathbf{m}(t) \times \mathbf{h}_{\rm a} \tag{3.1}$$

in which the gyromagnetic factor is denoted by

$$\gamma = g\frac{|e|}{m_{\rm e}} = \frac{1.602177330 \times 10^{-19}\ \mathrm{C}}{9.109389700 \times 10^{-31}\ \mathrm{kg}} = 1.758819617 \times 10^{11}\ \mathrm{T}^{-1}\ \mathrm{s}^{-1} \tag{3.2}$$

where g, the Landé splitting factor for electrons, has been set equal to 2, and thus $\mu_0\gamma = 2.210197915 \times 10^5$ m/As, $\mathbf{h}_{\rm a}$ is an effective magnetizing force caused by magnetic anisotropy [18]. For uniaxial magnetic anisotropy (Fig. 1),

$$\mathbf{h}_{\rm a} = \hat{\zeta}|\mathbf{h}_{\rm a}| = \hat{\zeta}h_{\rm a} \tag{3.3}$$

where $\hat{\zeta}$ is a unit vector along the Oζ axis fixed in the particle. The magnitude of $\mathbf{h}_{\rm a}$ is determined by equating the torque that such a magnetizing force would exert on $\mathbf{m}$, namely $\mathbf{m} \times \mathbf{h}_{\rm a}$, with the actual torque being created by magnetic anisotropy. Thus

$$\mathbf{m} \times \mathbf{h}_\mathrm{a} = -\hat{\mathbf{n}} \frac{\partial U(\theta)}{\partial \theta} \tag{3.4}$$

where

$$\begin{aligned} U(\theta) &= -\frac{1}{2}\kappa m_\zeta^2 \\ &= -\frac{1}{2}\kappa m^2 \cos^2\theta = -v_\mathrm{d} K_\mathrm{u} \cos^2\theta = -\frac{1}{2} v_\mathrm{d} K_\mathrm{u}(1 + \cos 2\theta) \end{aligned} \tag{3.5}$$

is the anisotropy energy of the domain when $\mathbf{m}$ makes an angle θ with the symmetry axis Oζ, K_u is the magnetic-anisotropy constant, and $\hat{\mathbf{n}}$ is a unit vector parallel to the torque $\mathbf{m} \times \mathbf{h}_\mathrm{a}$ (Fig. 2). Consequently,

$$mh_\mathrm{a} \sin\theta = 2K_\mathrm{u} v_\mathrm{d} \cos\theta \sin\theta \tag{3.6}$$

or

$$\mathbf{h}_\mathrm{a} = \hat{\boldsymbol{\zeta}} \frac{2K_\mathrm{u}}{M_\mathrm{S}} \cos\theta = \hat{\boldsymbol{\zeta}} \kappa m \cos\theta \tag{3.7}$$

Eq. (3.1), which may now be written in the form

$$\dot{\mathbf{m}}(t) = \frac{\mathrm{d}\mathbf{m}(t)}{\mathrm{d}t} = \omega_\mathrm{pr} \cos\theta \hat{\boldsymbol{\zeta}} \times \mathbf{m}(t) \tag{3.8}$$

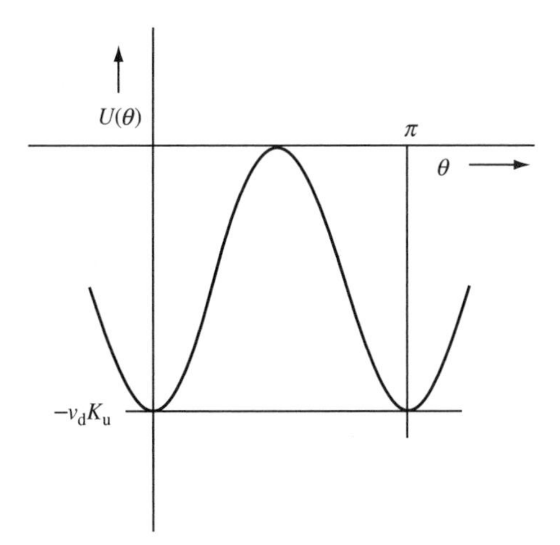

Figure 2. The variation with polar angle θ of the uniaxial magnetic-anisotropy energy $U(\theta) = -v_\mathrm{d} K_\mathrm{u} \cos^2\theta$ [Eq. (3.5)]. There are two energy minima: one at $\theta = 0$ and one at $\theta = \pi$.

indicates that when the direction of $\mathbf{m}$ deviates from the Oζ-axis, $\mathbf{m}$ precesses about this axis with Larmor angular velocity

$$\omega_{\text{pr}} \cos\theta = \hat{\zeta}\omega_{\text{pr}} \cos\theta \tag{3.9a}$$

with

$$\omega_{\text{pr}} = \mu_0 \gamma \kappa m = \mu_0 \gamma \frac{2K_{\text{u}}}{M_{\text{S}}} \tag{3.9b}$$

which is independent of the size of the spherical particle. For the materials of interest here [18–20], $K_{\text{u}} \approx 10^5$ J/m^3 and $M_{\text{S}} \approx 1$ Wb/m^2, so that $\omega_{\text{pr}}/2\pi \approx 10$ GHz.

It is clear from Figure 2, and from Eq. (3.5) that there are two antiparallel equilibrium orientations for $\mathbf{m}$. Therefore, at finite temperatures it is to be expected that $\mathbf{m}$ will reverse direction in a random manner and at a rate determined by the height $K_{\text{u}} v_{\text{d}}$ of the energy barrier that separates the two orientations. The significance of these fluctuations is that the domain will have a finite magnetic susceptibility along the Oζ-axis. This fact will be considered in more detail in Section VI.

Any uniformly magnetized body experiences a demagnetizing force arising from the uncompensated magnetic poles on the surface of the body. In the special case of a sphere, the demagnetizing force is

$$-\frac{1}{3\mu_0}\frac{\mathbf{m}}{v_{\text{d}}}.$$

Such a demagnetizing force cannot exert a torque on $\mathbf{m}$, since $\mathbf{m} \times \mathbf{m} = 0$. In the case of an ellipsoid, however, the demagnetizing force is no longer antiparallel to the magnetic moment and will, therefore, exert a torque on $\mathbf{m}$ [21,22].

Notice that it follows from Eq. (3.1) that the magnitude of $\mathbf{m}$ is a constant because

$$m\frac{\mathrm{d}m}{\mathrm{d}t} = \mathbf{m} \cdot \dot{\mathbf{m}} = -\mu_0 \gamma \mathbf{m} \cdot (\mathbf{m} \times \mathbf{h}_{\mathbf{a}}) \equiv 0 \tag{3.10}$$

The presence of a small, time-dependent external magnetizing force $\mathbf{h}(t)$ may be taken into account by adding $\mathbf{h}(t)$ to $\mathbf{h}_{\text{a}}$ in Eq. (3.1) with the result that

$$\dot{\mathbf{m}}(t) = \frac{\mathrm{d}\mathbf{m}(t)}{\mathrm{d}t} = \mu_0 \gamma \mathbf{f}(t) \times \mathbf{m}(t) \tag{3.11}$$

with

$$\mathbf{f}(t) = \mathbf{h}_{\text{a}} + \mathbf{h}(t) \tag{3.12}$$

Thermal agitation will tend to damp out any precessional motion. Landau and Lifshitz [9] proposed to take account of such damping by the empirical equation:

$$\dot{\mathbf{m}}(t) = \frac{\mathrm{d}\mathbf{m}(t)}{\mathrm{d}t} = \mu_0\gamma\mathbf{f}(t) \times \mathbf{m}(t) - \mu_0\gamma\frac{\lambda}{m^2}\mathbf{m}(t) \times [\mathbf{m}(t) \times \mathbf{f}(t)] \tag{3.13}$$

The term in Eq. (3.13) with the damping constant λ is so arranged as to be perpendicular to $\mathbf{m}$ and to the vector $\mathbf{m} \times \mathbf{f}$. Notice that the condition $\mathbf{m} \cdot \dot{\mathbf{m}} = 0$ holds for Eq. (3.13) and, therefore

$$\mathbf{m} \times (\mathbf{m} \times \dot{\mathbf{m}}) = \mathbf{m}(\mathbf{m} \cdot \dot{\mathbf{m}}) - m^2\dot{\mathbf{m}} = -m^2\dot{\mathbf{m}} \tag{3.14}$$

By means of repeated use of this result Gilbert and Kelly [38] have shown that, provided $\lambda < m$, the Landau–Lifshitz Eq. (3.13) may be transformed to the form

$$\frac{\mathrm{d}\mathbf{m}(t)}{\mathrm{d}t} = \mu_0\gamma\left(1 + \frac{\lambda^2}{m^2}\right)\mathbf{f}(t) \times \mathbf{m}(t) + \mu_0\gamma\frac{\lambda}{m^2}\mathbf{m}(t) \times \frac{\mathrm{d}\mathbf{m}(t)}{\mathrm{d}t} \qquad \lambda < m \tag{3.15}$$

On the basis of this result, Gilbert and Kelly proposed the following equation:

$$\frac{\mathrm{d}\mathbf{m}(t)}{\mathrm{d}t} = \mu_0\gamma\mathbf{f}(t) \times \mathbf{m}(t) + A\mathbf{m}(t) \times \frac{\mathrm{d}\mathbf{m}(t)}{\mathrm{d}t} \tag{3.16}$$

in which A is an empirical damping constant.

We shall confine our attention to solving Eq. (3.13) for the case of a small alternating external magnetizing force transverse to the magnetic axis of the particle. With this in mind we shall write

$$\mathbf{f}(t) = \mathbf{h}_\mathrm{a} + \mathbf{h}(t) = \hat{\zeta}h_\mathrm{a} + \hat{\xi}h_\xi(t) = \hat{\zeta}\kappa m_\zeta(t) + \hat{\xi}h_\xi(t) \tag{3.17}$$

with which we obtain from Eq. (3.13):

$$\dot{m}_\xi(t) = \mu_0\gamma\left\{-\kappa m_\zeta(t)m_\eta(t) + \lambda h_\xi(t) - \frac{\lambda}{m^2}[h_\xi(t)m_\xi(t) + \kappa m_\zeta^2(t)]m_\xi(t)\right\} \tag{3.18a}$$

$$\dot{m}_\eta(t) = \mu_0\gamma\left\{\kappa m_\zeta(t)m_\xi(t) - h_\xi(t)m_\zeta(t) - \frac{\lambda}{m^2}[h_\xi(t)m_\xi(t) + \kappa m_\zeta^2(t)]m_\eta(t)\right\} \tag{3.18b}$$

$$\dot{m}_\zeta(t) = \mu_0\gamma\left\{h_\xi(t)m_\eta(t) + \lambda\kappa m_\zeta(t) - \frac{\lambda}{m^2}[h_\xi(t)m_\xi(t) + \kappa m_\zeta^2(t)]m_\zeta(t)\right\} \tag{3.18c}$$

As already noted, the basic equation of motion proposed by Landau and Lifshitz ensures that $\mathbf{m}(t)$ is a vector of constant magnitude. Consequently, since

$$m_\zeta(t) = m\sqrt{1 - \frac{m_\xi^2(t)}{m^2} - \frac{m_\eta^2(t)}{m^2}} \tag{3.19}$$

and if $m_\xi(t)$ and $m_\eta(t)$ are small quantities, we may use the approximation:

$$m_\zeta(t) \approx m\left[1 - \frac{m_\xi^2(t)}{2m^2} - \frac{m_\eta^2(t)}{2m^2}\right] \tag{3.20}$$

The damping ensures that, in the absence of external fields, the vector **m** lies parallel to the magnetic Oζ-axis (Fig. 1). Thus to first order in $h_\xi(t)$, Eq. (3.18) reduces to:

$$\dot{m}_\xi(t) = \mu_0\gamma[-\kappa m m_\eta(t) + \lambda h_\xi(t) - \lambda\kappa m_\xi(t)] \tag{3.21a}$$
$$\dot{m}_\eta(t) = \mu_0\gamma[\kappa m m_\xi(t) - m h_\xi(t) - \lambda\kappa m_\eta(t)] \tag{3.21b}$$
$$\dot{m}_\zeta(t) = 0 \tag{3.21c}$$

Introducing the precession relaxation time $\tau_{\rm pr}$ defined by the equation

$$\tau_{\rm pr} = (\mu_0\gamma\lambda\kappa)^{-1} \tag{3.22}$$

and using the characteristic angular frequency $\omega_{\rm pr}$, defined by Eq. (3.9b), we may write Eq. (3.21) in the form:

$$\dot{m}_\xi(t) = \left[-\omega_{\rm pr} m_\eta(t) + \frac{1}{\kappa\tau_{\rm pr}} h_\xi(t) - \frac{1}{\tau_{\rm pr}} m_\xi(t)\right] \tag{3.23a}$$
$$\dot{m}_\eta(t) = \left[\omega_{\rm pr} m_\xi(t) - \frac{\omega_{\rm pr}}{\kappa} h_\xi(t) - \frac{1}{\tau_{\rm pr}} m_\eta(t)\right] \tag{3.23b}$$
$$\dot{m}_\zeta(t) = 0 \tag{3.23c}$$

If we eliminate $m_\eta(t)$ from these equations we obtain Eq. (1.11).

Suppose that

$$\mathbf{h}(t) = \hat{\xi} h_\xi(t) = \mathbb{R}\hat{\boldsymbol{\xi}} H_\xi(\omega)\exp(\imath\omega t) \tag{3.24}$$

and writing

$$\begin{aligned}\mathbf{m}(t) &= \hat{\boldsymbol{\xi}} m_\xi(t) + \hat{\boldsymbol{\eta}} m_\eta(t) + \hat{\boldsymbol{\zeta}} m_\zeta(t) \\ &= \mathbb{R}[\hat{\boldsymbol{\xi}} M_\xi(\omega) + \hat{\boldsymbol{\eta}} M_\eta(\omega) + \hat{\boldsymbol{\zeta}} M_\zeta(\omega)]\exp(\imath\omega t)\end{aligned} \tag{3.25}$$

we find that

$$\mathbf{m}(t) = \mathbb{R}[\hat{\boldsymbol{\xi}}\alpha_{\xi,\xi}(\omega) + \hat{\boldsymbol{\eta}}\alpha_{\eta,\xi}(\omega) + \hat{\boldsymbol{\zeta}}\alpha_{\zeta,\xi}(\omega)]H_\xi(\omega)\exp(\imath\omega t) \tag{3.26}$$

in which the complex magnetic polarizabilities are

$$\alpha_{\xi,\xi}(\omega) = \frac{M_\xi(\omega)}{H_\xi(\omega)} = \frac{1}{\kappa}\frac{1 + (\omega_{\rm pr}\tau_{\rm pr})^2 + \imath\omega\tau_{\rm pr}}{1 + (\omega_{\rm pr}\tau_{\rm pr})^2 - (\omega\tau_{\rm pr})^2 + \imath 2\omega\tau_{\rm pr}} \tag{3.27a}$$

$$\alpha_{\eta,\xi}(\omega) = \frac{M_\eta(\omega)}{H_\xi(\omega)} = \frac{1}{\kappa}\frac{-\imath\omega\omega_{\rm pr}\tau_{\rm pr}^2}{1 + (\omega_{\rm pr}\tau_{\rm pr})^2 - (\omega\tau_{\rm pr})^2 + \imath 2\omega\tau_{\rm pr}} \tag{3.27b}$$

$$\alpha_{\zeta,\xi}(\omega) = 0 \tag{3.27c}$$

These equations were first obtained by Landau and Lifshitz (1935). By changing from $\mathbf{h}(t) = \hat{\boldsymbol{\xi}} h_\xi(t)$ in Eq. (3.24) to $\mathbf{h}(t) = \hat{\boldsymbol{\eta}} h_\eta(t)$, it may also be shown that [23–26]

$$\alpha_{\xi,\eta}(\omega) = -\alpha_{\eta,\xi}(\omega) \tag{3.28a}$$

and that

$$\alpha_{\zeta,\xi}(\omega) = \alpha_{\xi,\zeta}(\omega) = \alpha_{\eta,\zeta}(\omega) = \alpha_{\zeta,\eta}(\omega) = 0 \tag{3.28b}$$

also, by symmetry,

$$\alpha_{\eta,\eta}(\omega) = \alpha_{\xi,\xi}(\omega) \tag{3.28c}$$

According to Eq. (3.21c), $\alpha_{\zeta,\zeta}(\omega)$ vanishes; we postpone a discussion of $\alpha_{\zeta,\zeta}(\omega)$ until Section VI.

An interesting result is obtained when the external magnetizing force is a rotating field of constant magnitude; consider, therefore, that in place of Eq. (3.24) we have

$$\mathbf{h}^{\pm}(t) = H(\omega)[\hat{\boldsymbol{\xi}}\cos\omega t \pm \hat{\boldsymbol{\eta}}\sin\omega t] = \mathbb{R}H(\omega)(\hat{\boldsymbol{\xi}} \mp \imath\hat{\boldsymbol{\eta}})\exp(\imath\omega t) \tag{3.29}$$

The $^+$ superscript indicates a right-handed rotation about the Oζ-axis, and the $^-$ superscript a left-handed rotation. From Eqs. (3.27) and (3.29) we find that

$$\begin{aligned} M_\xi^{\pm}(\omega) &= \frac{H(\omega)}{\kappa}[\alpha_{\xi,\xi}(\omega) \mp \imath\alpha_{\xi,\eta}(\omega)] \\ &= \frac{H(\omega)}{\kappa}\left\{\frac{[1 + (\omega_{\rm pr}\tau_{\rm pr})^2 + \imath\omega\tau_{\rm pr}] \mp \imath[+\imath\omega\omega_{\rm pr}\tau_{\rm pr}^2]}{1 + (\omega_{\rm pr}\tau_{\rm pr})^2 - (\omega\tau_{\rm pr})^2 + \imath 2\omega\tau_{\rm pr}}\right\} \\ &= \frac{H(\omega)}{\kappa}\left\{\frac{(1 + \imath\omega_{\rm pr}\tau_{\rm pr})(1 - \imath\omega_{\rm pr}\tau_{\rm pr}) + \imath\omega\tau_{\rm pr}(1 \mp \imath\omega_{\rm pr}\tau_{\rm pr})}{[1+\imath(\omega + \omega_{\rm pr})\tau_{\rm pr}][1 + \imath(\omega - \omega_{\rm pr})\tau_{\rm pr}]}\right\} \\ &= \frac{H(\omega)}{\kappa}\frac{1 \mp \imath\omega_{\rm pr}\tau_{\rm pr}}{1 + \imath(\omega \mp \omega_{\rm pr})\tau_{\rm pr}} = \pm\imath M_\eta^{\pm}(\omega) \end{aligned} \tag{3.30}$$

With the notation:

$$\frac{M_{\xi}^{\pm}(\omega)}{H(\omega)} = \pm\imath\frac{M_{\eta}^{\pm}(\omega)}{H(\omega)} = \alpha^{\pm}(\omega) = \frac{1}{\kappa}\Psi_{\mathrm{II}}^{\pm}(\omega, \omega_{\mathrm{pr}}, \tau_{\mathrm{pr}}) \tag{3.31}$$

it follows that

$$\alpha_{\xi,\xi}(\omega) = \frac{1}{2}[\alpha^{+}(\omega) + \alpha^{-}(\omega)] = \frac{1}{2\kappa}(\Psi_{\mathrm{II}}^{+} + \Psi_{\mathrm{II}}^{-}) = \frac{1}{2\kappa}\Psi_{\mathrm{II}}(\omega, \omega_{\mathrm{pr}}, \tau_{\mathrm{pr}}) \tag{3.32a}$$

$$\alpha_{\xi,\eta}(\omega) = -\alpha_{\eta,\xi}(\omega) = \imath\frac{1}{2}[\alpha^{+}(\omega) - \alpha^{-}(\omega)] = \imath\frac{1}{2\kappa}(\Psi_{\mathrm{II}}^{+} - \Psi_{\mathrm{II}}^{-}) \tag{3.32b}$$

It is sometimes convenient to introduce the dimensionless parameter σ, which is the ratio of the barrier height to the thermal energy, i.e.,

$$\sigma \equiv \frac{K_{\mathrm{u}}v_{\mathrm{d}}}{k_{\mathrm{B}}T} = \beta K_{\mathrm{u}}v_{\mathrm{d}} = \frac{1}{2}\beta\kappa m^2 \tag{3.33a}$$

so that

$$\frac{1}{\kappa} = \frac{1}{2\sigma}\beta m^2 = \frac{1}{2\sigma}\frac{m^2}{k_{\mathrm{B}}T} \tag{3.33b}$$

and hence Eq. (3.27a) can be written

$$\alpha_{\xi,\xi}(\omega) = \frac{1}{2\sigma}\frac{m^2}{k_{\mathrm{B}}T}\frac{1 + (\omega_{\mathrm{pr}}\tau_{\mathrm{pr}})^2 + \imath\omega\tau_{\mathrm{pr}}}{1 + (\omega_{\mathrm{pr}}\tau_{\mathrm{pr}})^2 - (\omega\tau_{\mathrm{pr}})^2 + \imath 2\omega\tau_{\mathrm{pr}}} \tag{3.34}$$

IV. SOME RESULTS FROM THE THEORY OF LINEAR SYSTEMS

According to linear system theory [16,27], to each of the normalized polarizability functions—$\Psi_{\mathrm{D}}(\omega, \tau_{\mathrm{D}})$, [Eq. (1.6)], $\Psi_{\mathrm{I}}(\omega, \omega_{\mathrm{L}}, \tau_{\mathrm{coll}})$ [Eq. (2.6)], and $\Psi_{\mathrm{II}}(\omega, \omega_{\mathrm{pr}}, \tau_{\mathrm{pr}})$ [Eq. (2.11)]—there corresponds a function of time, which is the response to a large, but very brief, pulse. More precisely, this function of time is the pulse, or delta-function, response. The pulse response is the time derivative of the unit step-function response.

We shall denote the pulse response corresponding to the dispersion, or frequency response, function $\Psi(\omega)$ by $\psi(t)$. For the moment we shall use a simplified notation that omits the subscripts (D, I, and II) and that omits reference to the constants τ_{D}, ω_{L}, τ_{coll}, ω_{pr}, and τ_{pr}). The pulse response and the dispersion function form a Fourier transform pair, i.e.,

$$\Psi(\omega) = \int_0^{\infty} \mathrm{d}t\, \psi(t) \exp(-\imath\omega t) \tag{4.1a}$$

$$\psi(t) = \frac{1}{2\pi}\int_{-\infty}^{\infty} \mathrm{d}\omega\, \Psi(\omega) \exp(\imath\omega t) \tag{4.1b}$$

The lower limit in the first integral is set to zero, because to satisfy causality, $\psi(t)$ must vanish for $t < 0$. We shall denote the unit step-function response by $s(t)$, which is related to $\psi(t)$ as follows:

$$\psi(t) = \frac{\mathrm{d}s(t)}{\mathrm{d}t} \tag{4.2}$$

Consequently,

$$\Psi(\omega) = \int_0^{\infty} \mathrm{d}t \frac{\mathrm{d}s(t)}{\mathrm{d}t} \exp(-\imath\omega t) \tag{4.3}$$

When we come to discuss fluctuation theory in Section V we shall need expressions for the decay, or aftereffect, function. The decay function, denoted by $b(t)$, is related to the unit step-function by the relation:

$$b(t) = s(\infty) - s(t) \qquad (t \geq 0) \tag{4.4}$$

From Eqs. (4.1) and (4.2), we deduce that

$$\Psi(0) = \int_0^{\infty} \mathrm{d}t\, \psi(t) = \int_0^{\infty} \mathrm{d}t \frac{\mathrm{d}s(t)}{\mathrm{d}t} = s(\infty) - s(0) = b(0) \tag{4.5}$$

Because no physical system can respond instantly to an external stimulus,

$$s(0) = 0 \tag{4.6a}$$

and since, by definition, $\Psi(0) = 1$ we see that

$$b(0) = 1 \tag{4.6b}$$

Hence we see from Eq. (4.5) that

$$s(\infty) = 1 \tag{4.7}$$

Using Eq. (4.4) and carrying out an integration by parts in Eq. (4.3), we obtain

$$\begin{aligned}\Psi(\omega) &= \int_0^\infty \mathrm{d}t \frac{\mathrm{d}[-b(t)]}{\mathrm{d}t} \exp(-\imath\omega t) \\ &= [-b(t)]_0^\infty - \imath\omega \int_0^\infty \mathrm{d}t\, b(t) \exp(-\imath\omega t) \\ &= 1 - \imath\omega \int_0^\infty \mathrm{d}t\, b(t) \exp(-\imath\omega t) \\ &= \left[1 - \omega \int_0^\infty \mathrm{d}t\, b(t) \sin\omega t\right] - \imath\left[\omega \int_0^\infty \mathrm{d}t\, b(t) \cos\omega t\right]\end{aligned} \tag{4.8}$$

We now list expressions for the step, aftereffect, and pulse functions for the three types of absorption considered in this Chapter.

Nonresonant Absorption [Eq. (1.6)]

$$s_\mathrm{D}(t, \tau_\mathrm{D}) = u(t) - b_\mathrm{D}(t, \tau_\mathrm{D}) = \left[1 - \exp\left(-\frac{t}{\tau_\mathrm{D}}\right)\right] u(t) \tag{4.9a}$$

$$\approx \frac{t}{\tau_\mathrm{D}} - \frac{1}{2}\left(\frac{t}{\tau_\mathrm{D}}\right)^2 + \cdots \quad \left(\frac{t}{\tau_\mathrm{D}}\right) \ll 1 \tag{4.9b}$$

$$b_\mathrm{D}(t, \tau_\mathrm{D}) = \exp\left(-\frac{t}{\tau_\mathrm{D}}\right) u(t) \tag{4.9c}$$

$$\psi_\mathrm{D}(t, \tau_\mathrm{D}) = \frac{\mathrm{d}s_\mathrm{D}(t, \tau_\mathrm{D})}{\mathrm{d}t} = \frac{1}{\tau_\mathrm{D}} \exp\left(-\frac{t}{\tau_\mathrm{D}}\right) u(t) \tag{4.9d}$$

Type I Resonance Absorption [Eq. (2.6)]

$$s_\mathrm{I}(t, \omega_\mathrm{L}, \tau_\mathrm{coll}) = u(t) - b_\mathrm{I}(t, \omega_\mathrm{L}, \tau_\mathrm{coll})$$

$$= \left\{1 - \left[\cos\omega_\mathrm{L} t + \frac{1}{\omega_\mathrm{L}\tau_\mathrm{coll}} \sin\omega_\mathrm{L} t\right] \exp\left(-\frac{t}{\tau_\mathrm{coll}}\right)\right\} u(t) \tag{4.10a}$$

$$\approx [1 + (\omega_\mathrm{L}\tau_\mathrm{coll})^2]\left[\frac{1}{2}\left(\frac{t}{\tau_\mathrm{coll}}\right)^2 - \frac{1}{3}\left(\frac{t}{\tau_\mathrm{coll}}\right)^3 \cdots\right] \left(\frac{t}{\tau_\mathrm{coll}}\right) \ll 1 \tag{4.10b}$$

$$b_\mathrm{I}(t, \omega_\mathrm{L}, \tau_\mathrm{coll}) = \left[\cos\omega_\mathrm{L} t + \frac{1}{\omega_\mathrm{L}\tau_\mathrm{coll}} \sin\omega_\mathrm{L} t\right] \exp\left(-\frac{t}{\tau_\mathrm{coll}}\right) u(t) \tag{4.10c}$$

$$\psi_{\mathrm{I}}(t, \omega_{\mathrm{L}}, \tau_{\mathrm{coll}}) = \frac{\mathrm{d}b_{\mathrm{I}}(t, \omega_{\mathrm{L}}, \tau_{\mathrm{coll}})}{\mathrm{d}t}$$
$$= \frac{[1 + (\omega_{\mathrm{L}}\tau_{\mathrm{coll}})^2]}{\tau_{\mathrm{coll}}^2} \exp\left(-\frac{t}{\tau_{\mathrm{coll}}}\right) \frac{\sin \omega_{\mathrm{L}} t}{\omega_{\mathrm{L}}} u(t) \tag{4.10d}$$

Type II Resonance Absorption [Eqs. (1.15) and (2.11)]

$$s_{\mathrm{II}}(t, \omega_{\mathrm{pr}}, \tau_{\mathrm{pr}}) = u(t) - b_{\mathrm{II}}(t, \omega_{\mathrm{pr}}, \tau_{\mathrm{pr}})$$
$$= \left[1 - \exp\left(-\frac{t}{\tau_{\mathrm{pr}}}\right) \cos \omega_{\mathrm{pr}} t\right] u(t) \tag{4.11a}$$
$$\approx \left(\frac{t}{\tau_{\mathrm{pr}}}\right) + [(\omega_{\mathrm{pr}}\tau_{\mathrm{pr}})^2 - 1]\frac{1}{2}\left(\frac{t}{\tau_{\mathrm{pr}}}\right)^2 \ldots \left(\frac{t}{\tau_{\mathrm{pr}}}\right) \ll 1 \tag{4.11b}$$
$$b_{\mathrm{II}}(t, \omega_{\mathrm{pr}}, \tau_{\mathrm{pr}}) = \left[\exp\left(-\frac{t}{\tau_{\mathrm{pr}}}\right) \cos \omega_{\mathrm{pr}} t\right] u(t) \tag{4.11c}$$
$$\psi_{\mathrm{II}}(t, \omega_{\mathrm{pr}}, \tau_{\mathrm{pr}}) = \frac{\mathrm{d}s_{\mathrm{II}}(t, \omega_{\mathrm{pr}}, \tau_{\mathrm{pr}})}{\mathrm{d}t}$$
$$= \frac{1}{\tau_{\mathrm{pr}}} \exp\left(-\frac{t}{\tau_{\mathrm{pr}}}\right) [\cos \omega_{\mathrm{pr}} t + (\omega_{\mathrm{pr}}\tau_{\mathrm{pr}}) \sin \omega_{\mathrm{pr}} t] u(t) \tag{4.11d}$$

These results have been obtained by standard Fourier transform techniques. Eq. (4.11d) was first given by Fröhlich [12]. The unit step-function $u(t)$ is defined by

$$u(t) = \begin{cases} 0 & t < 0 \\ 1 & t > 0 \end{cases} \tag{4.12}$$

Notice that for small values of t both $s_{\mathrm{D}}(t)$ and $s_{\mathrm{II}}(t)$ are proportional to t, whereas $s_{\mathrm{I}}(t)$ is proportional to t^2. This reflects the behavior at large values of ω, of $\Psi_{\mathrm{D}}(\omega, \tau_{\mathrm{D}})$, $\Psi_{\mathrm{II}}(\omega, \omega_{\mathrm{pr}}, \tau_{\mathrm{pr}})$, and $\Psi_{\mathrm{I}}(\omega, \omega_{\mathrm{L}}, \tau_{\mathrm{coll}})$, respectively (see Appendix A).

V. SOME RESULTS FROM FLUCTUATION THEORY

A. Basic Relations

The autocorrelation function $\rho_x(t)$ for some fluctuating physical variable $x(t)$ is defined [e.g., 15] as the time average of the product $x(t')x(t' + t)$, i.e.,

$$\rho_x(t) \equiv \lim_{\Theta \to \infty} \frac{1}{2\Theta} \int_{-\Theta}^{\Theta} \mathrm{d}t'\, x(t')x(t' + t) = \overline{x(t')x(t' + t)} = \langle x(t')x(t' + t) \rangle \tag{5.1}$$

The last equality indicates that time averages and ensemble averages (denoted by the angle brackets) are equal.

One of the consequences of the fluctuation–dissipation theorem [28] is that spontaneous fluctuations in physical variables in a linear system, such as magnetic and electric polarizations, decay in the same way as externally induced disturbances. Thus for $t > 0$, the time dependence of the autocorrelation function coincides with that of the decay function $f_x(t)$ for the particular variable; in fact [16]

$$\rho_x(t) = k_B T f_x(t) = \beta^{-1} f_x(t) \qquad (t > 0) \tag{5.2}$$

Even though $f_x(t)$ is defined only for $t > 0$ we may still calculate $\rho_x(t)$ for $t < 0$, because the autocorrelation function $\rho_x(t)$ is an even function of t, i.e.,

$$\rho_x(t) = \rho_x(-t) \tag{5.3}$$

The Fourier transform of $\rho_x(t)$ leads to the spectral density

$$\Phi_x(\omega) = \int_{-\infty}^{\infty} dt\, \rho_x(t) \exp(\imath\omega t) \tag{5.4}$$

and, in view of Eq. (5.3), it follows that $\Phi_x(\omega)$ is a real, even function of ω. Therefore

$$\Phi_x(\omega) = 2\int_0^{\infty} dt\, \rho_x(t) \cos \omega t \tag{5.5a}$$

which is the Wiener–Khinchin equation [15]. The inverse of Eq. (5.5a) is

$$\rho_x(t) = \frac{1}{2\pi}\int_{-\infty}^{\infty} d\omega\, \Phi_x(\omega) \exp(\imath\omega t) = \frac{1}{\pi}\int_0^{\infty} d\omega\, \Phi_x(\omega) \cos \omega t \tag{5.5b}$$

On combining Eqs. (5.2) and (5.5a), we obtain

$$\Phi_x(\omega) = 2k_B T \int_0^{\infty} dt\, f_x(t) \cos \omega t \tag{5.6}$$

or, with a normalized decay function

$$b_x(t) \equiv f_x(t)/f_x(0) \tag{5.7}$$

we have

$$\Phi_x(\omega) = 2k_B T\, f_x(0) \int_0^{\infty} dt\, b_x(t) \cos \omega t \tag{5.8}$$

In the previous section we obtained Eq. (4.8), which provides a relation between the complex normalized response function Ψ and the corresponding decay function $b(t)$. With the symbol $\mathbb{I}$ denoting "imaginary part of," we have from Eq. (4.8):

$$\Psi''(\omega) \equiv -\mathbb{I}\Psi = -\mathbb{I}[\Psi' - \imath\Psi''] = \omega \int_0^\infty d\omega\, b(t) \cos \omega t \tag{5.9}$$

consequently Eq. (5.8) becomes

$$\Phi_x(\omega) = 2k_B T\, f_x(0) \frac{\Psi''(\omega)}{\omega} = 2\rho_x(0) \frac{\Psi''(\omega)}{\omega} \tag{5.10}$$

The real and imaginary parts of a dispersion function are related to one another through the Kramers–Kronig relations [29,30], in particular [16]

$$\Psi'(0) = \frac{2}{\pi} \int_0^\infty d\omega \frac{\Psi''(\omega)}{\omega} \tag{5.11}$$

Therefore

$$\begin{aligned} \frac{1}{2\pi} \int_{-\infty}^{\infty} d\omega\, \Phi_x(\omega) &= 2\rho_x(0) \frac{1}{2\pi} \int_{-\infty}^{\infty} d\omega \frac{\Psi''(\omega)}{\omega} = 2\rho_x(0) \frac{1}{\pi} \int_0^\infty d\omega \frac{\Psi''(\omega)}{\omega} \\ &= 2\rho_x(0) \frac{1}{\pi} \frac{\pi}{2} \Psi'(0) = \rho_x(0) = \langle x^2 \rangle = \overline{x^2(t)} \end{aligned} \tag{5.12}$$

we used the fact that $\Psi'(0) = 1$. The equality between the first and last members of Eq. (5.12) is an example of the Parseval relation [15].

B. Particular Results

1. *The Kubo Relation*

The combination of Eqs. (4.8), (5.2), and (5.7) results in the Kubo [31] formula, namely

$$\Psi(\omega) = 1 - \imath \frac{1}{\rho(0)} \int_0^\infty dt\, \rho(t) \exp(-\imath\omega t) \tag{5.13}$$

It is important to realize that the range of integration is from zero to infinity, and thus the integral is a complex quantity.

2. *Correlation Function for a Random Telegraph Signal*

In a study of the statistical properties of telegraph signals, Kenrick [32] calculated the autocorrelation function for a random variable $r(t)$ that switches abruptly between the values +1 and −1. The instants at which these reversals occur is governed by a Poisson distribution [15] in which the reversals occur at a mean rate α, and the probability $p_k(t)$ that k (an integer) changes will occur in the interval between the times t' and $t' + t$ is given by

$$p_k(t) = \frac{(\alpha t)^k \exp(-\alpha t)}{k!}. \tag{5.14}$$

Kenrick showed that

$$\rho_k(t) = \langle r(t')r(t' + t)\rangle = \overline{r(t')r(t' + t)} = \exp(-2\alpha t) \qquad (t \geq 0) \tag{5.15}$$

To take account of the fact that $\rho_k(t)$ is an even function of t, it is usual to write this equation in the form

$$\rho_k(t) = \exp(-2\alpha|t|) \tag{5.16}$$

3. Correlation Function for a Sinusoid with Random Abrupt Changes in Phase

Consider the the random function

$$c(t) = \cos[\Omega t + \varpi(t)] \tag{5.17}$$

where $\varpi(t)$ is a randomly varying phase that changes abruptly at instants determined, as in the previous case, by a Poisson process. At each such instant, the actual change in phase may lie anywhere in the range from 0 to 2π. The autocorrelation function

$$\begin{aligned}\rho_c(t) &= \langle\cos[\Omega t' + \varpi(t')]\cos[\Omega t' + \Omega t + \varpi(t' + t)]\rangle \\ &= \langle\cos^2[\Omega t' + \varpi(t')]\cos[\Omega t + \Delta(t)]\rangle \\ &\quad - \frac{1}{2}\langle\sin[2\Omega t' + 2\varpi(t')]\cos[\Omega t + \Delta(t)]\rangle\end{aligned} \tag{5.18}$$

where $\Delta(t) = \varpi(t' + t) - \varpi(t')$ is the change in phase that occurs in the interval of time t. The changes in phase that occur before the beginning of this interval are independent of the changes that occur within the interval, therefore

$$\begin{aligned}\rho_c(t) =&\langle\cos^2[\Omega t' + \varpi(t')]\rangle\langle\cos[\Omega t + \Delta(t)]\rangle \\ &- \frac{1}{2}\langle\sin[2\Omega t' + 2\varpi(t')]\rangle\langle\cos[\Omega t + \Delta(t)]\rangle\end{aligned} \tag{5.19}$$

Because the initial phase $2[\Omega t' + \varpi(t')]$ may have any value between 0 and 2π,

$$\langle \sin[2\Omega t' + 2\varpi(t')] \rangle = 0 \tag{5.20}$$

and, therefore, Eq. (5.18) reduces to

$$\begin{aligned} \rho_c(t) &= \langle \cos^2[\Omega t' + \varpi(t')] \rangle \langle \cos[\Omega t + \Delta(t)] \rangle \\ &= \frac{1}{2} \langle \cos[\Omega t + \Delta(t)] \rangle \end{aligned} \tag{5.21}$$

If there is one, or more, change in phase in the interval t, the ensemble average in the second part of Eq. (5.21) will vanish. Only if there is no change in phase will there be a contribution to the autocorrelation function. The probability that no change will occur is, according to Eq. (5.14), $p_0(t) = \exp(-\alpha|t|)$; therefore [16]

$$\rho_c(t) = \frac{1}{2} \exp(-\alpha|t|) \cos \Omega t \tag{5.22}$$

VI. LONGITUDINAL AND TRANSVERSE POLARIZABILITIES FOR A FIXED, SPHERICAL, SINGLE-DOMAIN PARTICLE

A. Zero-Frequency Polarizabilities

As indicated, the model we use to describe the magnetization of a fixed, spherical, single-domain, uniaxial ferromagnetic particle consists of a magnetic moment vector **m** of constant magnitude that, at zero temperature, has two antiparallel equilibrium orientations along the, so-called, magnetic axis Oζ (Fig. 1). At finite temperatures, thermal agitation will cause **m** to deviate slightly from the Oζ-axis; as a result of magnetic anisotropy, a torque is exerted on **m**, which tends to return it to the Oζ-axis and causes it to precess about this axis. In Section III we sketched the equation of motion for **m** proposed by Landau and Lifshitz. The solution of this equation for a small, external, alternating magnetizing force, transverse to the magnetic axis, led to the equations for the transverse magnetic polarizabilities $\alpha_{\xi,\xi}$ and $\alpha_{\eta,\eta}$ [Eqs. (3.27a) and (3.28c)]. In the Landau–Lifshitz theory, the longitudinal polarizability $\alpha_{\zeta,\zeta}$ is zero. However, the finite height of the anisotropy–energy barrier [Eq. (3.5)] means that at nonzero temperatures $m_\zeta = m \cos\theta$, the component of **m** along the magnetic axis, will reverse direction in a random manner, giving rise to a finite value for $\alpha_{\zeta,\zeta}$. Clearly, these reversals will have some effect on the dynamics of **m**; in addition to the damping of the precessional motion introduced by Landau and Lifshitz, there will be damping arising from the fluctuations in m_ζ. We discuss below how this damping may be taken into account. First, however, we indicate how the equilibrium mean-square fluctuations $\langle m_\xi^2\rangle = \langle m_\eta^2\rangle$ and $\langle m_\zeta^2\rangle$ may be calculated.

By means of the Boltzmann distribution and Eqs. (3.5) and (3.33a), we have

$$\begin{aligned}\langle m_\xi^2\rangle &= m^2\langle \sin^2\theta\cos^2\theta\rangle \\ &= m^2\frac{\int_0^{2\pi} d\phi \int_0^\pi d\theta \sin\theta \sin^2\theta \cos^2\phi \exp[\frac{1}{2}\beta\kappa m^2\cos^2\theta]}{\int_0^{2\pi} d\phi \int_0^\pi d\theta \sin\theta \exp[\frac{1}{2}\beta\kappa m^2\cos^2\theta]} \\ &= \frac{1}{2}m^2\frac{\int_0^\pi d\theta \sin\theta\sin^2\theta\exp[\sigma\cos^2\theta]}{\int_0^\pi d\theta\sin\theta\exp[\sigma\cos^2\theta]} = \frac{1}{2}m^2\frac{\int_{-1}^1 dx(1-x^2)\exp[\sigma x^2]}{\int_{-1}^1 dx\exp[\sigma x^2]} \\ &= \frac{1}{2}m^2\frac{\int_0^1 dx(1-x^2)\exp[\sigma x^2]}{\int_0^1 dx\exp[\sigma x^2]} = \frac{1}{2}m^2\left[1-\frac{F'(\sigma)}{F(\sigma)}\right]\end{aligned} \tag{6.1}$$

By symmetry we must have

$$\langle m_\xi^2\rangle = \langle m_\eta^2\rangle \tag{6.2}$$

and since $\mathbf{m}$ is of constant magnitude,

$$\langle m_\zeta^2\rangle = m^2 - \langle m_\xi^2\rangle - \langle m_\eta^2\rangle = m^2 - 2\langle m_\xi^2\rangle \tag{6.3}$$

Hence it follows from Eq. (6.1) that

$$\langle m_\zeta^2\rangle = m^2\langle\cos^2\theta\rangle = m^2 - m^2\left(1-\frac{F'(\sigma)}{F(\sigma)}\right) = m^2\frac{F'(\sigma)}{F(\sigma)} \tag{6.4}$$

The functions

$$F(\sigma) \equiv \int_0^1 dx\exp(\sigma x^2) \tag{6.5a}$$

$$F'(\sigma) \equiv \frac{dF(\sigma)}{d\sigma} = \int_0^1 dx\, x^2\exp(\sigma x^2) \tag{6.5b}$$

were introduced by Raĭkher and Shliomis [33,34, p. 747] who obtained the following asymptotic expansions for large σ:

$$F(\sigma) \sim \frac{\exp\sigma}{2\sigma}\left(1+\frac{1}{2\sigma}+\frac{3}{4\sigma^2}+\cdots+\frac{1\cdot 3\cdot 5\cdots(2n-1)}{2^n\sigma^n}+\cdots\right) \tag{6.6a}$$

$$F'(\sigma) \sim \frac{\exp\sigma}{2\sigma}\left(1-\frac{1}{2\sigma}-\frac{1}{4\sigma^2}+\cdots\right) \tag{6.6b}$$

Therefore, for large σ we have the approximations:

$$\langle m_\xi^2 \rangle = \langle m_\eta^2 \rangle \approx \frac{1}{2\sigma} m^2 \qquad (\sigma \gg 1) \tag{6.7a}$$

$$\langle m_\zeta^2 \rangle \approx \left(1 - \frac{1}{\sigma}\right) m^2 \qquad (\sigma \gg 1) \tag{6.7b}$$

From these results we obtain, with the help of the Kirkwood–Fröhlich susceptibility theorem [35, p. 83], the following expressions for the zero-frequency polarizabilities:

$$\alpha_\perp(0) \equiv \alpha_{\xi,\xi}(0) = \alpha_{\eta,\eta}(0) = \frac{1}{2\sigma} \frac{m^2}{k_\mathrm{B} T} = \frac{\beta}{2\sigma} m^2 \qquad (\sigma \gg 1) \tag{6.8a}$$

$$\alpha_\parallel(0) \equiv \alpha_{\zeta,\zeta}(0) = \left(1 - \frac{1}{\sigma}\right) \frac{m^2}{k_\mathrm{B} T} = \left(1 - \frac{1}{\sigma}\right) \beta m^2 \qquad (\sigma \gg 1) \tag{6.8b}$$

A similar calculation would show that

$$\alpha_{\xi,\eta}(0) = \alpha_{\eta,\xi}(0) = \alpha_{\xi,\zeta}(0) = \alpha_{\zeta,\xi}(0) = \alpha_{\eta,\zeta}(0) = \alpha_{\zeta,\eta}(0) = 0 \tag{6.9}$$

Eq. (6.8a) coincides with Eq. (3.34) when $\omega = 0$, and Eq.(6.9) is consistent with Eq. (3.27). Notice that the solution of the Landau–Lifshitz equation proceeded on the basis that the deviation of **m** from the magnetic axis was small (in the sense that $1 - \cos\theta$ is small), which is tantamount to the assumption that σ, the ratio of the magnetic-anisotropy energy to the mean thermal energy, is large.

The calculation of the zero-frequency polarizabilities did not require a knowledge of the dynamics of the magnetization; this is entirely in accord with the principles of statistical mechanics.

It is clear from Eq. (6.8) that when σ is extremely large the only significant zero-frequency polarizability is the longitudinal polarizability $\alpha_\parallel(0)$. However, as we shall see presently, at frequencies near to resonance the frequency-dependent transverse polarizability $\alpha_\perp(\omega)$ can become significant if the damping is small, even though σ might be very large.

Eq. (3.33a) enables us to estimate the size of σ: with $K_\mathrm{u} \sim 10^5$ J/m^3 and $T = 300$ K, $\sigma \approx 12.6$ for a particle with a radius of 5 nm (50 Å). Knowing σ it is possible to estimate the magnitude of the angular deviation of **m** from the magnetic axis. From Eqs. (6.4) and (6.7b) we have

$$\langle \cos^2\theta \rangle \approx 1 - \frac{1}{2}\langle \theta^2 \rangle \approx 1 - \frac{1}{\sigma} \qquad (\sigma \gg 1) \tag{6.10}$$

so that

$$\theta_{\text{rms}} \equiv \sqrt{\langle \theta^2 \rangle} \approx \sqrt{\frac{2}{\sigma}} \tag{6.11}$$

For $\sigma \approx 12.6$, $\theta_{\text{rms}} \approx 23°$, and for $\sigma \approx 100$, $\theta_{\text{rms}} \approx 8°$; it is clear, therefore, that only for extremely large values of σ would it be correct to speak of them as blocked. Nevertheless, it is reasonable to regard particles with a value of σ in excess of about 12 as being blocked.

B. Frequency-Dependent Longitudinal Polarizability

On the assumption that σ is large, and that the time taken for $m_\zeta(t)$, the longitudinal component of $\mathbf{m}$, to change direction is very short, we may treat the random function $m_\zeta(t)$ as a random telegraph signal, switching between the values $\pm|m_\zeta(t)|$. From Eq. (5.16) we find that the autocorrelation function

$$\rho_{m_\zeta}(t) = m_\zeta^2 \exp\left(-\frac{|t|}{\tau_{\text{N}}}\right) \tag{6.12}$$

The time constant τ_{N} is called the Néel relaxation time [36]. Eq. (6.12) is analagous to the type of electric polarization arising from an electric dipole switching abruptly between two antiparallel orientations separated by a large energy barrier. This dielectric mechanism was first discussed by Debye [5, p. 105]. When we apply his reasoning to the magnetic case we find that

$$\tau_{\text{N}} = B(v_{\text{p}}, T) \exp \sigma \tag{6.13}$$

Although the factor $B(v_{\text{p}}, T)$ is a slowly varying function of particle volume and of temperature; the exponential function indicates that τ_{N} strongly depends on these two variables.

We shall replace m_ζ^2 in Eq. (6.12) by $\langle m_\zeta^2 \rangle$, and thus, in view of Eq. (6.7b), we obtain

$$\rho_{m_\zeta}(t) = m^2\left(1 - \frac{1}{\sigma}\right) \exp\left(-\frac{|t|}{\tau_{\text{N}}}\right) \tag{6.14}$$

which leads to a decay function

$$b_\zeta(t) = \beta m^2\left(1 - \frac{1}{\sigma}\right) \exp\left(-\frac{t}{\tau_{\text{N}}}\right) \tag{6.15}$$

With the help of the Kubo equation, Eq. (5.13), and Eqs. (4.8) and (4.9c), we get from Eq. (6.14) the following formula for the frequency-dependent longitudinal polarizability:

$$\begin{aligned}\alpha_{\|}(\omega) \equiv \alpha_{\zeta,\zeta}(\omega) &= \beta m^2\left(1 - \frac{1}{\sigma}\right)\frac{1}{1 + \imath\omega\tau_{\mathrm{N}}} \\ &= \beta m^2\left(1 - \frac{1}{\sigma}\right)\Psi_{\mathrm{D}}(\omega, \tau_{\mathrm{N}})\end{aligned} \tag{6.16}$$

C. Frequency-Dependent Transverse Polarizability

For this case, our main concern is to determine in what manner the abrupt reversals of the longitudinal component of **m** effect the damped precessional motion of **m**. Our basic assumption is that there is no correlation between azimuthal angle ϕ just before and just after a reversal of the longitudinal component (Fig. 1). We know from the expression for $\alpha_{\xi,\xi}(\omega)$ [Eq. (3.27a)], obtained from the Landau–Lifshitz equation of motion [Eq. (3.13)], that there is a corresponding decay function $b_{\mathrm{II}}(t, \omega_{\mathrm{pr}}, \tau_{\mathrm{pr}})$ [Eq. (4.11c)]. In an assembly of noninteracting particles, a portion of them, in any interval of time t, will not suffer reversal of the longitudinal component of **m**. The remaining members of the assembly will suffer at least one such reversal, which will destroy the coherence of the transverse components of **m.**

The damping process that leads to the time constant τ_{pr} is assumed to be statistically independent of the process that leads to the longitudinal reversals. The portion of the assembly that is free of reversals is $p_0(t)$ [Eq. (5.14)], the probability that no reversals occur in the interval of duration t. We conclude, therefore, that

$$\begin{aligned}\rho_{m_\xi}(t) \equiv \langle m_\xi(t')m_\xi(t'+t)\rangle &= \left[\langle m_\xi^2\rangle \exp\left(-\frac{|t|}{\tau_{\mathrm{pr}}}\right)\cos\omega_{\mathrm{pr}}t\right]p_0(t) \\ &= \langle m_\xi^2\rangle \exp\left[-\left(\frac{1}{\tau_{\mathrm{pr}}} + \frac{1}{2\tau_{\mathrm{N}}}\right)|t|\right]\cos\omega_{\mathrm{pr}}t \\ &= \frac{1}{2\sigma}m^2 \exp\left(-\frac{|t|}{\tau_1}\right)\cos\omega_{\mathrm{pr}}t\end{aligned} \tag{6.17}$$

which should be compared to Eq. (5.22). We see from Eq. (6.17) that the effect of the longitudinal fluctuations on the transverse fluctuations is to alter the time constant from τ_{pr} to

$$\tau_{\mathrm{I}} = \frac{1}{\dfrac{1}{\tau_{\mathrm{pr}}} + \dfrac{1}{2\tau_{\mathrm{N}}}} = \frac{2\tau_{\mathrm{pr}}\tau_{\mathrm{N}}}{2\tau_{\mathrm{N}} + \tau_{\mathrm{pr}}} \tag{6.18}$$

Using the same procedure as in the previous section we obtain the following expression for the frequency-dependent transverse polarizability:

$$\begin{aligned}\alpha_{\perp}(\omega) = \alpha_{\xi,\xi}(\omega) = \alpha_{\eta,\eta}(\omega) &= \frac{\beta m^2}{2\sigma}\frac{1+(\omega_{\text{pr}}\tau_1)^2 + \imath\omega\tau_1}{1+(\omega_{\text{pr}}\tau_1)^2 - (\omega\tau_1)^2 + \imath 2\omega\tau_1} \\ &= \frac{\beta m^2}{2\sigma}\Psi_{\text{II}}(\omega, \omega_{\text{pr}}, \tau_1)\end{aligned} \tag{6.19}$$

It may also be shown that the cross-polarizability [cf. Eq. (3.27b)] takes the form:

$$\alpha_{\xi,\eta}(\omega) = \frac{\beta m^2}{2\sigma}\frac{\imath\omega\omega_{\text{pr}}\tau_1}{1+(\omega_{\text{pr}}\tau_1)^2 - (\omega\tau_1)^2 + \imath 2\omega\tau_1} \tag{6.20}$$

An important difference between the approach adopted here and that of Brown [39] is that he used a single damping parameter whereas we have introduced two, independent, damping factors namely τ_{pr} and τ_{N}.

VII. CALCULATION OF THE COMPLEX, FREQUENCY-DEPENDENT SUSCEPTIBILITY OF A MAGNETIC FLUID

Consider a colloidal suspension of spherical, single-domain, uniaxial, ferromagnetic particles with particle number density N/V, and particle magnetic moment m. Let there be a coordinate system (x, y, z) fixed in space. As indicated in Section I, the particles will undergo rotational Brownian motion. The components of the magnetic moment vector **m** namely $m_x(t), m_y(t)$ and $m_z(t)$—will be random functions of time; the fluctuations in these components arise partly from the Brownian motion and partly from the internal fluctuations discussed in Section VI. If the magnetic interaction between the particles is very weak, we can calculate the susceptibility $\chi_{\text{mag}}(\omega)$ of the colloid if we can calculate the autocorrelation function $\langle m_x(t')m_x(t'+t)\rangle = \langle m_y(t')m_y(t'+t)\rangle = \langle m_z(t')m_z(t'+t)\rangle$ because, by the Kubo relation [Eq. (5.13)]

$$\chi_{\text{mag}}(\omega) = \frac{N}{\mu_0 V}\left[\langle m_x^2\rangle - \imath\int_0^\infty \text{d}t\langle m_x(t')m_x(t'+t)\rangle \exp(-\imath\omega t)\right] \tag{7.1}$$

Our first task is to relate the component $m_x(t)$ to the components $m_\xi(t), m_\eta(t)$, and $m_\zeta(t)$. The orientation in space of the axes of the coordinate system (ξ, η, ζ) fixed in the particle can be specified in terms of the direction cosines

$$l_{\xi,x}(t) = \cos[\gamma_{\xi,x}(t)] \qquad l_{\eta,x}(t) = \cos[\gamma_{\eta,x}(t)] \qquad l_{\zeta,x}(t) = \cos[\gamma_{\zeta,x}(t)] \tag{7.2}$$

the moving Oξ-, Oη-, and Oζ-axes make the angles $\gamma_{\xi,x}$, $\gamma_{\eta,x}$, and $\gamma_{\zeta,x}$, respectively, with the fixed Ox-axis. Therefore

$$m_x(t) = l_{\xi,x}(t)m_\xi(t) + l_{\eta,x}(t)m_\eta(t) + l_{\zeta,x}(t)m_\zeta(t) \tag{7.3}$$

so that

$$\begin{aligned}\langle m_x(t')m_x(t'+t)\rangle &= \langle l_{\xi,x}(t')l_{\xi,x}(t'+t)m_\xi(t')m_\xi(t'+t)\rangle \\ &+ \langle l_{\xi,x}(t')l_{\eta,x}(t'+t)m_\xi(t')m_\eta(t'+t)\rangle + \langle l_{\xi,x}(t')l_{\zeta,x}(t'+t)m_\xi(t')m_\zeta(t'+t)\rangle \\ &+ \langle l_{\eta,x}(t')l_{\xi,x}(t'+t)m_\eta(t')m_\xi(t'+t)\rangle + \langle l_{\eta,x}(t')l_{\eta,x}(t'+t)m_\eta(t')m_\eta(t'+t)\rangle \\ &+ \langle l_{\eta,x}(t')l_{\zeta,x}(t'+t)m_\eta(t')m_\zeta(t'+t)\rangle + \langle l_{\zeta,x}(t')l_{\xi,x}(t'+t)m_\zeta(t')m_\xi(t'+t)\rangle \\ &+ \langle l_{\zeta,x}(t')l_{\eta,x}(t'+t)m_\zeta(t')m_\eta(t'+t)\rangle + \langle l_{\zeta,x}(t')l_{\zeta,x}(t'+t)m_\zeta(t')m_\zeta(t'+t)\rangle\end{aligned} \tag{7.4}$$

If we assume that the bodily motion of the particle in no way affects the orientation of **m** relative to the axes Oξ, Oη, and Oζ fixed in the particle, then the components $m_\xi(t)$, $m_\eta(t)$, and $m_\zeta(t)$ will be statistically independent of the direction cosines $l_{\xi,x}(t)$, $l_{\eta,x}(t)$, and $l_{\zeta,x}(t)$. Consequently Eq. (7.4) reduces to

$$\begin{aligned}\langle m_x(t')m_x(t'+t)\rangle &= \langle l_{\xi,x}(t')l_{\xi,x}(t'+t)\rangle\langle m_\xi(t')m_\xi(t'+t)\rangle \\ &+ \langle l_{\xi,x}(t')l_{\eta,x}(t'+t)\rangle\langle m_\xi(t')m_\eta(t'+t)\rangle + \langle l_{\xi,x}(t')l_{\zeta,x}(t'+t)\rangle\langle m_\xi(t')m_\zeta(t'+t)\rangle \\ &+ \langle l_{\eta,x}(t')l_{\xi,x}(t'+t)\rangle\langle m_\eta(t')m_\xi(t'+t)\rangle + \langle l_{\eta,x}(t')l_{\eta,x}(t'+t)\rangle\langle m_\eta(t')m_\eta(t'+t)\rangle \\ &+ \langle l_{\eta,x}(t')l_{\zeta,x}(t'+t)\rangle\langle m_\eta(t')m_\zeta(t'+t)\rangle + \langle l_{\zeta,x}(t')l_{\xi,x}(t'+t)\rangle\langle m_\zeta(t')m_\xi(t'+t)\rangle \\ &+ \langle l_{\zeta,x}(t')l_{\eta,x}(t'+t)\rangle\langle m_\zeta(t')m_\eta(t'+t)\rangle + \langle l_{\zeta,x}(t')l_{\zeta,x}(t'+t)\rangle\langle m_\zeta(t')m_\zeta(t'+t)\rangle\end{aligned} \tag{7.5}$$

A particle has no preferred orientation in space; hence all the cross-correlations, such as $\langle l_{\xi,x}(t')l_{\eta,x}(t'+t)\rangle$, vanish. According to the theory of rotational Brownian motion [5], [7]

$$\begin{aligned}\langle l_{\xi,x}(t')l_{\xi,x}(t'+t)\rangle &= \langle l_{\eta,x}(t')l_{\eta,x}(t'+t)\rangle = \langle l_{\zeta,x}(t')l_{\zeta,x}(t'+t)\rangle \\ &= \frac{1}{3}\exp\left(-\frac{|t|}{\tau_B}\right)\end{aligned} \tag{7.6}$$

in which τ_B is the relaxation time owing to Brownian motion and defined by

Eq. (1.9). In the derivation of Eq. (7.5), the mechanical inertia of the particle is ignored.

We now make use of Eqs. (6.12) and (6.17), together with Eqs. (7.5) and (7.6), and get

$$\begin{aligned}
&\langle m_x(t')m_x(t'+t)\rangle \\
&= \frac{m^2}{3}\exp\left(-\frac{|t|}{\tau_B}\right)\left[\left(1-\frac{1}{\sigma}\right)\exp\left(-\frac{|t|}{\tau_N}\right)+\frac{1}{\sigma}\exp\left(-\frac{|t|}{\tau_1}\right)\cos\omega_{pr}t\right] \\
&= \frac{m^2}{3}\left[\left(1-\frac{1}{\sigma}\right)\exp\left(-\frac{|t|}{\tau_3}\right)+\frac{1}{\sigma}\exp\left(-\frac{|t|}{\tau_2}\right)\cos\omega_{pr}t\right]
\end{aligned} \tag{7.7}$$

in which the relaxation times τ_2 and τ_3 are defined as follows:

$$\frac{1}{\tau_2} = \frac{1}{\tau_B} + \frac{1}{\tau_1} = \frac{1}{\tau_B} + \frac{1}{2\tau_N} + \frac{1}{\tau_{pr}} \tag{7.8a}$$

$$\frac{1}{\tau_3} = \frac{1}{\tau_B} + \frac{1}{\tau_N} \tag{7.8b}$$

Notice that, as expected,

$$\langle m_x(t')m_x(t')\rangle \equiv \langle m_x^2\rangle = \frac{m^2}{3} \tag{7.9}$$

After inserting Eq. (7.7) in Eq. (7.1) and performing the integrations (using the results in Section IV) we finally obtain the equation

$$\begin{aligned}
\chi_{mag}(\omega) &= \frac{1}{3}\beta\frac{Nm^2}{\mu_0 V}\left[\left(1-\frac{1}{\sigma}\right)\Psi_D(\omega,\tau_3)+\frac{1}{\sigma}\Psi_{II}(\omega,\omega_{pr},\tau_2)\right] \\
&= \frac{1}{3}\beta\frac{Nm^2}{\mu_0 V}\left\{\left(1-\frac{1}{\sigma}\right)\frac{1}{[1+\imath\omega\tau_3]}+\frac{1}{\sigma}\frac{[1+(\omega_{pr}\tau_2)^2+\imath\omega\tau_2]}{[1+(\omega_{pr}\tau_2)^2-(\omega\tau_2)^2+\imath 2\omega\tau_2]}\right\}
\end{aligned} \tag{7.10}$$

There are thus three time constants that determine the frequency dependence of the susceptibility of a magnetic fluid. For particles with large values of σ, the Néel relaxation time τ_N is much greater than the Brownian relaxation time τ_B, which is much greater than the precession relaxation time τ_{pr}. In these circumstances $\tau_3 \approx \tau_B$ and $\tau_2 \approx \tau_{pr}$.

Eq. (7.10) is not inconsistent with some of the recent experimental data obtained by Fannin [3].

APPENDIX A. Comparison of the Functions Ψ_D, Ψ_I, and Ψ_{II}

To simplify the numerical analysis and comparison of the Debye, Lorentz, and Landau–Lifshitz equations it is convenient to introduce the following abbreviations.

Ψ_D, Nonresonant Absorption: Debye Eq. (1.6)

$$\Psi_D\left(\frac{x}{\tau_D}, \tau_D\right) = y_D(x) = y'_D(x) - \imath y''_D(x) = \frac{1}{1 + \imath x} \tag{A.1}$$

where

$$x = \omega\tau_D \tag{A.2}$$

Ψ_I, Type I Resonance Absorption: Lorentz Eq. (2.6)

$$\Psi_I\left(\frac{x}{\tau_{coll}}, \frac{a}{\tau_{coll}}, \tau_{coll}\right) = y_I(x) = y'_I(x) - \imath y''_I(x) = \frac{1 + a^2}{1 + a^2 - x^2 + \imath 2x} \tag{A.3}$$

where

$$x = \omega\tau_{coll} \quad \text{and} \quad a = \omega_L\tau_{coll} \tag{A.4}$$

Ψ_{II}, Type II Resonance Absorption: Landau and Lifshitz Eq. (1.15), Van Vleck and Weisskopf, and Fröhlich Eq. (2.11)

$$\Psi_{II}\left(\frac{x}{\tau_{coll}}, \frac{a}{\tau_{coll}}, \tau_{coll}\right) = y_{II}(x) = y'_{II}(x) - \imath y''_{II}(x) = \frac{1 + a^2 + \imath x}{1 + a^2 - x^2 + \imath 2x} \tag{A.5}$$

in which x and a are defined in Eq. (A.4).

A compact way of comparing Eqs. (A.1), (A.3), and (A.5) is by means of the Cole–Cole [37] plot in which, for example, $y''_D(x)$ is plotted as a function of $y'_D(x)$. In Figure 3 we show the Cole–Cole plots for $y_D(x)$, $y_I(x)$, and $y_{II}(x)$. Important features of the Cole–Cole plots include (Table I)

1. The slopes—$sl_D(0)$ and $sl_D(\infty)$, $sl_I(0)$ and $sl_I(\infty)$, $sl_{II}(0)$ and $sl_{II}(\infty)$—at which the plots intersect the horizontal axis at $x = 0$, and at $x = \infty$.

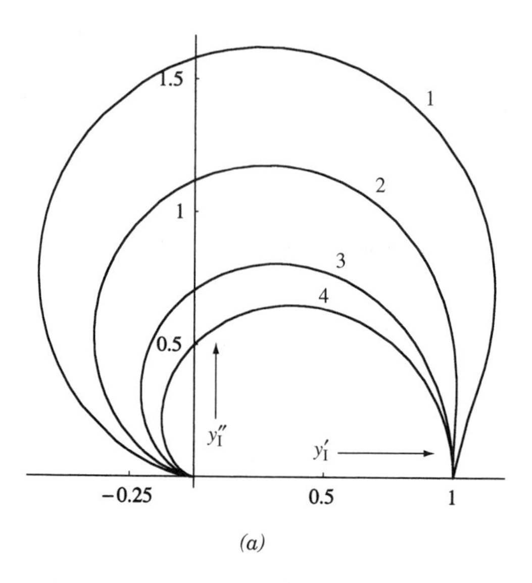

(a)

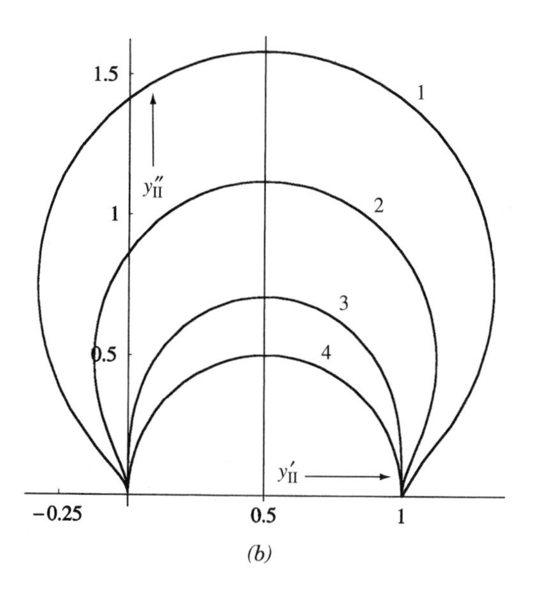

(b)

Figure 3. Continued opposite.

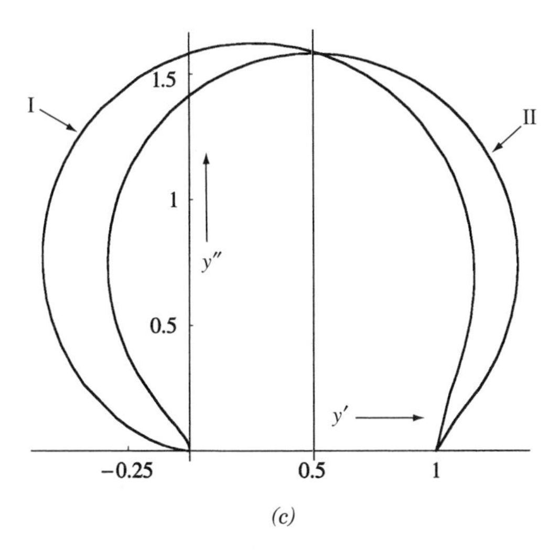

Figure 3. Cole–Cole plots in which the imaginary coordinate of a complex (vertical axis) function is plotted as a function of the real part (horizontal axis) of the function. (*a*) The function $y_{\mathrm{I}}(x) = y'_{\mathrm{I}}(x) - \imath y''_{\mathrm{I}}(x)$ for Lorentz Type I resonant absorption [Eq. (A.3)]. Plot 1, $a = 3$; plot 2, $a = 2$; plot 3, $a = 1$; plot 4, $a = 0$. (*b*) The function $y_{\mathrm{II}}(x) = y'_{\mathrm{II}}(x) - \imath y''_{\mathrm{II}}(x)$ for Landau–Lifshitz Type II resonant absorption [Eq. (A.5)]. Plot 1 $a = 3$; plot 2, $a = 2$; plot 3, $a = 1$; plot 4, $a = 0$. Plot 4 is also the plot for $y_{\mathrm{D}}(x) = y'_{\mathrm{D}}(x) - \imath y''_{\mathrm{D}}(x)$ for Debye nonresonant absorption [Eq. (A.1)], since, for $a = 0$, the two functions coincide. Notice that for $1 \geq a \geq 0$, $y'_{\mathrm{II}}(x)$ is never negative; whereas $y'_{\mathrm{I}}(x)$ will always become negative for sufficiently large values of x. (*c*) Comparison of plots for Type I and Type II absorptions. Plot 1 from part (*a*) is superimposed on plot 1 from part (*b*).

2. The value of x (denoted by x_{max}) at which the height of the plot is a maximum.
3. The value of x (denoted by x_{res}) at which the plot cuts the vertical axis.

Since the slope is the tangent of the angle of intersection, a slope of $+\infty$ corresponds to an angle slightly less than a right angle, a slope of $-\infty$ corresponds to an angle slightly greater than a right angle, and a slope of -0 corresponds to an angle slightly less than two right angles.

We determine the values of x_{max} by solving the equations:

$$\left[\frac{\mathrm{d}y''_{\mathrm{D}}(x)}{\mathrm{d}x}\right]_{x = x_{\mathrm{max}}} = 0, \text{ etc.,}$$

and we determine the values of x_{res} by solving the equations: $y'_{\mathrm{D}}(x_{\mathrm{res}}) = 0$, $y'_{\mathrm{I}}(x_{\mathrm{res}}) = 0$, and $y'_{\mathrm{II}}(x_{\mathrm{res}}) = 0$. Notice that the Cole–Cole plots for $y_{\mathrm{II}}(x)$ are symmetrical about the axis: $y'_{\mathrm{II}}(x_{\mathrm{max}}) = \frac{1}{2}$ [Fig. 3(*b*)].

TABLE I
Summary of the Cole–Cole Plots

Debye	Type I (Lorentz)	Type II (Landau, Lifshitz [9], Van Vleck, Weisskopf [10], Fröhlich [11])
$sl_D = -\infty$	$sl_I(0) = \begin{cases} -\infty, & a^2 < 3 \\ +\infty, & a^2 > 3 \end{cases}$	$sl_{II}(0) = \begin{cases} -\infty, & a^2 < 1 \\ +\infty, & a^2 > 1 \end{cases}$
$sl_D = +\infty$	$sl_I(\infty) = -0$	$sl_{II}(\infty) = \begin{cases} +\infty, & a^2 < 1 \\ -\infty, & a^2 > 1 \end{cases}$
$x_{max} = 1; y'_D(x_{max}) = y''_D(x_{max}) = \frac{1}{2}$	$x^2_{max} = \frac{1}{3}(a^2 - 1) + \frac{2}{3}\sqrt{1 + a^2 + a^4}$	$x^2_{max} = 1 + a^2; y'_{II}(x_{max}) = \frac{1}{2}$; $y''_{II}(x_{max}) = \frac{1}{2}\sqrt{1 + a^2}$
$x_{res} = \infty; y'_D(x_{res}) = y''_D(x_{res}) = 0$	$x^2_{res} = 1 + a^2; y'_I(x_{res}) = 0$; $y''_I(x_{res}) = \frac{1}{2}\sqrt{1 + a^2}$	$x^2_{res} = \frac{(1 + a^2)}{(a^2 - 1)}, a^2 > 1$; $y'_{II}(x_{res}) = 0$; $y''_D(x_{res}) = \frac{1}{2}\sqrt{a^2 - 1}$

This may be easily deduced by verifying that

$$y'_{II}(x) - y'_{II}(x_{max}) = y'_{II}(x) - \frac{1}{2} = -\left[y'_{II}\left(\frac{x^2_{max}}{x}\right) - \frac{1}{2}\right] \qquad x^2_{max} = 1 + a^2 \tag{A.6a}$$

and

$$y''_{II}(x) = y''_{II}\left(\frac{x^2_{max}}{x}\right) \tag{A.6b}$$

APPENDIX B. Different Forms of the Type II Resonance Equation

The equation associated with the names of Landau, Lifshitz, Van Vleck, Weisskopf, and Fröhlich, written in its simplest form, is

$$\begin{aligned}\Psi_{II}(\omega, \omega_L, \tau_{coll}) &\equiv \Psi'_{II}(\omega, \omega_L, \tau_{coll}) - \imath\Psi''_{II}(\omega, \omega_L, \tau_{coll}) \\ &= \frac{(1 + \omega_L^2\tau_{coll}^2 + \imath\omega\tau_{coll})}{(1 + \omega_L^2\tau_{coll}^2 - \omega^2\tau_{coll}^2 + 2\imath\omega\tau_{coll}}\end{aligned} \tag{B.1}$$

It may also be written in the form:

$$\Psi_{\rm II}(\omega, \omega_{\rm L}, \tau_{\rm coll}) = \frac{(1 + \omega_{\rm L}^2 \tau_{\rm coll}^2 + \imath\omega\tau_{\rm coll})}{[\omega_{\rm L}^2 \tau_{\rm coll}^2 + (1 + \imath\omega\tau_{\rm coll})^2]} \tag{B.2}$$

and in the form:

$$\Psi_{\rm II}(\omega, \omega_{\rm L}, \tau_{\rm coll}) = \frac{1}{2}\left\{\frac{(1 + \imath\omega_{\rm L}\tau_{\rm coll})}{[1 + \imath(\omega + \omega_{\rm L})\tau_{\rm coll}]} + \frac{(1 - \imath\omega_{\rm L}\tau_{\rm coll})}{[1 + \imath(\omega - \omega_{\rm L})\tau_{\rm coll}]}\right\} \tag{B.3}$$

The notation used in Eqs. (B.1)–(B.3) is that appropriate to the work of Lorentz, Van Vleck, Weisskopf, and Fröhlich for the collision-damped harmonic oscillator; Eqs. (B.2) and (B.3) were obtained by Fröhlich ([11] and [12], pp. 101 and 176); and in a different notation, Eq. (B.1) was obtained by Landau and Lifshitz [9] in their study of ferromagnetic resonance.

It is straighforward to separate $\Psi_{\rm II}(\omega, \omega_{\rm L}, \tau_{\rm coll})$ into its real and imaginary parts; thus, rationalizing Eq. (B.1), we obtain

$$\Psi'_{\rm II}(\omega, \omega_{\rm L}, \tau_{\rm coll}) = \frac{[(1 + \omega_{\rm L}^2 \tau_{\rm coll}^2)^2 + \omega^2\tau_{\rm coll}^2(1 - \omega_{\rm L}^2\tau_{\rm coll}^2)]}{[(1 + \omega_{\rm L}^2\tau_{\rm coll}^2 - \omega^2\tau_{\rm coll}^2)^2 + 4\omega^2\tau_{\rm coll}^2]} \tag{B.4a}$$

and

$$\Psi''_{\rm II}(\omega, \omega_{\rm L}, \tau_{\rm coll}) = \frac{2\omega\tau_{\rm coll}(1 + \omega_{\rm L}^2\tau_{\rm coll}^2 + \omega^2\tau_{\rm coll}^2)}{[(1 + \omega_{\rm L}^2\tau_{\rm coll}^2 - \omega^2\tau_{\rm coll}^2)^2 + 4\omega^2\tau_{\rm coll}^2]} \tag{B.4b}$$

The expressions corresponding to those derived by Van Vleck and Weisskopf are [10, Eq. 18]

$$\Psi'_{\rm II}(\omega, \omega_{\rm L}, \tau_{\rm coll}) = \frac{1}{\left(1 - \dfrac{\omega^2}{\omega_{\rm L}^2}\right)}\left\{1 - \frac{\frac{1}{2}(\omega/\omega_{\rm L}) + \frac{1}{2}(\omega/\omega_{\rm L})^2}{\tau_{\rm coll}^2[(\omega - \omega_{\rm L})^2 + (1/\tau_{\rm coll})^2]} + \frac{\frac{1}{2}(\omega/\omega_{\rm L}) - \frac{1}{2}(\omega/\omega_{\rm L})^2}{\tau_{\rm coll}^2[(\omega + \omega_{\rm L})^2 + (1/\tau_{\rm coll})^2]}\right\} \tag{B.5a}$$

and (Van Vleck and Weisskopf 1945, Eq. 17)

$$\Psi''_{\rm II}(\omega, \omega_{\rm L}, \tau_{\rm coll}) = \frac{1}{2}\left\{\frac{\omega\tau_{\rm coll}}{[1 + (\omega - \omega_{\rm L})^2\tau_{\rm coll}^2]} + \frac{\omega\tau_{\rm coll}}{[1 + (\omega + \omega_{\rm L})^2\tau_{\rm coll}^2]}\right\} \tag{B.5b}$$

It is a simple, though tedious, matter to verify that Eqs. (B.5a) and (B.5b) do reduce to Eqs. (B.4a) and (B.4b).

Acknowledgments

I express my warmest thanks to Professor William Coffey and to Dr Paul C Fannin for many stimulating and informative discussions.

REFERENCES

1. R. E. Rosensweig, *Ferrohydrodynamics.* Cambridge University Press, Cambridge, UK, 1985.
2. W. T. Coffey, ed., Relaxation phenomena in condensed matter. *Adv. Chem. Phys.* **87** (1994).
3. P. C. Fannin, "Wide band measurement and analysis techniques for the determination of the frequency-dependent, complex susceptibility of magnetic fluids," *Adv. Chem. Phys.*, **104**, 181–292 (1998).
4. L. D. Landau, E. M. Lifshitz, and L. P. Pitaevskiĭ, *Electrodynamics of Continuous Media*, 2nd rev. ed. Pergamon, Oxford, 1984.
5. P. Debye, *Polar Molecules.* Chem. Catalog Co., New York, 1929.
6. A. Einstein, *Investigation on the Theory of Brownian Movement* (R. Fürth, ed., A. D. Cowper, trans.). Methuen, London, 1926.
7. J. McConnell, *Rotational Brownian Motion and Dielectric Theory.* Academic Press, London, 1980.
8. H. Goldstein, *Classical Mechanics*, 2nd ed. Addison-Wesley, Reading, MA, 1980.
9. L. D. Landau, and E. M. Lifshitz, On the theory of the dispersion of magnetic permeability in ferromagnetic bodies. *Phys. Z. Sowjetunion* **8**, 153–169 (1935).
10. J. H. Van Vleck and V. F. Weisskopf, On the shape of collision-broadened lines. *Rev. Mod. Phys.* **17**, 227–236 (1945).
11. H. Fröhlich, Shape of collision-broadened spectral lines. *Nature (London)* **157**, 478 (1946).
12. H. Fröhlich, *Theory of Dielectrics*, 2nd ed. Clarendon Press, Oxford, 1958.
13. H. A. Lorentz, *The Theory of Electrons*, 2nd ed. Teubner, Leipzig, 1916.
14. R. G. Breene, *The Shift and Shape of Spectral Lines.* Pergamon, Oxford, 1961.
15. L. Maisel, *Probability, Statistics and Random Processes.* Simon & Schuster, New York, 1971.
16. B. K. P. Scaife, *Principles of Dielectrics*, rev. ed. Clarendon Press, Oxford, 1998.
17. S. V. Vonsovskiĭ, ed., *Ferromagnetic Resonance* (D. ter Haar, ed., H. S. H. Massey, trans.), Int. Ser. Solid State Phys. Pergamon, Oxford, 1966.
18. C. Kittel, *Introduction to Solid State Physics.* Chapman & Hall, London, 1953.
19. W. A. Yager, J. K. Galt, F. R. Merritt, and E. A. Wood, Ferromagnetic resonance in nickel ferrite. *Phys. Rev.* **80**, 744–748 (1950)
20. J. C. Anderson, *Magnetism and Magnetic Materials.* Chapman & Hall, London, 1968.
21. C. Kittel, Interpretation of anomalous Larmor frequencies in ferromagnetic resonance experiment. *Phys. Rev.* **71**, 270–271 (1947).
22. C. Kittel, On the theory of ferromagnetic absorption. *Phys. Rev.* **73**, 155–161 (1948).
23. D. Polder, On the theory of ferromagnetic resonance. *Philos. Mag.* [7] **40**, 99–115 (1949).
24. C. L. Hogan, The ferromagnetic Faraday effect at microwave frequencies and its application — the microwave gyrator. *Bell Syst. Tech. J.* **31**, 1–31 (1952).
25. C. L. Hogan, The ferromagnetic Faraday effect at microwave frequencies and its applications. *Rev. Mod. Phys.* **25**, 253–263 (1953).

26. R. A. Waldron, *Ferrites: An Introduction for Microwave Engineers.* Van Nostrand, London, 1961.

27. J. A. Aseltine, *Transform Method in Linear System Analysis.* McGraw-Hill, London, 1958.

28. H. B. Callen, and T. A. Welton, Irreversibility and generalized noise. *Phys. Rev.* **83**, 34–40 (1951).

29. H. A. Kramers, La diffusion de la lumière par les atomes. *Atti Congr. Int. Fisici, Como* **2**, 545–557 (1927).

30. R. de L. Kronig, Theory of dispersion of X-rays. *J. Opt. Soc. Am.* **12**, 547–557 (1926).

31. R. Kubo, Statistical-mechanical theory of irreversible processes. I. General theory and simple application to magnetic and conduction problems. *J. Phys. Soc. Jpn.* **12**, 570–586 (1957).

32. G. W. Kenrick, The analysis of irregular motions with applications to the energy frequency spectrum of static and of telegraph signals. *Philos. Mag.* [7] **7**, 176–196 (1929).

33. Yu. L. Raĭkher and M. I. Shliomis, Theory of dispersion of the magnetic susceptibility of fine ferromagnetic particles. *Sov. Phys. — JETP (Engl. Transl.)* **40**, 526–532 (1975).

34. Yu. L. Raĭkher and M. I. Shliomis, Effective field method. *Adv. Chem. Phys.* **87**, 595–748 (1994).

35. C. Kittel, *Elementary Statistical Physics.* Chapman & Hall, London, 1958.

36. L. Néel, Théorie du trainage magnétique des ferromagnétiques én grains fins avec applications aux terres cuites. *Ann. Géophys.* **5**, 99–136 (1949).

37. K. S. Cole, and R. J. Cole, Dispersion and absorption in dielectrics. I. Alternating current characteristics. *J. Chem. Phys.* **9**, 341–351 (1941).

38. T. L. Gilbert and J. M. Kelly, Anomalous rotational damping in ferromagnetic sheets, *Proceedings of the Conference on Magnetism and Magnetic Materials*, American Institute of Electrical Engineers pp. 253–263, 1955.

39. W. F. Brown, Jr., Thermal fluctuations of a single domain particle, *Phys. Rev.* **130**, 1677–1686 (1963).

SIMULATING MOLECULAR PROPERTIES OF LIQUID CRYSTALS

J. CRAIN

Department of Physics and Astronomy, The University of Edinburgh, Edinburgh, EH9 3JZ, Scotland, United Kingdom

A. V. KOMOLKIN

Institute of Physics, Saint Petersburg State University, Saint Petersburg, 198904, Russia

CONTENTS

Advances in Chemical Physics, Volume 109, Edited by I. Prigogine and Stuart A. Rice
ISBN 0-471-32920-7

I. INTRODUCTION

In condensed phases we can consider three types of molecular motion: translational motion of the molecular center of mass; rotational motion of the molecular tensor of inertia (its principal axes); and intramolecular, conformational motion. We can also envisage two types of molecular ordering: translational ordering of molecular positions and orientational ordering of molecular bonds.

The degree of translational order of molecules in condensed phases is determined simply by the thermodynamic state of the system (i.e., whether it is a fluid or a solid) however, orientational mobility is not. Translational order is defined by a function of the molecular center of mass density $\rho(\Delta\vec{r})$, where $\Delta\vec{r}$ is a vector from the center of mass of an arbitrary molecule and the function ρ is averaged over all molecules in a sample. In principle, ρ is a three-dimensional function of three spatial coordinates. In some types of materials, we can find different degrees of ordering in different directions. For example, perfect order can be seen in defect-free single crystals, where ρ is close to a periodic delta function $\rho \approx \delta(n \cdot \Delta\vec{r})$. In glasses and isotropic liquids there is no long-range translational order. This implies that ρ is a constant. It should be noted that on lengthscales of order one molecular radius, this function is equal to zero even in liquids because molecules cannot interpenetrate. At longer distances, the function has at least one maximum that corresponds to the first coordination shell. The number of minima and maxima of the function is a measure of the degree of translational ordering, with fewer maxima (i.e., faster relaxation of the function, to its average value) corresponding to less ordered substances. The term *long-range ordering* in this context means that ordering exists on distances longer than a few molecular lengths.

By contrast, long-range orientational order is described by a tensor quantity $Q_{\alpha\beta}^{ab}$

$$Q_{\alpha\beta}^{ab} = \left\langle \frac{3}{2}(\vec{n}_a \cdot \vec{n}_\alpha)(\vec{n}_b \cdot \vec{n}_\beta) - \frac{1}{2}\delta_{\alpha\beta}\delta_{ab} \right\rangle \tag{1.1}$$

where indices a and b denote laboratory frame axes X, Y, and Z and α and β correspond to molecular axes x, y, and z. The notation $(\vec{n}_a \cdot \vec{n}_\alpha)$ is a dot product of unit vectors of corresponding axes, that is, it is the cosine of the angle between the axes. If the orientation of molecules is homogeneous in space, then all elements of the tensor are equal to zero. If it is possible to determine a direction in a sample, along which molecules are oriented preferentially (usually, this direction is taken as laboratory frame axis Z), then, the only nonzero elements of Q are designated as $S_{\alpha\beta}$:

$$S_{\alpha\beta} = Q_{\alpha\beta}^{ZZ} = -2Q_{\alpha\beta}^{XX} = -2Q_{\alpha\beta}^{YY} \tag{1.2}$$

Two values, S_{zz} and $(S_{xx} - S_{yy})$, play the most important role in a description of orientational order in a sample. The first is the *orientational order parameter*, and the second is the *biaxiality*. The expressions are

$$S_{zz} = \frac{1}{2}\langle 3\cos^2\Theta - 1\rangle = \langle P_2(\cos\Theta)\rangle, \quad (S_{xx} - S_{yy}) = \frac{3}{2}\langle \sin^2\Theta \cos 2\varphi\rangle \tag{1.3}$$

Θ and φ are the polar angles of the laboratory frame axis Z in the molecular frame, and P_2 is the second Legendre polynomial. The orientational order parameter is equal to 0 in unoriented systems, and it is equal to 1 in perfectly oriented systems.

Liquid crystals are intermediate phases of matter, so called mesophases, which partially exhibit properties of both isotropic liquids (molecular mobility, fluidity) and crystals (long-range orientational ordering, anisotropy of electrical, optical, mechanical properties). The essential property of liquid crystals and, in principle, their definition is the presence of long-range orientational order in a fluid. The preferred direction in a sample, along which molecules orient, is called the *director*.

A. Liquid Crystal Types and Transitions

The term *liquid crystal* refers to both mesophases and substances that can show liquid crystalline phases under special conditions. Substances, for which mesophases are induced by temperature, are called *thermotropic* liquid crystals. These are usually pure organic compounds or their mixtures. *Lyotropic* liquid crystals are multicomponent systems, and they form mesophases owing to the influence of solvents. Classification of liquid crystals can be found in a variety of monographs [1,2]. We will confine the discussion in this article to thermotropic liquid crystals. These phases are usually formed by molecules of strongly anisotropic shapes resembling rods or disks. Of course, anisotropy of molecular shape is not the only requirement for liquid crystal phase formation, and some highly anisotropic molecules show no liquid crystal phase at all upon cooling. The nematic phase is the most symmetrical among the liquid crystalline phases exhibiting orientational order only and no long-range translational order. If a substance forms several mesophases, the nematic phase is usually found at the highest temperatures.

The isotropic–nematic transition itself occurs with small changes in density ρ, enthalpy H, and entropy S; and its thermodynamic character is only weakly first order. In *para*-azoxy-anisole (PAA), for example, $\Delta\rho/\rho = 0.0035$, $\Delta H = 0.14$ kcal/mol and $\Delta S = 0.4$ cal/mol $\cdot$ K [3]. Near the phase transition, the isotropic phase shows so-called pretransitional character, by which several properties exhibit critical behaviour. The origin of this has been linked to the existence of short-range orientational molecular order.

According to Landau–deGennes theory [1] the spatial decay of these orientational correlations is of the Ornstein–Zernike form

$$\langle S(0)S(R)\rangle = \frac{1}{R} e^{-R/\xi(T)} \tag{1.4}$$

where S is the order parameter and R is a distance. The correlation length $\xi(T)$ depends on temperature according to

$$\xi(T) = \xi_0 \left[\frac{T^*}{T - T^*}\right]^{1/2} \tag{1.5}$$

In this expression, ξ_0 is a molecular dimension that is comparable to the cube root of the molecular volume, and T^* is a temperature slightly lower (1–2°) than T_{NI}. Typical correlation lengths just below T_{NI} are of order 10^2 Å.

Pretransitional effects are most directly manifested in measurements of electric and magnetic birefringence and scattering intensity and in a variety of dynamic measurements. Chiral, optically active molecules can form cholesteric phases. This structure acquires a spontaneous periodic twist of the director about an axis normal to the local director.

Lower symmetry liquid crystalline phases are associated with the appearance (usually at low temperatures) of partial long-range positional order and an associated spatially modulated density of the form

$$\rho(z) = \rho_0 + \rho \cos\left(\frac{2\pi z}{d}\right) \tag{1.6}$$

which implies that the molecules segregate into diffuse layers. This so-called smectic ordering is described by an additional order parameter that represents the amplitude of the density modulation. A large number of smectic phases exist and are distinguished according to the average orientation the molecules adopt relative to the layers.

In smectic-A, -B, and -E phases, the director lies essentially parallel to the layer normal. There are many more smectic phases — denoted by the letters C, I, F, G, H, J, and K — in which the molecular axes are tilted with respect to the layer normal at an angle that usually increases with decreasing temperature. The low symmetry of these tilted phases permit ferroelectricity when the liquid crystal molecules are chiral and impose a long-range helical periodicity that is also temperature dependent. Schematic illustrations of the most common liquid crystalline phases are shown in Fig. 1. It is also possible to induce chiral phases by incorporating nonmesogenic chiral dopant molecules into liquid crystals. More than 10,000 compounds are reported as showing mesophase behavior, and their number is increasing rapidly.

B. Molecular Properties for Device Applications

From a technological point of view, liquid crystals are among the most pervasive of all molecular electronic materials in devices because the molecular orientation can be influenced by applied fields. This can be exploited in applications ranging from low-power, high-resolution, fast displays, optical filters, and switching systems. As a result, they have been the focus of sustained and widespread interdisciplinary attention for many years. Also central to the successful operation of these devices is the requirement that the alignment of LC molecules on surfaces be promoted and reproducibly controlled. With the current move toward submicron circuitry, structures for driving overlying LC layers can be envisaged as small as 1.0 μm^2, in which case the molecules will be arranged in layers of only a few hundred molecules [4]. In such structures, the details of molecular properties and the nature of the LC–surface interactions become particularly important. Such properties are, however, not easily probed in detail by experimental methods; and at present there

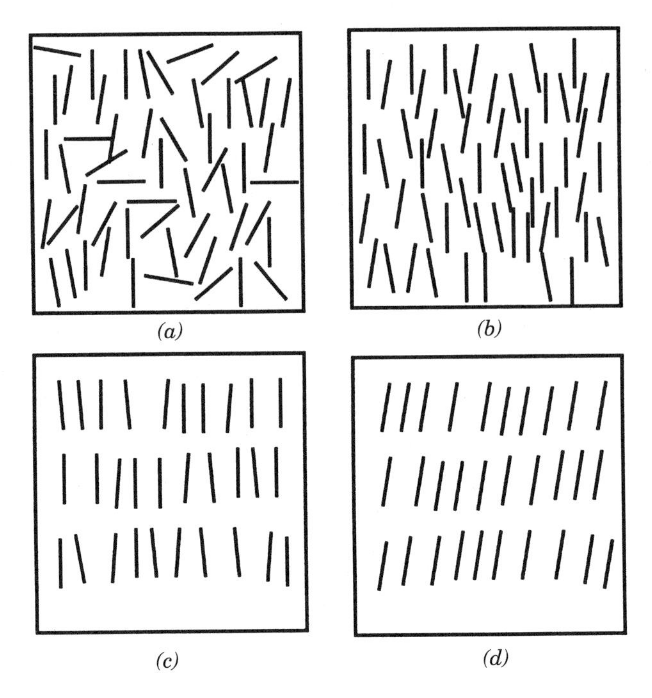

Figure 1. Illustration of the molecular orientational order defining several liquid crystalline phases: (*a*) isotropic, (*b*) nematic, (*c*) smectic-A, (*d*) smectic-C.

is little fundamental understanding of the nature of LC–surface interactions. It is clear, therefore, that molecular level modeling using powerful computational tools will become increasingly important in this field.

C. Role and Goals of Computer Simulations

Interestingly, some of the unique aspects of the liquid crystalline phase can be explored without *detailed* knowledge about the molecules that compose them. One of the most dramatic illustrations of this is Onsager's demonstration [5] that the competition between orientational entropy and free volume entropy leads to the formation of the orientationally ordered nematic phase in thin rod systems. Many reviews and seminal textbooks cover this and related areas in depth [1,2,6].

Computer simulations using simple molecular models have been applied with great success to explore the liquid crystal state. There are many reasons to study simple models. The most compelling of which is that large-scale, long-time molecular dynamics runs can be made. From these, important continuum properties, such as elastic constants and viscosities, are accessible from fluctuations and other dynamic behaviors of molecules. Furthermore, calculations based on model systems can be compared directly to the results of approximate, analytical theories, which are also based on idealized molecules. However, certain important liquid crystal properties (e.g., transition temperature, molecular order, phase stability, optical and interfacial properties, switching speed, and flow behavior) can be profoundly influenced by relatively small molecular-level modifications [1,6]. The simplest models cannot lead to realistic values for these thermodynamical quantities.

The urge to understand structure–property relationships in specific materials and to apply the power of computer simulation to predictive molecular design has begun to motivate the sacrifice of some of these advantages in favor of increased realizm of the molecular description. At this enhanced level of detail, one hopes to explore such issues as how liquid crystalline structure and properties are affected by molecular geometry, flexibility, and electronic structure. Further to this, there is the issue of how such properties can be calculated reliably either for isolated molecules or in the presence of others in condensed phases.

The importance of chemical detail is particularly significant in the emerging field of molecular electronics, which makes increasing demands for tightly prescribed functionality or multifunctional materials. These requirements are motivating efforts to understand the relationships between molecular structure and macroscopic properties and thus to improve molecular materials and device performance. In addition to advanced synthesis and characterization techniques, computer simulation is playing an increasingly valuable complementary role in the realization of molecular engineering as it moves closer to its goal of developing a realistic predictive capability.

The price paid for this increased specificity to real systems is that simulations are currently restricted to small sizes and relatively short time scales. Despite this, there have been several recent steps forward; and in view of its potential benefits and proliferation of computer power, this area of molecular simulation is likely to experience considerable expansion in the future.

In this chapter we review some of the recent advances in molecular-level simulation of liquid crystals. An attempt will be made to identify promising areas of future study, especially in light of forefront experimental capability. For completeness, we cover various types of molecular models in order of increasing molecular detail, but emphasis will be placed on so-called realistic models of liquid crystal phases and on new methods of determining intramolecular properties from parameter-free quantum mechanical methods.

We begin in Section II with examples of the relationship between molecular structure and material properties. In Section III we briefly review recent results obtained from idealized hard-body-excluded volume interactions and move on to consider the so-called soft anisotropic potentials and the influence of dipolar forces in Section IV. In Section V we consider the consequences of molecular detail and flexibility, with an emphasis on the use of atom–atom potentials to make connections to real molecular systems. Section VII is concerned with methods of extracting accurate intramolecular potentials ultimately for use in atomistic simulations of liquid crystals. For each of the different types of models, an attempt will be made to highlight recent examples of how it has been used to extract information on phase behavior, elasticity, dynamics, or intramolecular properties.

II. EMPIRICAL STRUCTURE–PROPERTY RELATIONSHIPS

A. Molecular Structure of Calamitic Liquid Crystals

Liquid crystal–forming compounds are not rare. Many thousands exist and are drawn from a wide variety of different chemical classes, although more than half contain derivatives of benzene. During the long history of liquid crystal physics and chemistry, a considerable amount of empirical information that relates molecular structure to material properties has emerged. Much of this work has in part been driven by the tight requirements for device applications. In this section, we give some examples of how chemical structure appears to affect material properties and introduce areas in which molecular modeling might deepen understanding and place existing empirical findings on firmer conceptual foundations. We begin by defining the basic building blocks of most calamitic (rodlike) liquid crystals and discussing the influence each of these mesogenic fragments appears to have on material properties.

In general, a liquid crystal molecule is composed of four basic structural units joined together to form an elongated molecule having typical overall dimensions of 5 × 20 Å. These constituent units are ring systems joined by a linking group that together form an essentially rigid core. At either end, various terminal groups are attached. At least one of these is usually a relatively long hydrocarbon chain. Some molecules may also have lateral substituents on the rings. An illustration of the general components of a calamitic liquid crystal is shown in Fig. 2.

The most common chain groups are

1. Alkyl groups ($C_n H_{2n+1}$)
2. Alkoxy groups ($C_nH_{2n+1}O$)
3. Alkenyl groups (C_nH_{2n-1})

The most common linking groups are ethylene, unsaturated double-bonded groups (ester, stilbene, azoy), and triple-bonded groups (acetylene and diacetylene). Examples of several linking groups are shown in Fig. 3.

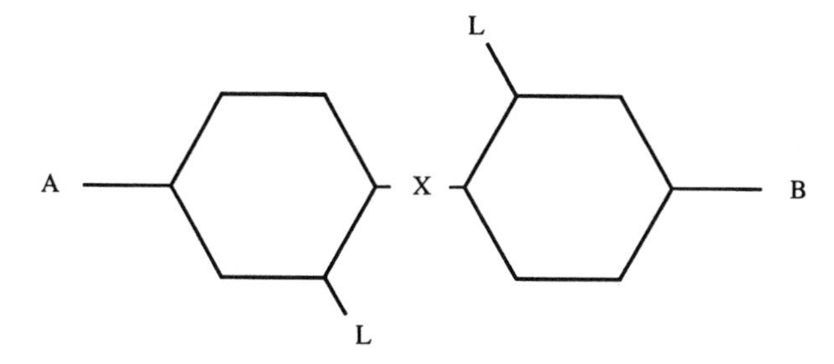

Figure 2. The main structural units of a rodlike calamitic liquid crystal showing the relative positions of the chain group (*R*), aromatic rings (*A* and *B*), a linking group (*Z*), and a terminal group (*X*).

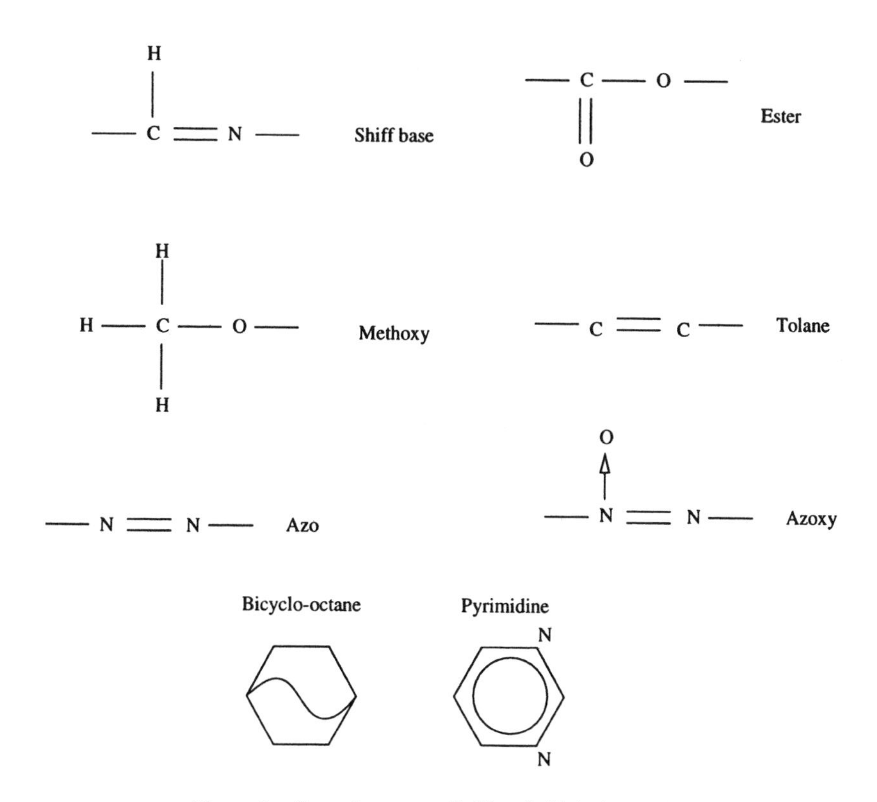

Figure 3. Several common linking (bridging) groups.

Examples of aromatic ring groups are normally cyclohexane; cyclocyclohexane; and phenyl, biphenyl, and terphenyl. Cyclohexane is a six-membered carbon ring with saturated σ bonds to 12 hydrogens. A phenyl ring is unsaturated and consists of mixed π and σ bonds to 6 hydrogens. Other heterocyclic systems have also been found to form liquid crystal phases. These include derivatives of thiophene, dioxane, quinoline, and benzothiazole.

Substitution of a chiral center for the terminal group can lead to chiral phases, as described in the previous section, as can the introduction of nonmesogenic chiral dopants [7].

Transition metal mesogenic complexes also exist, and the incorporation of paramagnetic species into liquid crystals can give rise to an important class of organometallic material that exhibits unusual magnetooptical properties [8]. Another class of liquid crystal includes those that are formed from dimers in which the molecular unit is composed of two (possibly different) mesogenic groups that are joined by a flexible spacer [9]. The phase behavior of these dimer systems has been shown to be sensitive to the chemical nature of the linking group and to possess properties different from its monomeric analogues. Computer simulation has just begun to explore this problem using simple molecular models [10], and there is considerable scope for future systematic studies.

B. Molecular Electronic and Optical Properties

Most electrooptical device applications of liquid crystals require a wide range of operating temperature, low operating voltages, fast switching times, steep voltage transmission response and sharp contrast between on and off states. These device attributes are ultimately related to molecular properties. For example, to operate over a range of temperatures, the liquid crystalline materials must exhibit a reasonably wide mesogenic range. To exhibit high contrast liquid crystals must be aligned well on solid substrates and be relatively free of defects. To minimize operating voltage, it is necessary to control combinations of dielectric and elastic properties. To reduce switching times, the elasticity, viscosity, and possibly spontaneous polarization must be optimized. The electron distribution in molecules and their long-range organization influence these properties and, therefore, the potential for exploitation in devices.

The electron distribution in mesogens can be quite different, depending (in most cases) on the spatial extent of delocalisation of π-bonded electrons (charge conjugation). In unsaturated systems characterized by delocalized electrons the optical spectrum is dominated by low-frequency π to π^* transitions, whereas saturated systems absorb light by σ to σ^* transitions. The latter occur at much higher energy, making the materials transparent in the visible part of the spectrum. Liquid crystals are also dichroic, meaning that optical absorption depends on the direction of the applied field relative to the molecular axis.

Optical properties can be tuned to some extent using different linking groups. The effect of this depends primarily on whether the linking group is saturated or unsaturated. In the latter case, electrons contribute to the effective conjugation length of the π bonds and thereby can change the optical properties. As a specific example of the relationship between electronic structure and physical properties, we consider the dielectric anisotropy $\Delta\varepsilon$, which determines the magnitude of the coupling between molecular orientation and applied fields. There have been a variety of models developed to link molecular structure and dielectric properties of anisotropic liquid crystals. Most of these lead to generalizations of the Lorentz–Lorenz relationship

$$\frac{\varepsilon - 1}{\varepsilon + 2} = \frac{4\pi N}{3}\alpha \tag{2.1}$$

which connects the average polarisability α to the dielectric permittivity of an isotropic fluid. N is the number density of molecules.

In the generalization to the case of liquid crystals, the dielectric anisotropy is related to two other important molecular electronic properties: the electric dipole moment and polarizability as follows

$$\frac{\varepsilon_{\|} - 1}{4\pi} = NhF\left[\langle\alpha_{\|}\rangle + \frac{F\langle\mu_{\|}^2\rangle}{kT}\right] \tag{2.2}$$

$$\frac{\varepsilon_{\perp} - 1}{4\pi} = NhF\left[\langle\alpha_{\perp}\rangle + \frac{F\langle\mu_{\perp}^2\rangle}{kT}\right] \tag{2.3}$$

where $\varepsilon_{\|}$ and $\varepsilon_{\perp}$ are the perpendicular and parallel components of the dielectric constant and h and f are called the cavity and reaction fields. Combining these two relations leads to an expression for the dielectric anisotropy

$$\frac{\Delta\varepsilon}{4\pi} = Nhf[(\alpha_l - \alpha_t) - F\frac{\mu^2}{2kt}(1 - 3\cos^2\beta)]S \tag{2.4}$$

Here, $(\alpha_l - \alpha_t)$ is the differential molecular polarizability defined such that $(\alpha_l - \alpha_t)S = \langle\alpha_{\|}\rangle - \langle\alpha_{\perp}\rangle$. This identifies α_l and α_t, respectively, as the longitudinal and transverse polarizabilities of the individual molecules. The quantity β is the angle between the dipole moment μ and the molecular axis, and S is the order parameter. The dielectric susceptibility itself is a tensor quantity that is described by its principal elements $\varepsilon_{\|} = \varepsilon_{zz}$ and $\varepsilon_{\perp} = \frac{1}{2}(\varepsilon_{xx} + \varepsilon_{yy})$. In biaxial phases, the polarizability (and any other second-rank tensor property) has three independent principal components. Other molecular theories for the dielectric permittivity of liquid crystals exist and have been discussed in references [11,12].

The complex bonding in liquid crystal molecules, in addition to their large size and low symmetry, makes it difficult to predict accurately the molecular electronic and dielectric properties of mesogens, although semiquantitative empirical relations do exist. Experimental results have been obtained for several liquid crystal-forming materials by extrapolation of refractive index measurements [13], use of depolarized light scattering [14], or the optical Kerr effect [15], although there are serious discrepancies in the results obtained by different methods in some cases [12]. As a result, molecular-level electronic structure calculations have much to offer in this important aspect of liquid crystal science (Section VII).

An extensive account of experimental results on the influence of molecular structure on dielectric permittivity in liquid crystals has been given by de Jeu [16]. In general, it is found that highly polar terminal substituents (cyano, fluoro, chloro) increase $\Delta\varepsilon$. Furthermore, the parallel component of the dielectric constant ($\varepsilon_{||}$) tends to increase with increasing alkyl chain length [17]. Variation of the linking group or the incorporation of an additional ring system (keeping terminal substituents fixed) can lead to large changes in the dielectric properties [16], depending on the symmetry and polarity of the bridge. Substituted phenylbenzoates have proved to show particularly complex behavior, owing to the presence of several sources of molecular dipole moments [16]. In materials having a strongly polar group, $\Delta\varepsilon$ can be either positive or negative, depending on the relative positions of the substituents. In the case where $\beta \approx 0$, the dipole moment lies along the long axis and $\Delta\varepsilon > 0$. The cyanobiphenyls are of this type and for 4-*n*-pentyl-4′cyanobiphenyl (5CB) for example, $\Delta\varepsilon = 10$ at a frequency of 1 KHz at room temperature. Laterally substituted species tilt the molecular dipole and lead to large values of β and a negative $\Delta\varepsilon$. In the case where $3\cos^2\beta = 1$ ($\beta = 55^\circ$), the molecular dipole makes an equal contribution to both $\varepsilon_{||}$ and $\varepsilon_{\perp}$. The dipolar contribution to $\Delta\varepsilon$ is positive for $\beta < 55^\circ$ and negative for $\beta > 55^\circ$. The overall sign of $\Delta\varepsilon$ is then determined by the relative magnitudes of the individual contributions.

For a nonpolar liquid crystal, the dielectric anisotropy is determined by the polarizability anisotropy and is expected to be relatively small. The dielectric anisotropy of liquid crystals ranges from -5 to about 25, and it usually increases with decreasing temperature. Of course, dielectric properties depend on the frequency of applied fields, and frequency-induced sign reversal of the dielectric anisotropy is also possible.

The influence of smectic order on dielectric permittivity can also be dramatic. In this case, the distribution of molecular centers of mass are no longer isotropic, and the dipolar interactions are different from those in the nematic phase. These interactions also depend on the precise origin of the dipole within a molecule (Section IV.B.1). For example, in a molecule with a central dipole, the distance between neighboring dipoles in different layers is much larger than the corresponding separation in the same layer. This usually results in a decrease in $\varepsilon_{||}$ and increase in $\varepsilon_{\perp}$ at the nematic smectic transition [16].

The macroscopic value of properties such as the dielectric anisotropy may differ from that expected from knowledge of the single molecule properties as a result of the relative orientation of molecules in condensed phases. For example, parallel alignment of molecules tends to enhance $\Delta\varepsilon$, whereas antiparallel arrangements reduce it. This local order has profound effects on the temperature dependence of the dielectric permittivity, which decreases with decreasing temperature in systems that associate in antiparallel, because the tendency to align in opposite directions increases at low temperature [18]. No simple rules exist for deciding on what the behavior will be for a specific molecule; however, the use of lateral substituents seems to suppress the tendency toward local antiparallal ordering in highly polar molecules.

The pressure-dependence of the dielectric anisotropy reflects changes in both molecular electronic structure and orientational order, and it has been measured for a few mesogens. For example, *n*-pentyl- and *n*-hexyl-cyanobiphenyl (5CB and 6CB) have been measured at constant temperature, and results were interpreted using the Maier–Meier equations [11] for the principal permittivity components in the nematic phase. The relation between $\Delta\varepsilon$ and the order parameter S, however, appears not to be properly described by these equations. This implies that the variation of the nematic order parameter cannot, by itself, account for the pressure dependence of the dielectric anisotropy [17] and changes to the molecular geometry or electronic structure must be considered.

In addition to molecular dipoles, liquid crystal molecules also posses higher electrostatic multipole moments (i.e., quadrupole moments), the magnitude of which also depends on the details of the molecular electronic structure. Little is currently known about empirical structure property relationships concerning higher-order moments, but accurate electronic structure calculations of higher-order moments may be an asset in this area.

C. Influence of Molecular Structure on Phase Stability

As stated previously, many of the most common mesogenic (liquid crystal forming) molecules comprise a cyclic or aromatic core unit with a single flexible chain at one end and possibly a polar substituent at the other end. The prototypical alkoxy-cyano-biphenyls (*n*CB) and alkoxy-benzyldiene-anilene (*n*OFBA) compounds are in this category as are members of the *n*OCB

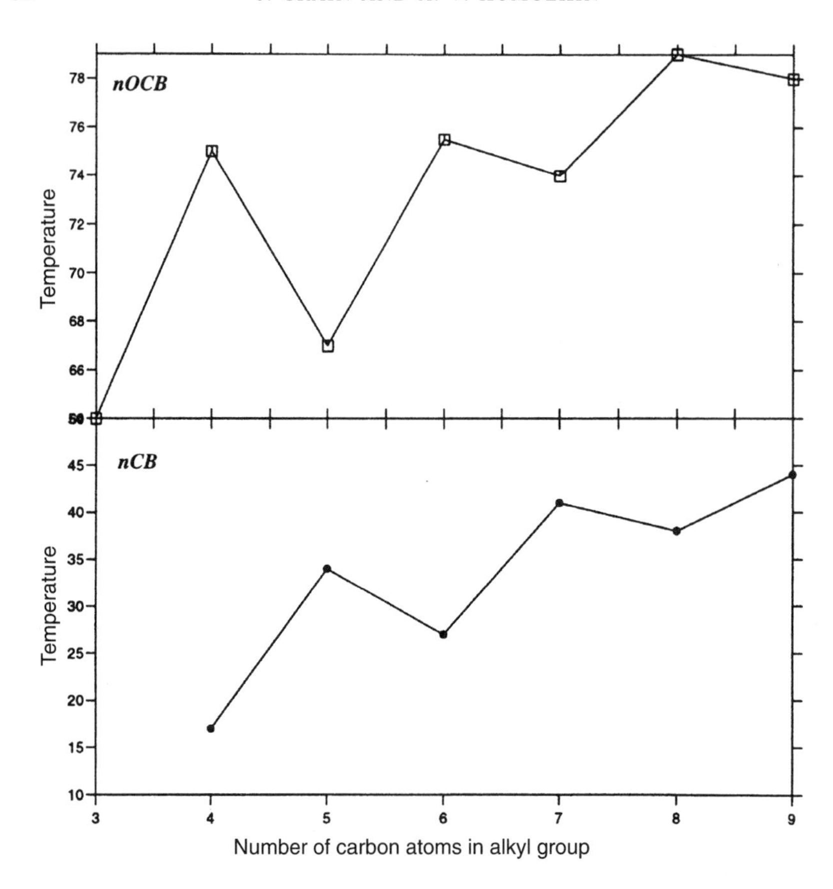

Figure 4. The alternation of the isotropic–nematic phase transition temperature with increasing length of the alkyl chain for the alkylcyanobiphenyl and alkoxycyanobiphenyl homologous series.

and *n*OFBA series. A variety of experiments have shown that these systems exhibit rich polymorphism simply as a function of chain length.

In fact, one of the most well documented properties of many liquid crystals is the alternation of the isotropic–nematic phase transition temperature with alkyl chain length. This so-called odd–even effect is illustrated for the cyanobiphenyl (*n*CB) homologous series in Fig. 4, where T_{NI} is consistently lower for relatively small even values of *n*. A partial explanation of this effect is that the molecules are more elongated and anisotropic in even homologues, thereby increasing the tendency to form the nematic phase. The extent of alternation decreases with increasing *n*, which is often attributed to the increased probability of nonextended conformers (i.e., less anisotropic molecules) [6].

However, the importance of this effect depends sensitively on the molecular flexibility. Smectic phases are common only for longer chain lengths, but the specific length at which smectic ordering is observed varies widely and appears to depend sensitively on the chemical structure of the other mesogenic fragments.

In the *n*OFBA series it is found that compounds with $n = 7$ or 8 show enantiotropic smectic-A and monotropic smectic-B phases. For $n = 6$, three enantiotropic liquid crystalline phases are formed (smectic-B and -A and nematic). For $n = 5$ and $n = 4$, only smectic-B and monotropic smectic-B, respectively, have been observed. No mesophases are observed for $n < 4$. This wide variety of phase behavior is strong testament to the importance of molecular detail in governing liquid crystal stability [19,20].

Empirical rules for the formation of tilted chiral phases also involve consideration of the chain structure: Tilted smectic phases are often found in nonlinear molecules containing two flexible alkyl chains of different lengths. Goodby [21] has suggested that dipoles associated with the core structure help drive the tilting. Branching of the terminal chains is also found to favor tilted phases. It is found, however, that phase behavior, helical sense, and spontaneous polarization are all sensitive to the location and configuration of the chiral center. Chiral centers located close to the aromatic groups tend to lead to large values of spontaneous polarization but narrow mesogenic range of the tilted phase [21]. In ferroelectric phases, it is generally found that the magnitude of the spontaneous polarization is determined by the dipole moment of the chiral species, although the core structure can have an influence. These ferroelectric liquid crystal materials have been exploited in devices such as displays, telecommunications, and advanced light modulators and are particularly important materials because they exhibit bistability and fast switching. The combination of two chiral units with a siloxane group has resulted in the formation of low-molar-mass organosiloxane antiferroelectric liquid crystal material [22]. The antiferroelectric ordering is attributed to competing intermolecular and intramolecular forces [22].

Lateral substitutions between rings can lower the conjugation of the electron distribution and thereby lower the polarizability. In this case, the contribution to LC stability arising from polarizability is also reduced. Lateral substitutions on the phenyl rings often lead to a larger effective width of the molecules, and this tends to reduce the stability of mesophases. In addition, ring systems may twist to reduce repulsive interactions between substituents. If, however, the substituent is capable of intramolecular hydrogen bonding, rotational motion may be suppressed and the nematic stability, enhanced [6]. The effect of lateral substitution can also be more subtle as in the case of difluorinated terphenyls. In this system, the relative position of the fluorine atoms dramatically affects phase behavior and transition temperatures [23] even for identical chemical compositions. Simulations on systems that differ from each other only by subtle chemical modifications have yet to be made.

It is not usually possible for a single liquid crystalline substance to satisfy all requirements for use in applications. As a result, mixtures of two or more liquid crystalline components can be used to tune certain properties (Section IV.A.2). In some cases the physical properties of eutectic mixtures (such as mesogenic range) can be reliably deduced from knowledge of the components [11]; however, this is not generally true. There may, for instance, be intervening phases in mixtures not present in the pure materials [21]. This is particularly problematic when tying to optimize dielectric anisotropy by mixing two materials having high and low $\Delta\varepsilon$. This often leads to the formation of smectic phases.

D. Polymorphism in Liquid Crystals at High Pressure

As was found for electronic and dielectric properties, pressure (density variation) can have a profound effect on phase stability of liquid crystals. In general, the effect of pressure on transition temperatures follows the Clausius–Clapyron relation

$$\frac{dT}{dP} = T\frac{\Delta V}{\Delta H} \tag{2.5}$$

where temperature T is related to volume ΔV and enthalpy change ΔH. This implies, for example, the isotropic–nematic transition will occur at higher temperatures as pressure is increased. In general, the pressure–temperature (P–T) curves will have different slopes; therefore, triple points are possible. A comprehensive discussion of pressure-induced effects can be found in [24]. A typical P–T phase diagram is shown for a liquid crystal [25] in Figure 5. High-pressure experiments have also revealed more complex behavior in phase stability. The compound 8OCB, for instance, shows a reentrant nematic phase upon cooling at high pressure [17]. In mean field theories of nematogens, a volume dependence of the molecular field potential is usually invoked and specified by an exponent γ. Experimentally, this quantity has been found to decrease with increasing chain length in the AOB homologous series. Moreover, it appears to depend on the position along the length of the alkyl chain being largest close to the aromatic core and smallest near the methyl group. Computer simulations of pressure-dependent properties using realistic molecular descriptions have so far been rare, although they are likely to be important in separating the effects of attractive and repulsive interactions in mesogens.

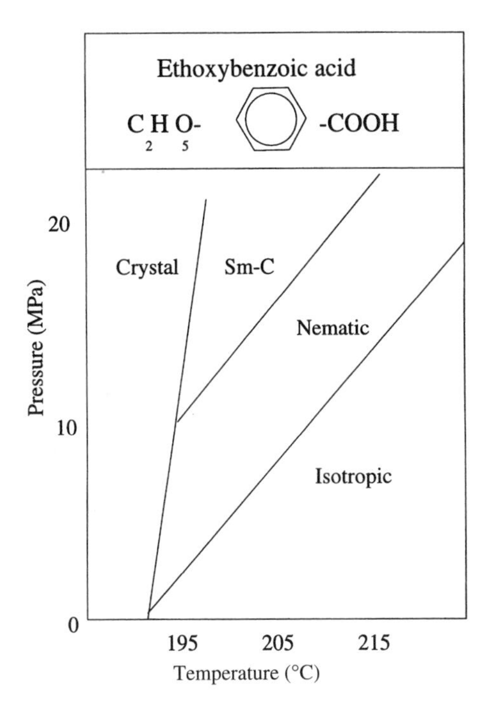

Figure 5. Pressure–temperature phase diagram characteristic of liquid crystals. Ethoxybenzoic acid is shown, illustrating two triple points and the regions of isotropic, nematic, smectic, and solid crystals.

E. Elasticity and Viscosity

By analogy to the case of solids, the elasticity of liquid crystals is a measure of the energy cost associated with deformations of the equilibrium director distribution. For most mesogens, elastic constants of order 10^{-6} dyn are typical. The elastic constants also depend sensitively on molecular structure and intermolecular interactions. The elastic constants are, therefore, also sensitive functions of temperature and order parameter. As yet molecular theories give insight into the behavior of elastic constants through order parameters, interaction potentials, and molecular shape but do not allow for an understanding of the influence of detailed molecular structure on elastic properties. In principle, computer simulation has a major role to play in this area.

For nematic liquid crystals, three types of independent distortions can be envisaged [Fig. 6]. These are the splay, twist, and bend deformations of the director, and a separate elastic constant (K_{11}, K_{22}, K_{33}) is associated with each of these.

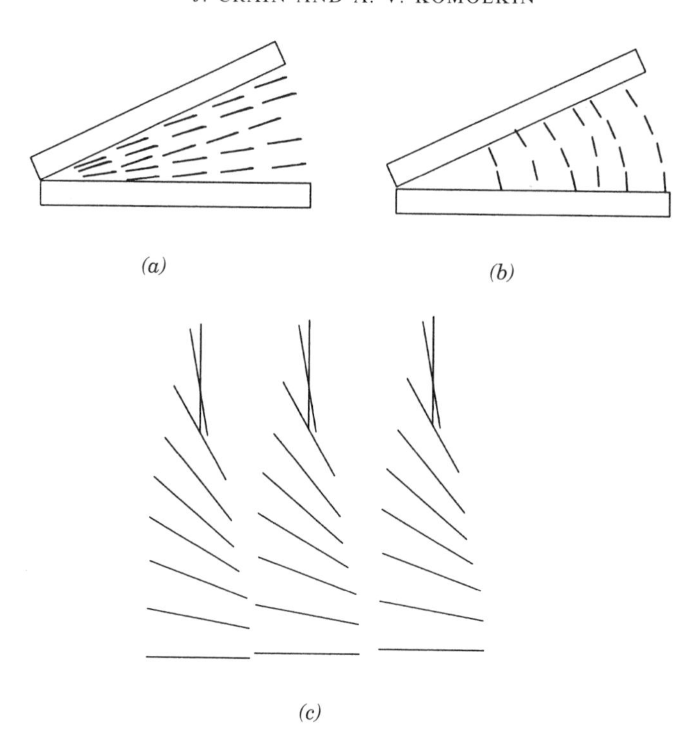

Figure 6. Independent deformations associated with the (*a*) splay, (*b*) bend, and (*c*) twist elastic constants for nematic liquid crystals.

The ratio $K_{33} : K_{11}$ is important in many display applications and tends to be between 0.5 and 3.0, depending on molecular type. Some evidence suggests that long chains give rise to low values of the bend : splay elastic constant ratio, which enhances the switching time of devices. A similar lowering of the bend : splay ratio can be achieved in materials having alkenyl tails. Mixtures of strongly and weakly polar mesogens have been found to bring about unusual nonmonotonic variation of the $K_{33} : K_{11}$ ratio with mole fraction.

Elastic constants of liquid crystals are often accessed experimentally by light scattering techniques that probe modulations in refractive indices induced by director fluctuations. In nematics, one has two fluctuation modes, which depend on combinations of elastic constants through

$$\langle n_1(\vec{q})^2 \rangle = \frac{kT}{V(K_{33}q_{\parallel}^2 + K_{11}q_{\perp}^2)} \tag{2.6}$$

$$\langle n_2(\vec{q})^2 \rangle = \frac{kT}{V(K_{33}q_{\|}^2 + K_{22}q_{\perp}^2)} \tag{2.7}$$

$$\tag{2.8}$$

where $q_{\|}$ and $q_{\perp}$ are wavevector components of scattered light relative to the director. Such fluctuations can also be exploited in simulations to extract elastic constants, as will be discussed in Section IV.A.3.

Even in relatively simple mesogenic systems, small modifications to the terminal groups influence both the bend and splay elastic constants as well as their temperature dependences [26]. Particularly striking is the observed nonmonotonic temperature dependence of the splay elastic constant for cyanobiphenyl ether with a butyl cyclohexane terminal group [26]. Additional elastic constants are needed to describe layer deformations in smectic liquid crystals.

Molecular dipolar properties, molecular shape, and elastic constants all enter the phenomenon of flexoelectricity in liquid crystals, which is similar to the piezoelectric effect in solids. Here, splay or bend deformations of the director in liquid crystalline phases of wedge-shaped or bent molecules, respectively, give rise to flexoelectric polarization [27] [Fig. 7]. For nematics this is quantified by flexoelectric moduli ε_{11} and ε_{33}

$$\varepsilon_{11} = 2\left(\frac{a}{b}\right)^{\frac{1}{3}} \mu_{\|} K_{11} \theta_0 N^{\frac{1}{3}}/kT \tag{2.9}$$

$$\varepsilon_{33} = 2\left(\frac{a}{b}\right)^{\frac{1}{3}} \mu_{\perp} K_{33} \theta_0 N^{\frac{1}{3}}/2kT \tag{2.10}$$

where $\mu_{\|}$ and $\mu_{\perp}$ are (as before) the parallel and perpendicular components of the molecular dipole moment, N is the number density, and a and b are parameters referring to molecular shape. The resulting polarization is then

$$\vec{P} = \varepsilon_{11}\vec{n}(\nabla \cdot \vec{n}) + \varepsilon_{33}(\nabla \times \vec{n}) \times \vec{n} \tag{2.11}$$

If a nematic material is placed in an electric field, there is a contribution to the free energy of the form $\vec{P} \cdot \vec{E}$.

The response times of liquid crystals are proportional to their viscosity. Nematic materials, however, require specification of six viscosity coefficients to determine fully their flow properties: There are three independent shear viscosities (v_1, v_2, and v_3), two bulk viscosities ($v_4 - v_2$ and v_5), and a rotational viscosity (γ_1). A complete measurement of all these coefficients exists only for a small number of materials. It more common to measure

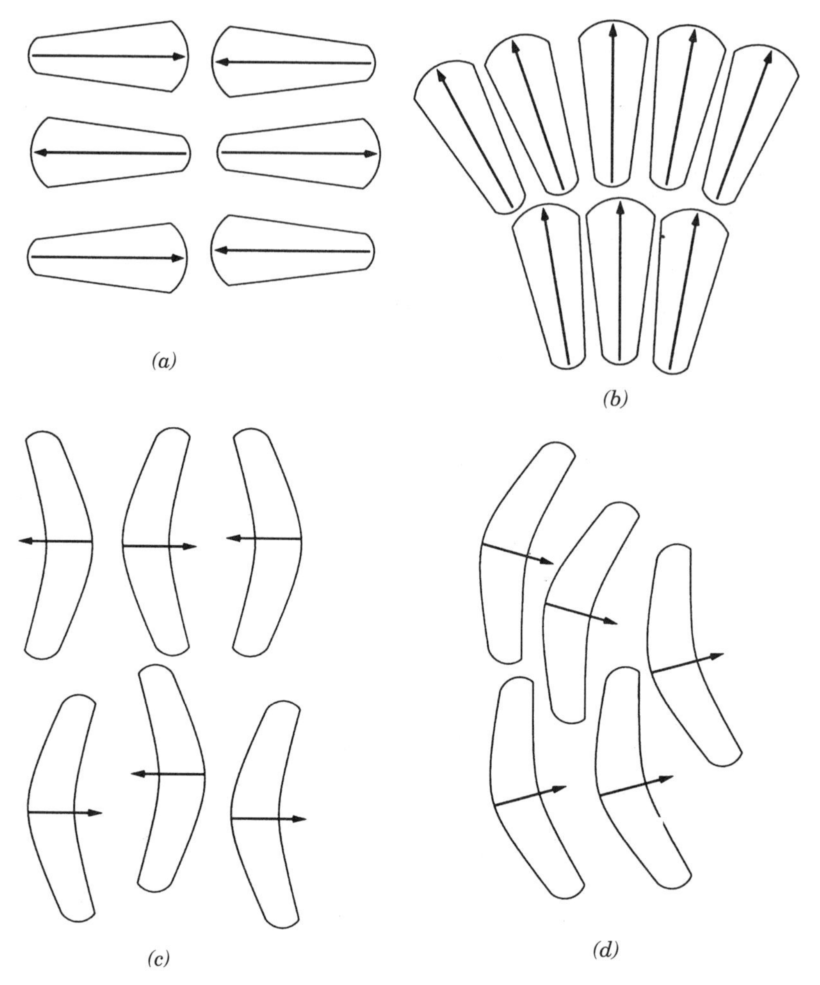

Figure 7. Generic shapes of mesogenic molecules that generate a spontaneous polarization (flexoelectricity) when the equilibrium director distribution is distorted as shown.

viscosity by techniques used for isotropic liquids. Such measurements normally access so-called Meisowicz viscosities [28], given by

$$\eta_1 = \nu_3 + \frac{1}{4}(1 - \lambda)^2 + \lambda\gamma_1 \tag{2.12}$$

$$\eta_2 = \nu_3 + \frac{1}{4}(1 - \lambda)^2 + \lambda\gamma_1 \tag{2.13}$$

$$\eta_3 = \nu_2 \tag{2.14}$$

Liquid crystal viscosities depend sensitively on molecular details and also show a strong temperature dependence generally of the form

$$\eta = \eta_0 \exp[E/kT] \tag{2.15}$$

where E ranges from 0.3 to 0.5 eV. However, nonmonotonic variation of shear viscosity with temperature has been observed in a few systems, such as PAA and MBBA [29], and some members of the alkyl cyano biphenyl series [30]. Longer or wider molecules (i.e., those with strongly polar lateral substituents) tend to have higher viscosities, as do those containing alkoxy tails. The position of certain linking groups (such as esters) is found to have dramatic effects on viscosity even in materials of identical composition. Related to viscosity are translational and rotational degrees of freedom, which will be explored in some detail later on. It is, however, appropriate to make the remark at this stage that molecular structure is closely connected to dynamical properties and orientational order; and in the case of translational diffusion anisotropy ($D_{\|}/D_{\perp}$), an explicit link between them [31] is often exploited:

$$\frac{D_{\|}}{D_{\perp}} = \frac{2\gamma(1-S) + 2S + 1}{\gamma(S+2) + 1 - S} \tag{2.16}$$

where $D_{\|}$ and $D_{\perp}$ are the parallel and perpendicular components of the translational diffusion coefficient relative to the director and $\gamma = 4l/\pi d$ involves the molecular length (l) and diameter (d).

It is clear from these examples that molecular structure and chemical detail exert major influences on nearly all measureable and exploitable material properties. The challenge facing computer simulation is to develop an understanding of the links between molecular properties and condensed phase behavior; computer simulation has yet to reach this stage, but the outlook is promising. In the next section, we outline several families of molecular models that have been developed to examine the influence of molecular properties on liquid crystallinity. We begin with the simplest and most familiar molecular models based on excluded volume interactions and progress to fully atomistic-level molecular descriptions.

III. NONPOLAR EXCLUDED-VOLUME MOLECULAR MODELS

A. Types of Excluded-Volume Models for Liquid Crystals

The hard sphere has long been recognized as a model system for studying phase behavior, crystallization, and nonequilibrium effects in atomic systems. Its simplicity and value have even motivated attempts to create the hard sphere

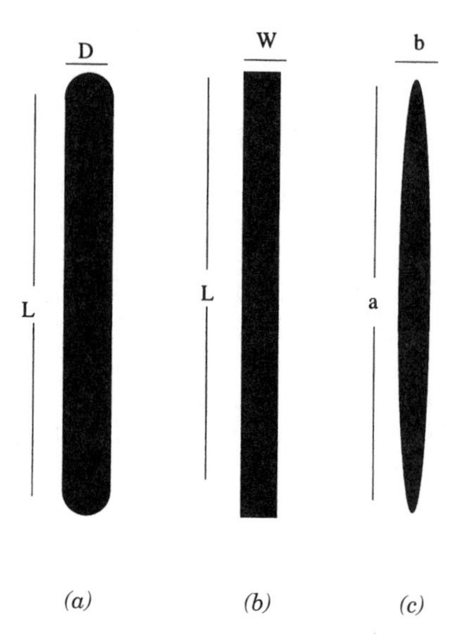

Figure 8. Excluded-volume, rigid molecular models of liquid crystals: (*a*) hard spherocylinder, (*b*) hard rod, and (*c*) hard ellipsoid.

system in the laboratory through the use of colloidal suspensions of submicron particles, which have short-ranged repulsive interparticle forces and, therefore, behave as nearly hard spheres [32]. The only interaction between particles is that which prevents interpenetration between the objects. This is called an *excluded volume* interaction.

Anisotropic molecular liquids can also be modeled using excluded-volume interactions by suitable generalizations of the hard sphere model. For the majority of liquid crystals, one of the simplest and most common molecular models is the spherocylinder, which is a rigid cylinder (of length L) terminated by a hemisphere (of diameter D) at each end [Fig. 8(*a*)]. In this system, the only relevant molecular property is the length : diameter ratio $L : D$.

Another model shape is the uniaxial ellipsoid of revolution [Fig. 8(*b*)], which has two semimajor axes of equal length ($a = b$) and a third axis (c) of a different length. Depending on these dimensions, molecules can be approximated either by prolate (rodlike) or (oblate) disklike objects. Details on simulation methods for these systems are given in the recent review by Allen et al. [33]. In view of the highly idealized nature of these models, direct comparison with experimental results on specific systems is not appropriate, although they do afford a valuable testing ground for theoretical predictions. Here as well, there have been experimental realizations of these models in the form of dispersions of rodlike particles [34] in which liquid crystalline phases have been observed. The important point is that there are no intramolecular degrees of freedom in these models, and no attractive interactions, and the only relevant thermodynamic variable is the density.

B. Phase Behavior and Material Properties

1. *Phase Diagrams*

Phase transitions in hard body systems are purely entropy driven, and temperature does not enter the phase diagram. Thermotropic phase transitions that are observed in real mesogenic systems are, therefore, inaccessible. Nevertheless, hard body models of anisotropic molecules exhibit considerable richness in their phase behavior, and there has been a great deal of recent effort and debate aimed at understanding the rigid rod phase diagram. It is now, however, well established that excluded volume models successfully account for the existence of orientationally ordered phases showing nematic [35,36], smectic [37,38], and columnar structures [39]. The tendency to form liquid crystal structures is influenced by the particular type of excluded volume molecular model chosen. For example, hard spherocylinders [Fig. 8(*a*)] tend to form smectics, whereas hard ellipsoids of revolution [Fig. 8(*b*)] tend to form nematics [37,38]. These trends have been explained on the basis of scaling arguments [40].

Veerman and Frenkel [39] made the first study of the hard spherocylinder phase diagram, although at that time it was not possible to survey a wide variety of L : D values. Since then, the phase diagram has been reexamined in greater detail. The shape anisotropy regime defined by $3 < L : D < 5$ has been explored most recently by McGrother and co-workers [41] using isothermal–isobaric NPT Monte Carlo simulations. This work has revealed that the smectic-A phase is the first mesogenic phase to form upon increasing the aspect ratio from the hard sphere limit. For $L : D = 3.2$, the observed phase sequence is I-SmA; nematic intervenes only for larger aspect ratios. For $L : D = 5$, isotropic, nematic, smectic, and solid phases are observed and are linked by first-order phase transitions. These results are in accord with earlier reports by Frenkel [36]. The phase diagram for hard spherocylinders with $3 < L : D < 5$ as a function of density and aspect ratio is shown in Fig. 9.

A wider range of L : D values and packing density has been investigated by Bolhuis and Frenkel [42]. These authors focused on identifying the orientational order–disorder transition in the crystalline solid phase, exploring the large $L : D$ regime (Onsager limit) and determining the location of triple points in the phase diagram.

Early computational studies aimed at exploring the nature of the isotropic–nematic transition in detail used several hundred hard ellipsoids and showed

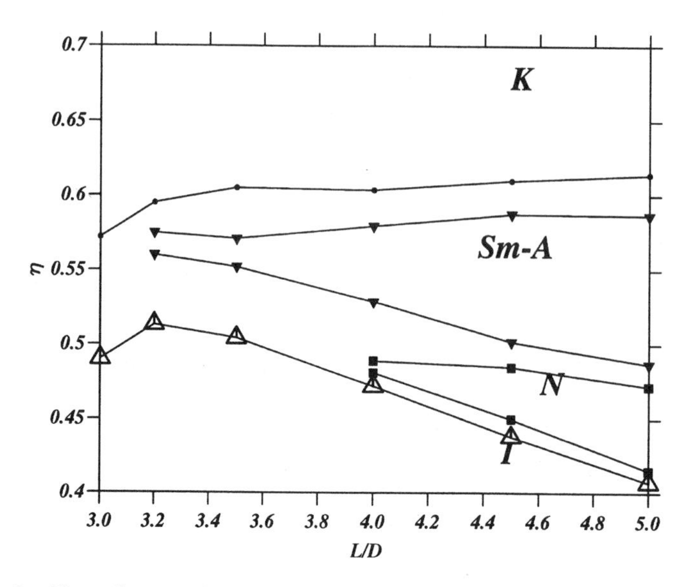

Figure 9. Phase diagram of nonpolar, rigid, hard spherocylinder molecules according to reference [41] as a function of density and the molecular aspect ratio $L : D$.

that collective reorientation slowed on approach to the transition [43]. These studies have now been extended considerably using other methods as will be outlined in Section IV.A.2.

In general, liquid crystal molecules are biaxial objects, which means that they are not cylindrically symmetric about their long axis. Theoretical investigations suggest that biaxiality weakens the N–I transition, and so-called biaxial phases and simulations using biaxial hard spheroids have begun to explore the effect [44]. In general, it has been shown that a hard ellipsoid [Fig. 8(*b*)] with three distinct axes exhibits a range of mesophase behavior, including nematic, discotic, and biaxial phases [45].

2. *Dynamic and Material Properties*

Hard particle models have also been used to study elastic [46,47] and dynamic [48,49] properties rather than simply phase behavior. The first of these calculations [46] led to the determination of Frank elastic constants for both prolate ellipsoids and ($L : D = 5$) spherocylinders. The order of magnitude and ratios of the calculated elastic constants were found to be reasonably close to expected values though the effects of boundary conditions were uncertain.

The translational self-diffusion coefficient is another important dynamic variable that is accessible to experiment through quasielastic neutron scattering (QENS) measurements, tracking of radioactive tracers, and nuclear magnetic resonance spin-echo techniques. Its component parallel to the nematic director ($D_{||}$) is defined by

$$D_{||} = \int_0^{\infty} < v_{||}(0)v_{||}(t) > dt \tag{3.1}$$

where the integrand is the velocity autocorrelation function.

In simulations of hard ellipsoids, this property has been observed to show an unusual nonmonotonic behavior with increasing density. It is found to increase initially and then to decrease, in contrast to the monotonic decrease found for $D_{\perp}$ [50]. This has been attributed to the corresponding density-induced increase in nematic order parameter, which suppresses the decorrelating effects of collisions parallel to the director.

Molecular diffusion in smectic phases is also possible, but dynamics of hard spherocylinders along the director are strongly confined, whereas diffusion perpendicular to the director seems to exhibit uncorrelated collisions [51]. Other features of molecular self-diffusion unique to smectic liquid crystals are the prospect of interlayer diffusion and the possibility of molecules moving to transverse interlayer positions [52,53]. The latter case corresponds to molecules existing between layers and having their long axes approximately normal to the director. Estimates of free energy barriers have revealed a considerable energy barrier to the formation of the transverse interlayer configuration [53].

IV. ATTRACTIVE INTERACTIONS AND ELECTROSTATIC FORCES

A. Anisotropic Attractive Interactions

1. *The Gay–Berne Model*

In addition to short-ranged, excluded-volume repulsive interactions, liquid crystal molecules may also interact through longer-range attractive forces that are ultimately electrostatic in origin. Recent surface freezing experiments on alkoxy-benzyldiene-aniline have revealed evidence for more exotic interactions, such as thermal casimir forces (which exhibit algebraic decay with distance) and short-range effective forces (which decay exponentially) [54]. The strength of these attractive interactions sets an energy scale by which temperature influences phase stability and gives rise to *thermotropic* transitions, which are absent in athermal excluded-volume systems. Also, as pointed out by deMiguel and co-workers [55], attractive interactions in general allow for gas–liquid phase separation, which is important in the study of interfacial properties.

One of the most common models by which to explore the influence of attractive forces is through the single-site Gay–Berne model [56]. The Gay–Berne (GB) potential can be regarded as an anisotropic version of the Lennard–Jones (LJ) potential normally suitable for uniaxial molecules, in which the strength and the range parameter depend on the orientations of the two particles and on their intermolecular vector. A detailed account of the development of the Gay–Berne model and its predecessors [57,58] is given by Luckhurst and Simmonds [59].

The form of the intermolecular interaction U is

$$U = 4\varepsilon(\hat{r}, \vec{u}, \vec{v})\left[\left(\frac{\sigma_s}{r - \sigma(\hat{r}, \vec{u}_i, \vec{u}_j) + \sigma_s}\right)^{12} - \left(\frac{\sigma_s}{r - \sigma(\hat{r}, \vec{u}_i, \vec{u}_j) + \sigma_s}\right)^{6}\right] \quad (4.1)$$

In this expression, r is the vector joining the center of masses of the molecules and $\hat{u}_i$ and $\hat{u}_j$ are unit vectors along the molecular axes. By analogy with the isotropic Lennard–Jones potential, σ and ε determine the range and strength of the interaction, respectively. The range, σ, depends on the molecular elongation $\kappa = \sigma_e/\sigma_s$, which is the ratio of the molecular end-to-end to side-by-side diameters. The orientation-dependent well depth ε is determined by two exponents μ and ν and the parameter $\kappa' = \varepsilon_e/\varepsilon_s$, which represents the ratio of the side-by-side and end-to-end well depths. Formally

$$\sigma(\hat{r}, \vec{u}_i, \vec{u}_j) = \sigma_s\left[1 - \frac{\chi}{2}\left(\frac{(\vec{u}_i \cdot \hat{r} + \vec{u}_j \cdot \hat{r})^2}{1 + \chi(\vec{u}_i \cdot \vec{u}_j)} + \frac{(\vec{u}_i \cdot \hat{r} - \vec{u}_j \cdot \hat{r})^2}{1 - \chi(\vec{u}_i \cdot \vec{u}_j)}\right)\right] \quad (4.2)$$

and $\chi = (\kappa^2 - 1)/(\kappa^2 + 1)$. Also

$$\varepsilon(\hat{r}, \vec{u}_i, \vec{v}_j) = \varepsilon_0 \varepsilon'^{\mu}(\hat{r}, \vec{u}_i, \vec{v}_j)\varepsilon'^{\nu}(\vec{u}_i, \vec{v}_j) \quad (4.3)$$

where $\varepsilon'(\vec{u}_i, \vec{u}_j) = [1 - \chi^2(\vec{u}_i \cdot \vec{u}_j)]^{-\frac{1}{2}}$ and

$$\varepsilon'(\vec{r}, \vec{u}_i, \vec{v}_j) = 1 - \frac{\chi'}{2}\left[\frac{(\vec{u}_i \cdot \hat{r} + \vec{u}_j \cdot \hat{r})^2}{1 + \chi(\vec{u}_i \cdot \vec{u}_j)} + \frac{(\vec{u}_i \cdot \hat{r} - \vec{u}_j \cdot \hat{r})^2}{1 - \chi(\vec{u}_i \cdot \vec{u}_j)}\right] \quad (4.4)$$

and $\chi' = (\kappa^{\frac{1}{\mu}} - 1)/(\kappa^{\frac{1}{\mu}} + 1)$. The function $\varepsilon(\vec{u}_i, \vec{u}_j)$ acts to favor parallel molecular alignment (nematic order), whereas $\varepsilon(\hat{r}, \vec{u}_i, \vec{u}_j)$ distinguishes between various parallel arrangements and may favor smectic arrangements. Despite the additional complexity of the GB description relative to the excluded volume models, there are still no intramolecular degrees of freedom present. There is only the geometrical molecular anisotropy. Further details

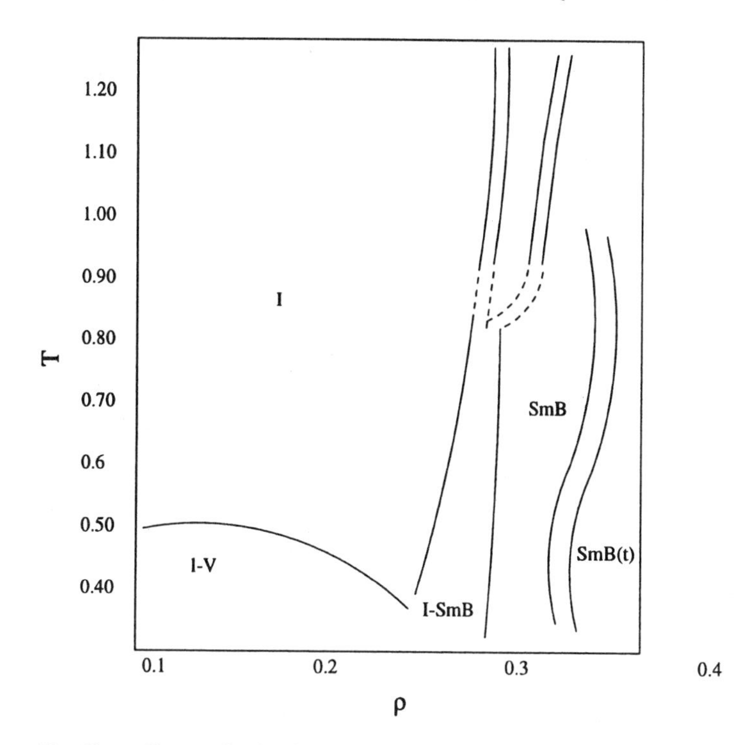

Figure 10. Phase diagram for the Gay–Berne fluid as a function of temperature and density.

concerning the GB potential as well as the explicit functional forms for the factors that enter the potential are reviewed by Rull [60] and Luckhurst and Simmonds [59].

2. *Phase Behavior*

As with all liquid crystal model potentials, one of the fundamental objectives is to understand how changes in the potential influence orientationally ordered phases and how the model parameters are related to molecular properties so that inference can be drawn for real systems.

The original parameterization of the Gay–Berne potential [56] was $\kappa = 3$, $\kappa' = 5$, $\mu = 2$, and $\nu = 1$; and it has been employed in this form in several investigations. Regarding phase behavior, it is now established that nematic and smectic-B phases exist for $\kappa = 3$ and $\kappa' = 5$ [55]. An approximate temperature–density phase diagram for the Gay–Berne fluid has been proposed by deMiguel and co-workers [55] for the $\kappa = 3, \kappa' = 5, \mu = 3,$ and $\nu = 1$ parameterization [Fig. 10]. This shows that the nematic phase is stable only at reduced temperatures $T^* > 0.8$ and that at lower temperatures a first order isotropic to smectic-B transition is observed. The presence of attractive interactions are found to be important in the formation of the smectic B phase. Additional complexity in the phase diagram may be introduced through variation of the energy exponents. The smectic-A phase appears to be particularly susceptible to variation of these parameters and can be eliminated completely in some cases [55,61,59]. Luckhurst and Simmonds [59] compared the phase behavior of the Gay–Berne model and hard ellipsoid system. They concluded that structures of the isotropic and nematic phases in the Gay–Berne sytsem are dominated by excluded volume effects. The smectic case is different. Below a critical value of the well depth anisotropy, no smectic phase is observed in the Gay–Berne system [59]. This implies that smectic ordering is not governed by excluded volume effects, as in the hard body case, but depends instead on attractive forces as well. As stated in the previous section, there is no smectic phase of hard ellipsoids [40], and the smectic phase of hard spherocylinders occurs only for relatively large aspect ratio.

DeMiguel and co-workers have identified the coexistence points of the isotropic–nematic transition in a Gay–Berne system [62] and have explored the influence of attractive interactions [63] by varying the parameter κ', keeping the values of κ, μ, and ν fixed at 3, 2, and 1, respectively. Isothermal compression of the isotropic phase results in successive transitions to the nematic and smectic phases for relatively large values of κ'. Decreasing T tends to destabilize the nematic phase until a direct isotropic to smectic B transition is observed [63]. Relatively low values of κ' are found to support coexistence between liquid crystalline and vapor phases [63].

So far it has proved difficult to make detailed comparisons of these results to those of real materials, because the potential parameter κ' cannot be directly associated with any specific molecular property. Also, attempts to parameterize simple phenomenologic models of liquid crystals to make them more representative of real molecular systems have been relatively rare. An exception is the recent effort to derive a Gay–Berne parameter set for *p*-terphenyl in which the molecular biaxiality was averaged out [59]. The strategy involved comparison of the Gay–Berne potential to a site–site interaction potential for the molecule. The parameters derived in this way were different from those used in other implementations of the Gay–Berne functional form; but nevertheless, they led to isotropic, nematic, and smectic ordering. The possibility of using information from molecular electronic structure calculations to parameterize such potentials is promising but has yet to be explored (Section VII).

Studies of Berardi and co-workers [64] of ensembles of GB particles with $\mu = 1$ and $\nu = 3$ show the existence of smectic, nematic, and isotropic phases. The behavior of the systems at the NI transition is consistent with a first-order transition. Calculated heat capacity, c_v, exhibits a well-defined peak at this point. Estimated values of $\langle P_2 \rangle$ and $\langle P_4 \rangle$ are greater than that expected from a simple Maier–Saupe model; however, the temperature dependence of $\langle P_2 \rangle$ is similar to that obtained in experiments for a large number of compounds.

A method of separating the influence of attractive and repulsive interactions on mesophase formation is to consider variations of orientational order P_2 with respect to either isothermal changes in volume $(\partial \langle P_2 \rangle / \partial V)_T$ or volume-conserving changes in temperature $(\partial \langle P_2 \rangle / \partial T)_V$.

These can be conveniently subsumed into a single parameter Γ [65] defined as

$$\Gamma = -(\partial \ln T / \partial \ln V) \tag{4.5}$$

The limiting cases occur when intermolecular interactions are either entirely repulsive or entirely attractive. In the former case, only the denominator influences orientational order, and Γ diverges. In the latter case, $\Gamma = 1$ is expected. The quantity Γ is experimentally accessible from combined isothermal and isobaric studies of orientational order, and results suggest that it depends sensitively on detailed chemical characteristics of liquid crystal molecules. The reported value of Γ for methoxyazoxybenzene is about 4.0 [65], whereas it is found to be only half this value for the hexylkoxy member of the homologous series [66]. Considerably higher values are found for 5CB [67,68].

The first simulations leading to reasonable estimates of Γ have now begun to emerge; the first of these are obtained using a novel parameterization of the Gay–Berne potential [69]. Here, a value of $\Gamma = 5.7 \pm 0.2$ was found. As this is rather high, there appears to be a slight tendency toward overestimation of the importance of attractive interactions. It is clear, however, that simulations aimed at estimating the value of Γ will be important in exploring the roles of attractive and repulsive forces.

Molecular simulation studies of the nematic–isotropic transition and associated pretransitional phenomena have been rare, owing the computational cost; however, there have been promising steps forward. In particular, Gay–Berne studies of orientational correlations have shown remarkable agreement with the predictions of the Landau–deGennes theory [70] and thereby have begun to provide insight into the nature of pretransitional phenomena at the molecular level.

The original Gay–Berne potential has now been modified to represent interactions between two different molecules [71] to handle mixtures. This generalized potential has been used to study equimolar mixtures of molecules with axial ratios 3.5 : 1 and 3 : 1 [72]. In this system, an unusual "presmectic" behavior was observed in simulations using the Gay–Berne potential in

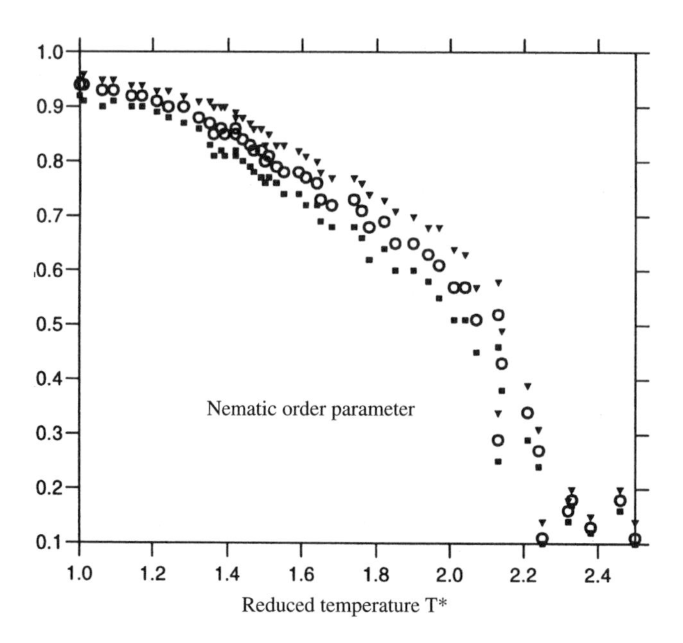

Figure 11. Variation with temperature of the orientational order parameters obtained for components of a mixture of Gay–Berne systems with differing aspect ratios.

which smectic density modulations for the different species are out of phase [72]. The phase difference decreases as the smectic transition is approached. The longer molecules are also observed to be more highly ordered than the shorter molecules, in accord with theoretical predictions [73,74]. Fig. 11 shows the nematic order parameter as a function of temperature for an equimolar mixture of GB particles. As with hard particle systems, there have also been efforts to generalize the GB model to include the effects of molecular biaxiality [71,75,76] and to adapt it for discotic molecules [77].

3. *Bulk Material Properties*

Elastic constants for the Gay–Berne model have been determined from long-wavelength bend, twist, and splay fluctuations [78,79] based on expressions similar to those in Eq. (2.8) and from direct correlation functions [80] in the nematic region of the phase diagram. Increasing density leads to an approximately uniform increase in the three elastic constants with $K_3 > K_1 \approx K_2$ [79].

Because elastic constants depend both on density and on order parameter, comparisons among different models may not be straightforward.

Nevertheless, the results obtained from orientational fluctuations appear to be in reasonable accord with earlier work on hard ellipsoids (despite the difference in interaction potential) [46] but are substantially different from results from a direct pair correlation function approach (obtained by inversion of the Ornstein–Zernike equation) for the same potential parameterization and identical state point in the nematic phase. The origin of the discrepancy is not clear, though some suggestions have been proposed [78,79]. There is also controversy surrounding surface elastic constants, which are found to be much smaller than bulk elastic constants [80]. Some surface elastic constants are found to be negative, which suggests surface instabilities in the weak-anchoring limit or in free-standing films. Previous GB simulations on liquid crystal surface elastic constants, however, showed that they were sensitive to the details of the intermolecular potential [81].

Single molecule dynamic properties have not been explored in the Gay–Berne system extensively; however, the self-diffusion coefficent $D_{||}$ shows a similar nonmonotonic variation with density [82] to that observed for hard ellipsoids [50].

Collective dynamic properties of the Gay–Berne nematic phase have been explored through simulations of viscosity coefficients [83], as obtained from time correlation functions of the stress tensor [84]

$$\sigma_{\alpha\beta} = \frac{1}{V}\left(\sum_i \frac{p^i_\alpha p^j_\beta}{m} + \sum_i \sum_{j>i} r^{ij}_\alpha f^{ij}_\beta\right) \tag{4.6}$$

where $\alpha, \beta = 1, 2, 3$; V is the volume; $\vec{p}^i$ is the linear momentum of molecule i; $\vec{r}$ and $\vec{f}$ are relative positions and forces between molecules i and j. These simulations account for the relative magnitudes of bulk and shear viscosities and for the observed temperature dependence of the shear and rotational viscosities and, in particular, reproduce the experimentally observed [29,30] nonmonotonic variation of one of the three shear viscosities (Section II.E). A variation of the GB potential consisting of only repulsive terms [85] was used to study the viscosity in the nematic phase for planar Couette flow: the Mięsovicz, shear, and twist viscosities have been evaluated from the nonequilibrium molecular dynamics [85], and the results agree well those obtained from Green–Kubo relations [84].

Also related to the molecular motion is the thermal conductivity $\lambda_{\alpha\alpha}$ given by

$$\lambda_{\alpha\alpha} = \frac{V}{kT^2}\int \langle J_{Q\alpha} J_{Q\alpha}\rangle \, dt \tag{4.7}$$

where $J_{Q\alpha}$ is the component of the heat flux in the α direction. The thermal conductivity in systems of prolate (3 : 1) and oblate (1 : 3) Gay–Berne systems has been demonstrated to be anisotropic [86]; the thermal conductivity parallel to the director is larger than that perpendicular to the director (i.e., $\lambda_{\|\|} > \lambda_{\perp\perp}$) in the prolate case. The opposite situation was found for the oblate system.

4. *Confined Geometries and Interfaces*

In addition to the obvious importance in devices, there is also great fundamental interest in the study of liquid crystals in a variety of confined geometries [87,88]. Droplets form one of the purest examples of a confined system and are appropriate models for polymer-dispersed liquid crystals [89]. Gay–Berne simulations reveal that at low temperatures a layer of radially aligned particles is formed in droplets, but this ordering does not propagate toward the center of the drop unless the drop is relatively large. Instead, concentric shells of aligned particles are formed, with a small smectic-like domain in the center. There is, however, some sensitivity to the form of the potential.

Interfaces between different condensed phases are also important areas of interest, but these have received relatively little attention at the molecular level. A notable exception in this area is the work of Bates and Zannoni [90], who explored orientational order at the nematic–isotropic interface of a Gay–Berne system and found planar as opposed to homeotropic or tilted alignment.

A more general confined geometry is one in which the liquid crystal is restricted to lie between two bounding surfaces, such as in most devices. Here the central quantity is the anchoring free energy, which is usually expressed according to the scheme of Rapini and Papoular [91] as

$$F = \frac{1}{2} A_\theta \sin^2(\theta - \theta_c) = \frac{1}{2} A_\theta [1 - (\hat{n} \cdot \hat{n}_c^2)] \tag{4.8}$$

for an anchoring geometry exhibiting cylindrical symmetry about the direction $\hat{n}_c$. In this expression, $\hat{n}$ is the surface director and θ and θ_c are the polar angles of $\hat{n}$ and $\hat{n}_c$, respectively. The Gay–Berne model has begun to be featured in the first simulations of LC–surface interactions [92,93]. In the recent work of Zhang and co-workers [92] a rubbed polymer alignment surface is modeled by an effective potential. This work suggests that the bulk pretilt angle is controlled by the surface through the orientation of the adsorbed liquid crystal monolayer. This is consistent with the results of recent experimental studies [92]. Further studies aimed at exploring the

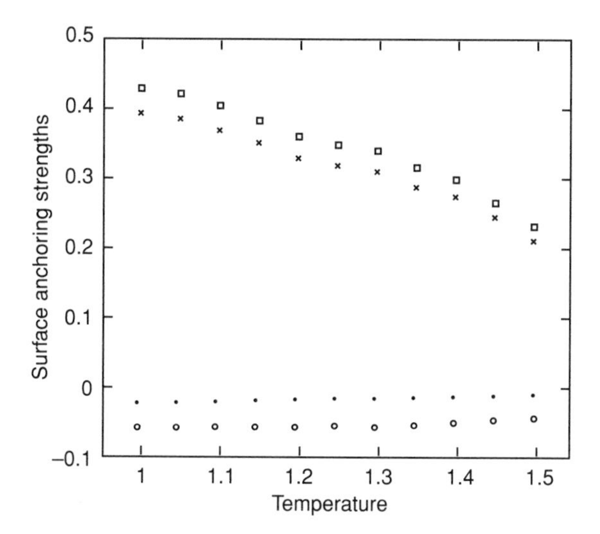

Figure 12. Temperature-dependent anchoring strengths for the Rapini–Papualar coefficients as reported in Ref. [93].

molecular origin of liquid crystal anchoring have been reported in the demonstration of homeotropic anchoring in a Gay–Berne system [93]. In addition to this, surface-induced smectic layers were observed to propagate into the bulk material with a temperature dependent coherence length [93]. Here, surface anchoring strengths were found to be about one order of magnitude smaller than bulk elastic constants and also temperature dependent, as shown in Fig. 12. These findings support the intuitive view that surface-induced orientational order propagates into the bulk material across over distances up to some macroscopic correlation length. However, a complete counterexample is found in Gay–Berne simulations in which molecular monolayer orientation at a graphite surface is uncorrelated with bulk orientation [94]. The intermolecular interactions are chosen to replicate the temperature dependence of order parameters and clearing temperature of 8CB. The molecule–surface interaction is formed by a combination of heteromolecular Gay–Berne interactions among mesogenic molecules and spherical carbon atoms (for the first two surface layers) and a Fourier series Steele potential [95] for lower-lying layers. The results imply that adsorption energy dominates over intermolecular interactions. Further systematic studies are certainly required to elucidate the mechanism of surface–molecule and monolayer–bulk interactions.

B. Influence of Dipolar and Quadrupolar Forces

1. Strength and Location of Molecular Dipoles in Hard and Soft Potential Systems

In general, the molecules that form liquid crystal phases are of low symmetry and as a result may posses a nonvanishing electric dipole moment. This represents an additional intermolecular interaction of considerable complexity, characterized by long-range and pronounced anisotropy. There is now evidence to suggest that highly polar molecules exhibit unusual properties, such as smectic bilayer formation [96] and reentrant phenomena [97]. Despite its importance, understanding the effect of molecular dipoles on mesogenic phase stability and orientational order has proved complex and sometimes controversial. Activity in this area has also served to highlight, in a general way, the importance of chemical detail in governing material properties.

Systematic experimental study of the influence of dipolar forces on mesophase properties is complicated, because it is not generally possible to tune the polarity of molecules while keeping all other intramolecular degrees of freedom unchanged. Computer simulation does not suffer from this limitation and, therefore, takes on added significance in this regard. Because dipolar forces are of long range, however, their inclusion in computer simulations is usually costly (requiring repeated evaluation of Ewald sums [98]); and this has limited the system sizes that can be studied.

Even in a simple molecular model of a liquid crystal, there are several scenarios to consider regarding location and direction of the dipole moment. For example, a point dipole in an anisotropic molecule may occupy a high symmetry position at the molecular center of mass (a central dipole) [Fig. 13(*a* and *d*)] or it may be displaced toward the end of the molecule (a terminal dipole) [Fig. 13(*b* and *c*)]. Moreover, the vector dipole moment may point either along the molecular long axis (longitudinal case) [Fig. 13(*a* and *b*)] or at right angles to it (transverse case) [Fig. 13(*c* and *d*)]. The main immediate questions are How do the magnitude, location, and direction of the molecular dipole influence stability and orientational order of the nematic and smectic phases? and How are the molecular dipoles arranged in mesophases?

Some of these issues have been explored in a variety of theoretical studies using methods such as a variational cluster expansion [99], a "scaled Onsager" formulation for the free energy [100], and a type of perturbation theory that incorporates many-body contributions [101]. All three methods lead to the conclusion that the location of the dipole in a molecule plays an important role in phase stability; however, the first two of these theoretical approaches to the problem suggest that central, longitudinal dipoles favor the

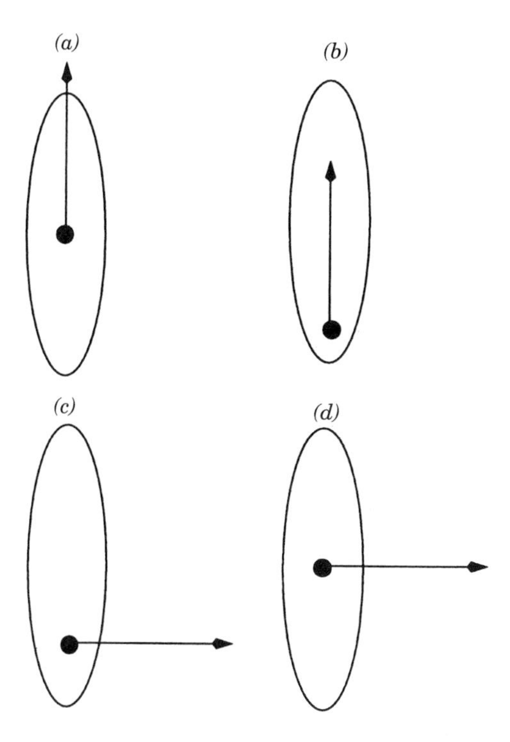

Figure 13. Possible combinations of point dipole position and orientation in a general calamitic liquid crystal, showing examples of (*a*) central longitudinal, (*b*) central transverse, (*c*) terminal transverse, and (*d*) terminal longitudinal cases.

nematic phase over the isotropic phase, whereas the latter method leads to the opposite conclusion. The difference is attributed to the presence of implicit frustrations that are absent in theories limited to two-body correlations [100].

We consider first the most symmetric case of an elongated molecule incorporating a central dipole directed along the long axis (the central longitudinal case shown in Fig. 13(*a*)). This scenario has been modeled recently [98] using hard spherocylinders ($L : D = 5$), by which dipolar interactions can be superimposed on the excluded volume repulsion through an extra term of the form

$$U_{\text{dipole}} = \frac{\mu_i \mu_j}{r^3}[\hat{\mu}_i \cdot \hat{\mu}_j - 3(\hat{\mu}_i \cdot \hat{\mu}_j)(\hat{\mu}_j \cdot \hat{\mu}_j)] \tag{4.9}$$

In the nonpolar case, this model exhibits isotropic, nematic and smectic-A mesophases [42,102]; but the addition of the central longitudinal dipole destabilizes the nematic phase (N) relative to both the isotropic (I) and smectic phases (Sm-A). Moreover, in the polar system, the nematic phase is eliminated completely below an $I - N -$ Sm-A triple point. Studies combining Monte Carlo simulation and thermodynamic perturbation also imply that the isotropic phase is stabilized by the presence of interactions between central dipoles [103]. These findings are in conflict with the theoretical results of Vega and Lago [100] and Williamson and Del Rio [101].

In soft-potential models (such as the Gay–Berne mesogen), dipolar interactions can also be added to the interaction potential in the same way as for excluded volume cases. In terms of phase behavior for the dipolar Gay–Berne system, the transition temperature from the isotropic to nematic phase is found to be insensitive to the value of the central, longitudinal dipole moment [104]; however, the nematic phase was found to be destabilized relative to the smectic phase [104]. This insensitivity of T_{NI} to the longitudinal central dipole is unlike the behavior of the dipolar hard spherocylinder case, in which a clear suppression of the nematic relative to the isotropic phase was found. The result is also unexpected from the theoretical results [100,101]. The dipole-induced instability of the nematic phase relative to Sm-A appears to be accounted for in both models.

Shifting the longitudinal dipole from a central to a terminal position has several consequences. Gay–Berne simulations [105] imply that the nematic phase is stabilized relative to the isotropic, whereas results on $(L : D = 5)$ spherocylinders with terminal dipoles [102] suggest that the nematic phase is slightly destabilized. The preference for isotropic relative to nematic order in hard particle systems with terminal dipoles may be the result of dipolar pairing, which leads to a reduction in effective aspect ratio of the molecules. The same argument applies in the case of central dipoles. This is perhaps overcome in the Gay–Berne systems by the presence of additional attractive interactions. In fact, there is some experimental evidence to support the results obtained on the Gay–Berne system based on the observation that T_{NI} for mesogenic derivatives of phenylpyrimidine is increased in compounds with terminal as opposed to central longitudinal dipoles [105].

Both hard [102] and soft [105] potential models imply that the terminal dipole influences the nematic–smectic transition. The most dramatic effect is seen in the hard spherocylinder simulations; in which the smectic-A phase is destabilized relative to the nematic. This behavior is also different from that found for central longitudinal dipoles, in which the smectic phase was stabilized relative to the nematic. The origin of the smectic instability may be owing to the dipole-induced local ordering, giving rise to staggered dimer configurations that cannot be packed efficiently into smectic layers.

Unlike the case of the symmetric longitudinal dipole, the presence of a transverse molecular dipole [Fig. 13(*c* and *d*)] leads to situations in which the directions of molecular orientational order and dipolar order are not colinear. This implies that, in the smectic-A phase for example, the dipole orientation is free to point in any direction in the plane normal to the director. Therefore, in addition to influencing phase behavior, the problem of collective organization of transverse dipoles becomes much richer than in the longitudinal case.

In hard spherocylinder systems, the effect of a transverse central dipole is similar to that for the longitudinal case, in which the smectic-A phase is stabilized relative to the nematic phase [106]. Soft potential Gay–Berne simulations comprising mesogens with transverse central dipoles have also been performed, but in this case a parameterization was chosen such that the nonpolar molecule exhibited a direct transition from isotropic to smectic-B. Therefore, no conclusions were drawn regarding the influence on nematic stability. It was, however, found that the N to Sm-B transition was not strongly influenced by the transverse dipolar interactions; although partial ordering before the transition was found, and orientational order in all phases was enhanced. There has been comparatively little work on elucidating the phase diagrams of dipolar rod-like molecules, although recent theoretical studies on hard rods with various longitudinal dipoles have begun to show promise in this area [107].

Collective organization of the dipoles depends on a balance between interactions favoring antiparallel orientation of dipole pairs and the formation of domains that have the same orientation. This balance also evidently depends in a nontrivial way on dipole position in the molecule [108]. For the central longitudinal case, the dipoles are arranged at random in a given smectic layer, and dipolar orientation in adjacent layers is uncorrelated [Fig. 14(*a*)]. In the shifted longitudinal case, large groups of common dipole orientation are found in a single smectic layer, and adjacent layers adopt the opposite orientation. The central dipole case also exhibits better defined layering with less interdigitation [Fig. 14(*b*)]. In the case of prolate spherocylinders with noncentral longitudinal dipoles [109,110], smectic phases form nonpolar monolayers with some interdigitation of the head groups [109]. In both soft [111] and hard potential [106] systems, the molecular transverse dipoles are confined in planes normal to the smectic layers. At high temperatures, the dipoles in the hard spherocylinder system are orientationally disordered; as the temperature is lowered, ringlike domains appear, and then elongated antiferroelectric chain-like domains form [106]. This transition from the high temperature disordered state of the quasi-2D dipolar system to the formation of dipolar aggregates has been compared to a Kosterlitz–Thouless type transition [106]. For off-center dipole positions, McGrother and coworkers [103] suggested that the influence of molecular dipoles on phase behavior may not be a monotonic function of dipolar strength [103], thereby introducing the possibility of reentrant behavior.

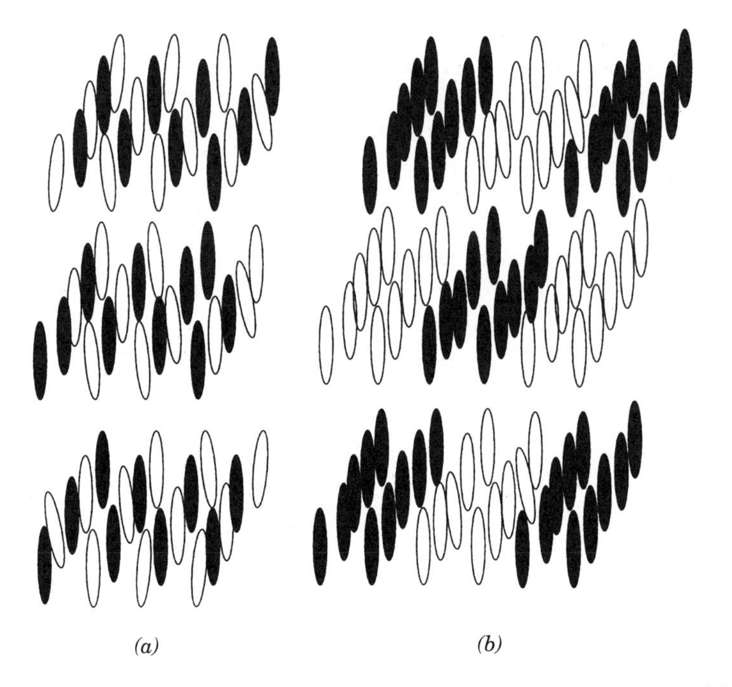

Figure 14. The molecular arrangement in the smectic phase for GB units with a central dipole (*a*) and a shifted dipole (*b*). The colors indicate up or down dipolar orientation.

As stated earlier, it is possible to determine the dielectric anisotropy if the molecular dipole moment and orientational order parameter are known. This relationship has led to estimates for the dielectric constants of hard spherocylinders with strong central dipoles. However, both $\varepsilon_{||}$ and $\varepsilon_{\perp}$ are much smaller than experimental values [109]. The difference is attributed to the neglect in the model of contributions from molecular polarizability. Moreover, strong antiparallel order of neighboring molecules is observed [109], which implies that models of dielectric properties must account for short-range correlations. Reaction field methods in conjunction with modified Ewald summation techniques [98] have also shown promise in evaluation of dielectric properties.

No ferroelectric phases consisting of prolate dipolar ellipsoids or spherocylinders have been reported. Interestingly, dipolar discotic particles do form ferroelectric and antiferroelectric phases, depending on temperature and aspect ratio [112]. This behavior shows some similarities to that found for strongly interacting dipolar spheres [113].

2. *Molecular Quadrupoles*

In addition to the influence of dipolar order on mesophase formation, there have been a number of attempts to assess the significance of quadruoplar effects as well. Although the nematic phase is found to be unstable if molecules interact only through quadrupolar forces [114] of the form,

$$V_q = D_{ij} \frac{x_i x_j}{2r^5} \tag{4.10}$$

such interactions are not negligible. In fact, theoretical studies have shown that the influence of quadrupolar forces is responsible for temperature-induced planar to homeotropic alignment transitions and for the appearance of a spontaneously polarized layer at the nematic–surface interface [115]. The effect of quadrupolar interactions has only begun to be explored in computer models. The addition of small ideal quadrupolar forces to otherwise hard body systems, however, was found to favor the nematic phase slightly, although multipole forces of large magnitudes favored more ordered phases [100]. The formation of certain columnar nematic phases (in mixtures of discotic multiynes and 2,4,7-trinitrofluorenone) has also been attributed to electrostatic quadrupolar interactions among the components, and the hypothesis is supported by Gay–Berne simulations [116] implemented to represent discotic molecules. The quadrupolar interaction appears to weaken the face-to-face attraction for like particles while strengthening it for unlike particles.

Nematic phase transition Gay–Berne simulations have also shown that the presence of model steric quadrupoles lowers the isotropic to nematic transition temperature [117]. The molecular model used in these simulations is illustrated in Fig. 15. The molecules interact through multisite Gay–Berne potentials, which are linked in various ways to give a composite unit representative of either a steric dipole [Fig. 15(*b*)] or steric quadrupole [Fig. 15(*c*)].

3. *Remarks*

This discussion has served to illustrate the complexity of incorporating point dipolar and quadrupolar interactions into otherwise relatively simple molecular systems and has demonstrated the richness of the resulting behavior. These model systems have, however, only just begun to reveal the complexity characteristic of real molecular systems. For instance, in real molecules, there may be several sources of strongly dipolar interactions. It may also not always be appropriate to refer to point dipoles; and in highly conjugated systems, such as cyano biphenyl compounds, the molecular dipole is normally considered to be delocalized. Moreover, dipoles may reside on flexible molecular segments, which have considerable conformational freedom, especially at

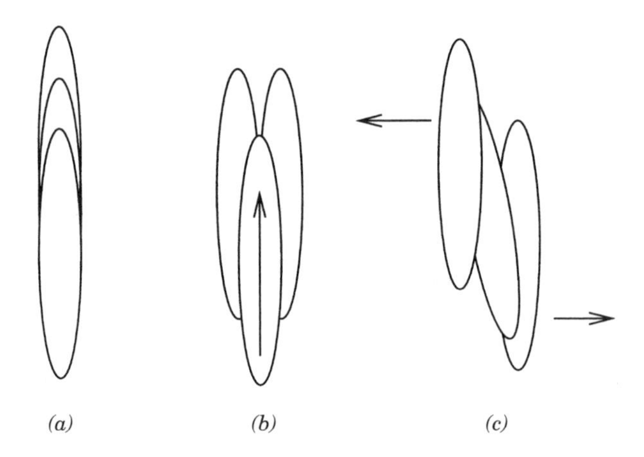

Figure 15. Three-site Gay–Berne molecular models representing (*a*) an ellipsoid, (*b*) a steric dipole, and (*c*) a steric quadrupole.

high temperature. These same considerations make it difficult to justify point–multipole and point–polarizability approximations.

It, therefore, becomes necessary to consider how molecular shape changes can be modeled and ultimately coupled with interactions that arise from molecular shape-dependent, electric multipoles. The issue of molecular flexibility is considered in the next section.

V. INFLUENCE OF MOLECULAR FLEXIBILITY

A. Background and Experimental Motivation

In addition to exhibiting complex electrostatic and steric interactions, the molecules that form liquid crystalline phases are not entirely rigid, and mesogens generally consist of both rigid and relatively flexible components. This additional structural freedom complicates substantially analytical theories based on statistical mechanics [118,119,120]. As a result, theoretical understanding of flexible mesogens lags behind that for rigid rod systems. Presently, no general theoretical formulation exists for the equation of state for flexible molecules; however, in the low-density limit, a virial expansion can be performed, yielding

$$\frac{p}{\rho kT} = 1 + B_2(T)\rho + B_3(T)\rho^2 + \dots \tag{5.1}$$

where ρ is the molecular density. In the case of nonspherical flexible molecules, B_2 (second virial coefficient) is of the form

$$B_2(T) = -\frac{1}{2}\frac{\sum_\lambda f_{12} e^{-\beta U_{\text{intra}}}}{\sum_\lambda e^{-\beta U_{\text{intra}}}} \tag{5.2}$$

The quantity $f_{12} = \text{Exp}(-\beta U_{\text{intra}}(R_{12}, \Omega_1, \Omega_2, \Gamma_1, \Gamma_2)) - 1$, where U_{intra} represents nonbonded intramolecular interactions and sums are taken over all configurations of two molecules with separation R_{12}, orienations Ω, and conformations Γ.

Despite the inherent complexity, the issue of molecular flexibility in liquid crystals remains an important technological issue, and it is essential to explore the consequences of molecular nonrigidity in the context of mesophase formation. It is equally important to predict and control the flexibility of individual molecules. At a molecular level, flexibility refers to the ease with which molecular segments can be rotated relative to each other, thereby giving rise to a variety of molecular conformations. Flexibility is usually quantified in terms of a *torsional potential*, which gives the molecular energy as a function of rotation angle. Such potentials are not easy to extract from experiment, although recent nuclear magnetic resonance experiments have begun to provide insight into molecular flexibility in mesogenic molecules. For example, it is possible to infer torsional potentials from observed dihedral angle distributions according to

$$P_T(\phi) = W \exp\left[\frac{-V(\phi)}{k_B T}\right] \tag{5.3}$$

In this expression, $P_T(\phi)$ is an observed distribution, V is the torsional potential associated with the torsional angle ϕ, and W is a normalization factor. Although the torsional potential is essentially a single-molecule property, it may be influenced by other molecules in condensed phases. We will return to this point later on in this section.

There is an increasing amount of experimental evidence that molecular flexibility plays an important role in mesophase behavior and dynamics. For example, the compound 1OCBF3 does not show a liquid crystalline phase, unlike its analogue, 4-cyano-4′-methoxybiphenyl (1OCB). Nuclear magnetic resonance has revealed that the major difference between the two molecules lies in the potential governing rotation about the ring-oxygen bonds [121]. Other examples in which the importance of flexibility has been identified include the phenomenon of nematic reentrance [122]. It has also been argued that a proper account of molecular internal flexibility is essential in the interpretation of quasielastic neutron scattering (QENS) [123,124] and Raman [125,126] spectroscopic data on liquid crystals. There have even been suggestions that non-negligible internal deformations occur in the solid phase of some mesogenic materials (such as TCDCBPh/di-(4-*n*-butyloxyphenyl) trans-cyclohexane-1,4-dicarboxylate) well before the melting transition [123].

Infrared spectroscopy on smectic-C* systems under switching conditions has shown that the influence of molecular flexibility in device environments is also significant in that certain molecular fragments are found to reach orientational equilibrium more quickly than others [127,128].

Computer simulations have provided a way of exploring generic consequences of molecular flexibility at different levels of approximation. One general point is that the order parameter is only uniquely defined for a rigid molecule. In the flexible case, deformed conformers do not possess a well-defined axis of symmetry, and interpretation of experimental results depends on the choice of reference frame. Therefore, a common procedure is to define the molecular axis of a flexible molecule as the principle axis of the tensor of inertia

$$I_{\alpha\beta} = \sum_i m_i (r_i^2 \delta_{\alpha\beta} - r_{i\alpha} r_{i\beta}) \tag{5.4}$$

The principal axis is the one having the smallest moment of inertia. Molecular order can also be defined in terms of the orientation of various molecular segments. An alternative is to use the biaxiality of the inertia tensor to describe orientational order. The order parameters are then defined as

$$S_{zz} = 1 - (A_x + A_y)/2Az \; ; \; S_{yy} = A_y/2A_z - \tfrac{1}{2} \tag{5.5}$$

$$S_{xx} = A_x/2A_z - \tfrac{1}{2} \tag{5.6}$$

where the semiaxes of the ellipsoids are defined as $A_\alpha = [(I_{\beta\beta} + I_{\gamma\gamma} - I_{\alpha\alpha})\frac{5}{2M}]^{\frac{1}{2}}$ and M is the molecular mass.

It is also equally valid to consider the polarizability tensor instead of the inerta tensor. Fuller discussions of methods for describing orientational order in flexible systems are given by Wilson [129] and Komolkin [130]. In most cases, simulations that incorporate flexibility are computationally demanding relative to the rigid-body case. The reason is that the introduction of flexible segments requires many interaction sites and additional potential terms. Moreover, intramolecular dynamics generally occur on relatively fast time scales, and the simulation time steps must be short enough to account accurately for internal dynamics. Normally, these are much shorter than those required to simply integrate classic equations of motion for rigid bodies.

In the remainder of this section we examine several ways in which molecular flexibility has been introduced into models of liquid crystals.

B. Incorporating Flexibility into Model Systems

1. *Flexibility in Excluded-Volume Models*

A simple method by which to introduce flexibility into excluded-volume models is to link spheres tangentially so that thay are bound together by narrow potential wells within which the spheres may move [Fig. 16(*a*)]. The potential parameters can be used to tune the flexibility. Intermolecular interactions still occur through hard-wall potentials. Such a model may be expected to

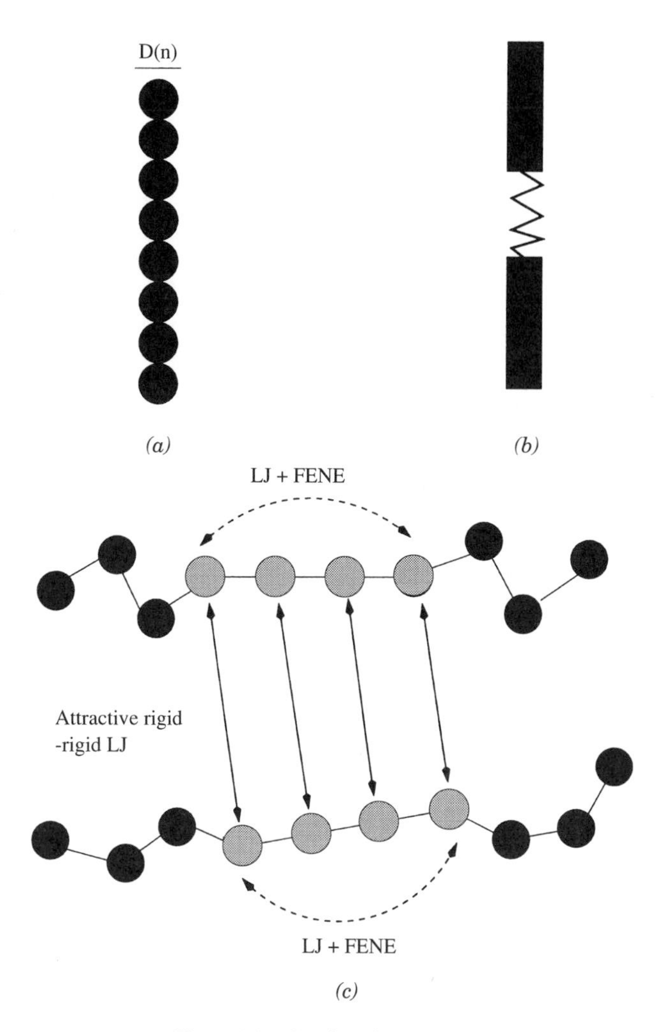

Figure 16. Continued on next page.

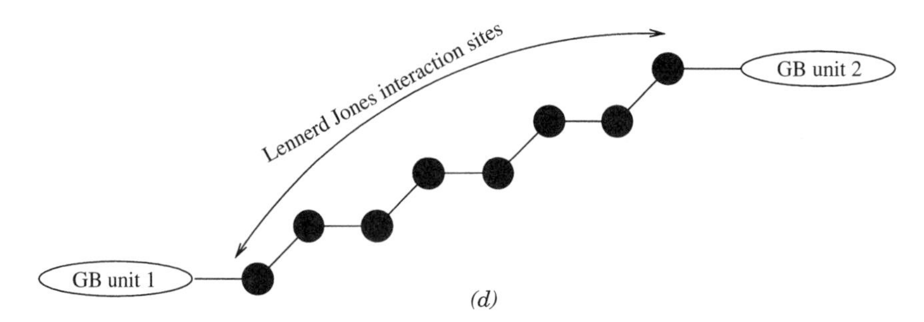

Figure 16. Flexible models of liquid crystal molecules, illustrating (*a*) the fused sphere model, (*b*) a model combining rigid segments joined by flexible spacers, (*c*) a variable-potential fused sphere model chosen to give flexible termini and a semi-rigid core, and (*d*) a flexible model mesogen similar to (*b*) but comprising combined Gay–Berne and Lennard–Jones units.

reveal only general consequences of flexibility, as it is not possible to choose parameters appropriate for specific real materials. A seven-site linked sphere model was found to exhibit smectic-A and nematic mesophases as well as an isotropic phase upon decreasing density [131]. No stable mesophases were identified in shorter molecules composed of five links, although a metastable smectic phase was observed for the most rigid five-member chain. The first-order transition from isotropic to nematic is accompanied by a small but discernible change in molecular shape in which the molecules become more linear [Fig. 17]. This observation is a simple example of the coupling between molecular structure and orientational order, which in this case corresponds to a decrease in the number of conformers that have segment bonds at an angle to the director [131]. Similar molecular shape changes at the nematic–isotropic transition have been reported in more realistic mesogenic models [132], which we discuss in a later section. Another way of incorporating flexibility into hard particle models is to add simple flexible tails (occupying zero volume) to otherwise rigid anisotropic units [133]. Their inclusion stabilizes the smectic phase relative to the nematic and increases the smectic layer separation.

2. *Flexibility in Soft Potential Models*

A variation on the fused sphere theme is to model mesogens using beads joined together, with interbead interactions chosen so that the central segment is relatively rigid while the termini are more flexible. This type of model is shown in Fig. 16(*c*), and it is the type used in the molecular dynamics simulations of Affouard and co-workers [134]. Here, the intramolecular potential (interaction between springs) is defined by so-called finitely extendible

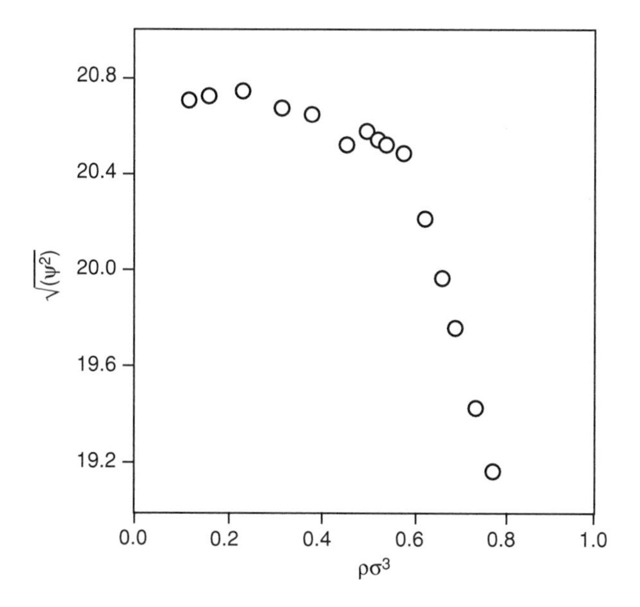

Figure 17. Density dependence of the angle between the head and the tail bonds.

nonlinear elastic springs, defined by

$$U_{\text{intra}} = \begin{cases} -\frac{1}{2}kR_0^2 \ln[1 - (r/R_0)^2] \text{ for } r \leq R_0 \\ \infty \text{ for } r > R_0 \end{cases} \tag{5.7}$$

Unlike the excluded volume case, the intermolecular potential (between beads) is taken to be of the Lennard–Jones type. The attractive part being

$$U_{\text{inter}}^{\text{att}} = \begin{cases} -\varepsilon_{\text{att}} \text{ for } r < 2^{\frac{1}{6}} \\ 4\varepsilon_{\text{att}}[r^{-12} - r^{-6}] \text{ for } 2^{\frac{1}{6}} < r \leq 2.5 \\ 0 \text{ for } r > 2.5 \end{cases} \tag{5.8}$$

whereas the repulsive part is

$$U_{\text{inter}}^{\text{rep}} = \begin{cases} 4[r^{-12} - r^{-6} + \frac{1}{4}] \text{ for } 2^{\frac{1}{6}} < r \leq 2.5 \\ 0 \text{ for } r > 2^{\frac{1}{6}} \end{cases} \tag{5.9}$$

The smectic phase for this model flexible system exists over a wide range of temperatures, but no nematic phase is observed. Decreasing the intermolecular attractive interaction does not lead to a nematic phase, although the smectic ordering does become less pronounced. It is clear that large changes

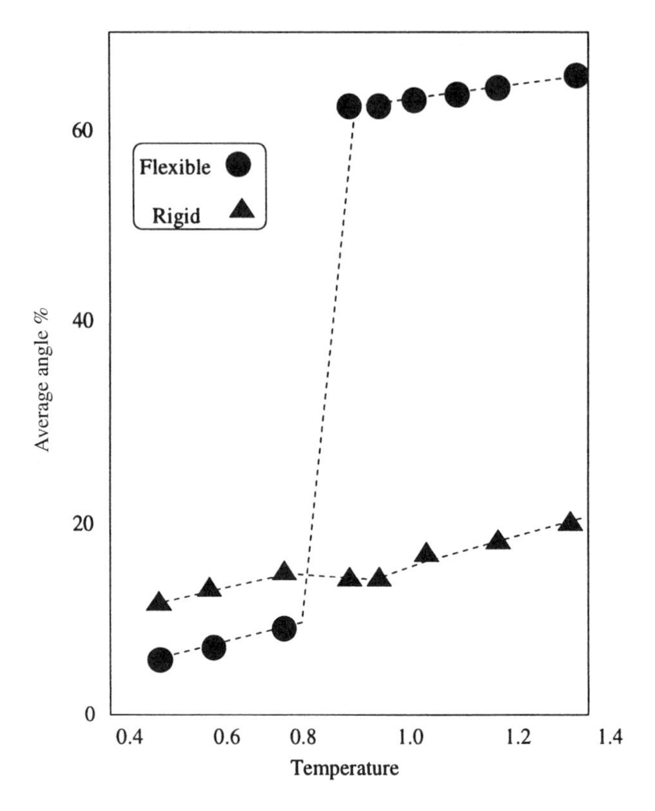

Figure 18. Angular disorder between two segments along the chains in the stiff and flexible parts. From Reference [134].

in conformational freedom of the flexible termini are observed at the clearing point, which is indicative of strong coupling between molecular structure and condensed-phase orientational order. This is quantified in terms of the temperature dependence of angular disorder shown in Fig. 18. Yoshida and Toriumi [135] used a simpler linked sphere model, in which each of four units acts as a center of LJ interactions. They concluded that LC formation at high density is dominated by repulsive interactions, whereas additional attractive interactions are necessary for mesophase formation at lower density [135].

In an attempt to study the influence of molecular shape and flexibility in more realistic systems without greatly increasing computational cost, it is possible to build up models that consist of rigid units joined by flexible spacers representative of realistic molecular fragments, such as alkyl tails [136]. A schematic illustration of this sort of flexible model mesogen is shown in

Fig. 16(*b*). In the model of Nicklas and co-workers [136], in which spherocylindrical fragments are linked by a flexible chain, the isotropic–nematic phase transition is found to occur with large conformational changes and to exhibit odd–even alternation of the transition temperature with chain length. Dynamic properties are not well described, as the model predicts both diffusion constants ($D_\perp$ and $D_{||}$) to be much higher than experimental values, which may be a symptom of the overly symmetric shape of the rigid components used in the model [136].

A similar strategy based on a combination of Gay–Berne units and Lennard–Jones interaction sites [137] has recently allowed the observation of spontaneous growth of the smectic-A phase from the isotropic liquid as well as a pronounced odd–even effect in the segmental order parameters of the chain bonds [Fig. 16(*d*)]. This system is intended to be similar to the α, ω-bis(4-4′cyanobiphenyloxy)hexane dimer, but the GB units replace the cyanobiphenyl moieties. The flexibility was defined by a torsional potential of the Ryckaert–Bellemans form [138], and the Gay–Berne parameterization used was that of de Miguel [55]. It should be noted, however, that only nematic phases are seen in the real homologous series and there is considerable scope to develop systemic methods for parameterizing these "hybrid" models for specific systems.

The importance of chiral liquid crystal phases in devices has motivated simulation studies aimed at exploring the relationship between molecular helicity and the handedness of chiral phases [7]. One such study considered a model of chiral atropisomers using Monte Carlo simulations in the NVT ensemble. The molecular chirality was introduced by joining two Gay–Berne particles through a bond with a fixed dihedral angle [139,140]. The transfer of chirality to rotamers was explored by introducing the possibility of barrier-less internal rotation about the bond axis as an internal degree of freedom for the guest molecules. Starting from an isotropic configuration, cholesteric phases were obtained on equilibrating the guest–host systems, whereby left-handed and right-handed cholesterics were formed, depending on the helicity of the atropisomers, respectively. The conformational distribution of the guest molecules in the cholesteric phase, shows an enantiomeric excess of rotamers of the guest molecules with the same helicity as the host molecules.

VI. INCORPORATING CHEMICAL DETAIL

B. Interaction Potentials for Realistic Molecular Models

A further step toward a completely atomistic description of mesophases involves potentials that represent interactions among all (or most) of the atoms in liquid crystal phases. In these atomistic models, the intermolecular and intramolecular interactions are usually expressed by a potential of the form

$$\begin{aligned} V(\vec{r}) = & \sum K_b(b - b_0)^2 \\ & + \sum K_\theta(\theta - \theta_0)^2 \\ & + \frac{1}{2}\sum V_n(1 + \cos\phi) \\ & + \sum_i \sum_j [A_{ij}/r_{ij}^{12} - C_{ij}/r_{ij}^6] \end{aligned} \tag{6.1}$$

where bond stretching and bending forces are described by the first two sums, respectively. The energy cost associated with torsional deformations is given by the third term (with V_n setting the barrier height). The remaining terms are the Lennard–Jones and electrostatic interactions.

To reduce the number of interaction sites, a common method of simplification is to treat common chemical groups (like $-CH_2-$) as superatoms, or united atoms. In this case, the form for the interaction potentials remains the same, but the parameters are modified to reflect composite interaction sites. Unlike simulations using highly idealized molecular models, the aims of realistic simulations are not generally to make connections to analytical theories, nor are they aimed at ultralarge simulation sizes. Rather the objective is to make detailed comparison to experiment, with an emphasis on predicting phase stability and revealing links between chemical structure and material properties.

B. Structure and Translational Diffusion

The prototypical nematic liquid crystal 5CB has been the focus of some of the first studies of mesophase order and dynamics using realistic atom–atom or united atom approaches. The effectiveness with which simplified united atom models reproduce quantitative static and dynamic properties for real systems was addressed by Cross and Fung [141,142] for the case of 5CB. The authors used a realistic atom–atom potential and a simplified model in which the phenyl rings were treated as simple spheres to extract information on molecular order parameters for the principal molecular axis and for mesogenic fragments. Illustrations of the two molecular structures are shown in Fig. 19.

The two models give similar results for the core order parameter and for some segmental order parameters in the alkyl chain. Anisotropic diffusion coefficients for the two models differ to a greater extent; the results for the simplified system are lower (for $D_{\parallel}$ and $D_{\perp}$) and in closer accord with

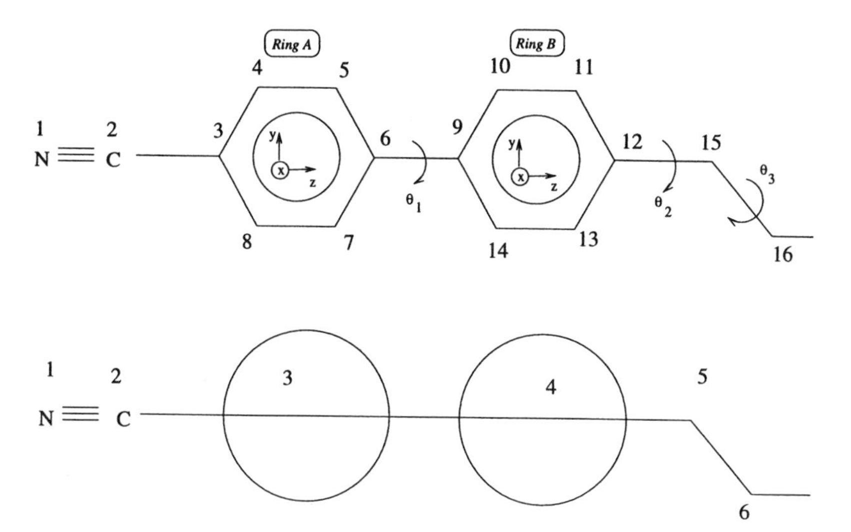

Figure 19. (*a*) Molecular models for 5CB, showing interaction sites and torsional angles. (*b*) Corresponding united atom representation.

the experiment than are those of the more detailed model. Molecular order and properties of nematic 5CB have also been studied by Komolkin and co-workers at 300 K using two interaction models [130]. The first of these comprises united atoms (UA) and the second includes the effect of hydrogen atoms explicitly. Molecular order in the nematic phase can be described in terms of partial intermolecular radial distribution functions $g(r_{ij})$, where r_{ij} is the separation between specific molecular sites on different molecules. In atomistic simulations of 5CB [130], for example, simulated radial distribution functions (RDFS) suggest that atomistic cores exist, on average, in an antiparallel local arrangement. Moreover, the average separation between methyl groups was found to be similar to that found in adsorbed *n*CB molecules. The distribution of dihedral angles in 5CB was found to depend on the type of model used to describe the molecule [Fig. 20]. A summary of the results on the translational diffusion constants is shown in Table I.

A model based on the Gay–Berne potential but combining a rigid anisotropic core and flexible united atom chain to represent alkoxy and methyl groups has yielded structural properties and segmental order parameters in good accord with experiments, however, dynamic properties are again not well described [19]. For example, the diffusion constants parallel and perpendicular to the director are 3×10^{-9} and 1×10^{-9} m^2/s. These values are larger than most experimental values obtained by NMR [143] and

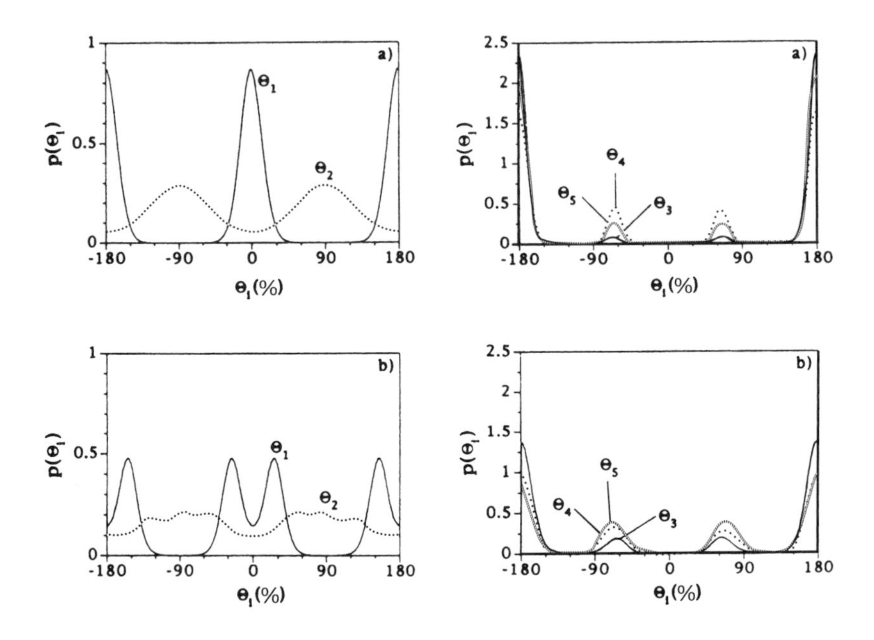

Figure 20. Simulated distribution of dihedral angles in 5CB.

TABLE I
Results on Simulated Translational Diffusion Coefficients as Obtained from Mean-Square Displacements (*MSD*) and Direct Velocity Autocorrelation Correlation Functions ($C_v(t)$), using either Full Atomic (*FA*) or United Atom (*UA*) Molecular Descriptions

Method[a]	$D_{\parallel} \times 10^{11}$	$D_{\perp} \times 10^{11}$
MSD (UA)	10(1)	25(2)
$C_v(t)$ (UA)	10(2)	30(2)
MSD (FA)	<3	<3
$C_v(t)$ (FA)	6(2)	26(2)
QENS[a] (296.5 K)	4.1	5.3

[a] Included for comparison. Data from Ref. [31].

QENS [144]. These are obtained from the components relative to the director of the mean square displacement shown in Fig. 21. The ratio $D_{\parallel} : D_{\perp} \approx 3$ is, however, typical of experimental results on calamitic nematic liquid crystals.

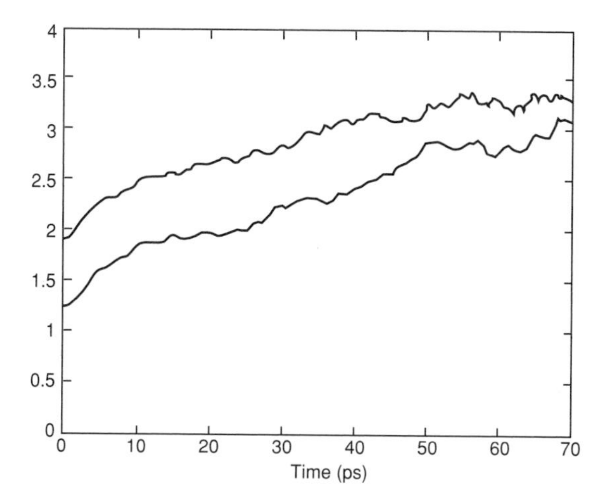

Figure 21. Mean square displacement for the model implemented by Lapenna [19].

C. Coupling between Molecular Structure and Orientational Order in Atomistic Models

Combined use of spectroscopic frequencies, linewidths, and shapes have recently been used to probe molecular flexibility of alkyl chains and thereby have become an important component in studying the relationships between molecular structure and orientational order of condensed phases [145,146]. In applying these techniques to thermotropic liquid crystals, several observations have been made, including that of a conformational collapse of the alkyl segment of 12OCB at the crystal–smectic transition. The smectic–isotropic transition, however, appears not to involve significant change in the relative conformer population [146]. In other systems, pretransitional structural effects appear to be evident even in single molecules in which the number of conformational kinks in an otherwise *trans*planar structure increases dramatically upon approach to a transition [147,148].

Emsley and co-workers considered the influence of the nematic mean field on molecular properties and proposed a form for the molecular energy in which the intermolecular and intramolecular contributions were separated according to

$$U(\Gamma, \omega) = U_{\text{intra}}(\Gamma) + U_{\text{nem}}(\Gamma, \omega) \tag{6.2}$$

where the first term is the intramolecular conformational energy and the second represents the influence of the nematic mean field. The effect of short-range intermolecular forces are not included. Originally, parameters for the nematic mean field were deduced by fitting to NMR data; however, a simplified form of the expression has been used in conjunction with MC simulations to infer the influence of the strength of the nematic mean field on molecular structure [149]. Using 5CB as a test case, the effect of the nematic mean field is to change the relative energies of the alkyl tail conformers (to favor elongated conformers) relative to the gas phase results obtained by using molecular mechanics force fields. The coupling between molecular structure and orientational order has been explored in only a few other cases that need realistic molecular models, but the degree of coupling appears to be system specific. In Lapennna's model [19] (combining rigid anisotropic sites and flexible isotropic sites), the chain conformation is observed to be strongly affected by intermolecular interactions but only weakly influenced by the intramolecular potential. In the smectic phase of hybrid Gay–Berne/LJ models [137], molecules appear to remain flexible and perform frequent *gauche* $\leftrightarrow$ *trans* interconversions. In this system, molecular structure is strongly coupled to orientational order, which tends to influence the effective torsional potentials to favor linear conformers.

There is also some evidence to suggest that torsional motions that do not alter molecular shape are not influenced appreciably by the liquid crystal environment [150]. This is illustrated in a molecular dynamics simulation in the nematic phase of hydroxybenzoic acid tetramers [Fig. 22(*a*)], which have revealed only a weak coupling between order and intramolecular structure. In this work, bond lengths were fixed, the force field was parameterized from quantum chemical simulations on dimers, and torsional potentials were corrected by a scaling factor. A high-order parameter ($S = 0.86$) was observed near experimental densities, but the liquid crystalline character was confirmed from anisotropic translational diffusion data. In this system, the nematic environment exerts only a marginal influence on the probability of conformers relative to the isolated molecule case [150]. This can be seen in Fig. 22, where the probabilities of torsional angles E, F, and G

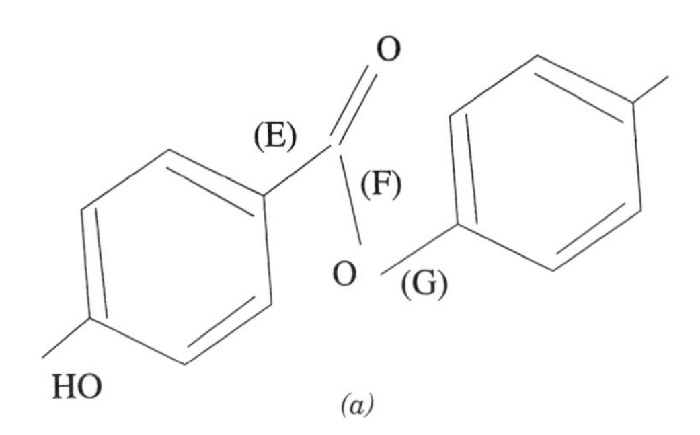

Figure 22. Continued opposite.

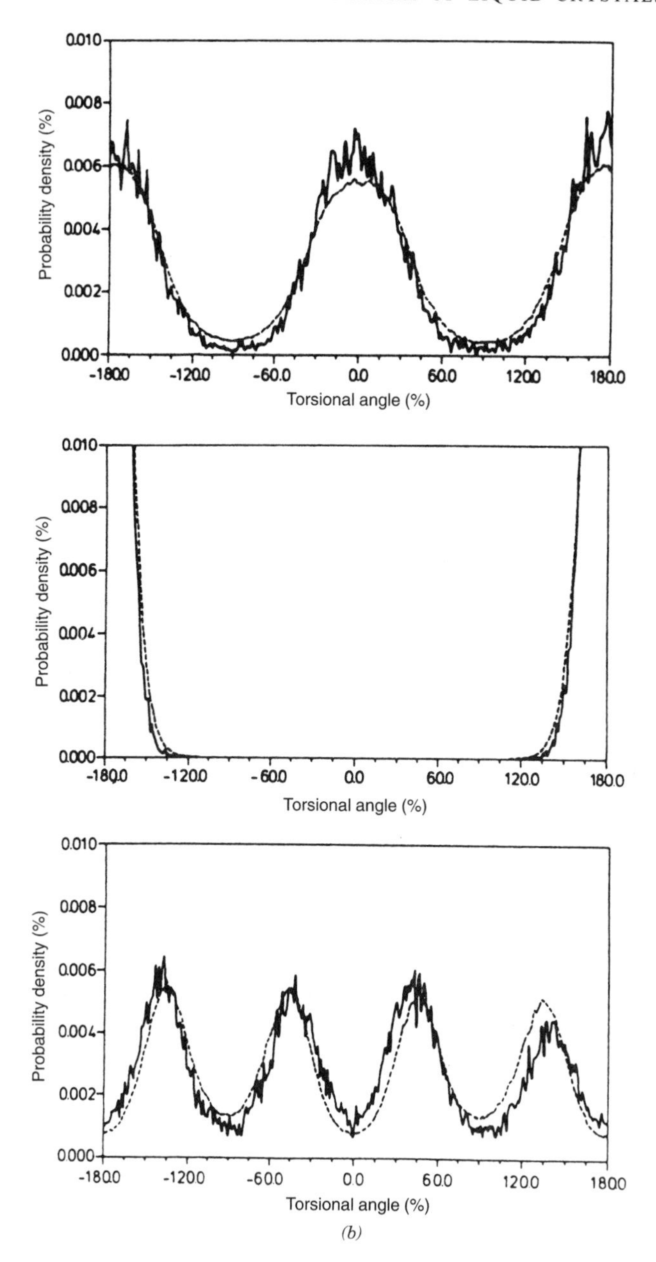

Figure 22. Dimer used for parameterization of the intramolecular potentials in *p*-hydroxybenzoic acid. (*a*) The probability densities for the torsional angles *E*, *F*, and *G* in the liquid crystalline phase at 475 K (*solid line*) are shown in the top, middle, and lower plots in (*b*). Also shown are the results obtained for the isolated molecule at the same temperature (*dotted line*).

are shown as obtained for isolated molecules (*dashed lines*) and in the nematic phase. A slight increase in the likelihood of the planar ring arrangement is found [150]. Another example of a case in which only weak coupling between molecular structure and orientational order has been found (but where associated shape changes are not small) is in atomistic MD simulations of 5OCB [151]. In the case of the torsional ring angle, for example, an equilibrium value of 35° is found in both liquid and liquid crystalline phases; however, the angle is near zero in the solid. Other changes in geometry occurring at the transition to the solid are found for certain torsions in the alkyl tail. The nonbonded interactions between hydrogen atoms in so-called *ortho* positions to the phenyl–phenyl unit were represented as a sum of Coulomb and Buckingham potentials

$$V_{nb} = \frac{1}{4\pi\varepsilon_0}\frac{q_i q_j}{r_{ij}} + A\left[\exp(-Br_{st}) - \left(\frac{C}{r_{st}}\right)^6\right] \tag{6.3}$$

Different methods of treating similar materials have produced some discrepant results regarding conformational statistics. For example, a flexible extended atom model of 8CB has been used to study the stability of the nematic phase by comparing the free energy with that in the nematic phase [152]. Throughout the nematic range, the molecules are found to exist in nearly all-*trans* conformations. This tendency is not reflected in the other atomic-level simulations of 5CB [130] and CCH5 [141,142]. The difference may lie in the fact that in the work of Levesque [152], attractive nonbonded interactions were not included, and these may play a role in determining chain conformations in condensed phases.

D. Reorientational Motion

In addition to anisotropic diffusional motion, the process of molecular reorientational motion has been of interest for several years. In liquid crystal applications, this problem is particularly relevant in connection with switching mechanisms. It also poses interesting fundamental questions connected with understanding

1. The influence of molecular anisotropy on dynamics.
2. The dynamic process that occurs on approach to an orientationally ordered phase.
3. The influence of local structure on reorientational dynamics [153].

The latter issue arises from the observation that in liquids exhibiting significant local structure orientational relaxation is nonhydrodynamic at short times and is decoupled from the bulk viscosity [153]. Therefore, molecular reorientational motion, even in the isotropic phase of liquid crystal–forming compounds, can be revealing. Equally interesting is the observation that the rate of molecular reorientation around the short axes does not change at the transition between the nematic and a liquid-like smectic phase (such as Sm-A and Sm-C), whereas it is considerably retarded at the nematic, solid-like Sm-B transition [154], as revealed by dielectric relaxation measurements. Generally, rotational diffusion models are applied to interpret experimental data obtained from a variety of spectroscopies. The central quantity that is accessible both from simulations and spectroscopic studies is the time autocorrelation function

$$C(t) = \langle P_l[\hat{\mu}(0) \cdot \hat{\mu}(t)] \rangle = \overbrace{\exp[-l(l+1)D_{\perp}t]}^{\text{diffusive limit}} \tag{6.4}$$

where μ is a unit vector that defines the molecular orientation and P_l is a low-order Legendre polynomial. Qualitatively, this function represents the average rate at which the molecules lose memory of their orientation. From these, it is possible to define rotational diffusion constants for tumbling and spinning motion. For example, in the long-time (rotational diffusion) limit, an exponential decay of the correlation function is expected with a characteristic diffusion constant ($D_{\perp}$) that measures the rate of end-over-end tumbling of the molecular long axis [Eq. (6.4)].

The most common spectroscopic techniques [155], such as Raman and infrared absorption, are restricted in the types of systems to which they can be applied, because they rely on the presence of isolated bands from which dynamic information can be obtained. Furthermore, reorientational information is limited to only a portion of the molecule that is responsible for the observed band, and, generally, full reorientational information is inaccessible. This limitation is not severe in rigid molecules but can seriously complicate analysis of reorientational dynamics in flexible systems. Moreover, the reorientational contribution to spectroscopic linewidths is generally small in large molecules compared to other broadening mechanisms, such as vibrational dephasing [156].

Ultrafast femtosecond measurements on cyanobiphenyls have been used to explore mesogen dynamics in the smectic phase [157,158]. These measurements revealed several characteristic fast relaxation time scales (in the picosecond to subpicosecond range), which have been attributed to molecular librational and reorientational diffusion processes. The coexistence of several relaxation time scales is taken as evidence for the presence of strong molecular interactions in the smectic layers [157].

There are also a few examples in which depolarized light scattering (DLS) [159] and transient grating optical Kerr effect (OKE) measurements have been made on liquid crystals in the pretransitional region above the nematic–isotropic transition [160,161] (Section I.A). The OKE response arises from fluctuations and relaxation of dielectric susceptibility; therefore, it allows access to collective dynamics and collision-induced phenomena.

Such measurements lead to a natural separation into several time ranges, which are interpreted as arising from [162].

1. Sub-picosecond (fast) librational motion.
2. Intradomain dynamics on time scales of 1 ps to 1 ns.
3. Slow (>1 ns) reorientation of pseudonematic domains.

The intradomain dynamics were observed to be of nonhydrodynamic character, exhibiting surprising insensitivity to temperature and viscosity η in a temperature range over which interdomain dynamics changed appreciably. For short times the dynamic response function $G(t)$ appears to follow a power law of the form

$$G(t) = G_0 t^{-\alpha} \tag{6.5}$$

with $\alpha \approx 0.63 \pm 0.03$ for both MBBA and 5CB, suggesting a universal character inherent in fast orientational dynamics in pretransitional liquid crystals [162]. Moreover, the slow dynamics of both systems show deviations from the predictions of the Landau–DeGennes theory at the same value for the correlation length, which is the same point at which the fast dynamics regime acquires a temperature dependence. For MBBA, this correlation length contains a small number (≈ 30) of molecules; yet it appears that such highly local order can have profound effects on the liquid state dynamics and so-called interaction-induced contributions to light-scattering spectra [163]. Such considerations suggest that simulations of dynamic processes in liquid crystals must account accurately for such correlations.

Computer simulation using molecular models is not subject to many of the practical restrictions that limit systematic experimental study, and in principle, atomic detail simulations open up the opportunity to study reorientational motion of molecular fragments in great detail. Simulations of reorientational dynamics using idealized rigid molecules have shown a large increase in reorientational relaxation time on approach to the nematic–isotropic transition [164], with considerable sensitivity to aspect ratio observed in both hard spherocylinder [164] and Lennard–Jones ellipsoids [165]. In Gay–Berne systems, observation of significant slowing down of orientational relaxation on approach to the nematic–isotropic transition has been reported [166]. Theoretical models based on reorientational diffusion or stochastic processes, however, have not led to a convincing description of single-particle reorientation dynamics in Gay–Berne systems

[82]. More flexible model systems have shown that segmental reorientational processes do not show single exponential charater [19].

At present, there are few investigations of reorientational motion in mesogens at near-atomistic levels of detail. However, Yakovenko and co-workers [167,163,168,125] have succeeded in separating reorientational dynamics of the *p*-*n*-pentyl-(*p*′-cyanophenyl)-cyclohexane (PCH5) into reorientational motions of the mesogenic bonds and fragments using united atom molecular dynamics in the nematic and isotropic phases. Considerable differences in the orientational autocorrelation functions for different fragments of the PCH5 molecule are found and attributed to the influence of molecular flexibility. Fig. 23 shows that the orientational autocorrelation functions decay most rapidly for the most flexible fragments, but some differences in rate are found even for bonds within a rigid core unit. Moreover, the rapidly fluctuating intermolecular forces give rise to conformational relaxations that occur on time scales that are much shorter than what would be expected based only on the molecular moment of inertia. The influence of molecular flexibility in this system has been proposed to account for the fast nonexponentional decay of the experimental autocorrelation functions for the cyano bond in PCH5 [167]. The general implication is that, although intermolecular torques may lead to a rather slow reorientation of the whole

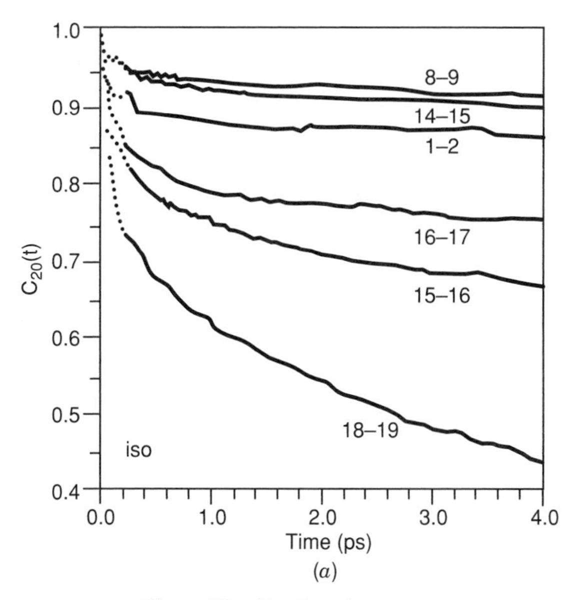

(*a*)

Figure 23. Continued on next page.

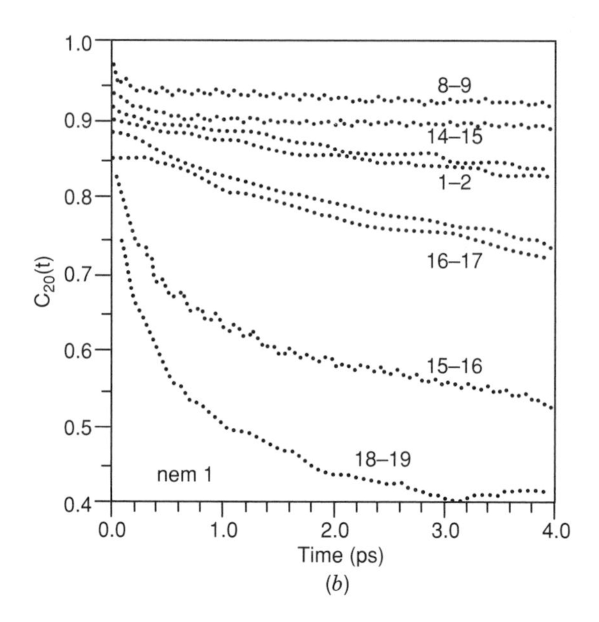

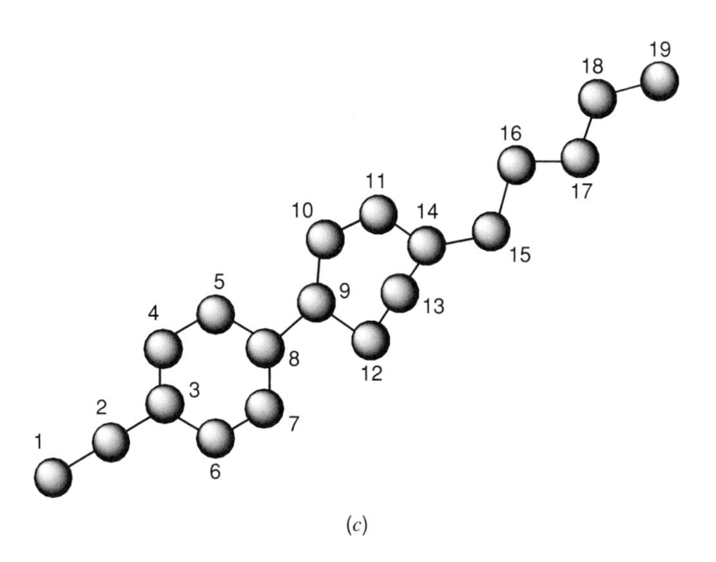

Figure 23. Reorientational autocorrelation functions for bonds in PCH5 in the isotropic (*a*) and nematic phases (*b*) and the location of the bonds in the molecule (*c*).

mesogenic molecule, the same torques may lead to rapid conformational changes that dominate experimentally accessible correlation functions. An important ingredient in these calculations is the molecular polarizability, which, in these cases, was constructed from anisotropic bond polarizabilities. Flexibility was also evident in simulations of isotropic PCH5 [163], in which fluctuations of the molecular transverse and longitudinal polarizability components of at least ± 8 and $\pm 3\%$, respectively, were observed in simulations of isotropic PCH5 [163].

Single molecule reorientational dynamics have been studied using realistic atomic and united atom models appropriate for 5CB in the nematic phase. The values obtained from the shorter full atom simulation are smaller than the united atom results by a factor of two; results are shown in Fig. 24. There is clearly need for further work combining spectroscopy and simulation in this area.

E. Interfacial and Confined Geometry Phenomena

Interfacial phenomena in thermotropic liquid crystals not only are central to the design and construction of electro-optic devices but also are of substantial fundamental interest. The central parameter that describes the liquid crystal–solid interface is the orientational molecular anchoring strength; and for a certain strength of surface interaction, a prewetting transition to a phase with an oriented boundary layer is expected. The importance of this issue has motivated a number of recent experiments that have illustrated the richness of surface-induced phenomena in real liquid crystal systems [169,170,171]. Despite its importance, little is currently known about the molecular nature of the LC–surface interface, especially in confined geometries such as those encountered in device environments. This is an area that is gaining in prominence, especially in view of the advances in so-called command surfaces [172]. There is also a growing body of evidence that suggests molecular conformation changes of alignment-promoting layers may lead to anchoring transitions between homeotropic and homogeneous alignment of liquid crystals [173].

The behavior of cyanobiphenyls adsorbed on a graphite surface have been studied using a 22-site united atom model and a LJ-type potential to model the surface [174]. The phenyl rings were constrained to be planar. Within these constraints, the minimum energy surface layer structures for 8CB are found to be consistent with scanning tunneling microscopy (STM) images, which reveal surface repeat units containing eight molecules (Fig. 25). Within a unit cell, interdigitation of the cyano groups is observed, but no interdigitation of the alkyl chains is found. These findings are not sensitive to electrostatic interactions among molecules; however, sensitivity of

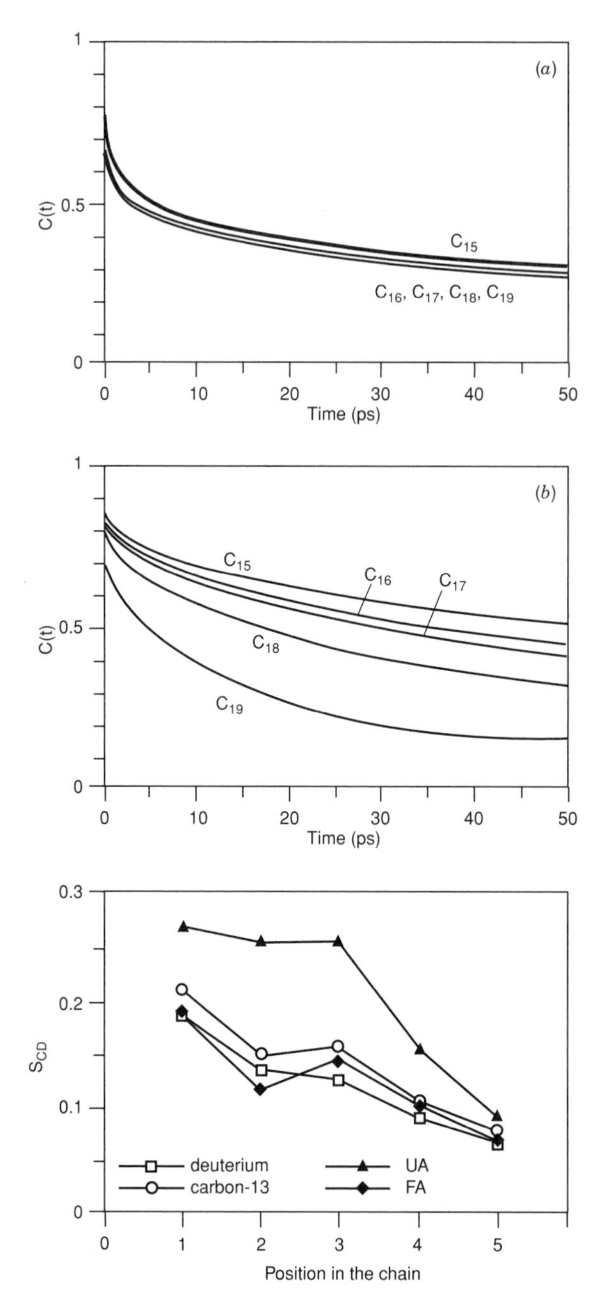

Figure 24. Segmental correlation function for C–H bond vectors in the alkyl chain as obtained by Komolkin and co-workers [130].

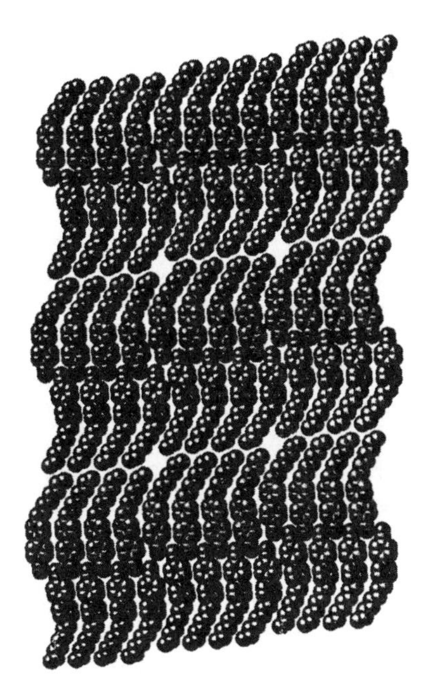

Figure 25. Structure of 8CB adsorbed onto a graphite surface.

surface order to molecular properties is reflected in these simulations, as evidenced by the fact that a different repeat unit containing six molecules is found for the longer 10CB molecule [174].

The first use of atomic-level detail simulations (using united atoms) on large ferroelectric liquid crystals have revealed that the tilted Sm-C structure develops spontaneously in initially untilted films of DOBAMBC [175]. Electrostatic interactions were not included, and the results suggest that anisotropic molecular shape is sufficient to cause the tilted phase. There is now also preliminary evidence to suggest that [176] ferroelectric liquid crystal alignment on polymer substrates is governed by interactions between the most flexible segments of LC molecules and the polymeric alignment-promoting layer.

Many of the simulation results using realistic models discussed so far depend on the accuracy of the interatomic and intraatomic potentials used. In the next section, we outline modern methods for determining such potentials and disuss their theoretical foundations.

VII. *AB INITIO* DETERMINATION OF INTRAMOLECULAR PROPERTIES AND POTENTIALS

A. Motivation

In all the examples cited so far it has been tacitly assumed that intramolecular properties of isolated molecules (such as shape, dipole moment, and flexibility) that are used to parameterize molecular potentials are known to arbitrary accuracy and that these potentials can be transferred from one molecule to another. In reality, however, these molecular properties are generally difficult to determine from experiment and require sensitive techniques and demanding sample preparation. Molecular geometry may be inferred from high-resolution diffraction or NMR, and the latter technique can also, in principle, provide information on flexibility, as discussed in the previous section. Dipole moments of isolated molecules are also difficult to deduce, and data analysis often relies on assumptions that are difficult to test. In addition to these problems in making accurate measurements of isolated molecular properties, it is also clearly advantageous to be able to predict the properties of mesophases *before* synthesis and without a specimen on which to perform experiments. To explore phase behavior or material properties of specific but unsynthesized mesogenic systems, it is necessary to develop the capacity to predict intramolecular properties to a high level of accuracy, without requiring input from experiment on the real material. In principal, such information is accessible from quantum mechanical computer simulations that solve the Schrodinger equation for the molecule. The difficulties are that liquid crystal molecules are generally large and the computational cost of quantum mechanical simulations for large low-symmetry molecules has precluded the application of the most reliable quantum chemical methods to mesogenic systems. Rapid increases in computational power and associated improvements in algorithms, however, have opened new opportunities for the detailed study of the structural, dynamic, and electronic properties of large molecules that form mesogenic phases.

There are a variety of computational chemistry strategies that are in principle available for the parameter-free determination of gas-phase, isolated molecule properties. In this section, we describe one of the most recent developments in this area and illustrate how it has overcome some of the problems associated with conventional quantum chemical techniques when applied to large, complex molecules and illustrate how it may prove to be a versatile approach for the study of liquid crystal physics. We also provide an outline of the background theory and methods of this emerging area along with examples of recent results.

B. Electronic Structure of Liquid Crystal Molecules from First Principles

1. Formulation

Recently, a method adapted from large-scale electronic structure calculations on periodic solids has been used to circumvent some of the problems associated with obtaining accurate intramolecular properties of liquid crystals [177,178]. An overview of the method is given here. The approach has as its foundation the so-called density functional principal [179,180], which is the two-part statement that

1. The energy of a molecule is a unique function of the electronic charge distribution.
2. The ground state electronic distribution is the one that minimizes the value of the energy functional.

Implementation of the scheme for a specific system (such as a liquid crystal molecule) requires explicit expressions for the individual contributions to the total energy.

For a molecule, the total energy $E[n(\mathbf{r})]$ can be formed from the sum of the following terms

$$E[n(\mathbf{r})] = T[n(\mathbf{r})] + \int V_{\text{ion}}(\mathbf{r})n(\mathbf{r})\,d\mathbf{r} + \tfrac{1}{2}\int\int \frac{n(\mathbf{r})n(\mathbf{r}')}{|\mathbf{r}-\mathbf{r}'|}\,d\mathbf{r}\,d\mathbf{r}' + G[n(\mathbf{r})] + E_{\text{ion}}[\{\mathbf{R_i}\}] \tag{7.1}$$

where the central quantity is the molecular electronic charge density $n(\mathbf{r})$ and where we assume that there exists a fixed set of atomic positions $\{\mathbf{R_i}\}$. The first term is the electronic kinetic energy. The second term is the Coulomb interaction among the electrons and ions. The third term represents the classical electrostatic interactions among electrons. These interaction terms are simply functions of the electronic charge density $n(\mathbf{r})$ The term $G[n(\mathbf{r})]$ represents all the quantum mechanical corrections to the classical electron–electron interactions arising from exchange and correlation effects. The last term is the classical electrostatic interaction between ions.

The relationship between this *ab initio* formulation and the *empirical* approach (discussed in Section VI.A) to molecular modeling is evident. The empirical parameters referring to bond extension–compression, bending, and torsion (K_b, K_θ, and V_n in Eq. 6.2) each have a well-defined molecular electronic origin related to the energy cost (measured by $E[n(\mathbf{r})]$) of deforming the electron charge density from its equilibrium configuration. The density functional technique provides a direct method for evaluation of the energy.

2. *Approximations*

As written, the above expression for the total energy is exact. Practical implementations, however, motivate the use of approximations to handle terms two and four. Specifically, the electron–ion interaction is normally handled by a *pseudopotential*. The validity of this approximation relies on the assumptions that electrons are separable into valance and core electrons and that only the valence electrons are active in bonding (the core electrons are inert). In this case, the valence electrons experience a relatively weak screened electrostatic potential owing to the ions. The advantage of a good pseudopotential is that it is transferable to a wide variety of different atomic environments. The pseudopotential method has been refined and applied to a variety of solid-state systems for several years [181,182,183,184,185]. Detailed references covering the generation and refining of pseudopotentials can be found in [186,187].

The many-body effects of exchange and correlation (term four) are incorporated through the local density approximation (LDA), which states that the exchange and correlation energy of an electron distribution at a given density can be treated as equivalent to that of a uniform electron gas at that density. The LDA method has been extended to include gradients of the charge density. Details of this generalized gradient approximation (GGA) can be found in [188].

3. *Basis Sets and Boundary Conditions*

Normally, it is necessary to expand the molecular electronic wave function in terms of a set of suitable basis functions. These are typically taken to be localized atomic orbitals or Gaussian functions centered on atomic sites, and some can be quite sophisticated and specialized. One of the simplest possible basis sets is the set of plane waves. In this basis, the molecular electronic wave function is expressed in the form

$$\psi_i(\vec{r}) = \Omega^{-\frac{1}{2}} \sum_{\vec{G}} C_i e^{\vec{k}_i + \vec{G} \cdot \vec{r}} \tag{7.2}$$

which is similar to a Fourier series expansion of the molecular electronic charge density.

This has several important advantages over traditional localized basis sets. First, completeness of the basis set can be achieved to arbitrary accuracy and convergence is easily tested. Second, because the basis functions are not localized on any atom and the set is orthogonal, there are no overlap integrals. Also, the delocalized character of the basis set makes the calculation of the force on the atoms simple via the Hellmann–Feynman theorem [189]. We can, therefore, perform efficient structural optimization and mode frequency calculation. Another important point is that, given the relevant pseudopotential, the delocalized plane wave basis set can be used for any atomic species [181,182,183,184,185], as it is guaranteed to span any possible wave function. This is in contrast to the commonly used localized sets that must be chosen carefully and that depend not only on atomic species but also on the type of bonding that is formed. This causes particular difficulty with carbon, owing to the vast range of bonding types that it forms.

To exploit the advantages afforded by the plane wave basis, the molecule to be simulated is usually placed in the center of a periodically repeating supercell, and periodic boundary conditions are enforced. The use of periodic boundary conditions allows for the expansion of the molecular electronic wave function in terms of the delocalized plane wave basis set. The coefficients of the plane waves in the expansion are then used as variational parameters until the lowest energy electronic configuration is found for a given set of ion positions. The method proceeds without the need for costly matrix diagonalization [190,191].

The problem then becomes equivalent to an electronic structure calculation for a periodic solid. In the limit of large separation between atoms in adjacent cells, the properties of the isolated molecules are described properly.

Only a single k-point (Brillouin zone center, by analogy with crystalline band structure terminology) is required as the sampling point ($\vec{k} = 0$). Finer sampling is not necessary, because electronic bands are dispersionless for isolated molecules. The size of the basis set (number of plane waves) required to span the molecular wave function is determined by the depth of the ionic pseudopotentials, which are used to represent the electron–ion interactions.

The ground state molecular electron distribution can be reconstructed directly from the the plane waves, which minimize the total energy from

$$\rho(\vec{r}) = \sum_{i=1}^{N} |\psi_i(\vec{r})|^2 \tag{7.3}$$

4. *Results for Mesogens and Fragments*

This approach has been used to calculate the three-dimensional valence electron charge density for 5CB (Fig. 26). The main molecular components (such as the biphenyl core, alkyl tal, and cyano terminal group) are easily visible.

Once the relaxed electronic structure for a given molecular conformation has been found, the calculation of the molecular multipole moments can be made. For example, the dipole can be determined by taking the vector

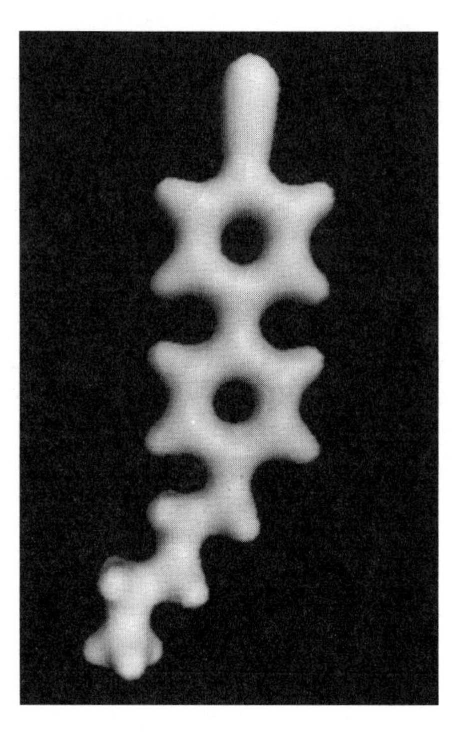

Figure 26. Electronic charge density for 5CB obtained from density functional calculations.

difference between the locations of the centroids of both the electronic and ionic charge distributions. We will return to this point below.

C. Molecular Structure and Flexibility

In the preceding sections a method of determining the total energy of a molecule for a given molecular structure was outlined. It is relatively simple, in principle, to repeat this procedure to map out the energy for different conformations. In real liquid crystals, there are, of course, several torsional angles; and the associated potentials are coupled to each other. The coupling between ring and tail angles in 5CB is an example. The density functional plane wave method has been used to map the conformational energy landscape for 5CB (Fig. 27). This is represented as a 2D potential for the relative ring angle and all-*trans* tail angle. It is found that a ring angle θ_r of 31° with an all-*trans* tail of 90° is the most favored conformer, which is similar to that found for solid 5CB by X-ray diffraction [192].

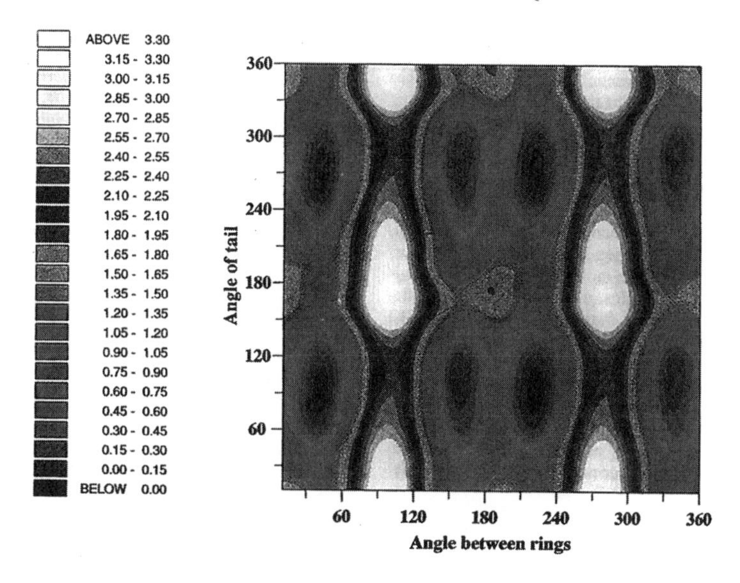

Figure 27. Conformational energy landscape for 5CB, as reported in Ref. [178].

D. Shape-Dependent Molecular Electronic Properties

It was emphasized that molecular structure, flexibility, and electronic properties are closely linked. This implies that molecular shape changes are expected to exert a major influence on electrostatic properties, such as the molecular dipole, quadrupole, and polarizability, which, together, govern many condensed-phase properties. In general, molecular dipoles can be defined either by a fixed molecular bond or by the relative position of the most electronegative species. In the former case, we expect a weak dependence of the molecular dipole on conformation and a more pronounced effect in the later situation. This issue has recently begun to be explored using quantum mechanical methods.

The mesogenic fragment 2-2′ difuluorobiphenyl represents a system in which molecular shape has a dramatic effect on overall dipole magnitude. This was revealed by density functional plane wave calculations (Fig. 28). The plot shows how the molecular dipole decreases from a value of about 2.5 D at a torsional angle of 0° (planar conformer) to a negligible value at 180°. The associated torsional potentials indicate that the favored torsional angle is about 60°, with a secondary minimum separated by a small barrier at 125°.

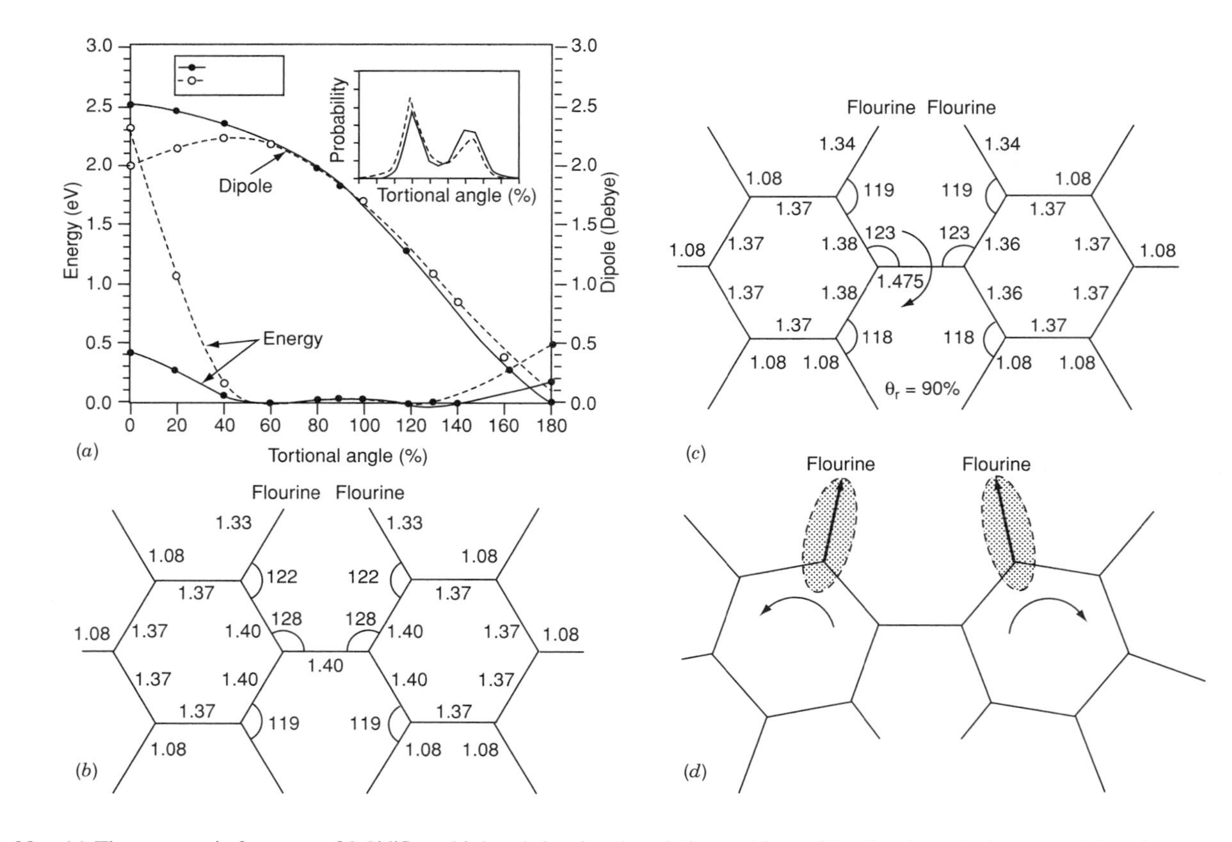

Figure 28. (*a*) The mesogenic fragment of 2-2′difluorobiphenyl showing the relative positions of the fluorine substituents and the relevant torsional angle. (*b*) The molecular dipole for 2-2′difluorobiphenyl as a function of torsional angle.

Another important issue from the point of view of simulations is the extent to which molecular dipoles can be deduced from knowledge of the dipoles of fragments. The largest contribution to the dipole moment of 5CB comes from the cyano group, because 0CB has a relatively large dipole of 5.83 D. The effect of the electron-donating tail is evident as the dipole moment of 5B is 0.55 D. The addition of these fragment dipoles is 6.38 D, which is marginally smaller than the dipole of 5CB (6.50 D). The results suggest that intramolecular charge transfer is not appreciable, implying that simple vector addition of dipole moments of fragments provides a good estimate of the total molecular dipole.

VIII. CONCLUSIONS

This review has several related objectives. Its principal goal is to motivate and stimulate interest in realistic molecular-level modeling of liquid crystals, the ultimate, long-range reward of which is to understand the relationships between molecular structure and condensed-phase behavior sufficiently well to predict properties before synthesis of materials. One of the further aims of this review is to illustrate through several examples how far molecular models and strategies, coupled with major advances in computing power, have brought physicists, chemists, materials scientists, and device engineers down this path.

In recent years, there have been considerable advances in our understanding of the structure, dynamics, and material properties of the liquid crystalline state, which have been made possible through combined use of sensitive experimental probes and simulations at the molecular level. It is also evident that the different levels of molecular realism are likely to remain essential, because the scope of problems is so diverse and there is no chance that fully atomistic or quantum mechanical simulations will be possible for the foreseeable future.

It has also emerged, however, that there are particularly promising opportunities for applying chemically detailed modeling to certain problems of relevance to condensed phases or interfacial behavior. For example, there appears to be a real prospect that accurate and predictive quantum mechanical methods will be useful in parameterizing potentials in a form that could be implemented efficiently using existing realistic models. It may be possible to modify density functional techniques to incorporate, in an approximate way, the effect of mean ordering fields to shed light on the the coupling between orientational order and molecular structure. Other areas include fully atomistic descriptions of liquid crystal–surface interactions in which the molecular basis for anchoring and alignment control might be explored at unprecedented levels of detail.

Acknowledgments

JC acknowledges support from the Royal Society of Edinburgh and the Royal Society of London. The authors are also grateful to S. J. Clark, D. J. Cleaver, M. R. Wilson, and S. Y. Yakovenko for helpful discussions.

REFERENCES

1. P. DeGennes and J. Prost, *The Physics of Liquid Crystals*, 2nd ed., Clarendon Press, Oxford, 1993.
2. S. Chandrasekhar, *Liquid Crystals*, 2nd ed., Cambridge University Press, Cambridge, 1992.
3. H. Arnold, *Z. Phys. Chem.* **226**, 146 (1964).
4. D.C. Burns, I. Underwood, J. Gourlay, A. Ohara, and D.G. Vass, *Optics Commun.* **119**, 62 (1995).
5. L. Onsager, *Ann. N.Y. Acad. Sci.* **51**, 627 (1949).
6. G.W. Gray and G. R. Luckhusrt, in *The Molecular Physics of Liquid Crystals, NATO Advanced Study Institute* (*ASI*), Academic Press, London, 1979.
7. A. Ferrarini and G. Moro and P. Nordio, *Mol. Phys.* **87**, 485 (1996).
8. J. Serrano, *Metallomesogens*, Wiley-VCH, London, 1998.
9. G.S. Attard, R.W. Date, C.T. Imrie, G.R. Luckhurst, S.J. Roskilly, and J.M. Seddon, *Liquid Crystals* **16**, 529 (1994).
10. S. McGrother, R. Sear, and G. Jackson , *J. Chem. Phys.* **106**, 731 (1997).
11. I. Khoo and S. Wu , *Optics and Nonlinear Optics of Liquid Crystals*, 1st ed., World Scientific, Singapore, 1993.
12. S. Elston and R. Sambles, *The Optics of Thermotropic Liquid Crystals*, Taylor and Francis, London, 1998.
13. A. Hauser, G. Pelzyl, C. Selbmann, D. Demus, S. Grande, and A.G. Petrov, *Mol. Cryst. Liq. Cryst.* **91**, 97 (1983).
14. P. Flory and P. Navard, *J. Chem. Soc. Faraday Trans.* **82**, 3381 (1986).
15. D. Dunmur and A. Tomes, *Mol. Cryst. Liq. Cryst.* **97**, 241 (1983).
16. W. de Jeu, *Solid State Physics Suppl. 14*, Academic Press, New York, 1978.
17. S. Urban and A. Würflinger, *Adv. Chem. Phys.* **96**, 143 (1997).
18. A. Drozd-Rzoska, S. Rzoska, and J. Ziolo, *Phys. Rev. E* **54**, 6452 (1996).
19. G. LaPenna, D. Catalano, and C. Veracini, *J. Chem. Phys.* **105**, 7097 (1996).
20. D. Grasso, C. Gandolfo, S. Fasone, and C. LaRosa, *Mol. Cryst. Liq. Cryst.* **221**, 85 (1989).
21. J. Goodby, *SPIE* **64**, 52 (1986).
22. W. Robinson, P. Kloess, C. Carboni, and H. Coles, *Liquid Crystals* **23**, 309 (1997).
23. L. Chan, G. Gray, and D. Lacey, *Mol. Cryst. Liq. Cryst.* **123**, 185 (1985).
24. S. Chandrashekar, *Advances in Liquid Crystals*, Academic Press, New York, 1976.
25. R. Sashidar and S. Chandrasekhar, *J. Phys. (Paris)* **36**, 1 (1975).
26. I. Molomiets, A. Ovsipyan, and S. Fillipov, *Mol. Cryst. Liq. Cryst.* **301**, 343 (1997).
27. A. Derzhanski and A. Petrov, *Acta Physica Polonica* **A55**, 747 (1979).
28. M. Miesowicz, *Nature* **158**, 27 (1946).
29. D. Langevin, *J. Phys. (Paris)* **33**, 249 (1979).
30. A. Chemielewski, *Mol. Cryst. Liq. Cryst.* **132**, 339 (1986).

31. A. Leadbetter, F. Temme, A. Heidemann, and W. Howells, *Chem. Phys. Lett.* **34**, 363 (1975).
32. W. C.-K. Poon and P. Warren, *Europhys. Lett.* **28**, 513 (1994).
33. M. Allen, G. Evans, D. Frenkel, and B. Mulder, *Adv. Chem. Phys.* **86**, 1 (1993).
34. M. Vanbruggen, F. Vanderkooij, and H. Lekkerkerkeer, *J. Phys. Cond. Mat.* **8**, 9451 (1996).
35. D. Frenkel and B. Mulder, *Mol. Phys.* **55**, 1171 (1985).
36. D. Frenkel, *J. Phys. Chem.* **92**, 3280 (1988).
37. D. Frenkel, H. Lekkerkerker, and D. Stroobants, *Nature* **332**, 822 (1988).
38. D. Frenkel, B. Mulder, and J. McTague, *Phys. Rev. Lett.* **52**, 287 (1984).
39. J. A. C. Veermen and D. Frankel, *Phys. Rev. A* **41**, 3237 (1990).
40. D. Frenkel, *Liquid Crystals* **5**, 929 (1989).
41. S. McGrother, D. Williamson, and G. Jackson, *J. Chem. Phys.* **104**, 6755 (1996).
42. P. Bolhuis and D. Frenkel, *J. Chem. Phys.* **106**, 666 (1997).
43. M. Allen and D. Frenkel, *Phys. Rev. Lett.* **58**, 1748 (1987).
44. P. Camp and M. Allen, *J. Chem. Phys.* **106**, 6681 (1997).
45. M. Allen, *Liquid Crystals* **8**, 499 (1990).
46. M. Allen and D. Frenkel, *Phys. Rev. A* **37**, 1813 (1988).
47. M. Allen and D. Frenkel, *Phys. Rev. A* **42**, 3641 (1990).
48. M. P. Allen, P. J. Camp, C. P. Mason, G. T. Evans, and A. J. Masters, *J. Chem. Phys.* **105**, 11175 (1996).
49. S. Tang, G. Evans, C. Mason, and M. Allen, *J. Chem. Phys.* **102**, 3794 (1995).
50. M. Allen, *Phys. Rev. Lett.* **65**, 2881 (1990).
51. D. Vaidya, D. Kofke, S. Tang, G. Evans, *Mol. Phys.* **83**, 101 (1994).
52. R. van Rioj, P. Bolhuis, B. Mulder, D. Frenkel, *Phys. Rev. E* **52**, 1277 (1995).
53. J. vanDuijneveldt and M. Allen, *Mol. Phys.* **90**, 243 (1997).
54. B. Swanson and L. Sorenson, *Phys. Rev. Lett.* **75**, 3293 (1995).
55. E. deMiguel, L. Rull, M. Chalam, and K. Gubbins, *J. Chem. Phys.* **74**, 405 (1991).
56. J. Gay and B. Berne, *J. Chem. Phys.* **74**, 3316 (1981).
57. J. Corner, *Phys. Rev. E* **192**, 275 (1948).
58. S. Walmsley, *Chem. Phys. Lett.* **49**, 320 (1977).
59. G. Luckhurst and P. Simmonds, *Mol. Phys.* **80**, 233 (1993).
60. L. Rull, *Physica A* **220**, 113 (1995).
61. G. Luckhurst, R. Stevens, and R. Phippen, *Liquid Crystals* **8**, 451 (1990).
62. E. DeMiguel, E. M. Delrio, J. T. Brown, and M. P. Allen, *Mol. Phys.* **72**, 593 (1991).
63. E. DeMiguel, E. Delrio, J. Brown, and M. P. Allen, *J. Chem. Phys.* **105**, 4234 (1996).
64. R. Berardi, A. Emerson, and C. Zannoni, *J. Chem. Phys.* **89**, 4096 (1993).
65. J. McColl and C. Shih, *Phys. Rev. Lett.* **29**, 85 (1972).
66. R. Tranfield and P. Collins, *Phys. Rev. A* **29**, 2744 (1982).
67. R. Horn and T. Faber, *Proc. Roy. Soc.* **25**, 199 (1979).
68. J. Emsley and G. Luckhurst, and B. Timini, *J. Phys.* **48**, 473 (1987).
69. M. Bates and G. Luckhurst, *Chem. Phys. Lett.* **281**, 193 (1997).
70. M. Allen and M. Warren, *Phys. Rev. E* **78**, 1291 (1997).
71. D. Cleaver, C. Care, M. Allen, and M.P. Neal, *Phys. Rev. E* **54**, 559 (1996).

72. R. Bemrose, C. Care, D. Cleaver, and M. Neal, *Mol. Phys.* **90**, 625 (1997).
73. H. Lekkerkerker, P. Coulon, R. van der Haegen, and R. Debliek, *J. Chem. Phys.* **80**, 3427 (1984).
74. T. Slukin, *Liquid Crystals* **6**, 111 (1989).
75. S. Sarman, *Computers Phys.* **104**, 342 (1996).
76. R. Berardi, C. Fava, and C. Zannoni, *Chem. Phys. Lett.* **236**, 462 (1995).
77. J. Stelzer , M. Bates, L. Longa, and G. Luckhurst, *J. Chem. Phys.* **107**, 7483 (1997).
78. M. P. Allen, P. J. Camp, C. P. Mason, G. T. Evans, and A. J. Masters, *J. Phys. Cond. Mat.* **8**, 9433 (1996).
79. M. P. Allen, M. A. Warren, M. R. Wilson, A. Sauron, and W. Smith, *J. Phys. Cond. Mat.* **105**, 2850 (1996).
80. J. Stelzer, L. Longa, and H. Trebin, *Mol. Cryst. Liq. Cryst.* **262**, 455 (1995).
81. P. Teixeira, V. Pergamenshchik, T. Slukin, *Mol. Phys.* **80**, 1339 (1993).
82. E. de Miguel, L. Rull, and K. Gubbins, *Phys. Rev. A* **45**, 3813 (1992).
83. A. Smondyrev, G. Loriot, and R. Pelcovitz, *Phys. Rev. Lett.* **75**, 2340 (1995).
84. S. Sarman, *Physica A* **240**, 160 (1997).
85. S. Sarman, *J. Chem. Phys.* **99**, 9021 (1993).
86. S. Sarman, *J. Chem. Phys.* **101**, 480 (1994).
87. S. Zumer, P. Ziherl, and M. Vilfan, *Mol. Cryst. Liq. Cryst.* **292**, 39 (1997).
88. G. Crawford and S. Zumer, *Liquid Crystals in Complex Geometries*, Taylor and Francis, London, 1996.
89. A. Emerson and C. Zannoni, *J. Chem. Soc. Faraday Transactions* **91**, 3441 (1995).
90. M. Bates and C. Zannoni, *Chem. Phys. Lett.* **280**, 40 (1997).
91. A. Rapini and M. Papoular, *J. Phys.* **30**, 4 (1969).
92. Z. Zhang, A. Chakrabarti, O. Mouritsen, and M. Zuckermann, *Phys. Rev. E* **53**, 2461 (1996).
93. J. Stelzer, L. Longa, and H. Trebin, *Phys. Rev. E* **55**, 7085 (1997).
94. V. Palermo, F. Biscarini, and C. Zannoni, *Phys. Rev. E* **57**, 2519 (1998).
95. W. Steele, *Surf. Sci.* **36**, 317 (1973).
96. A.J. Leadbetter, F.P. Temme, A. Heidemann, and W.S. Howells, *J. Phys.* **40**, 375 (1979).
97. P. Cladis, *Phys. Rev. Lett.* **35**, 48 (1975).
98. A. GilVillegas, S. McGrother, and G. Jackson, *Mol. Phys.* **92**, 723 (1997).
99. A. Vanakaras and D. Photinos, *Mol. Phys.* **85**, 1089 (1995).
100. C. Vega and S. Lago, *J. Chem. Phys.* **100**, 6727 (1994).
101. D. Williamson and F. DelRio, *J. Chem. Phys.* **107**, 9549 (1997).
102. S. McGrother, A. Gilvillegas, and G. Jackson, *J. Phys. Cond. Mat.* **8**, 9649 (1996).
103. S. McGrother, G. Jackson, and D. Photinos, *Mol. Phys.* **91**, 751 (1997).
104. K. Satoh, S. Mita, and S. Kondo, *Liquid Crystals* **20**, 757 (1996).
105. K. Satoh, S. Mita, and S. Kondo, *Chem. Phys. Lett.* **255**, 99 (1996).
106. A. Gilvillegas, S. McGrother, and G. Jackson, *Chem. Phys. Lett.* **269**, 441 (1997).
107. P. Teixeira, M. Osipov, and M. daGama, *Phys. Rev. E* **57**, 1752 (1998).
108. R. Berardi, S. Orlandi, and C. Zannoni, *Chem. Phys. Lett.* **261**, 357 (1996).
109. J. Weis, D. L. D, and Zarragoicoechea, *Mol. Phys.* **80**, 1077 (1993).

110. J. Weiss and Zarragoicoechea, *Phys. Rev. Lett.* **69**, 913 (1992).
111. E. Gwozdz, A. Brodka, and K. Pasterny, *Chem. Phys. Lett.* **267**, 557 (1997).
112. G. Ayton, D. Wei, and G. Patey, *Phys. Rev. E* **55**, 447 (1997).
113. D. Wei and G. Patey, *Phys. Rev. Lett.* **68**, 2043 (1992).
114. G. barbero, L. Evangelista, and S. Ponti, *Phys. Rev. E* **54**, 4442 (1996).
115. M. Osipov, T. Slukin, and S. Cox, *Phys. Rev. E* **55**, 464 (1997).
116. M. Bates and G. Luckhurst, *Liquid Crystals* **24**, 229 (1998).
117. M. Neal, A. Parker, and C. Car, *Mol. Phys.* **91**, 603 (1997).
118. A. Vanderschoot, *J. Phys.* **6**, 1557 (1996).
119. A. Tkachenko, *Phys. Rev. Lett.* **77**, 4218 (1996).
120. G. Evans, *Mol. Phys.* **87**, 239 (1996).
121. J.W. Emsley, G. Celebre, G. DeLuca, M. Longeri, and F. Lucchessini, *Liquid Crystals* **16**, 1037 (1994).
122. T. R. Bose, D. Ghose, C. D. Mokerjee, S. Saha, and M. Saha, *Phys. Rev. A* **43**, 4372 (1991).
123. R. Podsiadly, J. Janik, J. Janik, and K. Otnes, *Liquid Crystals* **14**, 1519 (1993).
124. R. Podsiadly, J. Mayer, J.A. Janik, J. Krawczyk, and T. Stanek, *Acta. Phys. Polonica* **91**, 513 (1997).
125. S. Yakovenko, G. Krömer, and A. Geiger, *Mol. Cryst. Liq. Cryst.* **275**, 91 (1996).
126. H. Hsueh, H. Vass, S. J. Clark, F. Y. Pu, W. C-K. Poon, and J. Crain, *Europhys. Lett.* **38**, 107 (1997).
127. F. Hide, N.A. Clark, K. Nito, A. Yasuda, and D.M. Walba, *Phys. Rev. Lett.* **75**, 2344 (1995).
128. A. L. Verma, B. Zhao, S. M. Jiang, J. C. Sheng, and Y. Ozaki, *Phys. Rev. E* **56**, 3053 (1997).
129. M. Wilson, *J. Mol. Liquids* **68**, 23 (1996).
130. A. Komolkin, A. Laaksonen, and A. Maliniak, *J. Chem. Phys.* **101**, 4103 (1994).
131. M. Wilson and M. Allen, *Mol. Phys.* **80**, 277 (1993).
132. M. Wilson and M. Allen, *Liquid Crystals* **12**, 157 (1992).
133. J. vanDuijneveldt and M. Allen, *Mol. Phys.* **92**, 855 (1997).
134. F. Affouard, M. Kroger, and S. Hess, *Phys. Rev. E* **54**, 5178 (1996).
135. M. Yoshida and H. Toriumi, *Mol. Cryst. Liq. Cryst.* **262**, 525 (1995).
136. K. Nicklas, P. Bopp, and J. Brickmann, *J. Chem. Phys.* **101**, 3157 (1994).
137. M. Wilson, *J. Chem. Phys.* **107**, 8654 (1997).
138. J. Ryckaert and A. Bellemans, *Chem. Phys. Lett.* **30**, 123 (1990).
139. R. Memmer, H. Kuball, and A. Schonhofer, *Mol. Phys.* **89**, 1633 (1996).
140. R. Memmer, H. Kuball, and A. Schonhofer, *Liquid Crystals* **19**, 749 (1995).
141. C. Cross and B. Fung, *Mol. Cryst. Liq. Cryst.* **262**, 50 (1995).
142. C. Cross and B. Fung, *J. Chem. Phys.* **101**, 6839 (1994).
143. F. Novak, *Mol. Cryst. Liq. Cryst.* **113**, 242 (1984).
144. R. Richardson, A. Leadbetter, and J. Frost, *Mol. Phys.* **45**, 1163 (1982).
145. G. Zerbi, P. Roncone, G. Longhi, and S. Wunder, *J. Chem. Phys.* **89**, 166 (1988).
146. E. Galbiati and G. Zerbi, *J. Chem. Phys.* **87**, 3653 (1987).
147. C. Almirante, G. Minoni, and G. Zerbi, *J. Phys. Cond. Mat.* **90**, 852 (1986).

148. G. Zerbi, in *Advances in IR and Raman Spectroscopy*, M. MacKenzie, ed., Heyden, London, 1984.

149. M. Wilson, *Liquid Crystals* **21**, 437 (1996).

150. J. Huth, T. Mosell, K. Nicklas, A. Sariban, and J. Brickmann, *J. Phys. Cond. Mat.* **98**, 7685 (1994).

151. S. Hauptmann, T. Mosell, S. Reiling, and J. Brickmann *Computers Phys.* **208**, 57 (1996).

152. D. Levesque, Mazars, and J. Weis, *Computers Phys.* **103**, 3820 (1995).

153. J. Stankus, R. Torre, and M. Fayer, *J. Phys. Cond. Mat.* **97**, 9478 (1993).

154. S. Urban, H. Kresse, and R. Dabrowski, *Zeit. fur Naturforsch. A* **52**, 403 (1997).

155. B. Berne and R. Pecora, *Dynamic Light Scattering*, 1st ed., Wiley, New York, 1976.

156. W. Rothschild, *Dynamics of Molecular Fluids*, Wiley-Interscience, New York, 1984.

157. A. Lecalvez, S. Montant, E. Freysz, A. Ducasse, X.W. Zhuang, and Y.R. Shen, *Chem. Phys. Lett.* **258**, 620 (1996).

158. A. Lecalvez, S. Montant, E. Freysz, R. Maleckrassoul, and A. Ducasse, *Annales de Physicque* **20**, 613 (1995).

159. T. Ueno, K. Sakai, and K. Takagi, *Phys. Rev. E* **54**, 6457 (1996).

160. J. J. Stankus, R. Torre, C. D. Marshall, S. R. Greenfield, A. Sengupta, and M. D. Fayer, *Chem. Phys. Lett.* **194**, 3503 (1992).

161. C. Flytzanis and Y. Shen, *Phys. Rev. Lett.* **33**, 14 (1974).

162. A. Sengupta and M. Fayer, *J. Chem. Phys.* **102**, 4193 (1995).

163. S. Yakovenko, A. Muravski, F. Eikelschulte, and A. Geiger, *Computers Phys.* **105**, 10766 (1996).

164. D. Thirumalai, *J. Phys. Cond. Mat.* **98**, 9265 (1994).

165. S. Ravichandran, A. Perera, M. Moreau, and B. Bagchi, *J. Chem. Phys.* **107**, 8469 (1997).

166. A. Perera, S. Ravichandran, M. Moreau, and B. Bagchi, *J. Chem. Phys.* **106**, 1280 (1997).

167. S. Yakovenko, A. Muravski, G. Krömer, and A. Geiger, *Mol. Phys.* **86**, 1099 (1995).

168. S. Yakovenko, A. Minko, G. Krömer, and A. Geiger, *Liquid Crystals* **17**, 127 (1994).

169. Cagnon, *Phys. Rev. Lett.* **70**, 2742 (1993).

170. R. Ondris-Crawford, G. Crawford, S. Zumer, and J. Doane, *Phys. Rev. Lett.* **70**, 194 (1993).

171. T. Moses and Y. Shen, *Phys. Rev. Lett.* **67**, 2033 (1991).

172. M. Büchel, B. Weichart, C. Minx, H. Menzel, and D. Johannsmann, *Phys. Rev. E* **55**, 455 (1997).

173. Y. Zhu and Y. Wei, *Computers Phys.* **101**, 10023 (1994).

174. D. J. Cleaver, M. J. Callaway, T. Forester, W. Smith, and D. J. Tildesley, *Mol. Phys.* **81**, 781 (1994).

175. M. Glaser, R. Malzbender, N. Clark, and D. Walba, *J. Phys.* **6**, A261 (1994).

176. D. Binger and S. Hannah, *Mol. Cryst. Liq. Cryst.* **302**, 63 (1997).

177. S. J. Clark, C. J. Adam, J. A. White, G. J. Ackland, and J. Crain, *Liquid Crystals* **22**, 469 (1997).

178. S. Clark, C. Adam, D. Cleaver, and J. Crain, *Liquid Crystals* **22**, 475 (1997).

179. W. Kohn and L. Sham, *Phys. Rev.* **140**, 1133A (1965).

180. E. Kryachko and E. Ludena, *Energy Density Functional Theory of Many Electron Systems*, Klewer Academic, Boston, 1990.

181. R. M. Wentzcovich, W. W. Schultz, and P. B. Allen, *Phys. Rev. Lett.* **72**, 3389 (1994).

182. H. C. Hsueh, H. Vass, S.J. Clark, G.J. Ackland, and J. Crain, *Phys. Rev. B* **51**, 16750 (1995).
183. H. C. Hsueh et al., *Phys. Rev. B* **51**, 12216 (1995).
184. K. Refson et al., *Phys. Rev. B* **52**, 10823 (1995).
185. I. Stich et al., *Phys. Rev. B* **49**, 8076 (1994).
186. J. S. Lin, A. Qteish, M. C. Payne, and V. Heine, *Phys. Rev. B* **47**, 4174 (1993).
187. G. Bachelet, D. Hamman, and M. Schlüter, *Phys. Rev. B* **26**, 4199 (1982).
188. J. P. Perdew et al., *Phys. Rev. B* **46**, 6671 (1992).
189. R. Feynman, *Phys. Rev.* **56**, 340 (1939).
190. M. C. Payne et al., *Rev. Mod. Phys.* **64**, 1046 (1992).
191. M. Schlüter and L. J. Sham, *Physics Today* **36** (1982).
192. T. Hannemann, W. Haase, I. Svomoda, and H. Feuss, *Liquid Crystals* **19**, 669 (1995).

MOLECULAR-BASED MODELING OF WATER AND AQUEOUS SOLUTIONS AT SUPERCRITICAL CONDITIONS

ARIEL A. CHIALVO AND PETER T. CUMMINGS

Department of Chemical Engineering, University of Tennessee, Knoxville, TN 37996-2200 and Chemical Technology Division, Oak Ridge National Laboratory, Oak Ridge, TN 37831-6181

CONTENTS

Advances in Chemical Physics, Volume 109, Edited by I. Prigogine and Stuart A. Rice
ISBN 0-471-32920-7

I. INTRODUCTION

Water, crucial for the existence of life and an essential component in many industrial and biochemical processes, is without a doubt the most intriguing and least understood fluid in nature. For a low molecular weight, nonionic, nonmetallic compound, it exhibits a striking number of unusual properties: [1] volume contraction upon melting, extrema in density (maximum) and isothermal compressibility (minimum) at normal conditions, at least nine crystalline polymorphs, and on anomalously large dielectric constant and critical temperature [2–4]. These properties have been ascribed to water's ability to hydrogen bond and form three-dimensional networks with a rather small number of nearest neighbors, (about 4) [3,5] compared to more normal liquids (10–12) [6].

Most biochemical and industrial processes occur in solution, making it crucial to understand the participation of the solvent (i.e., solvation effects) in determining chemical reactivity [7,8]. Numerous recent developments have increased interest in supercritical water (SCW) and aqueous solutions [9–11]. In particular, supercritical water has become an attractive environmentally friendly reaction medium (solvent) for a variety of chemical processes and technologic applications in which water participates as a solvent, catalyst, and reactant, including selective synthesis [12], coal conversion [13], deuteration of simple organic compounds [14], and conversion of organic waste to light feedstocks [15]. Perhaps the most promising application in this area is the destructive oxidation of biochemical and pharmacologic hazardous wastes, known as the supercritical water oxidation (SCWO) process [16–20].

Some of the properties that make SCW a powerful solvating agent for separation purposes, apart from its high compressibility and mass transfer characteristics, also make it an unusually advantageous medium to conduct chemical reactions [9,11,21]. This also implies that its solvation power can be changed continuously, from liquid-like to gas-like by modest changes in temperature and/or pressure; thus SCW exhibits gas-like viscosity [22], liquid-like diffusivities [23], and an eight-orders-of-magnitude variation in the ion product [24]. This combination of features makes it possible for SCW to dissolve polar and nonpolar organic compounds, [25,26] as well as gases, [27–31] resulting in homogeneous systems capable of reacting through polar [32], ionic [33,34], and free-radical [35] mechanisms.

Although chemistry in SCW has great potential, it also involves fundamental and practical (i.e., technological) challenges. Among them, corrosion [36] is probably one of the most important, not only because it has the potential to ultimately destroy a reactor, but also because the resulting insoluble salts can clog reactor pipes [37,38]; furthermore, the remaining traces of solvated ion might catalyze unwanted reactions [39]. Here we are especially concerned with the fundamental difficulties characterized by our poor microscopic understanding of high-temperature solvation of species with a wide range of molecular asymmetries, such as ionic, polar, and nonpolar compounds. Important aspects of supercritical solvation are associated with, and enhanced by, the pronounced local density changes around charged species, which translate into electrostriction and dielectric saturation [40,41,42]. These local changes play a central role in the solvation process at high temperature, especially in compressible solvents.

No less important are the potential applications of near and supercritical aqueous solutions, conditions at which newly discovered forms of life can thrive (such as that in deep sea hydrothermal vents [43]), creating opportunities in high-temperature microbiologic research [44,45]. Although the recent rush of potential applications of SCW chemistry appears encouraging, their exploitation is still uncertain owing to the lack of a precise microscopic (molecular-level) interpretation of the observed behavior. In other words, while our understanding of the physicochemical properties of water and aqueous solutions at normal conditions has improved significantly in the last two decades [46,47], it is far from sufficient to even conjecture the behavior of supercritical aqueous systems, which are currently of great scientific and practical interest [48,49]. The failure of conventional solvation approaches for near-critical and supercritical conditions hinges on the coexistence of mechanisms with two rather different length scales — short-range (solvation) and long-range (compressibility driven) phenomena [50,51] — making these systems extremely challenging to study by simulation, theory, or experiment.

II. SUPERCRITICAL WATER

B. Microstructural Analysis, Hydrogen-Bonding Characterization, and the Interplay between Experiments and Molecular Simulation

For decades researchers have debated over the nature of the hydrogen bond interactions and its microstructural manifestation. Molecular simulation, X-ray, and neutron scattering techniques have been the primary tools used by researchers to determine the spatial local structure of water and thus to extract valuable information on the relevant molecular configurations (geometry and energy). Currently, more sophisticated experimental techniques, including NMR [52–54], Raman scattering [55], and microwave spectroscopy, [56] have become available to obtain additional insight into the hydrogen-bonding (HB) interactions.

For more than a quarter of century, attention has been mainly focused on ambient water (driven in part by the need for an adequate water model for simulations of biochemical interest [57]); however, the accuracy of existing diffraction results has been the subject of controversy [58], even though a particular data set [59] had become the *de facto* standard until recently. The renewed interest in this subject was motivated by the results for water at ambient and high temperatures from neutron diffraction with isotope substitution (NDIS) by Postorino and co-workers [60]. Because these results were somewhat controversial, they generated an enthusiastic and healthy scientific discussion, which is still ongoing [61].

In this chapter, we present an account of the significant progress made in the elucidation of water microstructure and its changes with state conditions, the concomitant modeling effort, and the fruitful interplay between experiment and simulation (see Section II.B) [62]. We first review the experimental results, including those from neutron and X-ray diffraction, NMR, and Raman spectroscopy. In the following sections, we discuss the theoretical and simulation studies carried out either to support or to challenge the experimental findings and, more important, to place emphasis on the resulting interplay between them, which encouraged an insightful effort toward the microscopic characterization of water in any state condition.

As noted, the first real task before any serious analysis of HB can be made is to set up a working definition, one that is consistent with the accepted (conventional wisdom) picture of a hydrogen-bonded pair [5]. Although there is general consensus that a proper description of HB must involve some positional and orientational constraints [such as the extent of the collinear O–H···O configurations and the relative orientation of the pair of water molecules (dihedral angle δ_D in Fig. 1)], the actual assessment of the hydrogen bond strength is still a matter of discussion [61]. In fact, there are many ways to identify an HB configuration, depending on how we approach it [63]. For example, a scatterer who determines the radial distribution function $g_{OH}(r)$ will likely choose the first peak of this quantity to describe the strength of the HB configuration [60], whereas a spectroscopist might define it in terms of the observed shift in the frequency spectrum [52,55].

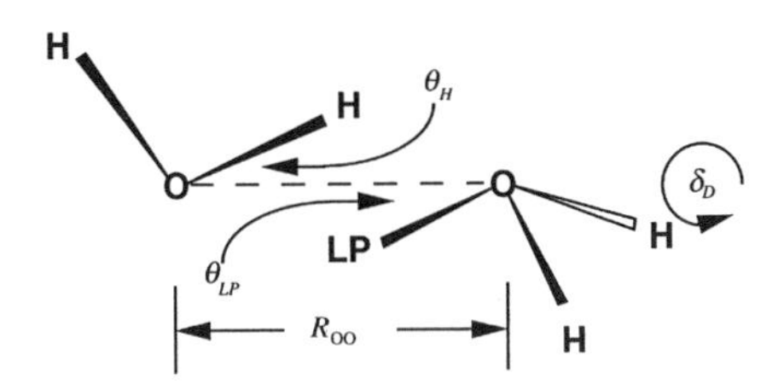

Figure 1. Definition of hydrogen bond parameters.

There are two widely used definitions to describe hydrogen-bonding configurations—geometric [57,64,65] and energetic criteria [66,67]—as well as a gamut of variations and combinations [68–78]. The main issue with either criterion is the lack of consistency with the same physical picture of hydrogen bonding; for example, for some water configurations, one criterion might define HB whereas the other might not [79,80].

Another interesting aspect of the controversy concerns the maximum temperature at which HB might disappear, which has changed since earlier estimations. For example, based on the approach of significant structures, Murchi and Eyring [81] suggested the disappearance of HB at about $T = 523$ K, a condition at which the water dimers behave as free rotators. Soon after, Luck [82] extended this temperature up to 647 K (the critical point), based on the study of IR absorption in liquid water. Later high-temperature spectroscopic work pushed this temperature still higher [83–85]. Finally, X-ray scattering results for high-temperature water suggested a non-negligible probability of tetrahedral hydrogen-bonded water configurations at $T = 773$ K [86,87].

Some progress on the subject has been achieved through the development of theoretical techniques to systematically analyze the angular and spatial local structure of water [75,80,88]. A particular application of this orientational-dependent microstructural analysis was performed with supercritical water [72,89,90] to dispute the claims of the disappearance of HB made by Postorino and co-workers [60]; this orientational approach appears to be a special case of a newly published analysis [75,88].

As we will see in Section II.B, the great challenge is to interpret the experimental results and put them on common ground with the theoretical developments, without introducing (implicitly) any preconceived argument into the manipulation of the raw data. Undoubtedly, this is a situation in which computer simulation of model fluids may help in the microscopic interpretation of experimental data, because it can deal with precisely defined systems with complete control of and full access to the molecular environment.

Neutron diffraction scattering with isotope substitution is considered the only means by which we can extract direct information about all the site–site pair correlation functions of liquids and amorphous materials [91–94]. What makes this approach possible is the justifiable assumption that light (H_2O) and deuterated (D_2O) water exhibit the same structural features [95] and allow the extraction of the three site–site pair correlation functions from the diffraction data of three isotopically different samples (see below for more details). Although this technique was used to study the structure of water at nonambient conditions [96], the first complete study of supercritical water, and comparison with new results for ambient water results, was reported by Soper and collaborators in late 1993 [60], followed shortly by two other contributions [97,98].

These researchers studied water at three state conditions — ambient ($P = 0.1$ MPa and $T = 298$ K), $T = 573$ K and $\rho = 0.72$ g/cm^3, and $T = 673$ K and $\rho = 0.66$ g/cm^3 — and compared their diffraction results (hereafter referred to as NDIS-93) directly with the molecular simulation data of Cummings and co-workers [99] for the SPC water model [100] at rather different state conditions. Based on this comparison and on the amplitude and position of the first peak of the O···H radial pair distribution function $g_{OH}(r)$ as a measure of hydrogen bonding, they

1. Concluded that almost all hydrogen bonding is broken down at $T = 673$ K.
2. Cast some doubt on the capability of current water models to describe properly the high-temperature structure of water.
3. Claimed that their new structural information should allow theoreticians to adjust their models of water.

These results, and the associated claims were surprising, since at criticality the kinetic energy of water is just about one-third of the assumed energy for the hydrogen bond (i.e., 2.6 kcal/mol [55,101]), making it rather difficult to break the hydrogen-bonding structure. Beyond the validity of the authors' claims, the newly determined structural information exhibited several rather controversial features. The first and most obvious was the appearance of an unphysical bump on the left-hand side of the O···O radial pair distribution function $g_{OO}(r)$ at ambient conditions (Fig. 2). Because this is the least accurately determined function (in that it contributes only $\sim$10% to the total scattering pattern in two of the three NDIS experiments [92]), there was reason to suspect that this bump was not real but the result of the numerical manipulation of the raw scattering data (see the analysis of Löffler and co-worker's results). This feature, not present in any earlier diffraction experiment [59,102,103], degenerates into a pronounced shoulder at higher temperatures (Fig. 2). The second obvious controversial feature concerns the behavior of the hard-core diameter for the $g_{OH}(r)$, which increases with temperature, i.e., is at odds with the typical behavior of a molecule's excluded volume in which increasing temperatures translates into effective decreasing hard-core diameter. Surprisingly, the new ambient microstructural data were remarkably different from the accepted standard (hereafter referred to as NDIS-86 [59]), a feature not discussed in the Postorino and co-workers paper.

These claims (the elimination of hydrogen bonding by $T = 673$ K and the inadequacy of current water models) were immediately challenged through molecular dynamics simulation by Chialvo and Cummings [89] and by

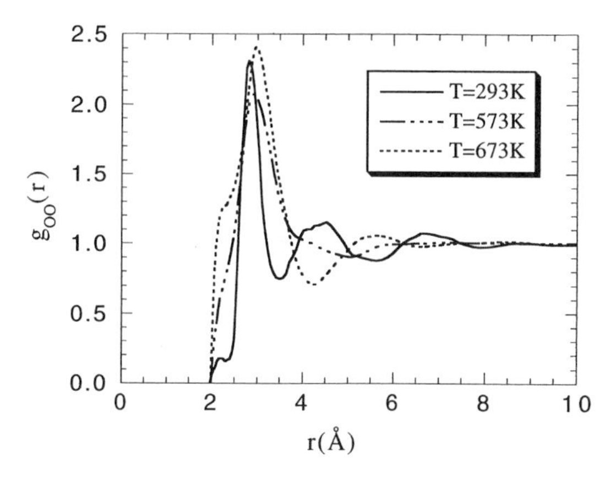

Figure 2. Temperature dependence of the NDIS-93 O···O radial distribution functions of water at $T = 293$ K and $\rho = 1$ g/cm^3; $T = 572$ K and $\rho = 0.72$ g/cm^3; and $T = 673$ K and $\rho = 0.64$ g/cm^3.

Löffler and co-workers [104], and at once it became clear that the NDIS-93 data were seriously flawed. Given the complexity of the NDIS experiments and the numerous possible sources of errors in the processing of the raw data, it was reasonable to suspect the presence of artifacts [104]. Yet, initially, it was also unclear whether the disagreement between the predicted and measured correlation functions was a reflection of unrealistic intermolecular models, inadequate methods of processing the diffraction raw data, or a combination of both.

Molecular simulations gave the first clues concerning the anomalies in the NDIS-93 data. Chialvo and Cummings [89] pointed out the unphysical features of the NDIS-93 data and the unusual isothermal density dependence of the amplitude of the first peak of the oxygen–hydrogen radial distribution function, $g_{OH}(r)$; i.e., the amplitude of its first peak decreases with increasing density and the corresponding coordination number increases [89,104]. By analyzing an orientationally dependent version of $g_{OH}(r)$, Chialvo and Cummings [72] also concluded that neither the height of the first peak of $g_{OH}(r)$ nor the corresponding coordination number, $n_{OH}(r) = 4\pi\rho_H \int_0^r g_{OH}(r)r^2\,dr$, adequately measures the strength of the hydrogen bonding in water (see Section II.B). Thus, even if the NDIS-93 were correct, one could not conclude on the basis of the NDIS-93 data that hydrogen bonding was essentially nonexistent by $T = 673$ K.

Moreover, Löffler and co-workers [104] suggested that the observed unphysical features were the result of an improper correction for inelasticity and incoherent effects in the scattering data from light water. This conclusion was based on the analysis of the behavior of the Fourier transform of the structural factors $G(r)$,

$$G(r) = (b_O^2 g_{OO}(r) + 4b_O b_{HD} g_{OH}(r) + 4b_{HD}^2 g_{HH}(r))/(b_O + 2b_{HD})^2 \quad (2.1)$$

where b_O(5.8 fm) denotes the oxygen's scattering length, and b_{HD} is that corresponding to the hydrogen–deuterium mixture:

$$b_{HD} = x_H b_H + (1 - x_H) b_D \quad (2.2)$$

where b_H(−3.74 fm) and b_D(6.67 fm) are the corresponding scattering length for hydrogen and deuterium, respectively. Thus by performing three scattering experiments with different isotopic compositions, say, one with fully deuterated water ($x_D = 1 - x_H \approx 0.997$),

$$G_1(r) = 0.092 g_{OO}(r) + 0.422 g_{OH}(r) + 0.486 g_{HH}(r) \quad (2.3)$$

another with half-deuterated water ($x_D = 1 - x_H \approx 0.5$),

$$G_2(r) = 0.441 g_{OO}(r) + 0.446 g_{OH}(r) + 0.113 g_{HH}(r) \quad (2.4)$$

and the final one with light water ($x_D = 1 - x_H \approx 0.0$),

$$G_3(r) = 11.92 g_{OO}(r) - 30.74 g_{OH}(r) + 19.82 g_{HH}(r) \quad (2.5)$$

The $g_{OO}(r)$, $g_{OH}(r)$, and $g_{HH}(r)$ are obtained by matrix inversion of Eqs. (2.3)–(2.5):

$$g_{OO}(r) = -1.101 G_1(r) + 2.086 G_2(r) + 0.015 G_3(r) \quad (2.6)$$
$$g_{OH}(r) = 0.663 G_1(r) + 0.355 G_2(r) - 0.018 G_3(r) \quad (2.7)$$
$$g_{HH}(r) = 1.690 G_1(r) - 0.703 G_2(r) + 0.013 G_3(r) \quad (2.8)$$

According to this formalism, Löffler and co-workers were able to study the weighted contribution of individual scattering experiments to the shape of each of the site–site distribution functions. For example, they [104] determined the three $G_i(r)$ values through the simulated site–site pair correlation functions of the extended simple point change (SPC/E) water model [i.e., Eqs. (2.3)–(2.5); Fig. 3], and from them they concluded

1. $G_3(r)$ plays a key role in determining the final shape of $g_{OO}(r)$ and $g_{OH}(r)$, especially around the first peak of both correlations, where the NDIS-93 data show the most controversial features.
2. Although $G_3(r)$ contributes little to the shape of $g_{HH}(r)$, it simultaneously enhances the first peak of $g_{OH}(r)$ and annihilates the first peak of $g_{OO}(r)$ (Fig. 3), based on our own simulations of SPC water.

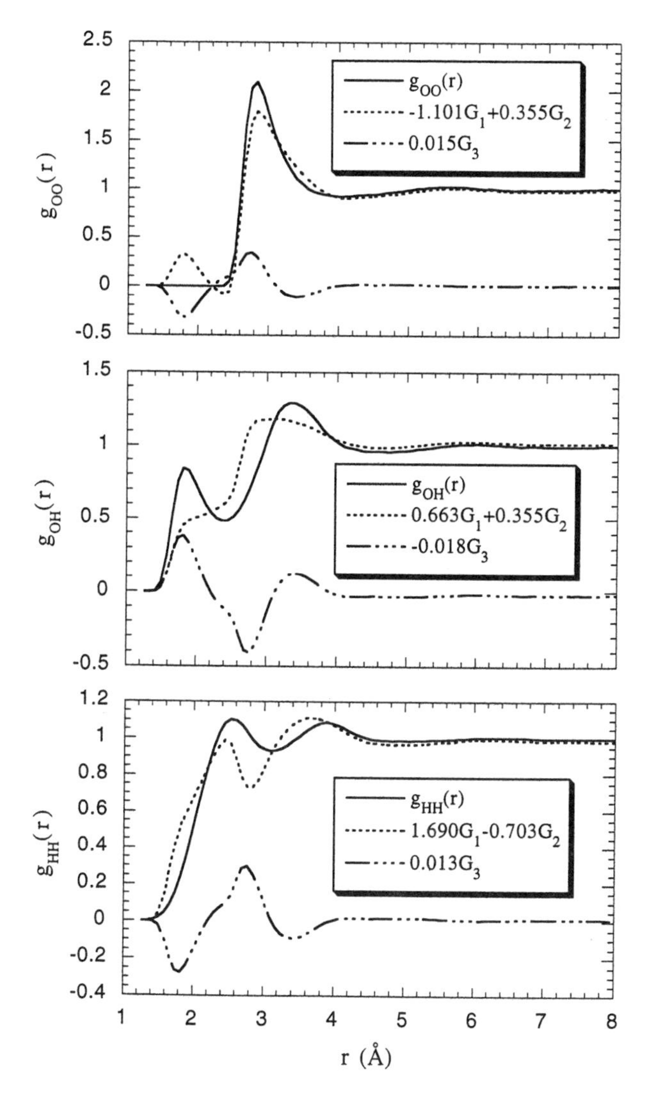

Figure 3. Comparison of the three contributions to the radial distribution functions for the SPC water model at $T = 572$ K and $\rho = 0.72$ g/cm^3, according to Eqs. 2.3–2.5. Modified from [104].

In summary, there is a correlation between the appearance of the bump on $g_{OO}(r \approx 1.9$ Å) and the attenuation of the first peak (the so-called HB peak) of $g_{OH} r \approx 1.9$ Å). Interestingly, $G_3(r)$ is a troublesome quantity because it arises from the scattering of light hydrogen, which usually carries high incoherent and inelasticity effects [105]. The correction of these effects becomes a large source of uncertainty at high temperature, because they are remarkably difficult to estimate for light isotopes [106]. Therefore, any uncertainty in $G_3(r)$, owing to either neglecting these effects or improperly correcting for them, will result in a partial annihilation of the first peak on $g_{OO}(r \approx 1.9\ \text{Å})$ (by showing up as a bump) and, simultaneously, in a partial enhancement of the first peak of $g_{OH}(r)$ (partial neglecting of the proper hydrogen bonding).

Another convincing indication of inconsistencies in the NDIS-93 data came from the comparison of *ab initio* [107] and classical simulations of the microstructure of supercritical water at $T = 730$ K and $\rho = 0.64$ g/cm^3 (the state conditions at which the *ab initio* calculations have been performed) contrasted with Fois and co-workers' comparison of *ab initio* at $T = 730$ K and $\rho = 0.64$ g/cm^3 and the NDIS-93 at $T = 653$ K and $\rho = 0.66$ g/cm^3. *Ab initio* simulations provide a description of the fluid behavior not by preassigning an interaction potential but by determining the interactions from electronic structure calculations within the Born–Oppenheimer approximation during the simulation (see Section II.B). Tromp and co-workers [98] claimed

> It is a straightforward matter to compare the results with those of computer simulation. Indeed in our previous letter [60], we showed that in so doing there exists a significant difference between SPC model and the experimental results.... In passing we note that in private communication with the group of Parrinello [107], an *ab initio* computer simulation based on the Car–Parrinello method [108] gives good agreement with our results.

Their comparison with SPC, however, was not at the same state conditions (Figs. 4–6), nor were the *ab initio* results in any acceptable agreement with the NDIS-93 data. In fact, a direct comparison of classical and *ab initio* simulations at precisely the same state conditions [62]—unlike Fois and co-workers' [107] comparison of their *ab initio* and the NDIS-93 data at rather different state conditions—actually supports the accuracy of the prediction of the classical water models instead of the NDIS-93 data.

According to Figures 4–6, the hard-core diameters (excluded volumes) from the *ab initio* radial distribution functions $g_{OO}(r)$ and $g_{OH}(r)$ (i.e., ~2.5 and ~1.5 Å, respectively) compare quite well to ~2.4 and ~1.5 Å as predicted by most water models, and they are in sharp contrast to ~2 and ~2 Å, respectively, found in the NDIS-93 data. Moreover, the classical simulations predict $g_{OO}(r)$ with a first-peak height of ~1.8 compared to the *ab initio* value of ~1.6 and the NDIS-93 value of ~2.4. Note also the formation of a "flat" first peak for the *ab initio* $g_{OH}(r)$ at $2\ \text{Å} < r < 2.5\ \text{Å}$, in remarkable agreement with the classical water model (especially the polarizable model). Most significant, the *ab initio* total dipole moment (i.e.,

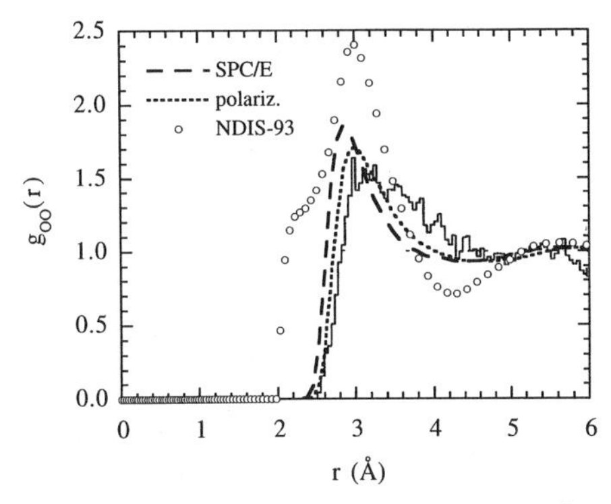

Figure 4. Comparison of the NDIS-93 data [60] at $\rho = 0.66$ g/cm^3 and $T = 673$ K, the *ab initio* (*choppy line*) data [107], and the classical simulated O–O radial distribution function (*solid line* and *dotted line*) of water at $\rho = 0.64$ g/cm^3 and $T = 730$ K.

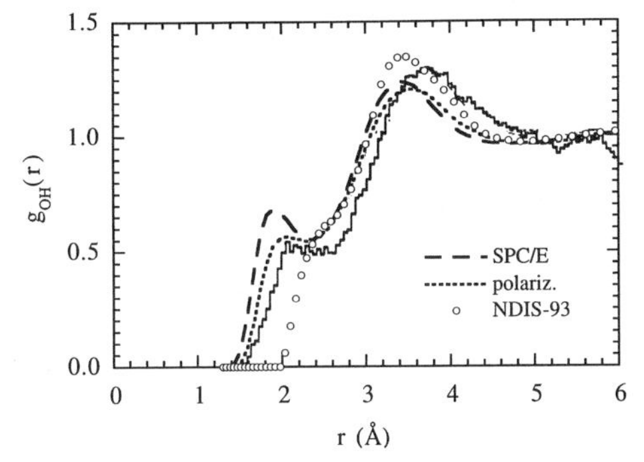

Figure 5. Comparison of the NDIS-93 data [60] at $\rho = 0.66$ g/cm^3 and $T = 673$ K, the *ab initio* data (*choppy line*) [107], and the classical simulated intermolecular O–H radial distribution function (*solid line* and *dotted line*) of water at $\rho = 0.64$ g/cm^3 and $T = 730$ K.

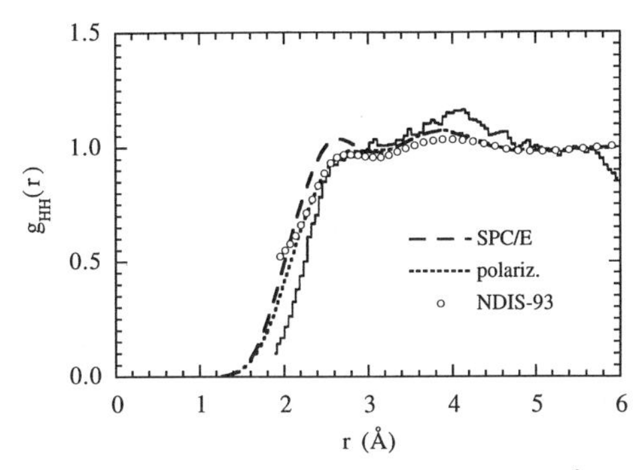

Figure 6. Comparison of the NDIS-93 data [60] at $\rho = 0.66$ g/cm^3 and $T = 673$ K, the *ab initio* data (*choppy line*) [107], and the classical simulated intermolecular H–H radial distribution function (*solid line* and *dotted line*) of water at $\rho = 0.64$ g/cm^3 and $T = 730$ K.

2.3 ± 0.2 D) is in remarkably good agreement with that from the polarizable models (2.27 ± 0.01 D) [62] (note that the SPC/E model has an effective permanent dipole moment of 2.35 D).

In a separate effort, Yamanaka and co-workers [109] developed a new X-ray diffractometer with an imaging plate area detector that makes fast determination of high-temperatute and -pressure water microstructure, i.e., in the range of $300 \leq T \leq 649$ K and $0.1 \leq P \leq 98$ MPa, corresponding to $0.7 \leq \rho \leq 1.0$ g/cm^3. Because their scattering experiments did not allow any direct determination of O$\cdots$H correlations, resulting from the low X-ray scattering factor of the hydrogen atom, they were not able to conclusively establish strong hydrogen bond correlations. Their analysis of the radial distributions at $T = 649$ K and $\rho = 0.7$ g/cm^3 in terms of Gaussian functions [110], however, indicates that the first peak cannot be reproduced accurately by a single function, suggesting the existence of two types of hydrogen bonds in dense supercritical water [109], which amounts to 30–40% of those present at ambient conditions [111,112].

Postorino and co-workers' claims were also challenged by Gorbaty and Kalinichev [113], who presented an alternative experimental study of hydrogen bonding in supercritical water based on a combination of IR absorption spectroscopic and X-ray diffraction techniques. After discussing the semantics behind previous HB definitions, they offered a rather strong case for the persistence of HB to temperatures beyond 673 K, based on the measure of the integral intensity of the valence band B of H–O–D ($B = \int_{v_1}^{v_2} \varepsilon(v)\, dv$; where $\varepsilon(v)$ is the wave number v dependence of the molar absorption coefficient) for the v_{OH} ($B_{v_{\mathrm{OH}}}$) compared to the corresponding for the v_{OD} ($B_{v_{\mathrm{OD}}}$) from Franck and Roth [83]. According to its isobaric temperature dependence, $B(T)$ exhibits a monotonic decrease from $B(293\text{ K}) \approx 160$ km/mol to $B(823\text{ K}) \approx 30$ km/mol; and considering that the value for an isolated (nonbonded) water molecule is $B_{\mathrm{isolated}} \approx 11$ km/mol [114], it follows that the strength of the hydrogen bond is still nonnegligible even at $T = 823$ K since $B(823\text{ K})/B_{\mathrm{isolated}} \approx 3$.

An additional study of the temperature dependence of the hydrogen bond was performed by considering the short-range portion of the molecular pair correlation function $g_m(r)$ determined by X-ray diffraction (which for all practical purposes is equal to $g_{\mathrm{OO}}(r)$ [115]) as a sum of three Gaussian functions centered at r_i ($i = 1, 3$). The first two Gaussian functions encompass the first peak of $g_m(r)$ and its shoulder, and the third describes the contributions from the second and higher coordination shells [103,110]. Thus, by assuming that the first and second nearest neighbors form the first coordination shell with a coordination number $Z = \alpha_1 r_1 + \alpha_2 r_2$ (where α_i denotes the area under the Gaussian function centered at r_i), the overall degree of HB χ can be defined as $\alpha_1 r_1 / Z$. According to X-ray and IR results, this quan-

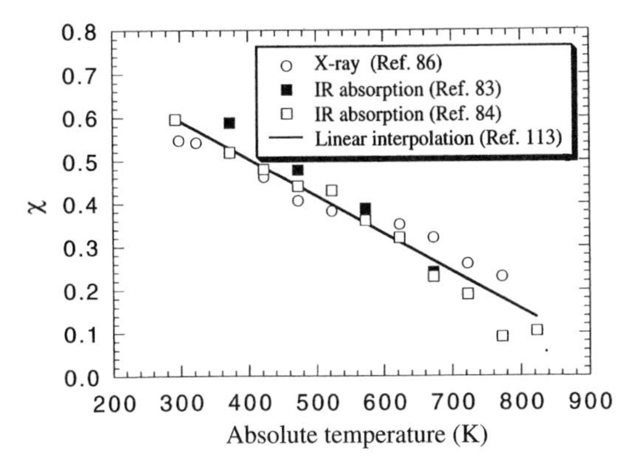

Figure 7. Temperature dependence of χ derived from X-ray data (*open circles*), IR absorption by ν_{OH}HDO (*open squares*), and IR absorption by ν_{OD}HDO (*filled squares*).

tity exhibits a linear temperature dependence for $293 \leq T \leq 923$ K within the density range of $0.7 \leq \rho \leq 1.1$ g/cm^3 (Fig. 7) where χ can be considered only a function of temperature, i.e., $\chi \approx 0.851 - 8.68 \cdot 10^{-6} T$. Although it is expected that χ will show a strong density dependence for $\rho < 0.7$ g/cm^3, χ should approach 0 asymptotically in a similar trend as per the integral of the intensity band B [113]. Within this context, the analysis suggests that the degree of HB will not become negligible until an even higher temperature, i.e., ~ 1000 K. These two experimental approaches indicate that the degree of hydrogen bonding cannot be taken as nonexistent in near-critical water, contrary to Postorino and co-workers' claim [60].

Further evidence of these unphysical features were provided by Jedlovszky and Vallauri [116] through their reverse Monte Carlo simulation study, which did not rely on any intermolecular potential model [117]. Their results showed that the NDIS-93 correlation functions could not possibly represent any physically realizable geometrical arrangement of real water molecules.

More recently, Hoffmann and Conradi [52] measured the NMR proton chemical shift owing to hydrogen bonding in subcritical and supercritical water. The authors measured the proton shift in water, within the range of $298 \leq T \leq 873$ K and $0.1 \leq P \leq 40$ MPa, using dilute benzene as an internal chemical shift reference [118]. The chemical shift σ was correlated linearly with the degree of hydrogen bonding η, following the prediction of the two-state (bonded–nonbonded) model of hydrogen bonding:

$$\eta = 0.2439\sigma + 1.610 \qquad (2.9)$$

such that ambient water exhibits $\eta = 1$ with $\sigma = -2.5$ ppm. Thus, according to this correlation, the extent of the HB in water at 673 K and $\rho = 0.52$ g/cm^3 is

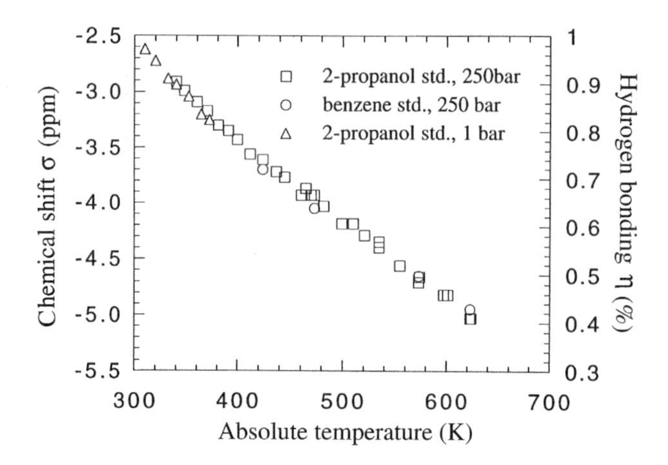

Figure 8. Chemical shift of water relative to dilute internal benzene in the lower temperature (high-density) region.

ca. 30%, in sharp contrast to that claimed from NDIS-93 [60] for water at 673 K and $\rho = 0.66$ g/cm^3 (Fig. 8).

In related work, Matubayasi and co-workers [53,54] measured the proton shift of water using a capillary method that allowed them to use a very low density state of water as an external reference. By comparing the chemical shifts of water in neat fluids at high temperature to those in organic solvents at ambient conditions, they concluded that hydrogen bonding persists in supercritical water in the density region of $0.41 \leq \rho \leq 0.60$ g/cm^3.

Bellissent-Funel and co-workers [77] determined the neutron structure factor of supercritical heavy water at $0.2 \leq \rho \leq 0.7$ g/cm^3 and $653 \leq T \leq 773$ K to investigate the effect of pressure and temperature on the pair correlation function $d_{\mathrm{L}}(r)$, defined as

$$d_{\mathrm{L}}(r) = 4\pi\rho r[g_{\mathrm{L}}(r) - 1] \tag{2.10}$$

where $g_{\mathrm{L}}(r)$ is the weighted sum of the site–site pair correlation functions:

$$g_{\mathrm{L}}(r) = 0.489 g_{\mathrm{DD}}(r) + 0.421 g_{\mathrm{OD}}(r) + 0.090 g_{\mathrm{OO}}(r) \tag{2.11}$$

Based on a simple geometric criterion, one in which a hydrogen bond between a pair of water molecules exists if the O$\cdots$H intermolecular distance $r \leq 2.4$ Å; its signature in $d_{\mathrm{L}}(r)$ is a peak at $r \approx 1.9$ Å. According to their experimental data [77], the mentioned peak (even though less intense and widened) is still observed at $\rho = 0.23$ g/cm^3 along the isotherm $T = 653$ K.

These experiments were complemented with molecular dynamic simulations of the SPC/E water model [119] by determining the $g_L(r)$ at $T = 653$ K and $\rho = 0.73$ g/cm^3 and comparing it to the diffraction results at precisely the same state conditions and to those from Postorino and co-workers [60] obtained at $T = 673$ K and $\rho = 0.73$ g/cm^3. The simulation results for $g_L(r)$ are in remarkable agreement with their corresponding diffraction data, especially in the region $1.5 \leq r \leq 2.1$ Å, i.e., the so-called HB distance. In contrast, Postorino and co-workers' data indicate a steeper fall in $g_L(r)$ for $r \leq 2.1$ Å, a feature that became the source of much controversy. An additional comparison between SPC/E simulation results and diffraction data at $T = 723$ K and $\rho = 0.35$ g/cm^3, as well as $T = 773$ K and $\rho = 0.55$ g/cm^3, indicates reasonable agreement between simulation and experiment, a rather surprising behavior, given that the model was parameterized at ambient conditions.

B. Simulation Approaches to Microscopic Behavior

1. *Classical Molecular Simulations*

Numerous rigid and flexible intermolecular potential models for ambient liquid water have been proposed since the pioneering work of Ben-Naim and Stillinger [120]. A common and somewhat remarkable feature of all these models is that they have succeeded, up to differing extents, in reproducing adequately the thermophysical, microstructural, and dynamical properties of water at ambient conditions (Table I) [121–127] even though none of them includes explicitly the orientational features associated with hydrogen bonding. Considering that their parameterization was based on thermophysical properties at ambient conditions, an obvious question is whether their success at ambient conditions extrapolates to near and supercritical state conditions. The rationale is that the success of a model typically hinges on capturing the proper effective dipole moment through the adjustment of the electrostatic interactions of the model. For the case of rigid models (such as SPC, ST2, and TIP4P) this effective dipole moment represents an enhanced value over that of the bare dipole moment of water (or the gas phase value), which accounts for all polarization many-body effects in the condensed phase. As the state conditions change, the water polarization and related quantities change accordingly; therefore, we expect a failure of the model parameterization as state conditions approach and go beyond the critical point of the model [128].

Most frequently, simulation of model fluids at near-critical and supercritical conditions have been performed without the knowledge of the model vapor–liquid equilibrium envelope, i.e., by simply taking the state conditions from the phase diagrams of the real substances. Consequently, it becomes

TABLE I
Thermophysical Properties of Water at Ambient Conditions

Model[a]	U_c, Kcal/mol	ΔH_{vap},[b] Kcal/mol	C_p,[b] Kcal/(mol·K)	$10^5\kappa$,[b] MPa	Dielectric Constant ε	Diffusion Constant D, 10^5 cm^2/s
SPC	−9.89	10.77	23.4	27	63 ± 31[c]	4.3[d]
SPC/E	−9.80	—	—	—	81.3[e]	2.4[d]
TIP4P	−9.79	10.66	19.3	35	29 ± 13[c]	—
TJE[f]	−9.8	—	—	—	84 ± 5	4.1
ST2	−9.85	10.96	22.2	63	78.3[g]	2.9[g]
Expt.	−9.92[d]	10.51	17.99	45.8	77.7[h]	2.3[i]

[a] TIP4P ≡ transferable intermolecular potential, 4 point charges; TJE ≡ Teleman-Jönsson-Engström; ST2 ≡ Stillinger-2; Expt ≡ experimental.
[b] Ref. [121].
[c] Ref. [122].
[d] Ref. [119].
[e] Ref. [123].
[f] Refs. [126] and [127].
[g] Interpolated value [124].
[h] Ref. [4]
[i] Ref. [125].

rather difficult to make unambiguous comparisons between the model and real behavior, since we cannot locate the state conditions in relation to the critical state conditions of the model under study. Fortunately, with the recent development of powerful new simulation techniques for the direct determination of phase equilibria, such as the Gibbs Ensemble Monte Carlo [129], the Gibbs–Duhem integration [130], and the Grand Canonical Monte Carlo with histogram reweighting method [131], we are now able to determine the phase diagram of water models with unprecedented accuracy and computational efficiency [132] and to make some accurate prediction of their critical points.

a. Intermolecular Potentials. Most simulation work performed on high-temperature water in the last 15 years has been based on a wide variety of simple molecular force fields described by arrangements of point charges located in well-defined molecular geometries [57]. The first obvious and common defect to all these models is that their Coulombic charges cannot change in response to the changing local electric field; therefore, they are unable to describe the considerable polarization effects occurring in condensed phases. In fact, the many-body polarization contributions are consequently described in a pairwise-additive fashion through the interaction of enhanced dipole moments. These enhanced (over the corresponding gas-phase values) dipoles are the result of fitting the model's charges to the fluid thermodynamic properties of the real system.

Note that, even though the assumption that the potential energy can be expressed as a sum over pair interactions is most frequently used, it has also been recognized that many-body polarization effects are responsible for a substantial contribution to the properties of the condensed phase of highly polar fluids such as water [133], especially on the dielectric constant [134]. Although the pairwise assumption might work well for treating many-body interactions as effective two-body interactions in fluids exhibiting quasi-isotropic polarization effects [134] (such as in pure water in which the charge distribution can be considered centered at the oxygen), this approximation breaks down for systems with anisotropic polarization (such as in the neighborhood of an ion in solution), i.e., when the direction of the average dipole moment does not coincide with the bisector of the HOH bond angle.

A second serious defect common to all nonpolarizable water pair potential models is their inability to describe both the two-body interactions of an isolated molecular pair (i.e., the behavior at low density) and the many-body interactions of the condensed phase without running into undesired state-dependent force-field parameters [128,135]. Thus nonpolarizable water models suffer from the lack of transferability; therefore, in principle, they should not be able to describe accurately the phase behavior at conditions other than that used in the parameterization.

Among the effective pairwise potential models with rigid geometry, TIP4P [121] and its predecessors TIPnP ($n = 2, 3$), SPC [100], SPC/E [119] and its newly proposed modified version MSPC/E [136], ST2 [124] and its modified version ST2M [137], and some variations on the SPC such as the simple point change gas phase dipole (SPCG) [89] have been most frequently used in the study of supercritical water. All these models are described as effective pair potentials composed of Lennard–Jones (LJ) and Coulombic terms. In addition to their geometry, they differ from one another in the number and the strength of their Lennard–Jones and Coulombic sites.

Specifically, the SPC water model (three-point charge) is given by [100]

$$u_{SPC}(\mathbf{r}_{12}\omega_1\omega_2) = \sum_{\alpha=1}^{3}\sum_{\beta=1}^{3}\frac{q_1^{\alpha}q_2^{\beta}e^2}{r_{12}^{\alpha\beta}} + 4\varepsilon_{OO}\left[\left(\frac{\sigma_{OO}}{r_{OO}}\right)^{12} - \left(\frac{\sigma_{OO}}{r_{OO}}\right)^{6}\right] \tag{2.12}$$

where the subscript OO represents oxygen–oxygen interactions, r_{12} is the vector joining the centers of molecules 1 and 2, ω_i is the vector containing the Euler angles describing the orientation of molecules i, q_i^{α} is the charge on site α on molecule i, e is the electronic charge, $r_{12}^{\alpha\beta}$ is the distance between site α on molecule 1 and site β on molecule 2, and ε and σ are the Lennard–Jones's energy and size parameters, respectively. For the SPC and related models, the negatively charged site (site 1) is coincident with the center of the oxygen, so that $r_{12}^{11} \equiv r_{\mathrm{OO}}$. The SPC model was the first for which we have the vapor–liquid phase envelope determined by Gibbs Ensemble Monte Carlo (GEMC) [138], which allowed one to estimate the model's critical point: $T_c \approx 587$ K and $\rho_c \approx 0.27$ g/cm^3.

The SPC/E water model differs from the SPC only by a reparametrization of the site electrostatic charges to account for the missing polarization correction $E_{\mathrm{pol}} = 0.5(\mu - \mu_o)^2/\alpha$, where μ and μ_o are the model's and the gas phase's dipole moments, respectively, and α is the molecular polarizability [119]. The vapor–liquid phase envelope was determined through an indirect molecular dynamics approach [123], resulting in a better agreement with the phase diagram of real water in that the critical point of the SPC/E model is $T_c \approx 651$ K and $\rho_c \approx 0.326$ g/cm^3.

Three other variations of the SPC water model are the SPCG, SPC0, and MSPC/E. The SPCG water model differs from the original SPC model in that the coulombic charges are scaled to reproduce the gas-phase dipole moment of 1.85 D. [Note that in Ref. 139 a typographical error results in the statement that the charge scaling is $\sqrt{1.8/2.24}$, whereas in fact the scaling is 1.85/2.27 since $\mu_{\mathrm{SPC}} \equiv 2q_{\mathrm{H}} \cos(109.5°/2) = 2.27$ D [89,139].] Similarly, the SPC0 denotes a SPC model for which the dipole moment is small but nonzero, approximately 300 times smaller than that of SPC [72]. Finally, the MSPC/E water model, proposed recently by Boulougouris and co-workers [136], describes the thermodynamics of water more accurately within a wide range of state conditions. Although the new model improves the representation of the phase envelope over that of the SPC model, the SPC/E representation is still the best among all nonpolarizable water models; however, the MSPC/E model gives a better prediction of the vapor pressure and partial improvements on the enthalpy of vaporization (at low temperature) and second virial coefficients (at high temperature).

The TIP4P water model is a four-site rigid model (three Coulombic sites and a LJ site) [121] in which the negative charge is located in the bisectrix of the HOH angle and 0.15 Å away from the Lennard–Jones oxygen site, i.e., $r_{12}^{11} \neq r_{\mathrm{OO}}$.

The ST2 water model is a five-site rigid model with four charged sites arranged tetrahedrally around the LJ oxygen site [124]:

$$u_{\mathrm{ST2}}(\mathbf{r}_{12}\omega_1\omega_2) = S(r_{\mathrm{OO}}) \sum_{\alpha=1}^{4} \sum_{\beta=1}^{4} \frac{q_1^{\alpha} q_2^{\beta} e^2}{r_{12}^{\alpha\beta}} + 4\varepsilon_{\mathrm{OO}} \left[\left(\frac{\sigma_{\mathrm{OO}}}{r_{\mathrm{OO}}} \right)^{12} - \left(\frac{\sigma_{\mathrm{OO}}}{r_{\mathrm{OO}}} \right)^{6} \right] \tag{2.13}$$

where $s(r_{OO})$ is the switching function introduced to prevent the catastrophe $r_{12}^{\alpha\beta} = 0$ for attractive electrostatic interactions that might arise at small center-to-center distances [72].

A few models have flexible geometry including the TJE [126] and the Bopp–Jancsó–Heinzinger (BJH) [140]. In addition, at least three types of polarizable models have been used for supercritical water: polarizable point charge (PPC) [141], a polarizable TIP-type model [142], and a few variations of SPC-type models with either point or smeared electrostatic charges [143–145].

Attempts to improve the description of water behavior within a classical approach have followed several routes, such as the introduction of additional interaction sites to handle the electrostatic distribution [146], the account of nonadditivity effects via charge or dipole polarizabilities [141,143,147–152], the use of flexible geometries [153–155], or a combination of flexibility and polarizability [156–158]. All these options frequently imply a substantial increase in CPU-requirement and a not so obvious improvement in the accuracy of the resulting model. With the development of multiple time-step integration methods, such as r-RESPA [159], models with flexible geometry have become increasingly popular under the assumption that flexibility should improve the model's ability to describe more realistically the equilibrium and the dynamic properties of water [160–162].

Although the use of a flexible geometry allows the calculation of intramolecular modes, it is unclear whether a classical approach may be able to give a proper account of their dynamics when weakly coupled modes of rather different frequencies are involved. This feature results in a slow energy distribution between modes and, consequently, can lead to non-equilibrium energy distribution between the intramolecular and the intermolecular (roto-translational) degrees of freedom unless, the corresponding forces are determined with comparable accuracy [163]. The literature on the relative merits of the flexible water models over their corresponding rigid counterparts is rather controversial and somewhat contradictory [126,161,164–166], because the parameterization of the models to be compared is not based on the same set of experimental data and fitting criteria (as a case in point, contrast the rather distinct conclusions drawn by Tironi and co-workers [163] and by Mizan and co-workers [166] for the same comparison between the SPC and the TJE models).

Typically, the popular route to molecular flexibility is the introduction of harmonic [126] or Morse-type [153] intramolecular potentials to describe oscillatory motion of classical point masses. The flexible TJE water model [126] is based on the rigid SPC water model [167] for the intermolecular

TABLE II
Potential Parameters of Rigid Water Models

Model	σ_{OO}, Å	ε_{OO}/k, K	q_H^a	Number of Sites	μ, D
SPC	3.166	78.23	0.4100	3	2.27
SPC/E	3.166	78.23	0.4328	3	2.35
SPCG	3.166	78.23	0.3348	3	1.85
SPC0	3.166	78.23	0.00133	3	0.0074
MSPC/E	3.116	74.20	0.4108	3	2.24
TIP4P	3.153	78.08	0.5200	4	2.17
ST2	3.100	38.23	0.2357	5	2.35

[a] $\sum_{\alpha} q_{\alpha} = 0$.

interactions:

$$u_{\text{TJE}}^{\text{inter}}(\mathrm{r}_{12}\omega_1\omega_2) = \sum_{\alpha=1}^{3}\sum_{\beta=1}^{3}\frac{q_1^{\alpha}q_2^{\beta}e^2}{r_{12}^{\alpha\beta}} + 4\varepsilon_{OO}\left[\left(\frac{\sigma_{OO}}{r_{OO}}\right)^{12} - \left(\frac{\sigma_{OO}}{r_{OO}}\right)^{6}\right] \quad (2.14)$$

where the Coulombic charges and the Lennard–Jones parameters are those of the SPC model (Table II) and the intramolecular interactions are given by harmonic bond and angle vibration around the SPC geometry:

$$u_{\text{TJE}}^{\text{intra}}(\mathrm{r}_{OH}\vartheta) = 0.5k_{\text{b}}\sum_{\text{all bonds}}(\mathrm{r}_{OH} - \mathrm{r}_{OH}^{\text{SPC}})^2 + 0.5k_{\vartheta}\sum_{\text{all angles}}(\vartheta - \vartheta_{\text{SPC}})^2 \quad (2.15)$$

where the force-field parameters are given in Ref. [126]. Following a similar approach to the one for the SPC/E model [124], Mizan and co-workers [168] determined the coexistence curve and the critical point of the TJE model, which shows a slight improvement over its rigid counterpart (SPC), as we might have expected as the result of its variable total dipole moment, with a critical point at $T_c \approx 604$ K and $\rho_c \approx 0.27$ g/cm^3.

The BJH water model [140] is a modification of the so-called central force model of Stillinger and Rahman [169], in which the intermolecular interactions are given by

$$u_{\mathrm{OO}}^{\mathrm{inter}}(r) = (604.6/r) + (111889/r^{8.86}) - 1.045\{\exp[-4(\mathrm{r} - 3.4)^2] + \exp[-1.5(\mathrm{r} - 4.5)^2]\} \tag{2.16}$$

$$u_{\mathrm{OH}}^{\mathrm{inter}}(r) = -(302.2/r) + (26.07/r^{9.2}) - \{41.79/(1 + \exp[40(\mathrm{r} - 1.05)])\} - \{16.74/(1 + \exp[5.493(\mathrm{r} - 2.2)])\} \tag{2.17}$$

$$u_{\mathrm{HH}}^{\mathrm{inter}}(r) = (151.1/r) + \{418.33/(1 + \exp[29.9(\mathrm{r} - 1.968)])\} \tag{2.18}$$

where all distances are in angstroms, and the potentials in kilojoules per mole. The corresponding intramolecular potential is based on the formulation of Carney and co-workers [170]:

$$u_{\mathrm{intra}} = \sum_{i,j} L_{ij}\rho_i\rho_j + \sum_{i,j,k} L_{ijk}\rho_i\rho_j\rho_k + \sum_{i,j,k,l} L_{ijkl}\rho_i\rho_j\rho_k\rho_l \tag{2.19}$$

where $\rho_1 = 1 - (r_{\mathrm{OH}}^{\mathrm{e}}/r_{\mathrm{OH}_1})$, $\rho_2 = 1 - (r_{\mathrm{OH}}^{\mathrm{e}}/r_{\mathrm{OH}_2})$, $\rho_3 = \theta - \theta_e$, and r_{OH_i} is the O–H bond length for hydrogen i, θ is the HOH bond angle, subscript e denotes equilibrium values, and the L coefficients are given in [140].

To tackle the mentioned defects of simple pairwise effective potential, numerous polarizable water models have recently been developed [143,149–151,171–174]. These developments have not targeted the proper description of water at high temperature but rather at ambient conditions. With a few exceptions [148,152,156] these models are typically built on successful nonpolarizable counterparts, by scaling the Coulombic charges to match the gas-phase dipole moment and by including either a polarizable point charge or a point dipole to account for the many-body polarization contributions. Moreover, sometimes the permanent dipole moment is set a bit larger than the gas-phase value of 1.85 D to obtain better agreement with experimental data at ambient conditions [143,174].

Thus the polarizable models used in the simulation of supercritical water typically involve the rigid geometry of either the SPC or the TIP4P water model [141,143–145,173], i.e., a planar configuration with an HOH angle of either 109.5° (tetrahedral) or 104.5° (gas phase) and an OH bond length of either 1.0 or 0.9572 Å (gas phase).

For the case of models with point dipole polarizabilities, the permanent Coulombic charges are located on the three SPC or TIP4P sites, with their magnitudes constrained by electroneutrality and the permanent dipole moment of the isolated water molecule, i.e., 1.85 D [3]. With this geometry, the models consist of a Lennard–Jones OO pair, the charge–charge electrostatic pair interactions among permanent charges, and the isotropic–linear point dipole polarizability at the center of mass (or at the O-site) to account for the many-body polarizability effects, so that the induced dipole moment on the center of mass of molecule i is given by,

$$\begin{aligned} \mathrm{p}_i &= \alpha \mathrm{E}_i \\ &= \alpha(\mathrm{E}_i^q + \mathrm{E}_i^p) \end{aligned} \tag{2.20}$$

where α is the scalar molecular polarizability, E_i is the total electric field on the center of mass of molecule i, whose contribution from the permanent charges of sites γ on molecule j is given by

$$\mathrm{E}_i^q = \sum_{j \neq i}^{N} \sum_{\gamma=1}^{s} q_j^\gamma \frac{\mathrm{r}_{i,j\gamma}}{|\mathrm{r}_i - \mathrm{r}_{j\gamma}|^3} \tag{2.21}$$

while the corresponding polarization contribution at the center of mass of molecule i is given by,

$$\mathrm{E}_i^p = \sum_{j \neq i}^{N} \mathrm{T}_{ij} \mathrm{p}_j \tag{2.22}$$

In the preceding equations we used the following notation: $\mathrm{r}_{i,j\gamma} = \mathrm{r}_i - \mathrm{r}_{j\gamma}$, where $\mathrm{r}_{j\gamma} = \mathrm{r}_j + \mathrm{r}_j^\gamma$ and r_j are the vector positions of site γ and the center of mass of molecule j, respectively. In Eq. (2.22) T_{ij} is the symmetric dipole tensor:

$$\mathrm{T}_{ij} = \frac{1}{r_{ij}^3} \left(\frac{3\mathrm{r}_{ij}\mathrm{r}_{ij}}{r_{ij}^2} - \mathrm{I} \right) \tag{2.23}$$

and I is the unit tensor.

The total electrostatic energy for this system is given by [175]

$$\begin{aligned} U &= U_{\mathrm{qq}} + U_{\mathrm{pol}} \\ &= U_{\mathrm{qq}} + U_{\mathrm{qp}} + U_{\mathrm{pp}} + U_{\mathrm{self}} \end{aligned} \tag{2.24}$$

with the charge–charge contribution U_{qq}

$$U_{\mathrm{qq}} = 0.5 \sum_{j \neq i}^{N} \sum_{\beta,\gamma}^{s} \frac{q_i^\beta q_j^\gamma}{|\mathrm{r}_{i\beta} - \mathrm{r}_{j\gamma}|} \tag{2.25}$$

the charge–dipole contribution U_{qp}

$$U_{\text{qp}} = -\sum_{i=1}^{N} \mathrm{p}_i \mathrm{E}_i^q \tag{2.26}$$

the dipole–dipole contribution U_{pp}

$$U_{\text{pp}} = -0.5 \sum_{j \neq i} \mathrm{p}_i \mathrm{T}_{ij} \mathrm{p}_j \tag{2.27}$$

and the self-polarizability term U_{self}

$$U_{\text{self}} = 0.5\alpha^{-1} \sum_{i=1}^{N} \mathrm{p}_i^2 \tag{2.28}$$

Now, from Eqs. (2.20)–(2.28) and recalling Eq. (2.20), we have

$$\begin{aligned} U_{\text{pol}} &= -\sum_{i=1}^{N} (\mathrm{p}_i \mathrm{E}_i^q - 0.5\mathrm{p}_i^2/\alpha) - 0.5 \sum_{i \neq j}^{N} \mathrm{p}_i \mathrm{T}_{ij} \mathrm{p}_j \\ &= -0.5 \sum_{i=1}^{N} \mathrm{p}_\mathrm{i} \mathrm{E}_\mathrm{i}^\mathrm{q} \end{aligned} \tag{2.29}$$

Therefore, from Eq. (2.29) and by recalling Eq. (2.20), we have

$$\left(\frac{\partial U_{\text{pol}}}{\partial \mathrm{p}_i} \right) = -\mathrm{E}_i^q - \sum_{i \neq j}^{N} [(\mathrm{T}_{ij} \mathrm{p}_j - \mathrm{p}_i/\alpha)] = 0 \tag{2.30}$$

i.e., the induced dipoles adjust themselves to minimize U_{pol} [150]. This also means that the contributions from induced dipoles do not depend on derivatives of dipole moments (they depend on explicit functions of distances).

Finally, the total potential energy for a system of n water molecules described by the self-consistent point dipole polarizability (SCPDP) model becomes

$$\begin{aligned} U_{\text{SCPDP}} &= 0.5 \sum_{i,j=1}^{N} \left[\sum_{\beta=1}^{3} \sum_{\gamma=1}^{3} \frac{q_i^\beta q_j^\gamma}{|\mathrm{r}_i^\beta - \mathrm{r}_j^\gamma|} - \mathrm{p}_i \mathrm{E}_i^q \right] \\ &\quad + 4\varepsilon_{\text{OO}} \sum_{i \neq j}^{N} \left[\left(\frac{\sigma_{\text{OO}}}{|\mathrm{r}_i^4 - \mathrm{r}_j^4|} \right)^{12} - \left(\frac{\sigma_{\text{OO}}}{|\mathrm{r}_i^4 - \mathrm{r}_j^4|} \right)^{6} \right] \end{aligned} \tag{2.31}$$

where the oxygen site is taken as the fourth site in the model's geometry, i.e., $\beta = \gamma = 4$ for the non-Coulombic OO interactions.

A variation on this model involves the replacement of the point charges by smeared (Gaussian distributed) charges and the use of an exponential-6-potential type model for the non-Coulombic interactions in place of the traditional Lennard–Jones model [145]. An elegant way to introduce the charge smearing was given by Thole [176] and applied to the modeling of water by Bernardo and co-workers [177,178]. Unfortunately, Bernardo and co-workers used an *ad hoc* modification of Thole's formalism to avoid a discontinuity in the original expression; this modification appears to be incorrect in that the dipole tensor $T_{ij} \neq \partial^2\phi_s/\partial r_i \partial r_j$ as required by Thole's formalism, where $\phi_s(r)$ is the smeared electrostatic potential.

The polarizable point charge (PPC) model of Kusalik and co-workers [141,174] retains the simplicity of most nonpolarizable three-site models while incorporating the nonadditivity polarization through polarizable point charges that fluctuate in response to the local electric field. The novelty in this model is that the electric field dependence of the point charges has been determined by quantum-chemical calculations using a commercial package. The model involves three sites in a rigid geometry with an OH bond length of 0.943 Å and a HOH angle of 106°. At the hydrogen sites, the Coulombic positive charges are given by

$$q_{\mathrm{H}}(e) = 0.486 \pm 0.03E_x + 0.02E_z \tag{2.32}$$

where E_x and E_z are the components of the local electric field (in volts per angstroms) along the corresponding directions, and the change of sign in the E_x coefficient takes care of the symmetry of the HOH angle along the z-direction [141]. Note that the polarization takes place only along the x and z directions, characterized by the polarizabilities $\alpha_x = 1.01$ Å^3 and $\alpha_z = 0.66$ Å^3. The fluctuating oxygen charge is determined by electroneutrality and located according to the following electric field dependence:

$$l_z(\text{Å}) = 0.11 - 0.03E_z \tag{2.33}$$

and

$$l_x(\text{Å}) = -0.025E_x \tag{2.34}$$

where l measures the distance with respect to the Lennard–Jones oxygen site, placed at the vertex of the HOH angle, with $\sigma_{\mathrm{OO}} = 3.234$ Å and $\varepsilon/k = 72.18$ K. With this parameterization, the model predicts a rather high permanent dipole moment for the water dimer (for $E = 0$, $\mu = 2.14$ D).

As per most polarizable models, the self-consistent electric field equations are determined by an iterative approach. A second-order predictor approach, based on the three previous self-consistent electric field values, is used to determine the instantaneous electric field around the oxygen site:

$$\mathrm{E}(t_n) = 3.0\mathrm{E}(t_{n-1}) - 3.0\mathrm{E}(t_{n-2}) + \mathrm{E}(t_{n-3}) \tag{2.35}$$

where t_n is the current integration time step. With this prediction, the location of the negative charge is determined using Eqs. (2.33) and (2.34); and the force calculation for the electrostatic interactions is performed until convergence, given by the criterion of $(\delta E)^2/\langle E^2 \rangle < 0.0003$, typically achieved after two or three iterations.

Given the implicit nature of the expression for induced dipoles [Eq. (2.20)], several methods have been proposed for their efficient calculation. Although the matrix solution is the most straightforward from the algebraic viewpoint, its computer implementation becomes rather expensive, at least compared to other available options, unless a parallel implementation is available [178]. The most popular of them is the combination of a predictive scheme with the traditional self-consistent iterative procedure for the calculation of nonadditive effects [150] and variations around [179] which typically scales as the number of iterations per time step required to attain self-consistency (usually two to five iterations, depending on the required accuracy).

Yet, there is a molecular dynamics option that completely avoids either the iteration approach or the matrix inversion required to solve the equations to self-consistency. This method, proposed by Sprik and Klein [148], follows closely the extended Lagrangian scheme for isothermal [180] and isobaric simulations [181] in that it treats the induced dipole in each molecule as an additional degree of freedom. Thus an extended Lagrangian scheme can be set up, including a fictitious kinetic energy associated with the induced dipoles:

$$K_p = 0.5 m_p \sum_{i=1}^{N} \dot{\mathrm{p}}_i^2 \tag{2.36}$$

where $\dot{\mathrm{p}}_i$ denotes the time derivative of the induced dipole moment of molecule i, and m_p is the fictitious inertial mass associated with the additional dynamical variable (with units of mass $\times$ charge^{-2}). Then the equations of motion are derived following standard procedures [182]:

$$m_p \ddot{\mathrm{p}}_i = \mathrm{E}_i - \alpha^{-1} \mathrm{p}_i \tag{2.37}$$

where $\ddot{\mathrm{p}}_i$ indicates the second time-derivative of the induced dipole, and α is the molecular polarizability. By comparing Eqs. (2.20) and (2.37), it becomes clear that the self-consistency is recovered when $m_p \to 0$, i.e., the dipole's inertial mass determines the time scale for the reorientation of the dipoles owing to the local fluctuations of the electric field. In other words, the average induced dipole [183]

$$\langle \mathrm{p}_i \rangle = \alpha \mathrm{E}_i \tag{2.38}$$

follows the direction of the electric field, with a fluctuation about this average given by

$$\langle \mathrm{p}_i^2 \rangle = 3\alpha k_\mathrm{B} T + \alpha^2 \mathrm{E}_i^2 \tag{2.39}$$

These expressions indicate that, in contrast to the actual behavior of induced dipoles that orient along the direction of the local electric field, this integration scheme (model) reduces to the actual behavior in the limit of zero temperature. In a simulation, this is typically achieved by coupling the system to a thermal bath of an arbitrarily low temperature, and its accuracy can be checked by monitoring the polarization contribution to the total energy U_{pol}, which should be equal to half the charge dipole contribution U_{qp} [see Eqs. (2.26) and (2.29)]. This can be efficiently accomplished by using separate thermostats for the roto-translational and the polarization degrees of freedom in a Nosé-type formalism, as recently described by Mountain [184].

Finally, a few words about the parameterization of these models. Although it is not always clear how the model parameterization has been done, there are some common features in the fitting process. Typically, the portion of the model to be fitted involves the nonelectrostatic interactions, i.e., the pairwise repulsion and attraction, which is most frequently modeled by a Lennard–Jones potential for the OO interactions. The typical empirical parameterization procedure is based on constraining as many of the model's thermophysical properties as needed (depending on the number of force-field parameters) to the experimental values at ambient conditions by adjusting those parameters in a series of short simulations [163,173,185] or during a rather longer simulation [128,144]. The choice of the property to be targeted in the parameterization depends strongly on its sensitivity to small perturbations of the force fields to be adjusted and on the accuracy with which it can be obtained by simulation and measured experimentally. The most convenient choices are the configurational internal energy and the pressure to fit the energy and size force field parameters for the pairwise repulsion–attraction terms. Sometimes the choice is based on performing short simulations, so that configurational internal energy and microstructure are the most frequently targeted properties [185]. The choice of force field to be fitted depends on whether the model is polarizable or nonpolarizable and rigid or flexible, in that several of the models' force fields are already fixed by the corresponding experimental properties (such as polarizabilities, equilibrium bond lengths, and bond angles). Consequently, the nonelectrostatic part of the potential is typically used to achieve agreement between the model predictions and the targeted thermophysical properties.

b. Simulation Results. In what follows we analyze separately the three site–site radial distributions for the water models and compare them with NDIS-97 data at ambient and high temperature. A rather complete discussion of the NDIS-93 data is given in Ref. [90].

I. AMBIENT CONDITIONS. Figures 9–11 display the microstructural results from the simulation of the nonpolarizable SPC, re-engineered simple point change (RSPC), TIP4P, and ST2 models [72] and the polarizable

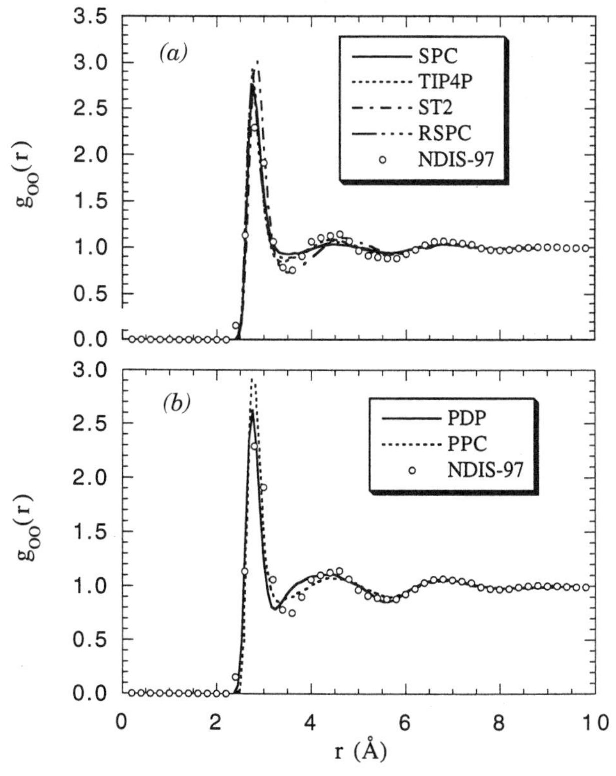

Figure 9. Comparison of the NDIS-97 data and the simulated O···O radial distribution functions of several water models at ambient conditions: (*a*) nonpolarizable and (*b*) polarizable models.

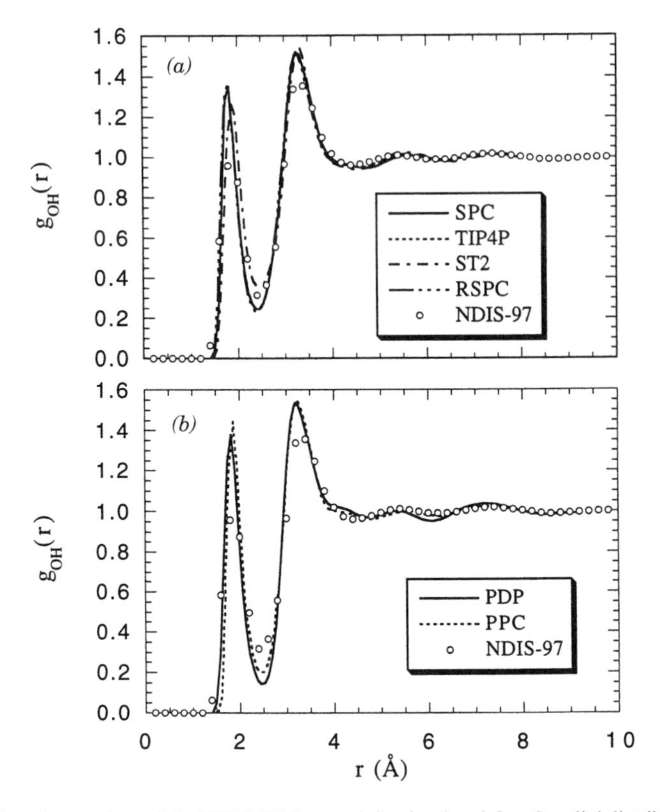

Figure 10. Comparison of the NDIS-97 data and the simulated O· · ·O radial distribution functions of several water models at ambient conditions: (*a*) Nonpolarizable and (*b*) polarizable models.

point charge (PPC) [141] and SCPDP [145] models. In general, these models predict accurately the position of the first peak of the distribution functions, although they overpredict the strength of the correlations; the SPC model is slightly better. The descriptions of $g_{OO}(r)$ by the SPC/E and TIP4P models are undistinguishable, with the first peak ~30% higher than that from the NDIS-97 data. In addition, even though the ST2 model consistently gives higher peaks than those from the NDIS-97 data, it captures the main features of the three-pair correlation functions.

It is not surprising to find that these models, including most flexible [127] and polarizable [143,150,173,174] models, overpredict the first peak of the site–site correlations, because their parameterization has been (directly or indirectly) guided by structural information from the *de facto* standard NDIS-86 data, which are now known to be inaccurate (see Section II.A).

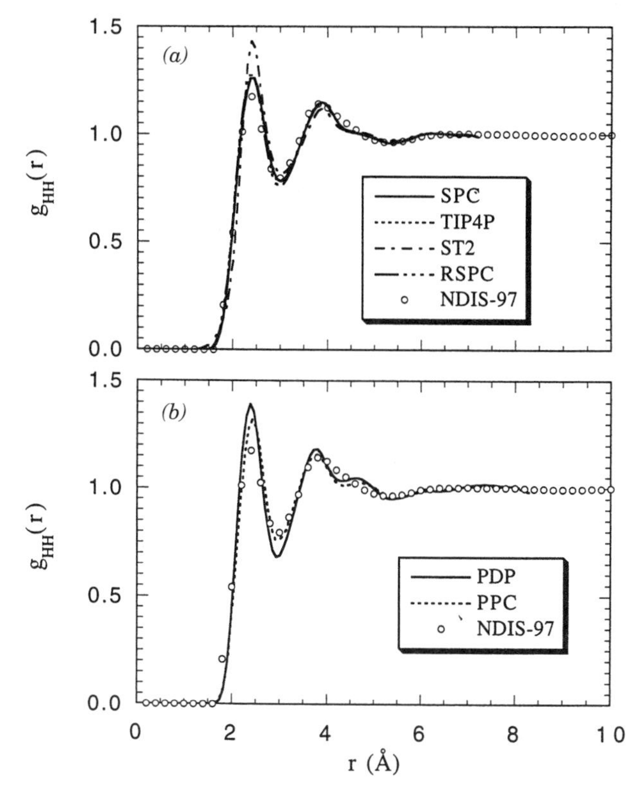

Figure 11. Comparison of the NDIS-97 data and the simulated H···H radial distribution functions of several water models at ambient conditions: (*a*) Nonpolarizable and (*b*) polarizable models.

II. HIGH-TEMPERATURE CONDITIONS. Here we consider mainly the two state conditions studied originally by NDIS [60]: $T = 573$ K and $\rho = 0.72$ g/cm^3, and $T = 673$ K and $\rho = 0.66$ g/cm^3. All polarizable and nonpolarizable models, with the exception of the RSPC model, capture qualitatively the main features of the site–site distributions for real water; the main departure is in the representation of the $g_{OH}(r)$ and thus in the representation of the HB interactions.

It is quite interesting to see that the re-engineered SPC water model, RSPC [128]—whose force field parameters are state dependent to achieve a more accurate fit of the thermophysical properties of water along the coexistence curve—fails dramatically to predict the structure at high temperature [135]. In contrast, the RSPC predicts the first two peaks to be shifted ~0.5 Å to the left, to be more structured, and to have a rather slow decay, as if the system exhibited a near criticality (Figs. 12–14).

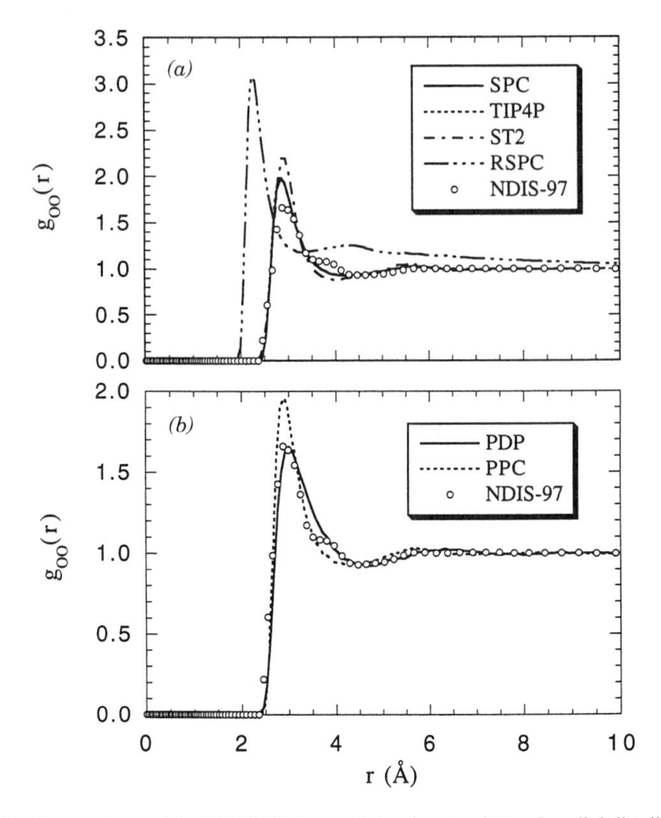

Figure 12. Comparison of the NDIS-97 data and the simulated O···O radial distribution functions of several water models at $T = 573$ K and $\rho = 0.72\,\mathrm{g/cm^3}$: (*a*) nonpolarizable models and (*b*) polarizable models.

Next, we analyze each site–site pair correlation function in more detail.

(a) Oxygen–Oxygen Radial Distribution Functions

At $T = 573$ K the SPC, ST2, and TIP4P models predict essentially the same structure, with a first peak 15–30% higher than that given by the NDIS-97 data [Fig. 12(*a*)]. Between the two polarizable models, the PDP model predicts the correct location and strength of the peaks, whereas the PPC overpredicts the first peak by 25% (not surprising, being that it also overpredicted the strength of this peak at ambient conditions) [Fig. 12(*b*)]. Yet, neither model is able to describe the shoulder right before the first valley.

At $T = 673$ K the three nonpolarizable models show only a slight decrease (<5%) in the strength of their first peak with respect to that at $T = 573$ K

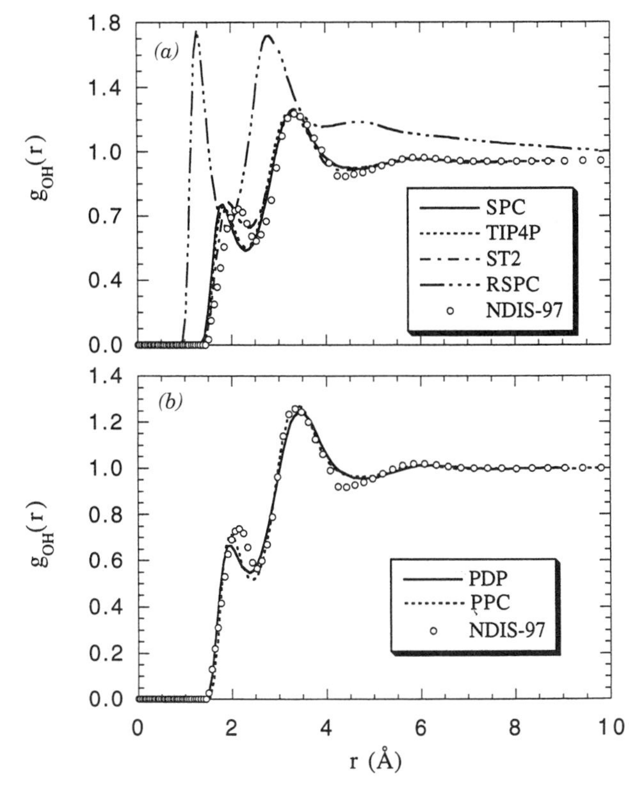

Figure 13. Comparison of the NDIS-97 data and the simulated O···H radial distribution functions of several water models at $T = 573$ K and $\rho = 0.72\,\mathrm{g/cm^3}$: (*a*) Nonpolarizable and (*b*) polarizable models.

(Figs. 12 and 15). An interesting feature in the NDIS-97 data is the disappearance of the shoulder and the simultaneous deepening of the first valley with increasing temperature, a behavior that is not captured by any of the models tested.

(b) Oxygen–Hydrogen Radial Distribution Functions

According to Figure 13(*a*), the SPC, TIP4P, and ST2 models capture the main features of the site–site structures measured by NDIS-97. The only noticeable difference is the shift in the location of the first peak at $T = 573$ K, by 0.1–0.15 Å. There is a clear improvement in the prediction of this correlation by the polarizable models, especially in the first peak; both models give essentially the same description. None of the nonpolarizable models captures the disappearance of the first peak at $T = 673$ K (Fig. 16) which may be an arti-

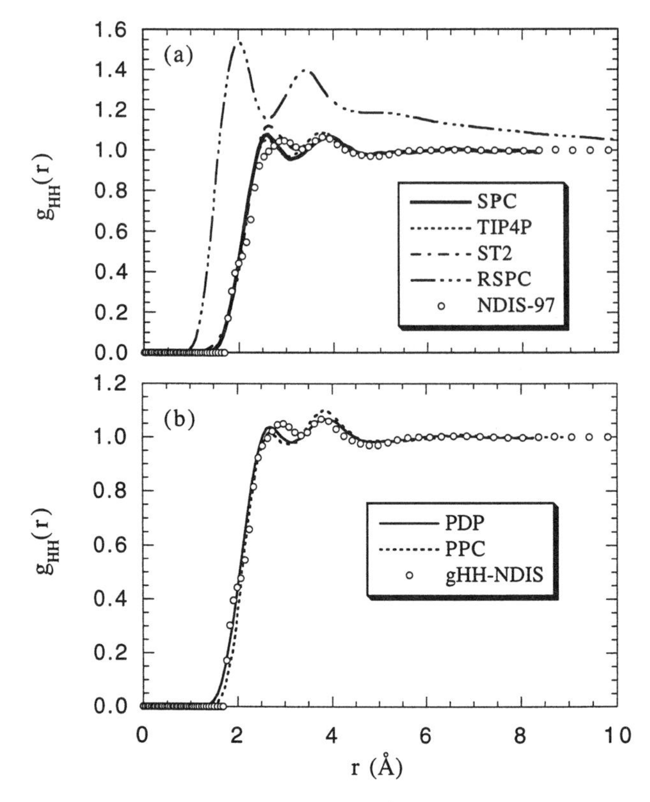

Figure 14. Comparison of the NDIS-97 data and the simulated H···H radial distribution functions of several water models at $T = 573$ K and $\rho = 0.72\,\text{g/cm}^3$: (*a*) Nonpolarizable and (*b*) polarizable models.

fact of the processing of the raw data [77]. However, the models predict with remarkable accuracy the position and strength of the second peak. No predictions were available for the polarizable models.

(c) Hydrogen–Hydrogen Radial Distribution Functions

At these supercritical conditions, the SPC, TIP4P, and ST2 models qualitatively predict the position and shape of the first peak [Fig. 14(*a*)]; however, the three models predict the second peak accurately. In contrast, the polarizable models significantly improve the description of the first two peaks at $T = 573$ K [Fig. 14(*b*)]. Regarding $T = 673$ K, the nonpolarizable models overpredict the strength of the first peak (actually, this becomes a shoulder); yet they predict the second peak in good agreement with the NDIS-97 data (Fig. 17).

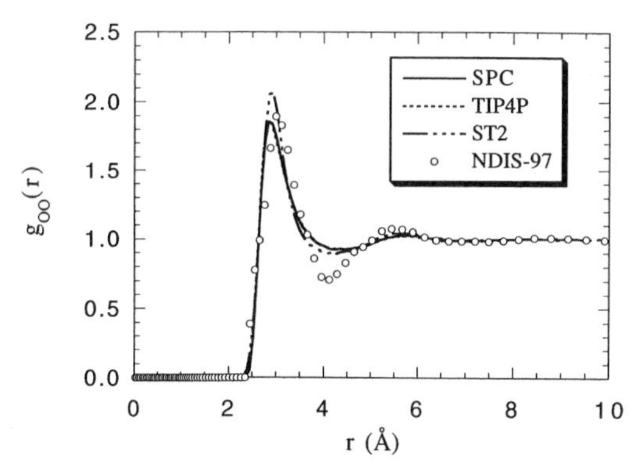

Figure 15. Comparison of the NDIS-97 data and the simulated O···O radial distribution functions of several nonpolarizable water models at $T = 673$ K and $\rho = 0.66$ g/cm^3.

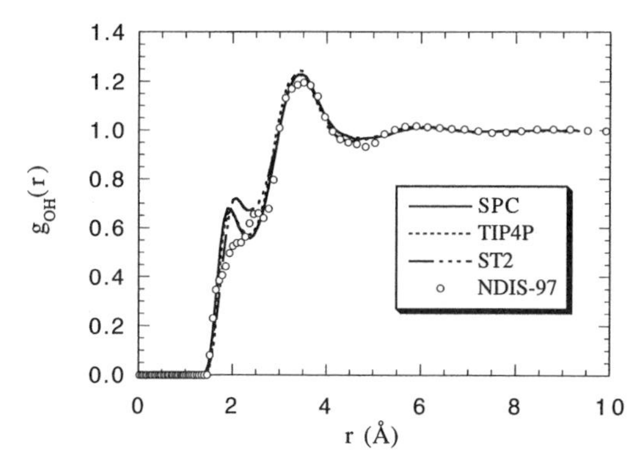

Figure 16. Comparison of the NDIS-97 data and the simulated O···H radial distribution functions of several nonpolarizable water models at $T = 673$ K and $\rho = 0.66$ g/cm^3.

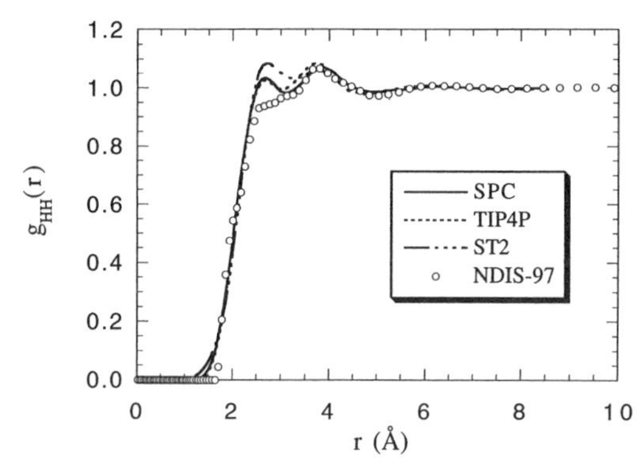

Figure 17. Comparison of the NDIS-97 data and the simulated H···H radial distribution functions of several nonpolarizable water models at $T = 673$ K and $\rho = 0.66$ g/cm^3.

c. Hydrogen Bonding. One of the first extensive molecular dynamics studies on the microstructures features of high-temperature water was performed by Mountain [69] using the TIP4P model [186]. This study covered a wide range of state conditions, with temperatures from the coexistence curve to about 1130 K and densities from vapor-like to liquid-like ($0.1 \le \rho \le 1.0$ g/cm^3). It focused attention on the short-range behavior of $g_{\mathrm{OH}}(r)$ as the signature of hydrogen bonding in water.

More precisely, Mountain used the conventional geometric definition of hydrogen bonding [i.e., coordination number or volume integral over $g_{\mathrm{OH}}(r)$ and within a solvation shell of radius 2.4 Å] as a measure of the HB strength. According to this definition, the TIP4P model predicts a disappearing HB structure [in that the first peak of $g_{\mathrm{OH}}(r)$ at $r \approx 1.8$ Å becomes a shoulder] at $T \approx 780 \pm 40$ K within the range of densities studied, suggesting a shortening of the hydrogen bond lifetime. In addition, the study indicated that the quantity obtained by scaling the hydrogen bond strength with the system density exhibits a single temperature dependence for the liquid-like densities (i.e., $\rho \ge 0.6$ g/cm^3) and not for vapor-like densities; the transition is around the critical density. Note also that the behavior breaks down for $\rho \ge 1.0$ g/cm^3 at which the HB strength saturates at $N_{\mathrm{B}} \approx 2$, independent of the prevailing density. The interpretation of these results can be extracted from the analysis of the mentioned scaling:

$$N_{\mathrm{B}}/\rho = 4\pi(\rho_{\mathrm{H}}/\rho) \int_0^{2.4\,\text{Å}} g_{\mathrm{OH}}(r) r^2 \, dr \tag{2.40}$$

where $(\rho_{\mathrm{H}}/\rho) = 2$. According to Eq. (2.40), the explicit density dependence has disappeared and what is left is an implicit dependence through $g_{\mathrm{OH}}(r)$. The fact that the above integral exhibits temperature—but not density—dependence in the range of $0.6 \le \rho \le 1.0$ g/cm^3 hinges on the peculiar behavior of water, in that an increase of density results in a lowering of the first peak of $g_{\mathrm{OH}}(r)$ (unlike the behavior of most other fluids at liquid-like densities) [72,89,104]. Mountain's findings on the temperature dependence of the HB strength confirm and complement earlier Monte Carlo results by Kalinichev [187].

During the investigation of the solvation behavior of infinitely dilute solutes in supercritical water, Cummings and collaborators [188,189] used a more restrictive geometric definition of HB: the so-called minimal geometric criterion of Mezei and Beveridge [64,65]. This criterion allows the study of the radial dependence of the number of neighboring water molecules whose configurations with respect to a central one are consistent with the qualitative picture of HB, defined by four internal coordinates between the pair of water molecules (Fig. 1): the interoxygen separation R_{OO}, the angle θ_H between the H–O and the O···O bonds, the angle θ_{LP} between the LP–O and the O···O bonds, and the dihedral angle δ_D between the planes H–O···O and the LP–O···O. In this definition, LP is an imaginary "pseudoatom" on the water molecule corresponding to the location of tetrahedrally oriented lone pair (LP) orbitals, i.e., the LP–O–LP triangle is of the same dimensions as, but perpendicular to, the H–O–H triangle. For each water molecule, the atom or pseudoatom H-bonded to another water molecule is the one on the donor water closest to the oxygen atom of the acceptor water that satisfies the following geometrical constraints: $R_{OO} \leq 3.3$ Å, $\theta_H \leq 70.53°$, $\theta_{LP} \leq 70.53°$, and $\delta_D \leq 180.0°$.

Based on this definition, Cummings and co-workers [188–190] determined the radial dependence (with respect to a central water molecule or solute species) of the number of hydrogen-bonded water molecules in ambient and supercritical SPC water and the effect of the presence of different types of infinitely dilute solutes. The remarkable finding is the lack of any significant effect of the nature of the solute on the strength of the observed hydrogen bonding (see Section III.A). A close analysis of the raw data also indicates the need for a more precise sampling of the hydrogen bonding at short distances from the central species to make more definite claims about the effect of the nature of the solute species. According to the predictions of the SPC water model, the strength of the hydrogen bond at near critical conditions decreases by a factor of two from the corresponding value at ambient conditions (Fig. 18).

Subsequently, Chialvo and Cummings [72,89] suggested that, within the spirit of the geometric definition, a better interpretation for the strength

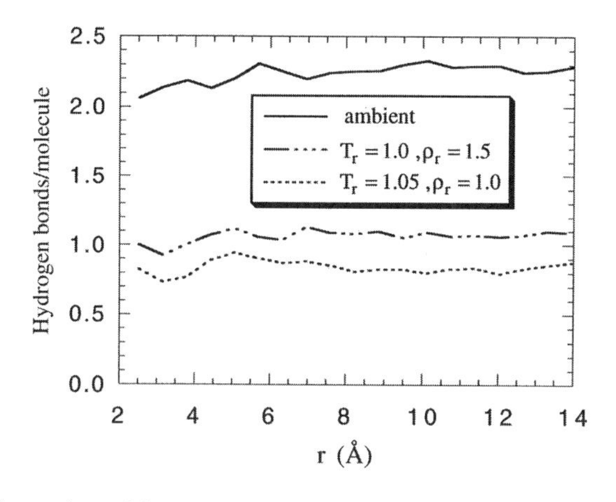

Figure 18. Comparison of the supercritical and the ambient behavior of the average number of water molecules hydrogen-bonded to a water molecule at a distance r from its center of mass.

of hydrogen bonding could be achieved by splitting the $g_{\mathrm{OH}}(r)$ into two complementary orientational pair distribution functions—$g_{\mathrm{OH}}^{\mathrm{HB}}(r)$ and $g_{\mathrm{OH}}^{\mathrm{NHB}}(r)$—associated with the geometry of bonding and nonbonding water configurations. Within this framework, $g_{\mathrm{OH}}^{\mathrm{HB}}(r)$ gives the relative to bulk probability of finding two water molecules separated by a distance r in a hydrogen-bonding configuration. Likewise, $g_{\mathrm{OH}}^{\mathrm{NHB}}(r)$ gives the relative to bulk probability of finding two water molecules separated by a distance r in any configuration other than that for hydrogen bonding. Thus $g_{\mathrm{OH}}^{\mathrm{HB}}(r)$ involves a specific angular slice of the total pair correlation function $g_{\mathrm{OH}}(r)$, whose configurational space volume [64]

$$\begin{aligned}\Omega_{\mathrm{HB}}(r) &= 4\pi \int_0^r R_{\mathrm{OO}}^2\, dR_{\mathrm{OO}} \int_0^{70.53^\circ} \sin\theta_{\mathrm{H}}\, d\theta_{\mathrm{H}} \int_0^{70.53^\circ} \sin\theta_{\mathrm{LP}}\, d\theta_{\mathrm{LP}} \int_0^{180^\circ} d\delta_{\mathrm{D}} \\ &= 16\pi^2 r^3/27 \end{aligned} \tag{2.41}$$

is a fraction of the total

$$\begin{aligned}\Omega(r) &= 4\pi \int_0^r R_{\mathrm{OO}}^2\, dR_{\mathrm{OO}} \int_0^{180^\circ} \sin\theta_{\mathrm{H}}\, d\theta_{\mathrm{H}} \int_0^{180^\circ} \sin\theta_{\mathrm{LP}}\, d\theta_{\mathrm{LP}} \int_0^{180^\circ} d\delta_{\mathrm{D}} \\ &= 16\pi^2 r^3/3 \end{aligned} \tag{2.42}$$

According to Eqs. (2.41) and (2.42) it becomes clear that the radial O···H pair distribution function will be composed of the two contributions: $g_{\mathrm{OH}}^{\mathrm{HB}}(r)$ and $g_{\mathrm{OH}}^{\mathrm{NHB}}(r)$:

$$g_{\mathrm{OH}}(r) = (\tfrac{1}{9}) g_{\mathrm{OH}}^{\mathrm{HB}}(r) + (\tfrac{8}{9}) g_{\mathrm{OH}}^{\mathrm{NHB}}(r) \tag{2.43}$$

Therefore, we can split the total coordination number

$$n_{\mathrm{OH}}(r) = 4\pi\rho_{\mathrm{H}} \int_0^r g_{\mathrm{OH}}(r) r^2\, dr \tag{2.44}$$

[for $n_{\mathrm{OH}}(r < 2.4\ \text{Å}) = N_{\mathrm{B}}$ in Mountain's definition [191]] into hydrogen-bonding coordination

$$n_{\mathrm{OH}}^{\mathrm{HB}}(r) = (4\pi\rho_{\mathrm{H}}/9) \int_0^r g_{\mathrm{OH}}^{\mathrm{HB}}(r) r^2\, dr \tag{2.45}$$

and non-hydrogen-bonding coordination contributions

$$n_{\mathrm{OH}}^{\mathrm{NHB}}(r) = (32\pi\rho_{\mathrm{H}}/9) \int_0^r g_{\mathrm{OH}}^{\mathrm{NHB}}(r) r^2\, dr \tag{2.46}$$

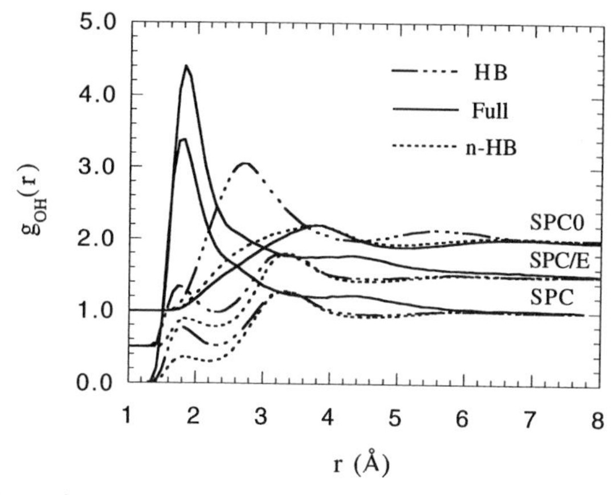

Figure 19. Comparison of the simulated orientational O···H radial distribution functions for the SPC, SPC/E, and SPC0 water models, each shifted up by 0.5 unit for clarity, at $T = 573$ K and $\rho = 0.72$ g/cm^3.

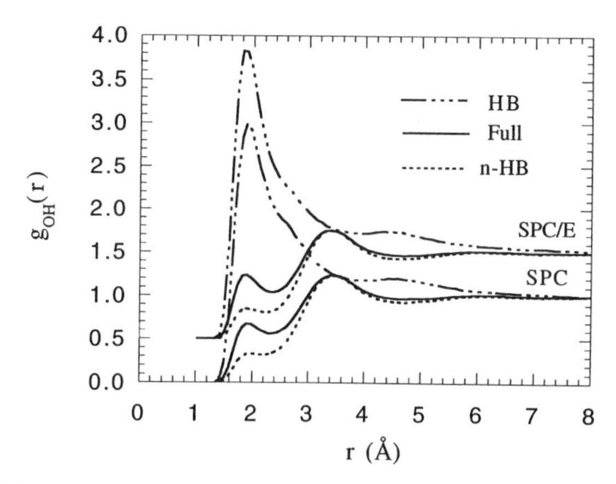

Figure 20. Comparison of the simulated orientational O···H radial distribution functions for the SPC and SPC/E water models, each shifted up by 0.5 unit for clarity, at $T = 673$ K and $\rho = 0.66$ g/cm^3.

Obviously, in the Mezei–Beveridge definition [64], $g_{\text{OH}}^{\text{HB}}(r)$ for $r \leq R_{\text{OO}}$ defines the hydrogen-bonding structure around a molecule so that $n_{\text{OH}}^{\text{HB}}(R_{\text{OO}})$ gives the number of hydrogen-bonded water molecules.

The behavior of the correlation functions $g_{\text{OH}}^{\text{HB}}(r)$, $g_{\text{OH}}^{\text{NHB}}(r)$, and $g_{\text{OH}}(r)$ as described by the SPC, SPC/E, TIP4P, ST2, and SPC0 models at $T = 573$ K and $\rho = 0.72$ g/cm^3 and (except for the SPC0 model) at $T = 673$ K and $\rho = 0.66$ g/cm^3 is shown in Figures 19 and 20. The outstanding feature in these results is given by the relative strength, measured as the size ratio between the first peaks of $g_{\mathrm{OH}}^{\mathrm{HB}}(r)$ to $g_{\mathrm{OH}}(r)$, which remains approximately the same for all models (4.0–4.5) and does not change from ambient to supercritical conditions. In addition, note that even though the hydrogen-bonded pairs contribute only one-ninth of the total $g_{\mathrm{OH}}(r)$, according to Eqs. (2.45) and (2.46), these pairs account for about 50% of the total coordination number: $n_{\mathrm{OH}}^{\mathrm{HB}}(r = 2.4\ \text{Å}) \approx n_{\mathrm{OH}}(r = 2.4\ \text{Å})$. This feature appears to show little dependence on state conditions from ambient to near critical conditions. The choice of the distance $r \approx 2.4$ Å is based on the location of the first valley of the $g_{\mathrm{OH}}(r)$. Note that this radius also covers the first peak of $g_{\mathrm{OH}}^{\mathrm{HB}}(r)$; although, because of the skewness of the latter, its first valley is not clearly defined.

Recently, Kalinichev performed a series of isobaric–isothermal Monte Carlo simulations of TIP4P water model along the supercritical isotherm of $T = 773$ K and a density range of $0.03 \leq \rho \leq 1.67$ g/cm^3 to study the hydrogen bond distribution, based on an intermolecular energy distance distribution [74]. Thus the analysis is based on a three-dimensional surface that depicts a measure of hydrogen bond strength (z coordinate) as a function of the intermolecular O$\cdots$H distance (y coordinate) and the pair-interaction energy (x coordinate). Therefore, both a purely geometric (G) and the usual energetic (E) criteria are represented simultaneously, a feature that allows Kalinichev to study the usefulness of each criterion at any state condition. Note that the geometric criterion here is simple (based only on the O$\cdots$H distance). It should not be confused with the much more stringent geometric definition of the Mezei–Beveridge definition [64] used by Cummings and co-workers in the studies described above. As clearly indicated in [74], either criterion successfully predicts hydrogen bonding at normal conditions; however, at supercritical conditions, neither criterion by itself is able to give an accurate account of HB, since some of the pairs that satisfy the simple geometric criterion might represent repulsive pair interactions.

According to Figure 21, the energetic and simple geometric criteria predict approximately the same hydrogen-bonding behavior along the isotherm $T = 773$ K and up to a density of $\rho \approx 0.77$ g/cm^3 ($P \approx 300$ MPa). For denser systems, both criteria overpredict the strength of hydrogen bonding: the geometric by as much as 100% and the energetic by approximately 25% at $\rho \approx 1.0$ g/cm^3 ($P \approx 1000$ MPa), with respect to the prediction of the G + E criterion. An interesting feature of Kalinichev's simulation results is the lack of density dependence for the average values of the O–H$\cdots$O and H$\cdots$O–H angles (Fig. 1) and a rather weak density dependence for the average r_{OH} distance along the isotherm $T = 773$ K. These results clearly indicate that the temperature effect (from ambient to $T = 773$ K) is by far

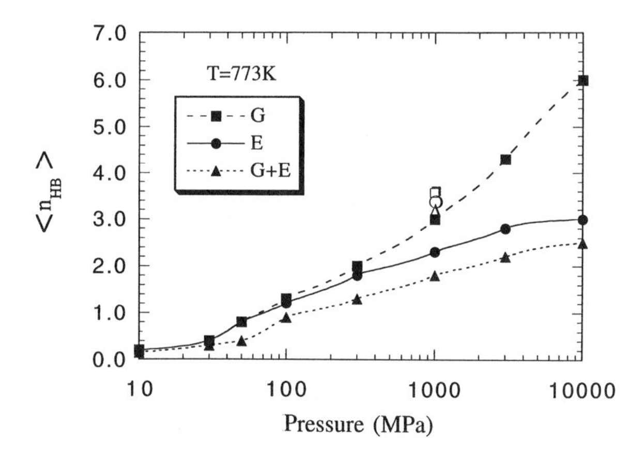

Figure 21. Pressure dependence of the average number of hydrogen bonds per water molecule according to the geometric (G), energetic (E), and combined (E+G) criteria. The corresponding *open symbols* for ambient water are plotted at $P = 1000$ MPa, at which the supercritical density is ≈ 1 g/cm^3.

larger than the pressure effect (from ambient to $P \approx 1000$ MPa) on the hydrogen-bonding behavior in water.

Additional simulations of the ST2 water model were performed by Mountain [191], who also studied the so-called RPOL polarizable water model of Dang [143] to compare the predicted microstructures with those from the NDIS-93 data [60]. His comparison resulted in similar conclusions to those of other authors regarding the accuracy of the NDIS-93 data. He also analyzed the strength of hydrogen bonding based on the geometric criterion of Mezei and Beveridge [64], although under the assumption that the dihedral constraint $\delta_D \leq \pi$ is always satisfied. Under this angular condition, Mountain defined $g_B(R_{OH})$ as the bonded-pair distribution, with $R_{OH} \leq R_B \approx 3.1$ Å. Note that this function differs from the $g_{OH}^{HB}(r)$ as defined by Chialvo and Cummings [72,89] in two aspects: although there is a restriction on R_{OH}, there is no restriction on δ_D [see Eqs. (2.41)–(2.46)].

In a follow-up to Mountain's work on the TIP4P water model [69], Mizan and co-workers [71,192] studied the temperature and density dependence of hydrogen bonding in the flexible TJE water model [126]. The analysis was performed along the supercritical isotherm $T = 773$ K within the density range of $0.115 \leq \rho \leq 0.659$ g/cm^3, and along the isochores $\rho = 0.257$ (subcritical) and 0.659 g/cm^3 (supercritical) within the temperature range of $773 \leq T \leq 1073$ K. Based on the energetic criterion [66], they studied the hydrogen-bonding strength, the hydrogen bond cluster size distribution, the hydrogen bond persistence time, and its autocorrelation function.

Their results indicate that the hydrogen bond strength decreases with increasing temperature along both isochores; the cluster size distribution appears to be weakly dependent on temperature; and according to the behavior of the hydrogen bond persistence and the autocorrelation function, the hydrogen bond rupture is essentially a temperature-driven process.

More recently, Kalinichev and Bass [193] completed their studies on the comparison of the NDIS-97 data [76] and computer simulation results for the TIP4P and BJH water models. Although their comparison at $T = 573$ K and $\rho = 0.72$ g/cm^3 revealed essentially the same agreement found by others [145], the most interesting features were noted at $T = 673$ K and $\rho = 0.66$ g/cm^3 where the NDIS-97 data predict a disappearing of the first peak of $g_{OH}(r)$ at $r \approx 2$ Å. Even though their comparison was done at rather different state conditions (i.e., NDIS-97 data and TIP4P results at 673 K and 0.66 g/cm^3, *ab initio* results [107] at 730 K and 0.64 g/cm^3, and BJH results at 630 K and 0.69 g/cm^3), it indicates a remarkably different behavior for the flexible BJH water model. The most evident disagreement between NDIS-97 and the simulations is in the hard-core diameter for the H$\cdots$H interactions, which the BJH model predicts to be ~ 0.5 Å larger than that observed by NDIS-97 and predicted by other models. In general, the BJH model predicts more structured features for the three site–site distribution functions. On the other hand, the TIP4P model gives a rather good representation of all site-distribution functions, even though it does not predict the disappearing of the first peak of $g_{OH}(r)$. The first peak of $g_{OH}(r)$ becomes a plateau. This plateau is not reproduced even by the newest experiments [77].

Based on their previous work [113], Kalinichev and Bass made contact between the experimentally overall degree of hydrogen bonding χ (see Section II.A) and the simulated average number of hydrogen-bonded water molecules, $\langle n_{HB} \rangle$, according to the G + E criterion:

$$\langle n_{HB} \rangle = 4\chi \tag{2.47}$$

This relation is based on the fact that the maximum number of ideal hydrogen bonds is four (i.e., in ice-I). Moreover, by performing a few straightforward scaling transformations, they were able to plot most of the available data from simulation and experiments (Fig. 22). The remarkable feature in this comparison is the degree of agreement among the rather different approaches and the fact that even at high temperatures low-density water still exhibits a noticeable degree of HB, most probably represented by small clusters of a few water molecules (as opposed to the percolating network at ambient conditions).

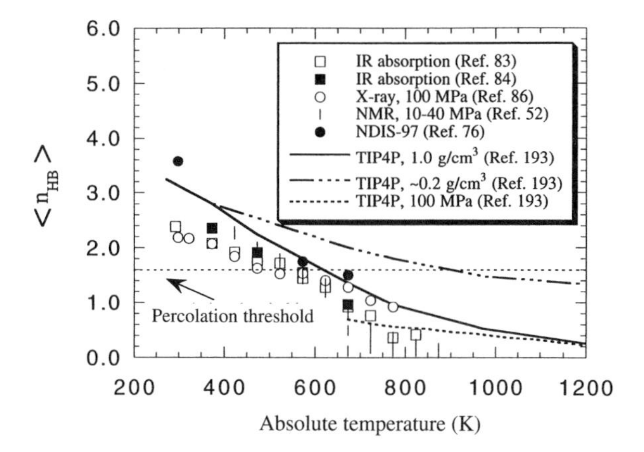

Figure 22. Temperature dependence of the average number of hydrogen bonds per molecule obtained by X-ray data, IR absorption by ν_{OH}HDO, IR absorption by ν_{OD}HDO, neutron diffraction, proton NMR shift, and Monte Carlo simulations.

In a second round of reverse Monte Carlo (RMC) simulations, Jedlovszky and co-workers [194] analyzed the newest and expanded neutron diffraction data set of NDIS-97 [76] for ambient and supercritical water and performed new molecular dynamics simulations matching eight state conditions from the NDIS-97 data. To determine the extent of the hydrogen bonding from the experimental site–site correlation functions, they took advantage of the configurational information generated in the RMC simulations; therefore, they used a geometric criterion for their definition of a hydrogen-bonded water pair (note that there is no energetic information involved in the RMC technique). Thus they consider two molecules to be hydrogen bonded if their O···O separation is less than the distance at which $g_{OO}(r)$ exhibits the first minimum, their closest O···H separation is less than the distance at which $g_{OH}(r)$ exhibits the first minimum, and the H–O···O angle (γ) is smaller than some arbitrarily fixed value. They also analyzed the geometry of the hydrogen bonds in terms of the angle distribution $P(\cos\gamma)$ at all state conditions and found that it peaked at 1, which is indicative of a preferred colinear hydrogen bond geometry. The sharpness of this peak, however, decreases with the increase in temperature; so that at $T = 573$ K, this distribution becomes barely dependent on density.

These authors also investigated the topology of the hydrogen bond structure, i.e., the properties of the hydrogen bond clusters. Because ambient water forms a space-filling percolating hydrogen-bonded network [195–197] and the extent of hydrogen bonding decreases with increasing temperature and decreasing density [52,113], they concluded that this network should cross a percolation transition, i.e., a condition at which the space-filling hydrogen-bonded network breaks down to form smaller hydrogen-bonded clusters of water molecules.

2. *Ab Initio Simulation Approach*

Recent progress in the calculation of electronic structure via density functional theory (DFT) has made it possible to perform *ab initio* molecular dynamics simulation of ambient and supercritical water [107,198–200]. The *ab initio* molecular simulation (often known as the Car–Parrinello (CP) method, after its developers [108]) is based on the use of a DFT description of the electronic structure of molecules with dynamical finite-temperature disorder, such as molecular liquids including water. Thus the DFT gives the instantaneous (adiabatic) forces acting on the atomic nuclei (within the Born–Oppenheimer approximation), which are then used to perform the typical time-step integration in the classical molecular dynamics. In this sense, the DFT calculation step replaces the intermolecular potential model interaction in a classical simulation, and for that reason *ab initio* methods are often labeled "parameter-free" simulations [107,201].

The popularity of the *ab initio* method hinges on its computational efficiency for large periodic systems and the fact that a relatively simple local density approximation (LDA) of the energy functional is often a good guess for many system properties [202]. The calculation of electronic structure for molecular liquids is rather demanding, however, because it requires interaction energies with uncertainties smaller than 10^{-3} au, otherwise even these small inaccuracies translate into large perturbations in the resulting fluid's structural and dynamical behavior. Precisely for this reason, much of the development in DFT has focused on improving the energy evaluation by extending the LDA with nonlocal terms, such as gradient corrections (GC) [203]. Currently, the *ab initio* structural and dynamical results for water indicate that

1. The LDA corrected for exchange effects according to Becke [204] yields a rather weak hydrogen bonding.
2. The addition of a GC for correlation effects according to Perdew [205] results in too strong effective hydrogen bonding
3. The combination of the Becke correction and a semilocal correlation function according to Lee and co-workers [206] renders the best agreement with experimental data of water at ambient condition [199].

An alternative *ab initio* simulation method was recently proposed by Ortega and co-workers [207] in which the water geometry is fixed at the beginning of the simulation, based on the assumption that the water behavior is essentially determined by the intermolecular (as opposed to intermolecular and intramolecular) interactions. According to these authors, the rationale behind the simplification is to direct all the effort toward properly describing the weak interactions and make the calculation computationally efficient. For that purpose, they constructed an approximate many-body Hamiltonian function based on a linear combination of atomic orbitals (LCAO) [208], and kept only the intermolecular interactions that contribute to the total energy up to the second order in the overlap between atomic orbitals on different molecules. Following a similar approach as Kohn–Sham and Hartree–Fock, they solve many-body Hamiltonian using a complementary one-electron Hamiltonian [209].

As in the CP *ab initio* approach, Ortega and co-workers' approach also relies on corrections to the exchange and correlation contributions to the interaction energy (see Ref. [207] for details). The bottom line of this approach is that all interactions involve only two center integrals (i.e., higher-order integrals are neglected), a feature that makes it possible to store the integrals as a function of the corresponding pair distances during the simulation.

It is important to keep in mind that *ab initio* methods are considered parameterization-free approaches, in that they do not rely on a fitted intermolecular potential model; but that the suitability of the approximations introduced into the DFT or LCAO calculations — and, therefore, the accuracy of the simulation results — are known only after these results are compared with experimental data (usually on a case-by-case basis, such as in the study of water by Sprik [200]). For instance, Ortega and co-workers [210] applied their first principle simulation method [207] to the determination of the structural properties of water at ambient and high temperature. They computed the radial distribution function for the O···H interactions, $g_{OH}(r)$, for the state condition of $T = 580$ K and $\rho = 0.72$ g/cm^3, which corresponds closely to that of the NDIS-93 experiments ($T = 573$ K and $\rho = 0.72$ g/cm^3) [60,97]. Although the corresponding NDIS-93 data were not plotted together with their simulation results, the authors claimed that their simulated $g_{OH}(r)$ predicted the correct location of the first peak, 2.02 Å, and its position shift from 1.91 to 2.02 Å as the system conditions move from $T = 300$ K and $\rho = 1.0$ g/cm^3 to $T = 580$ K and $\rho = 0.72$ g/cm^3. In addition, they pointed out the occurrence of a substantial reduction in the strength of the first peak of $g_{OH}(r)$ and argued that "classical computer simulations fail to describe this behavior, and in fact show that the hydrogen-bond peak survives up to $T = 1000$ K and $\rho \cong 0.5$ g/cm^3," by quoting earlier simulation work on the SPC water model [99].

The comparison of their simulation results and the NDIS-93 data failed to mention one of the most striking and controversial features of the NDIS-93 structural behavior: the $g_{OH}(r)$ at $T = 573$ K and $\rho = 0.72$ g/cm^3 exhibits a hard-core diameter of ~1.8 Å in contrast to ~1.5 Å for ambient water. Moreover, the authors contended that, according to Cummings and co-workers' earlier water simulations [99], the hydrogen bond peak survives up to $T = 1000$ K and $\rho \cong 0.5$ g/cm^3. Unfortunately, this statement was not made or suggested in the Cummings and co-workers paper, in which the reported simulations involved the SPC model at two supercritical conditions: $T/T_c = 1.0, \rho/\rho_c = 1.5$ and, $T/T_c = 1.05, \rho/\rho_c = 1.0$ relative to the SPC critical conditions. As shown in Section II.B.1.b, classical water models can, and indeed do, properly describe the decreasing strength of the O···H pair correlation, as long as the models are properly parameterized.

A second controversial assertion made by Ortega and co-workers concerns the $g_{OO}(r)$ used in their figures. A prominent feature of the simulated $g_{OO}(r)$ is the lack of a well-defined second peak at ambient conditions. Typically, this peak is located at ~4.5 Å and represents the signature of the tetrahedral water structure at ambient conditions [86,103,109]. The fact that this peak does not appear in the simulations at ambient conditions, and does appear at $T = 350$ K and $\rho = 0.72$ g/cm^3, raises questions about the authors' model and/or simulation methodology. Moreover, the lack of agreement between the $g_{OO}(r)$ from first principle simulations and the NDIS-93 data may not be ascribed to "an insufficient accuracy in the inversion of the experimental data from ***k***-space to real space" as suggested by Ortega and co-workers after invoking the *ab initio* work of Fois and co-workers [107]. In fact, even though there are inherent difficulties in handling the Fourier transformation of the NDIS raw data [92] (in addition to the formidable task of assessing the inelasticity corrections [105,106], the suitable analysis of which resulted in the NDIS-97 study [76]), the NDIS-97 $g_{OO}(r)$ values are in remarkable quantitative agreement with earlier X-ray results [86,211], which more directly yield $g_{OO}(r)$ [109] (Fig. 23).

In addition, even though the water's $g_{OO}(r)$ value from X-ray experiments is a molecular rather than atomic distribution function (in that it contains about 12% contributions from O···H interactions and less than 2% contributions from H···H interactions [102]), the existence of the second and third peaks in $g_{OO}(r)$ does not appear to be associated with those contributions. If this were the case, we should also be able to reproduce the molecular distribution function by a linear combination of the three radial distribution functions [$g_{HH}(r), g_{OH}(r)$, and $g_{OO}(r)$] from either simulation or NDIS, using a $g_{OO}(r)$ that exhibits only the first peak and is equal to 1 for larger distances [this will approximately describe the $g_{OO}(r)$ of Ortega and co-workers). Thus the molecular radial distribution function would be given by $g_m^{lc}(r) \approx 0.86 g_{OO}(r) + 0.12 g_{OH}(r) + 0.02 g_{HH}(r)$ and should exhibit the second and third peaks at 4.5 and 7.6 Å, respectively. According to Figure 24, however—where $g_m(r)$ from X-ray diffraction is compared with that

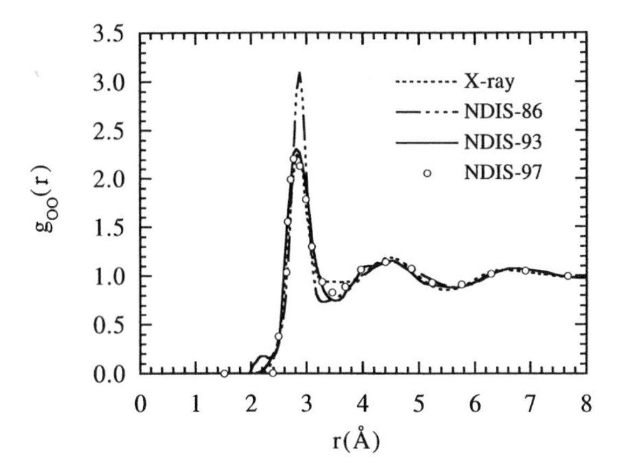

Figure 23. Comparison between scattering results from Soper and Phillips [59] (NDIS-86), Postorino and co-workers [60] (NDIS-93), Soper and co-workers [76] (NDIS-97), and X-ray results from Gorbaty and Demyanets [103] for the O···O radial distribution function of water at ambient conditions.

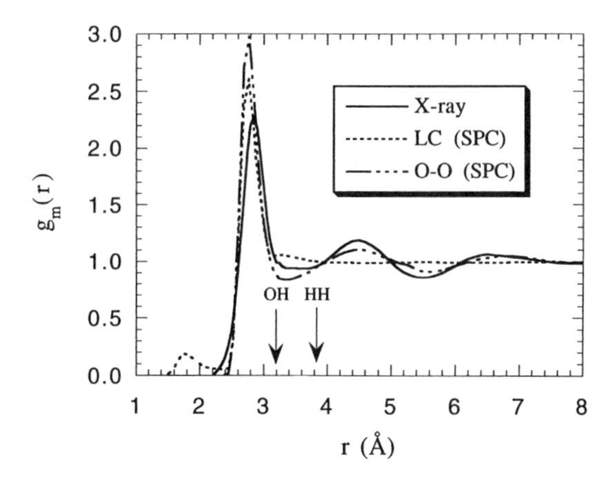

Figure 24. Simulated $g_m^{lc}(r)$ and $g_{OO}(r)$ comparison to the experimental $g_m(r)$ from X-ray diffraction [86]. *Arrows* indicate the location of the second peaks of the O···H and H···H correlation functions.

from a linear combination of SPC microstructures [with the mentioned requirement for $g_{OO}(r)$] and the actual SPC $g_{OO}(r)$—this is not the case. Clearly, $g_m^{lc}(r)$ shows no peaks beyond the first coordination shell because neither $g_{HH}(r)$ nor $g_{OH}(r)$ exhibits any substantial structure for $r > 4$ Å.

Moreover, even though earlier *ab initio* simulation results of $g_{OO}(r)$ for ambient water by Laasonen and co-workers [198] were at odds with the

NDIS-86 [59], in a later publication these authors [199] studied the effect of different gradient-corrected density functionals in their *ab initio* approach and reached a remarkable agreement between the simulated and the experimental $g_{OO}(r)$ values; their best *ab initio* approach predicts a $g_{OO}(r)$ with a well-defined peak at around 4.4 Å.

Although *ab initio* methods for the study of ambient and supercritical water show considerable future promise and are able to study many phenomena generally inaccessible to classical simulations (such as dissociation of the water molecule and speciation of the hydroxyl and hydronium ions), the long calculation times and the limited predictive capability (functional results still need to be compared to experimental results) suggest that classical models and simulations will continue to be used for a long time into the future.

C. Discussion

For three decades, molecular models for water have been developed and refined by fitting to structures measured by neutron scattering. The decade-old widely accepted structure of water at room temperature and pressure was recently revised as a by-product of attempts to understand the structure of high-temperature, high-pressure water for which, in a remarkable reversal of roles, molecular models successfully pinpointed inaccuracies in scattering data. This remarkable interplay between molecular modeling and experiment suggests that molecular methods can effectively complement experimental approaches in the quest for an accurate microscopic picture of the structure of water and indicates the increasing reliability of the intermolecular potential models and the accuracy of the simulation results. The substantial agreement between models and experiments that resulted from this interplay gives greater confidence in our abilities to measure and predict the microstructural properties of water at all conditions.

The revised NDIS results have a few unexpected consequences. The comparison of earlier NDIS structural data for ambient water by Soper and Phillips [59], NDIS-86, and the newest NDIS-97 data [76] indicates that the NDIS-86 results actually overpredicted the amplitude of the $g_{OO}(r)$ and $g_{OH}(r)$ correlations, even though they satisfied some essential thermodynamic constraints [58]. This means that most current water models included in commercial simulation packages, whose parametrization was guided by the Soper and Phillips data (the *de facto* standard NDIS-86), might also fail to reproduce the actual structure of water. In addition, as shown in Figure 23, the NDIS-97 results suggest that earlier X-ray and neutron-scattering determinations of the $g_{OO}(r)$ values of water at ambient conditions by Narten [211] and Gorbaty [110] are in fact more accurate than has been generally accepted over the past decade.

The comparison of simulation results and experimental data, an essential validation approach for any molecular-based modeling effort, indicates a few important points. Because the experimental complexity and likelihood of undesirable numerical artifacts in raw data processing, it appears imperative to have independent verification of the current NDIS data. Fortunately, work is already in progress in that direction [77]. In addition, it seems desirable to have alternative experimental methods to determine microstructures to avoid the use of isotope substitution and its troublesome inelasticity corrections. Thus it is reasonable to seek alternative or complementary molecular-based methods to efficiently check the reliability of the experimental structural results, before they are used in model parameterization. Work in that direction, such as new developments on the RMC method [212,213], is already under way.

III. SUPERCRITICAL AQUEOUS SOLUTIONS

The physical chemistry of supercritical aqueous solutions has become a subject of wide-ranging interest because of their potential applications in emerging new technologies [214]. Progress has been made in the understanding of the fundamental role played by the solute-induced structural changes of the solvent in the vicinity of a species in solution, i.e., the solvation structure, in determining the thermophysical behavior of solutions [90,99,215–219]. This feature is most evident in (although not exclusive to) aqueous solutions, in which the microstructural changes of water around solutes have been linked to a wide variety of solvation-related phenomena, such as ion mobility and conductivity [220,221], solvent electrostriction and dielectric saturation [222], salt solubility and ion speciation [223,224], electrode processes [225], salting-in and -out in mixed-solvent electrolytes [42,226], antifreeze activity of certain agents [227], the response of taste chemoreceptors [228–230], the hydrophobicity of gas hydration [231], and protein folding [232–234].

Solute solvation in a supercritical solvent can be formally described as a two-step process involving two rather different length scales (see Section III.B):

1. The insertion of a solute into the system that induces a finite density perturbation of its average local density around a solvent molecule [50,51,235].
2. Its subsequent propagation across the medium by a distance given by the prevailing correlation length (note that in the case of infinitely dilute solutions this length, as well as all the solute's mechanical partial

molar properties, will diverge with the isothermal compressibility as the solvent approaches its critical point [236]).

Yet, the solvation process at high temperature can be fully characterized, from a microscopic viewpoint, in terms of the local density perturbation (solute-induced effects [50,90,235]), because the (compressibility driven) propagation plays an accessory role; the large compressibility allows the tuning of the solvation power (large density changes) by small pressure changes.

Even though the coexistence of these two effects makes dilute near-critical solutions extremely challenging to model properly, it also provides a useful way to characterize their thermophysical properties in terms of solute-induced and compressibility driven contributions. For modeling purposes, this characterization usually takes the form of asymptotic expressions for the mixture's Helmholtz free energy and its temperature, volume, and composition derivatives around the solvent's critical point. The usefulness of the resulting asymptotic expressions [from the residual Helmholtz free energy expansion $A^r(T, \rho, x)$] resides in their mathematical simplicity. They require only the pure solvent thermophysical properties and the critical value of the derivative $(\partial P/\partial x_U)^\infty_{T,\rho}$—Krichevskii's parameter [236]—with a clear microscopic meaning, where P is the total pressure, x_U is the mole fraction of solute, T is the absolute temperature, ρ is the solvent's density (see Section III.B for a complete analysis). Typical examples of this asymptotic expression are given in the literature for the solute distribution factor K^∞ [237,238], Henry's constant [237–239], and solute solubility enhancement [50,240].

The key to the successful modeling of dilute near-critical mixtures is the microscopic understanding of the solvation phenomena; a convenient starting point to gaining such an understanding is the microscopic interpretation of the driving force for the solvation processes, i.e., the species-residual chemical potentials and their related quantity $(\partial P/\partial x_U)^\infty_{T,\rho}$ [50,238]. According to the behavior (magnitude and sign) of $(\partial P/\partial x_U)^\infty_{T,\rho}$, the infinitely dilute solutes can be classified into nonvolatiles [240] (attractive [241]) and volatiles [240] (repulsive [241]), which typically correspond to the solvation of electrolyte- and nonelectrolyte-condensed solutes and gases, respectively.

Accurate gas solubility data are still scarce, despite the intense activity in gas solubility and its impact in several application-oriented fields, such as biomedical technology (solubility of oxygen in perfluorinated hydrocarbons as blood substitutes) [242], anesthesia [243], environmental pollution control (oxygen and ozone in waste water treatment), geochemistry (undersea gas hydrates) [244], and chemical process design (enhanced oil recovery, gas sweetening, sludge oxidation). Nonpolar gases are typically one to three orders of magnitude less soluble in water than in liquid hydrocarbons [245]; and the thermodynamics of dissolution indicates some uncommon solvation features, such as the temperature dependence of gas solubility in water (and other solvents). For example, at low and moderate pressures, gas solubility (which is directly connected to Henry's constant) decreases with increasing temperature, exhibits a minimum, and then increases steeply as approaching the solvent's critical temperature [246]. This behavior, which appears to be a universal feature for aqueous and nonaqueous near-critical solvents [247], encouraged some researchers to find some phenomenologic justification [237,247,248]. In particular, Japas and Levelt Sengers [237] derived the limiting temperature dependence of Henry's constant and the solute distribution factor K_U^∞ and proposed useful linear correlations for these quantities in terms of the so-called Krichevskii's parameter and the solvent's orthobaric density in the vicinity of the solvent's critical point. Later, Wilhelm [249] suggested an alternative asymptotic linear correlation involving Ostwald's coefficient L_U^∞ instead of K_U^∞, arguing that the former might show a wider linear range than the latter.

Among the relevant supercritical aqueous-solute solvations, electrolytes are perhaps the most challenging to study experimentally and are currently the focus of renewed interest because of the significant role played by ion speciation in new environmentally relevant technologies (such as the supercritical water oxidation of toxic wastes [16,17,19]) and significant complications with current technologies (such as solid deposition [250] and metal corrosion [225] in electric power generators).

The experimental determination of equilibrium properties of dilute aqueous electrolyte solutions is notoriously difficult because of the simultaneous occurrence of long-range density–composition fluctuations and long-range Coulombic interactions. Considering that the dielectric constant regulates the water's solvation power through the screening of the solute's electrostatic interactions [222], we expect ion speciation as the dielectric screening of the electrostatic interactions changes dramatically in the highly compressible medium [48,251,252] where the dielectric permittivity of water is small and changes dramatically with small perturbations of temperature and pressure [4], as well as by the large electric field gradients around charged species (in the form of dielectric saturation and electrostriction [42]). The microscopic manifestation of these phenomena can be probed in terms of the local properties of the solvent in the vicinity of the species in solution, i.e., the local properties of water relative to the unperturbed water (ideal solution) (see below). Thus the occurrence of speciation at near-critical conditions adds another complication to the already problematic modeling of dilute aqueous electrolyte systems [253,254].

A. Molecular Simulation of Supercritical Aqueous Solutions

Typically, theoretical studies of supercritical aqueous solutions focus on the connection between the solute partial molar properties at infinite dilution and the solvent microstructure in the vicinity of the solute. Although this connection can be obtained rigorously from statistical mechanics (see Section III.B), the usefulness of this interpretation is not guaranteed, unless we are able to obtain valuable microscopic information regarding the link between solute–solvent molecular asymmetry and the resulting structure, from molecular simulation or other molecular-based means (such as accurate integral equation approximations [255]). It is here that molecular-based simulation becomes a powerful tool for probing the solvation process at a microscopic level, as long as we are aware of the intrinsic limitations of the technique when used at near critical conditions [256].

In the following sections we review relevant classical molecular simulations of supercritical aqueous solutions involving a wide variety of solutes (including noble gases and organic and ionic species) and solvation phenomena.

1. *Intermolecular Potentials and Structure of Infinitely Dilute Solutions*

Guillot and Guissani [257] performed molecular dynamics simulation studies of the temperature dependence of nonpolar gas solubility in water through the use of cumulant expansions in the canonical and isobaric–isothermal ensembles [258,259]. Based on statistical mechanical arguments, the authors were able to show that a low-order truncated cumulant expansion, generated from a single reference state point, can describe accurately the temperature dependence of the solubility (Henry's constant) over a wide range of temperatures along the coexistence curve.

To test the validity of the hypothesis behind the linear (asymptotic) correlations for the solubility coefficients proposed by Levelt Sengers and colleagues [237], Guillot and Guissani [231,260] also studied the solubility of noble gases and methane, via the test particle method [261,262], along the saturation curve (from the triple to the critical point) of SPC/E water. Their calculations comprised the solubility parameter $\gamma_l = \exp(-\mu_U^{r(l)}/k_B T)$, and Ostwald coefficient $L_U \equiv \rho_U^{(l)}/\rho_U^{(g)} = \gamma_l \exp(\mu_U^{r(g)}/k_B T)$, where $\mu_U^{r(l)}$ and $\mu_U^{r(g)}$ are the residual chemical potentials of the solute in the liquid and gas phases, respectively. According to their results, $T\ln(H_U/f_V)$ and $T\ln K_U^{\infty}$ exhibit well-defined linear dependences with the solvent density in a rather wide range of orthobaric conditions, i.e., $\rho - \rho_c \leq 0.35\ \text{g/cm}^3$, even though the observed linearity is less pronounced for $T\ln K_U^{\infty}$ [260].

Moreover, their calculations indicated that the solute–solvent interaction energy is responsible for the observed solubility hierarchy of noble gases in water, whereas the peculiar temperature dependence of and minimum in the solubility along the water coexistence curve can be attributed to an entropic contribution related to the so-called cavity formation. The microstructural analysis allowed them to establish a connection between the increase of solubility with a decrease in temperature in cold water and the formation of clathrate-type of cages around the solutes. They also concluded that as the temperature increases beyond the boiling point, the first solvation shell restructures so that the water molecules do not orient tangentially to the solute but reorient toward the bulk. Moreover, as the temperature approaches criticality, the structure indicates the lowering of the water local density around the solutes, which translates into an increase of the corresponding partial molar volumes.

Lin and Wood [263] performed molecular dynamic simulation of light hydrocarbons (methane, ethane, and propane described by an united atom model) at infinite dilution in TIP3P water at high temperature ($600 \leq T \leq 1200\ ^\circ\mathrm{C}$) in a wide range of densities ($0 \leq \rho \leq 1.0\ \mathrm{g/cm^3}$) to estimate the solute chemical potentials in aqueous solution. The simulation results were then used to develop a seven-parameter equation of state for the quantity $\Delta_h^V G^\theta \equiv kT \ln(\phi_U^\infty P/kT\rho)$ [which, according to our nomenclature, is simply the infinite dilute solute residual chemical potential $\mu_U^{r\infty}(T, \rho)$, based on the ideal gas properties at the same temperature and density [50]] as a function of temperature, water density, and number of carbon atoms, which was tested against the predictions of other equations of state, including the SUPCRT92 [202] and the SUPERFLUID [264,265].

As part of an ongoing molecular dynamic simulation investigation of supercritical aqueous solutions, Chialvo and Cummings [90] and Cummings and co-workers [99,188–190] analyzed the solvent microstructural behavior around a variety of solutes, including noble gases (He, Ar, Xe, and Ne), organics (methanol, benzene, toluene), and isolated ions and ion pairs (Na^+, Li^+, Cl^-) along the near-critical residual isotherm $T_r = 1.05$ and the supercritical residual isochor $\rho_r = 1.5$ of SPC water. According to the behavior of the radial distribution functions of water surrounding the infinitely dilute He and Ne water forms a cavity around these solutes, i.e., water is locally less dense around each solute than around any water molecule (ideal solution) (Figs. 25 and 26). Moreover, because of the closeness to the water critical conditions, these distribution functions exhibit the characteristic slow-decaying (compressibility driven) tails, from above unity for water and from below unity for the noble gases. This behavior has been rigorously characterized by the derivative $(\partial P/\partial x_U)_{\rho,T}^\infty$ (see Section III.B), which becomes Krichevskii's parameter at the solvent's critical point, a finite quantity that plays a central role in the asymptotic behavior of the solubility coefficients [240]. In fact, at constant temperature and solvent

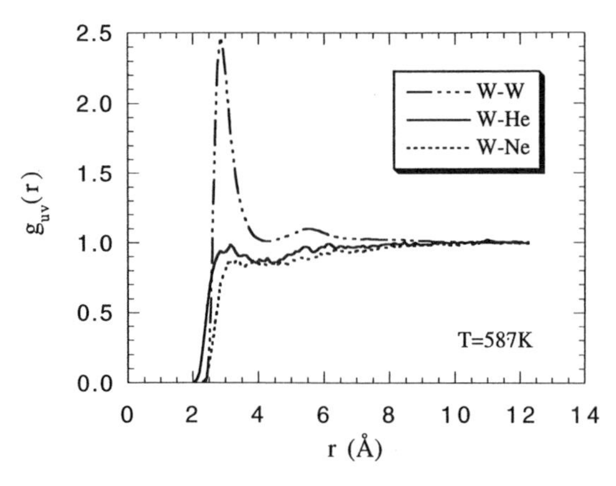

Figure 25. Radial distribution functions for the water–water (ideal solution), helium–water, and neon–water interactions at infinite dilution at $T_r = 1.0$ and $\rho_r = 1.5$.

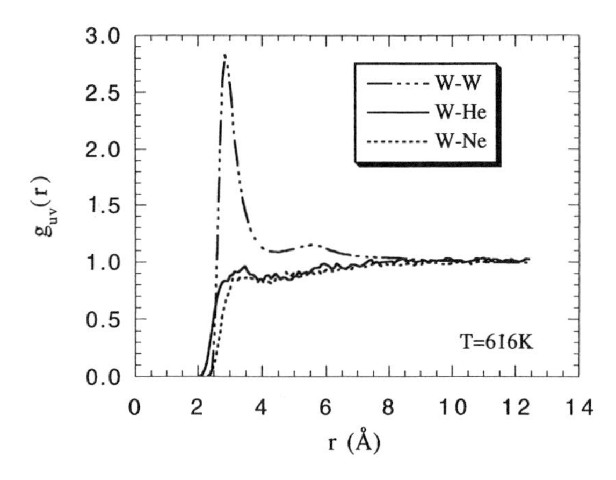

Figure 26. Radial distribution functions for the water–water (ideal solution), helium–water, and neon–water interactions at infinite dilution at $T_r = 1.05$ and $\rho_r = 1.0$.

density, the presence of a noble gas solute induces a local density depletion with a concomitant increase of $(\partial P/\partial x_U)^{\infty}_{\rho,T}$ over (both) that of the ideal solution $[(\partial P/\partial x_U)^{\infty}_{\rho,T} = 0]$ and that of an ideal gas solute $[(\partial P/\partial x_U)^{\infty}_{\rho,T} = kT\rho - \kappa^{-1}]$ [266]. This behavior (volatile or repulsive solute) is clearly seen in Figures 27 and 28 where we display the corresponding excess function $N^{\infty}_{\text{ex}}(R)$ (defined by Eq. 3.12, below), which show negative values within the first solvation shells and give rise to positive values for the corresponding partial molar volumes of the solutes at infinite dilution.

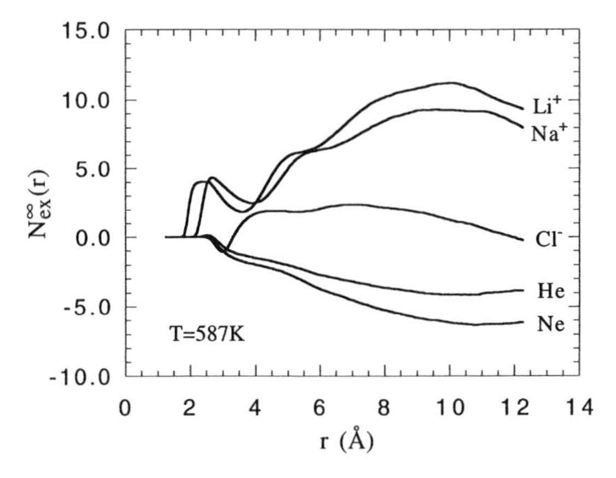

Figure 27. The number of water molecules around each infinite dilute solute in excess to that for the ideal solution as a function of the radial distance from the center of mass of the solute at $T_r = 1.0$ and $\rho_r = 1.5$.

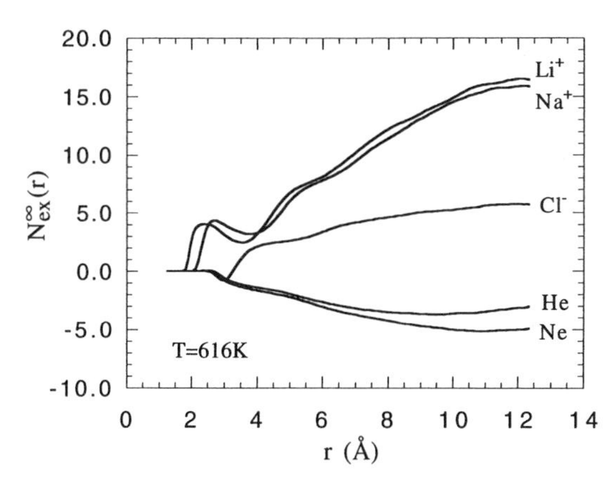

Figure 28. The number of water molecules around each infinite dilute solute in excess to that for the ideal solution as a function of the radial distance from the center of mass of the solute at $T_r = 1.05$ and $\rho_r = 1.0$.

In contrast to the behavior of noble gases, the radial distribution functions of water around the infinitely dilute anion (Cl^-) and cations (Na^+ and Li^+) exhibit a rather strong restructuring; i.e., relative to pure (unperturbed) water there is a substantial increase in the local water density around the ions as a result of strong ion–dipole interactions (Figs. 29 and 30). In other words, the presence of an ion induces an increase in the solvent local density, relative to the pure solvent, with a consequent decrease of

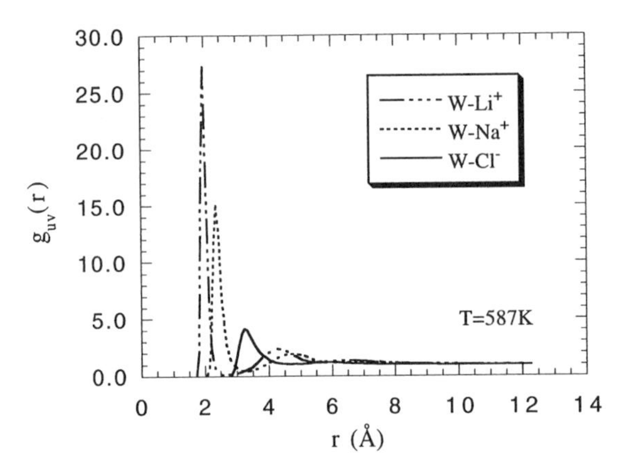

Figure 29. Radial distribution functions for the water–cation and water–anion interactions at infinite dilution at $T_r = 1.0$ and $\rho_r = 1.5$.

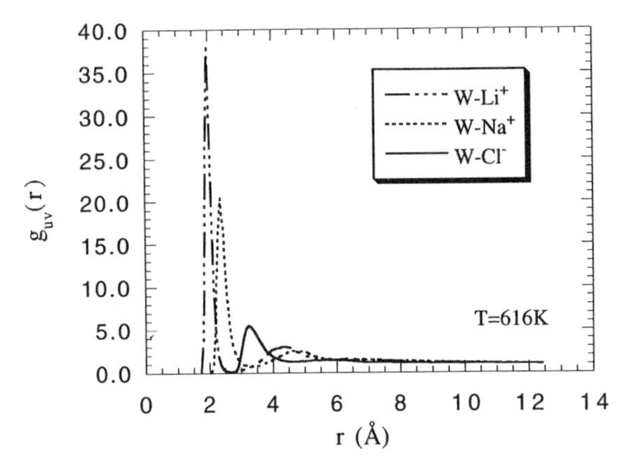

Figure 30. Radial distribution functions for the water–cation and water–anion interactions at infinite dilution at $T_r = 1.05$ and $\rho_r = 1.0$.

$(\partial P/\partial x_U)^{\infty}_{\rho,T}$ (nonvolatile or attractive solute). In turn, the $N^{\infty}_{\text{ex}}(R)$ indicates that the cation first solvation shell contains an excess, over the ideal solution, of at least four water molecules and gives rise to a negative $(\partial P/\partial x_U)^{\infty}_{\rho,T}$. Note that, because of its larger size, the anion's $N^{\infty}_{\text{ex}}(R)$ shows smaller values at both state conditions.

Regarding the degree of water-to-water hydrogen bonding, simulation results from Chialvo and Cummings [90] indicate that the average number of hydrogen bonds for pure water (ideal solution) decreases from ~1.1 (at

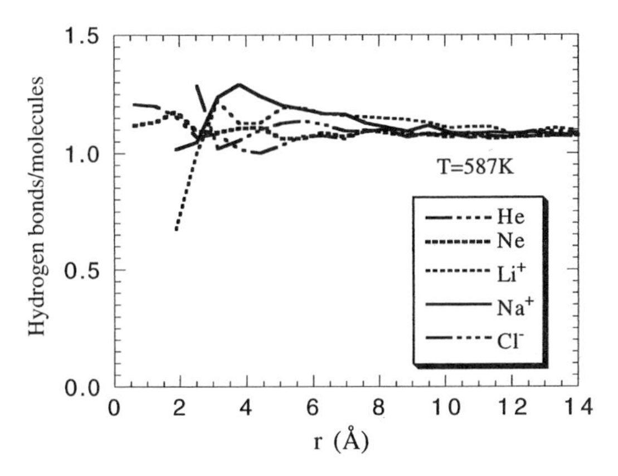

Figure 31. Average number of water molecules hydrogen bonded to a water molecule at a distance r from the solute's center of mass at $T_r = 1.0$ and $\rho_r = 1.5$.

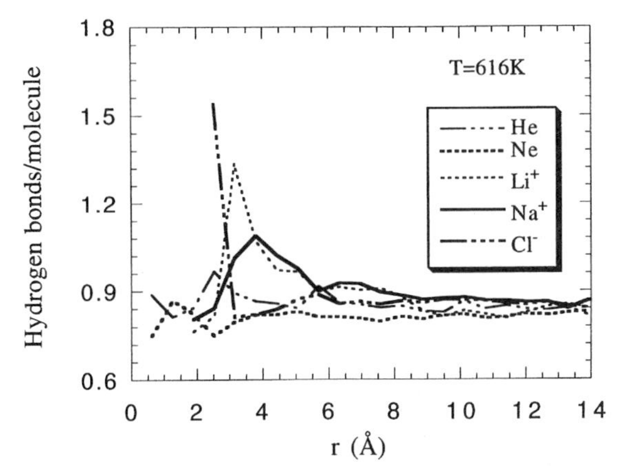

Figure 32. Average number of water molecules hydrogen bonded to a water molecule at a distance r from the solute's center of mass at $T_r = 1.05$ and $\rho_r = 1.0$.

$T_r = 1.05$ and $\rho_r = 1.0$) to ~0.8 ($T_r = 1.0$ and $\rho_r = 1.5$) at all distances (Fig. 18). Moreover, the presence of a volatile solute shows little effect on these numbers at both state conditions. Although the picture stays the same for other volatile and nonvolatile solutes, the (strong) solvation of small ions induces a large distortion of the water structure in the first solvation shell and, consequently, a small but noticeable effect on the degree of hydrogen bonding (see the small peaks in Figs. 31 and 32). Owing to poor sampling, little can be said about the HB in the close vicinity of the solutes (especially the ions).

Driesner and co-workers performed a molecular dynamics study of the hydration and ionic association in the archetypal NaCl aqueous system (based on GROMOS force fields [267]), from infinite dilution to 1 molal and from ambient to supercritical conditions [268]. They concluded that, for infinitely dilute solutions, the hydration number for Na^+ remains approximately constant around 5.5 from ambient to supercritical temperatures along the $\rho = 1.0$ g/cm^3 isochore. For Cl^-, the geometrical definition of the hydration shell becomes difficult, because the location of the first minimum in the $g_{OCl}(r)$ changes dramatically from ambient to supercritical conditions. For the case of 1 molal solutions, the hydration number decreases from 4.9 to 2.8 as a result of strong ion association. For the same reason as for the infinitely dilute case, the hydration number for Cl^- becomes more difficult to define than before. These results highlight once more the drawbacks of a geometric definition of hydration shells and numbers as descriptors of the solvation process and suggest the need for additional or alternative ways to analyze the solvation phenomenon (see Section III.B).

Potential of mean force (PMF) calculations have been frequently used in the study of solvation at ambient and supercritical conditions [217,269–275], especially when researchers are interested in the behavior of solutes at high dilution, conditions at which solute–solute interactions become rare events and, consequently, cannot be accounted for by conventional distribution function calculations. The resulting PMF are in fact free energy profiles, which can be used to determine the corresponding association constants and the (transition state theory) kinetic rate constants governing the conversion of different solute-pair configurations separated by energy barriers.

The two approaches frequently used to determine PMF [i.e., $W(r) = -k_B T \ln g(r)$] are the constraint dynamics route, which allows the direct determination of the solvent mean force on the constrained solute-pair configuration [276], and the thermodynamic free energy perturbation route [277], which directly gives the change of free energy $W(r)$ relative to a reference value at a determined pair distance (typically a few molecular diameters). Although these methods usually appear as two rather different approaches, the latter is actually a special case of the former in that the free energy is determined via a thermodynamic integration (the coupling parameter is simply the radial distance), as opposed to a perturbation approach [278]. In that sense, the method of constraints might exhibit slightly better statistics than the perturbation method, for reasons discussed by Mezei and Beveridge [279]. Yet, if properly performed, both methods must give the same result [275].

The thermodynamic free energy perturbation route was used by Gao [271] in the analysis of the benzene dimer formation in supercritical water at $T = 673$ K and $P = 35$ MPa, via NPT Monte Carlo simulations of TIP4P water model and optimized potentials for liquid simulations (OPLS) benzene. The resulting PMF indicated no local water density increase around benzene, as if benzene behaved as a weakly repulsive solute [241]. In fact, the calculated association (binding) constant was approximately three times smaller than that corresponding to water at ambient conditions.

A similar approach was used by Johnston and co-workers [280–282] who studied the S_N2 reaction of methyl chloride with a chloride ion in supercritical water, via molecular dynamic simulations and continuum electrostatic calculations. They examined the solvation of species along the reaction coordinate through the determination of the free-energy profile (PMF) via a molecular dynamic perturbation approach, in which water was described by the SPC model [100], along three reduced isotherms ($1.0 \leq T_r \leq 1.3$) and within the density range of $0.5 \leq \rho_r \leq 2.0$.

Their simulations [283] provided structural information on the distribution of water around the reacting species, including the transition state complex, such as radial distribution functions, coordination numbers, local densities, and pair distribution energies. Based on this structural information and on the determination of the number of hydrogen bonds between the water and the nucleophile Flanagin and co-workers [283] were able to draw some conclusions about the 9 to 12 orders of magnitude increase of the supercritical reaction rate over that at ambient conditions.

Finally, Johnston and co-workers [280–282] used the simulation results to test a continuum electrostatics model in which the system is represented by a set of discrete charges inside a low dielectric cavity embedded in a polarizable continuum solvent. The solution of the corresponding Poisson's equation was obtained through the program *DelPhi* [284], using two sets of parameters to define the atomic radii and cavities: In the first set, the radii were taken directly from the solute model used in the simulations (OPLS-based parameters), whereas the second set was based on optimized radii to match experimental free-energy data at ambient conditions.

According to their account [282], the continuum model made a reasonable prediction of the free energy along the reaction coordinate, with the exception of the middensity region. The success of the model appears closely tied to the choice of the cavity radius (an already well-documented finding from earlier studies based on Born-type calculations [41,285–287], and the disagreement with simulation is ascribed to the fact that the model cannot account for electrostriction around charged species in the solvent. This comparison highlights an important drawback often found (and overlooked) in the molecular-based study of solutions, in which the source of success or failure cannot be traced to specific hypotheses (approximations) in the theory, because it usually involves too many factors with compensating effects.

In a related work, Johnston and collaborators [288–290] performed molecular dynamic simulations of aqueous electrolyte solutions to determine the solvation Helmholtz free energy of Cl^-, OH^-, Na^+, HCl, and H_2O at ambient and supercritical conditions (up to $T = 673$ K and $\rho = 0.087$ g/cm^3). In addition, they characterized the microstructure of ion–water interactions, in which water was described by the SPC/E model [119], and the force fields for the solutes were taken from the available literature [291–293], based on the corresponding radial distribution functions, coordination numbers, a geometric definition of the hydrogen bond, and the pair energy distributions. According to their analysis, the ions show high coordination numbers even at $T = 673$ K and $\rho = 0.087$ g/cm^3, due to the strong ion–water interactions (not a surprise taking into account that the dielectric screening of the Coulombic interactions is drastically reduced because the dielectric constant at this conditions is approximately 1.5 [4]). This is not the case for the neutral HCl solute, whose coordination number decreases dramatically from ambient to supercritical conditions (note, however, that this calculation was based on the coordination with the Cl site of the molecule, not the center of mass).

In a subsequent paper, Johnston and collaborators [289] studied the acidity of HCl relative to H_2O, by a hybrid route based on thermodynamic cycles, free-energy calculations from molecular simulation, and experimental data for the change of free energy between the acid and water dissociation in the gas phase:

$$\begin{aligned}
&\Delta G^{\text{diss}}_{\text{HCl}}(aq, T, P) - \Delta G^{\text{diss}}_{\text{H}_2\text{O}}(aq, T, P) = 2.3K_B T(pK_a - pK_w) \\
&\quad = [\Delta G^{\text{diss}}_{\text{HCl}}(g, T, P \to 0) - \Delta G^{\text{diss}}_{\text{H}_2\text{O}}(g, T, P \to 0)] \\
&\qquad + [\Delta G^{\text{solv}}_{\text{Cl}^-}(g, T, P) - \Delta G^{\text{solv}}_{\text{OH}^-}(g, T, P)] - [\Delta G^{\text{solv}}_{\text{HCl}}(g, T, P) - \Delta G^{\text{solv}}_{\text{H}_2\text{O}}(g, T, P)]
\end{aligned} \tag{3.1}$$

where $\Delta G^{\text{diss}}_{\text{HCl}}(aq, T, P) = 2.3k_\text{B}T\ \text{pK}_\text{a}$, and $\Delta G^{\text{diss}}_{\text{H}_2\text{O}}(aq, T, P) = 2.3k_\text{B}T\ \text{pK}_\text{w}$. Thus the first line of this equation gives the HCl acidity with respect to that of H_2O (the solvent) at the same state conditions, which is calculated by using experimental data for the second line, and free energy simulation calculations for the two mutation processes indicated in the third line of Eq. (3.1).

Unfortunately, the molecular dynamic free-energy perturbation calculations for the mutation were performed in the canonical rather than the isobaric–isothermal ensemble as required by Eq. (3.1), i.e., the calculated and reported quantities were $\Delta A^{\text{diss}}_U(aq, \rho, T)$ and $\Delta A^{\text{solv}}_U(aq, \rho, T)$ rather than $\Delta G^{\text{diss}}_U(aq, P, T)$ and $\Delta G^{\text{solv}}_U(aq, P, T)$ for the solute U. In addition, the mixing of experimental data with simulation results toward the final determination of a free-energy change [Eq. (3.1)] masks the intrinsic value of the final results, in that real and SPC/E water do not behave equally at any (T, P) or (T, ρ) conditions, especially in the highly compressible near-critical and supercritical regions [124,132]; consequently, those results may be compromised by fortuitous cancellation or magnification of inaccuracies.

a. Ion Speciation in High-Temperature Electrolyte Solutions. Speciation at high temperature has been typically analyzed through conductimetric measurements in dilute aqueous alkali halides [294–296] by invoking the definition of the association constant [48]. Even though a conductance measurement at near-critical conditions is a formidable task, this technique provides a direct measure of the degree of association as well as useful information for rationalizing experimental data from volatility [297], isopiestic [298], and calorimetric [48,299] measurements. An important drawback of all these techniques is their inability to offer any insight into the solution structure and dynamics of the ion pairs, i.e., they render little detail on the mechanism of the speciation process. This shortcoming suggests that, although supercritical fluids are widely used in a variety of processes, a molecular-based understanding of the mechanism underlying supercritical solvation (especially for aqueous–electrolyte solutions) is still in its infancy [10,300].

In search for that understanding, Cui and Harris [217] pioneered the study of ion association in dilute supercritical NaCl aqueous solutions, through PMF calculations by constraint dynamics, and the characterization of the resulting microstructure at the extreme conditions typically found in supercritical water oxidation reactors [38], i.e., $700 \leq T \leq 1000$ K and $P \approx 25$ MPa ($\rho \leq 0.1$ g/cm^3). Using the SPC water model, the Pettitt–Rossky model for the ion–water interactions [301], and the Fumi–Tosi–Huggins–Mayer model for the ion–ion interactions [302,303], Cui and Harris performed molecular dynamics simulations to determine the association constant of NaCl and the detailed analysis of the water structure around the ions, including the radial dependence of the corresponding coordination numbers. Their comparison of simulation results and extrapolated association constants [294] indicated that the models underpredict the extrapolated values from conductimetric measurements by one to two orders of magnitude.

In addition, Cui and Harris decomposed the PMF into its energetic and entropic contributions, i.e., $W(r) = E(r) - TS(r)$, to assess their relative role in the association process. With that purpose they carried out the following splitting:

$$E(r) = U(r) + [E(r_o) - U(r_o)] \tag{3.2}$$

where $U(r)$ is the total potential energy for the ion–ion interaction for a separation r, and r_o is the separation at which the PMF is anchored to the reference $W(r_o)$. Note that Eq. (3.2) suggests that $E(r) - U(r) =$ a constant that is independent of r, an assumption that may be unwarranted, especially for pair configurations at small distances.

In a subsequent work, Cui and Harris [304] studied the solubility of NaCl in supercritical water, as described in their previous work, in the range of $723 \leq T \leq 823$ K and $10 \leq P \leq 30$ MPa. To achieve that, they used Kirkwood coupling parameter integration supplemented by Widom's test particle method [262] for the determination of the residual chemical potentials for the species in solution, which were taken to be neutral ion pairs (dumbbells), and the quasiharmonic approximation for the determination of the residual chemical potential of the NaCl in the solid phase. The solubility results from the free-energy calculations were in remarkably good agreement with experimental data [305] and indicated that simple approaches in which the ions are treated as hard spheres in a continuum dielectric are not able to predict the observed behavior because they neglect the pronounced structural changes of the solvent around the charged solutes.

As part of their molecular-based investigation of solvation at supercritical conditions, Chialvo et al. [273] performed molecular dynamic simulations of supercritical electrolyte solutions with three different ion–water models to determine the association constant for the ion pair Na^+–Cl^- and the constant of equilibrium between the solvent-separated and the contact ion pairs at a near critical state condition. According to the chemical equilibrium between the free anion A, cation C and the possible associated neutral pair AC configurations in a dilute aqueous electrolyte solution of a 1 : 1 salt (for which we can assume that the association process does not involve neutral polyionic clusters [306]) the following multistep mechanism [307] can be assumed:

$$\begin{array}{ccccccc} \text{A} + \text{C} & \overset{K_1}{\Leftrightarrow} & \text{AC} & \overset{K_2}{\Leftrightarrow} & \text{A}|\text{C} & \overset{K_3}{\Leftrightarrow} & \text{A}\|\text{C} \\ \text{free ions} & & \text{CIP} & & \text{SSHIP} & & \text{SSIP} \end{array} \tag{3.3}$$

where SSIP is solvent-separated ion pairs, SSHIP is solvent-shaped ion pairs, and CIP is contact ion pairs. Thus the equilibrium association constant can be expressed as

$$K_a(T, \rho) = K_1(1 + K_2 + K_2K_3) \cong K_1(1 + K_2) \tag{3.4}$$

which can be microscopically interpreted as the following simple integral [273]:

$$K_a = 4\pi \int_0^{r_2} g_{\mathrm{AC}}^{\infty}(r) r^2 \, dr \tag{3.5}$$

where ∞ denotes an infinite dilution property, and r_2 indicates the position of the second valley of the $g_{\mathrm{AC}}^{\infty}(r)$ (i.e., it encompasses the first and second peak, or the CIP and SSHIP configurations). In addition, within the range of dilution in which the involved activity coefficients can be taken as effectively unity [273], the degree of dissociation α becomes

$$\alpha = (\tfrac{1}{2} c_o K_a)[(1 + 4c_o K_a)^{0.5} - 1] \tag{3.6}$$

Similarly, the constant of equilibrium between the SSHIP and CIP species, usually referred to as K_e, can be determined as

$$K_e = K_2 = \int_{r_1}^{r_2} g_{\mathrm{AC}}^{\infty}(r) r^2 \, dr \Big/ \int_0^{r_1} g_{\mathrm{AC}}^{\infty}(r) r^2 \, dr \tag{3.7}$$

where r_i denotes the location of the first valley of the $g_{\mathrm{AC}}^{\infty}(r)$.

Therefore, the evaluation of the association and equilibrium constants via molecular simulation hinges on the nontrivial determination of the anion–cation radial distribution function at infinite dilution as discussed elsewhere [273]. The authors assessed the realism of three models by comparing the simulated association constants with those from high-temperature conductance measurements [295]. They noted that before drawing any conclusions from the comparison of the simulation results and experimental values of the association constant, it is necessary to assess the impact and validity of the assumptions involved in the working expressions, especially that associated with the reference PMF. Typically, this reference is approximated by the dielectric attenuated electrostatic ion–ion interaction, i.e., $W(r_o) \approx -q^2/r_o\varepsilon$ [308,309]. In fact, because this reference term contributes exponentially to the expressions for the association constant (through the corresponding radial distribution function), any uncertainty in the dielectric constant will be greatly magnified in the determination of the association constant. This is precisely the case of using, as a common practice, the experimental value of the dielectric constant, instead of that corresponding to the model solvent.

According to the simulation results of Chialvo et al. [224,273,310] (along the reduced isotherm of $T_r = 1.05$ in the density range of $1.0 \leq \rho_r \leq 2.0$, and along the reduced density of $\rho_r = 1.5$ within the temperature range of $1.05 \leq T_r \leq 1.4$), temperature and density exhibit opposite (compensating)

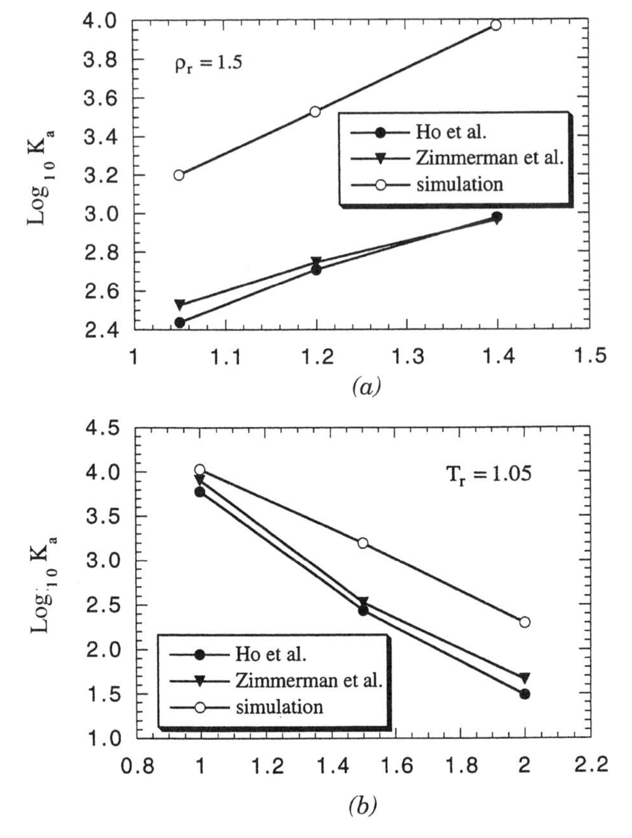

Figure 33. Simulation results for the isothermal density dependence, reduced temperature (*a*) and isochoric temperature dependence, reduced density (*b*) of the ion pair association constant compared to the corresponding experimental data from conductance experiments of Ho and co-workers [295] and Zimmerman and co-workers [296].

effects on K_a (Fig. 33), similar to the experimentally observed behavior from conductimetric measurements [295,296].

A few points about these results are worth highlighting. Note the strong isothermal (near-critical) density and isochoric (supercritical) temperature effects on the degree of dissociation (Fig. 34). This steep increase of the dissociation at a constant total salt concentration c_o is a clear manifestation of the dielectric screening effect of the solvent, because the solvent's dielectric constant shows a threefold increase as the density doubles. Furthermore, the isochoric temperature effect on the degree of dissociation is also pronounced, even though the range of temperature is much smaller. Based on the determination of the constant of equilibrium K_e between CIP and SSSIP, the con-

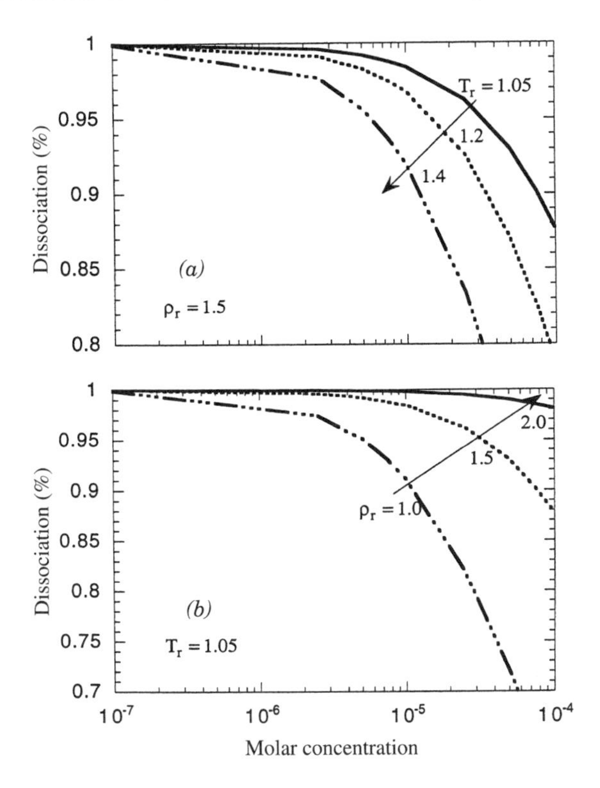

Figure 34. (*a*) Isochoric temperature and (*b*) isothermal density effects on the degree of dissociation as a function of the salt concentration.

centration of CIP decreases by only 10% when the solvent density is increased by a factor of 2, from the critical value, along an isothermal path [Fig. 35(*a*)]. This contrasts with the 25% increase when the temperature is increased by a factor of 1.4, from the critical value, along the isochoric path. In fact, at $T_r = 1.4$ and $\rho_r = 1.5$, ~95% of all ion pairs are in the CIP configuration [Fig. 35(*b*)]. This behavior contrasts with that at ambient conditions, at which water dissociates more than 99% of the ion pairs [311], and it indicates that water is still a powerful solvation agent at these near critical conditions.

b. *Solvent Properties in the Vicinity of an Ion.* A key ingredient in the solvation of ions is the ability of the solvent to attenuate the long-range electrostatic interactions through dielectric screening. In other words, the solvation power of a solvent can in principle be tailored through its dielectric constant by

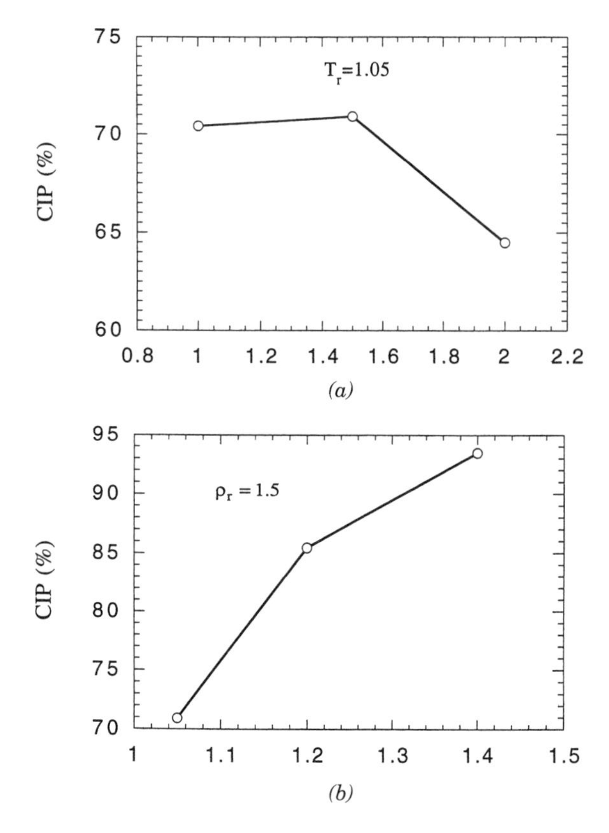

Figure 35. Density and temperature effects on the relative concentration of ion pairs in the CIP configuration along (*a*) at reduced density, the isotherm $T_r = 1.05$, and (*b*) at reduced temperature, the isochore $\rho_r = 1.5$.

manipulating the solvent's state conditions. This possibility is more dramatically manifested in the highly compressible region of the solvent phase diagram in which small isothermal pressure perturbations give rise to large density changes and, concomitantly, to large gradients of electric field and dielectric screening.

Although the above argument might describe an accurate picture for a solvent that behaves as a continuum dielectric, the specific short-range repulsive and dispersive ion–ion and ion–solvent interactions in real systems induce pronounced changes in the solvent's properties, especially in the vicinity of the charged species and in the strength of the dielectric screening (e.g., through electrostriction and dielectric saturation). Specifically, the large gradients in the solvent's electric field associated with the enhanced local density surrounding the ions become the dominant factors governing the local behavior of the solvent's properties and render the continuum dielectric as limited help for additional insights [41].

A great deal of effort has been made to determine the solvation (hydration) structure of ions (i.e., the solvent structure in their vicinity) from a variety of spectroscopic techniques (NMR, XAFS [312–314], Mössbauer, IR, Raman [315]), scattering techniques (X-rays, electron and neutron diffraction [315,316]), electrochemical techniques [8,42], and simulation methods [315]. The rationale behind these studies hinges on the idea that a realistic description of the thermophysical properties of electrolyte solutions must take into account the ion-induced local distortion of the solvent properties, i.e., it should go beyond the so-called continuum or primitive models. The challenge resides in our ability to probe the properties of the solvent in the vicinity of ions, and then to make explicit contact with meaningful solvation-related macroscopic properties.

Although the local-density perturbation is common to electrolyte and nonelectrolyte solutions, it becomes rather large for ion solvation, because of the resultant increased dipole concentration in the region of highest field strength. Typically, the perturbation process takes place within the first hydration shells [42] and affects the solvent's electric field and dielectric behavior. Perhaps for this reason great attention has been focused on the behavior of the first solvation shell of species in solution, determined either by experimental or theoretical means [312–314,317]. Yet molecular simulation offers the best tool for probing the properties of the solvent (water) in the vicinity of the ions, because it provides total control and full manipulation of the variables involved and allows us to make precise cause and effect connections.

An example of this type of approach is the study performed by Chialvo and co-workers [318] on the water local environment around species in the archetypal infinitely dilute NaCl aqueous solution (i.e., Na^+, Cl^-, and H_2O) at supercritical conditions. The authors determined the water properties within the ion's hydration shells and compared them to the average local environment around any water molecule. These properties included the radial profiles for the water local density (and the resulting coordination numbers), the local pressure, the local electric field, and a measure of the local dielectric constant around the species Na^+, Cl^-, and H_2O. Their simulation results (Figs. 36–39) provide rather compelling evidence for the pronounced distortion of the water properties around the species in solution and suggest the need for unambiguous molecular-based interpretation of the local properties, their connection to experimentally measured quantities, and their usefulness as descriptors of nonidealities for modeling purposes.

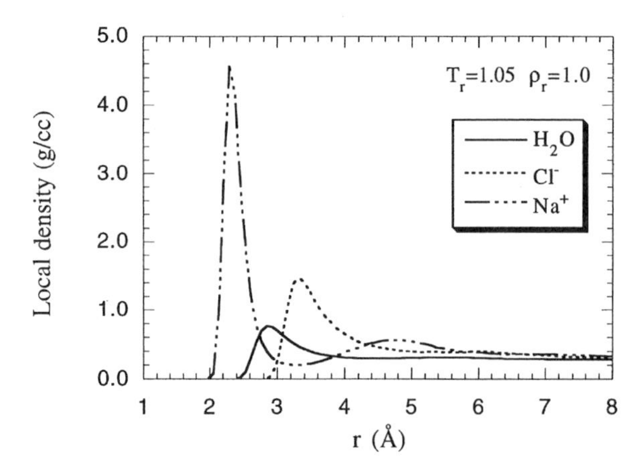

Figure 36. Comparison of local density profiles around species in solution for an infinite dilute NaCl aqueous solution at $T_r = 1.05$ and $\rho_r = 1.0$.

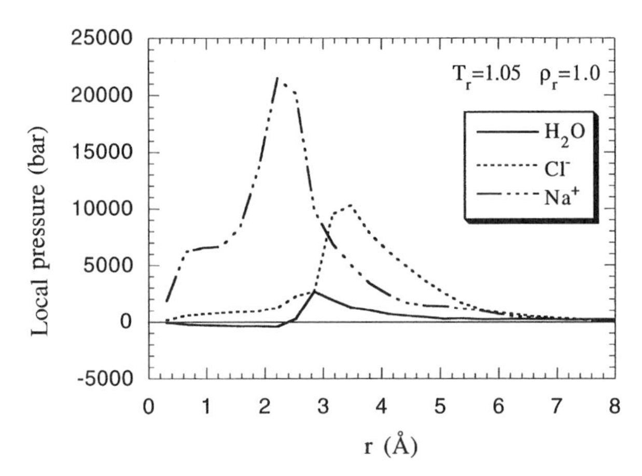

Figure 37. Comparison of local pressure profiles around species in solution for an infinite dilute NaCl aqueous solution at $T_r = 1.05$ and $\rho_r = 1.0$.

In particular, owing to the large electric field gradient in the vicinity of the ions (Fig. 38), resulting in exceptionally compact first solvation shells at high temperature (Fig. 40), the absolute coordination numbers exhibit weak dependence on state conditions, rendering them poor descriptors of ion solvation. For precisely the same reasons, the simulation results (Fig. 38) highlight the failure of the Born equation to describe the electric field (and derived properties) in the vicinity of a charged species, i.e., within

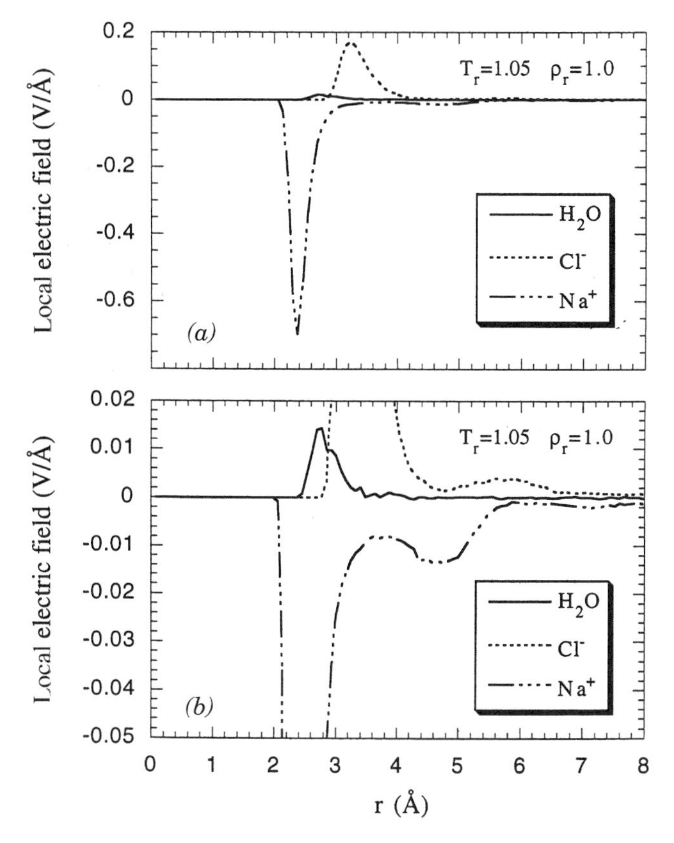

Figure 38. Comparison of local electric field profiles (*a*) around species in solution for an infinite dilute NaCl aqueous solution at $T_r = 1.05$ and $\rho_r = 1.0$ and (*b*) at magnification.

the first two solvation shells and the need for a spatially dependent dielectric solvent permittivity [40,41,287].

The behavior of ions in solutions has been traditionally associated with the idea of coordination (solvation in general, or hydration in particular) number [222,319], the quantities that played a significant role in theoretical discussions concerning the short- and long-range interactions of the ions with the solvent [42,319,320]. The usefulness of the coordination numbers as descriptors of ionic behavior in solution is rather controversial [42,222,315,319,321], and the simulation results of Chialvo and co-workers [318] cast additional doubts on the application of these quantities at supercritical conditions and point again to the need for a more appropriate way to describe the behavior of electrolyte solutions on a molecular-based level.

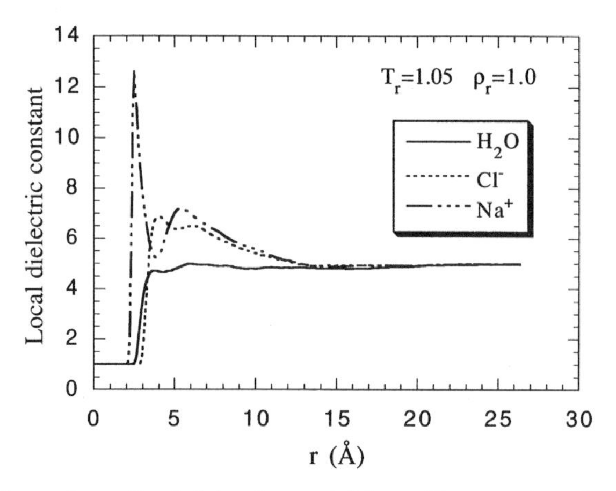

Figure 39. Comparison of local dielectric permittivity profiles around species in solution for an infinite dilute NaCl aqueous solution at $T_r = 1.05$ and $\rho_r = 1.0$.

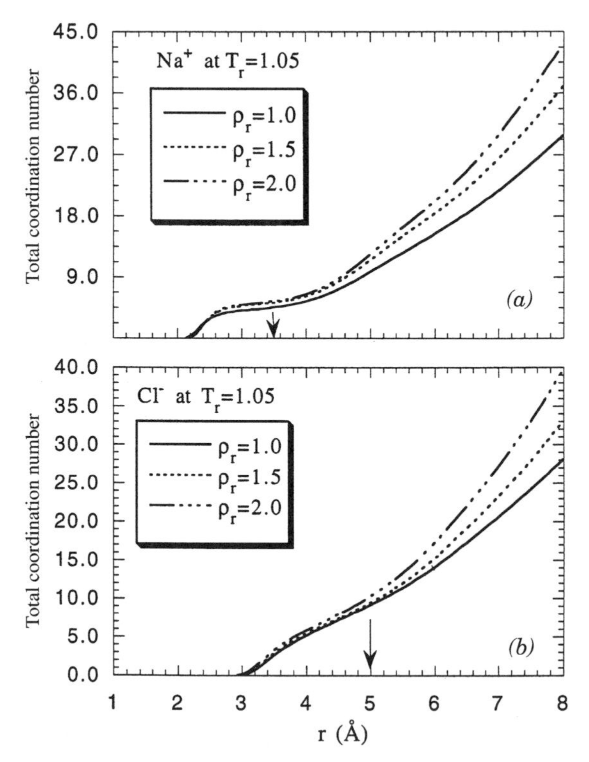

Figure 40. Radial profile of coordination numbers around species in solution for an infinite dilute NaCl aqueous solution at $T_r = 1.05$ and $\rho_r = 1.0$: (*a*) cation and (*b*) anion. *Arrows* indicate the radius of the first solvation shell.

B. Solvation Formalism

The main thrust behind these developments is the formulation of unambiguous statistical mechanical connections between the structural changes of the near-critical solvent around dilute species and the corresponding macroscopic analogues. The resulting formalism involves a clear-cut separation between the structural changes (short-ranged phenomena) associated with the solvation driving force (difference of chemical potentials) and the coexistent slow-decaying structural correlation (long-range) associated with the divergent solvent isothermal compressibility. In fact, the formalism differentiates the local density perturbation (i.e., the solvent's microstructural changes, resulting from mutating a solvent particle by a solute particle and occurring within a few molecular diameters around the solute molecule) from its propagating effect (up to a distance given by the solvent's correlation length), which diverges at the solvent's critical point.

1. *Solvation Thermodynamics*

We have recently developed a statistical mechanical formalism to accomplish the separation between solvation and compressibility driven phenomena, based on the natural splitting of the total correlation functions into their corresponding direct and indirect contributions [50,235], according to the Ornstein–Zernike (OZ) equation [6]. The derived formalism was based on the Kirkwood–Buff fluctuation theory of nonelectrolyte mixtures [322], which defines exact relations between the microscopic details of the system and integrals over its microstructure characterized by the pair correlation functions. The formalism was successfully applied to study the microscopic mechanism of supercritical solubility enhancement in nonelectrolyte solutions [50], to interpret gas solubility [238], and to determine the solvation effects on the reaction kinetic rate [266] and other solvation effects [323].

While Kirkwood–Buff's fluctuation formalism can be straightforwardly applied to nonelectrolyte mixtures [324–327], its formal application to electrolyte systems is not trivial (in that individual ions must obey the electroneutrality condition, i.e., their concentration in solution cannot be changed independently), because correlation functions between dependent species render indeterminate all thermodynamic properties [328,329].

An analogous approach to the solvation quantities derived for nonelectrolyte systems [50,235], based on Kusalik and Patey's version of the Kirkwood–Buff fluctuation theory of mixtures [328], was developed [51] to make explicit contact not only between the solvation structure of individual ions and its corresponding macroscopic properties but also between the individual ions' and the salts' properties without invoking any extrathermodynamic assumption [222,330].

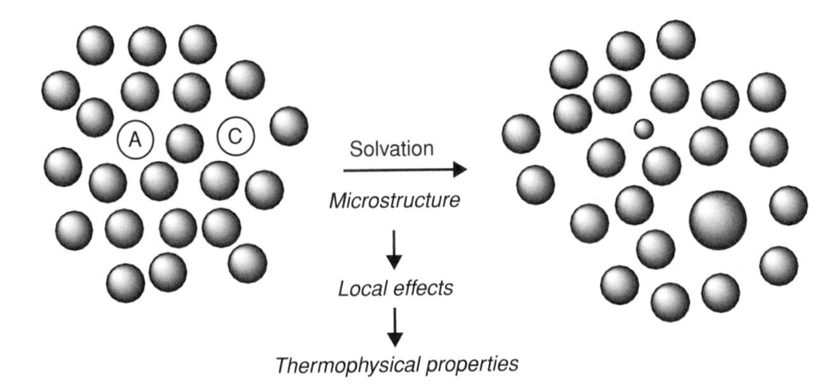

Figure 41. Representation of the solvation process. On the left, A and C are two solvent molecules that will be mutated to the anion and cation in solution, respectively. On the left is pure solvent, "ideal solution;" on the right is infinitely dilute solution, "nonideal solution."

Essentially, the formalism is based on the imaginary process that takes place when a solvent particle is mutated to a solute one (i.e., according to its residual properties, the initial system is considered to be an ideal solution in which the solute is initially a solvent molecule that differs from the others by its label only [331]), in terms of well-defined short-range quantities that are fundamentally connected to the system's macroscopic properties (Fig. 41). The driving force for this thought experiment can be envisaged as the difference of residual chemical potentials between the infinitely dilute solute and the pure solvent [50,235]:

$$\mu_U^{r\infty}(T,P) - \mu_V^{ro}(T,P) = \int_0^{\rho(P)} (\partial P/\partial x_U)_{\rho,T}^{\infty} \frac{d\rho}{\rho^2} = k_B T \ln[H_U(T,P)/f_V^o(T,P)] \tag{3.8}$$

where k_B is Boltzmann's constant, T is the absolute temperature, H_U is the solute's Henry constant, and f_V^o is the pure solvent's fugacity. Subscripts U and V denote solute and solvent species, respectively. Superscripts O and ∞ indicate pure component and infinitely dilute conditions, respectively. By expressing the integrand of Eq. (3.8) in terms of the solute–solvent and solvent–solvent (either total or direct) correlation function integrals [50,235], we relate the solvation phenomenon (microscopic solvent structural changes around solutes) and the (macroscopic) change of pressure on addition of an infinitely dilute solute at constant density and temperature:

$$\left(\frac{\partial P}{\partial x_U}\right)_{\rho,T}^{\infty} = \frac{\rho}{\beta}(C_{VV}^o - C_{UV}^{\infty}) = \frac{\rho^2 k_B T(G_{VV}^o - G_{UV}^{\infty})}{1+\rho G_{VV}^o} \tag{3.9}$$

where the total correlation function integrals (TCFI) are given by [332]

$$G_{ij} = \int h_{ij}(r)\, d\mathbf{r} = \hat{h}_{ij}(\mathbf{0}) \tag{3.10}$$

and the direct correlation function integrals (DCFI) defined as [325,329]

$$C_{ij} = \rho \int c_{ij}(r)\, d\mathbf{r} = \rho \hat{c}_{ij}(\mathbf{0}) \tag{3.11}$$

where $\hat{h}_{ij}(\boldsymbol{k})$ is the Fourier transform of the total ij pair correlation function, $h_{ij}(r) \equiv [g_{ij}(r) - 1]$, $g_{ij}(r) = \langle g_{ij}(r_{ij}, \omega_i, \omega_j) \rangle_{\omega_i, \omega_j}$ is the spatial (unweighted angle average) center-to-center ij pair distribution function, and $\hat{c}_{ij}(\boldsymbol{k})$ is the Fourier transform of the direct ij pair correlation function. Note that at the solvent's critical conditions, this pressure derivative becomes Krichevskii's parameter, a finite quantity that plays an important role in describing near critical solvation phenomena, such as the supercritical solubility enhancement [50], gas solubility [238], mixed-solute and cosolvent effects [216], and solvent effects on kinetic rate constant of reactions at supercritical conditions [266]. For electrolyte solutions Eq. 3.9 reads [51]

$$\left(\frac{\partial P}{\partial x_U}\right)_{\rho,T}^{\infty} = \frac{\nu\rho}{\beta}(C_{VV}^{o} - C_{UV}^{\infty}) \tag{3.12}$$

where $\nu = \nu_+ + \nu_-$ is the sum of the stoichiometric coefficients of the ion pair and $C_{uv}^{\infty} = (\nu_+ C_{+v}^{\infty} + \nu_- C_{-v}^{\infty})/\nu$ [328], so that we can also write the corresponding expression for the individual ions [51]:

$$\left(\frac{\partial P}{\partial x_i}\right)_{\rho,T}^{\infty} = \frac{\rho}{\beta}(C_{VV}^{o} - C_{iV}^{\infty}), \qquad i = +, - \tag{3.13}$$

In addition, $(\partial P/\partial x_U)_{\rho,T}^{\infty}$ can be macroscopically interpreted in terms of a difference of partial molar volumes:

$$\left(\frac{\partial(\mu_U^{r\infty}(T,P) - \mu_V^{ro}(T,P))}{k_B T\, \partial \rho}\right)_T = \frac{1}{k_B T \rho^2}\left(\frac{\partial P}{\partial x_U}\right)_{\rho,T}^{\infty} = (\bar{v}_U^{\infty}(SR) - \bar{v}_V^{o}) \tag{3.14}$$

where $\bar{v}_V^{o} = \rho_V^{-1}$, which becomes the formal definition of the short-range finite (solvation) contribution to the solute partial molar volume at infinite dilution. Thus the solute partial molar properties at infinite dilution can be split into the solvation (finite) and compressibility driven (divergent) contributions.

For example, for the partial molar volume $\bar{v}_U^\infty$ of a nonelectrolyte we have

$$\begin{aligned}\bar{v}_U^\infty &= \bar{v}_V^o + G_{VV}^o - G_{UV}^\infty = [(1 - C_{UV}^\infty)/\rho(1 - C_{VV}^o)] \\ &= \underbrace{\bar{v}_V^o(1 + C_{VV}^o - C_{UV}^\infty)}_{\bar{v}_U^\infty(\mathrm{SR})} + \underbrace{[C_{VV}^o(C_{VV}^o - C_{UV}^\infty)/\rho(1 - C_{VV}^o)]}_{\bar{v}_U^\infty(\mathrm{LR})}\end{aligned} \tag{3.15}$$

Note that, as required by its physical meaning, the solute-induced effect (upon adding an infinitely-dilute solute) on the system's microstructure and thermodynamics approaches zero as the mixture approaches ideality, i.e., as the solute molecule becomes a solvent molecule. Furthermore, this splitting also allows us to define and analyze the entire solvation process without invoking the compressibility driven phenomena, i.e., simply in terms of $(\partial P/\partial x_U)_{\rho,T}^\infty$ and its temperature derivative so that

$$\begin{aligned}\Delta G_{\mathrm{solvation}} &= [\bar{h}_U^{r\infty}(\mathrm{SR}) - h_V^{ro}]_{P,T} - T[\bar{s}_U^{r\infty}(\mathrm{SR}) - s_V^{ro}]_{P,T} \\ &= \mu_U^{r\infty}(T, P) - \mu_V^{ro}(T, P) = \mu_U^{*ro}(T, \rho) - \mu_V^{*ro}(T, \rho)\end{aligned} \tag{3.16}$$

where the enthalpic and entropic terms in parentheses have been derived elsewhere [50,51], and the asterisks indicate that the residual properties are defined at constant density and temperature (as opposed to pressure and temperature). Therefore, the key to the understanding of the solvation phenomenon in dilute supercritical solutions is the study of the microstructural solvent changes around the infinite dilute solute, whose macroscopic manifestation is $(\partial P/\partial x_U)_{\rho,T}^\infty$. The case of electrolyte solutions is developed in detail in Ref. [51].

According to this formalism, the solute-induced effect on the solvent structure can be studied through the definition of the excess quantity $N_{\mathrm{ex}}^\infty(R)$ [51,235],

$$N_{\mathrm{ex}}^\infty(R) = 4\pi\rho \int_0^R (g_{UV}^\infty(r) - g_{VV}^o(r))r^2\, dr \tag{3.17}$$

which measures the excess number of solvent molecules from what we would obtain if the solute molecule behaved as another solvent molecule (ideal solution), where R denotes the radius of the first few solvation shells. This removes from the definition of $N_{\mathrm{ex}}^\infty(R)$ the buildup of solvent molecules that would take place in the absence of the solute (better yet, when the solute is just another solvent molecule) primarily due to the diverging correlation length as the solvent approaches its critical point. Note that the sign of $N_{\mathrm{ex}}^\infty(R \to \infty) \equiv N_{\mathrm{ex}}^\infty$ is intimately connected to the sign of the partial molar volume of the solute at infinite dilution $\bar{v}_U^\infty$:

$$N_{\text{ex}}^\infty = -\kappa\left(\frac{\partial P}{\partial x_U}\right)_{\rho,T}^\infty = v - \rho\bar{v}_u^\infty \qquad (3.18)$$

and, therefore, to the corresponding Krichevskii's parameter. Thus the behavior of the supercritical solute can be characterized entirely by the sign of N_{ex}^∞ or $(\partial P/\partial x_U)_{\rho,T}^\infty$. In fact, solutes in supercritical fluids can be classified into nonvolatile [240] (or attractive [241]) if $(\partial P/\partial x_U)_{\rho,T}^\infty$, corresponding to $N_{\text{ex}}^\infty > 0$, and volatile [240] (or repulsive [241]) if $(\partial P/\partial x_U)_{\rho,T}^\infty > 0$, corresponding to $N_{\text{ex}}^\infty < 0$.

2. *Solvation Effects on Kinetic Rate Constants*

Solvation affects the solute reaction rates by altering the free energy activation profile along the solute's reaction path and the free energy barrier $\Delta G^\ddagger$ that goes exponentially in the definition of the reaction rate constant according to the transition state theory (TST) formalism [7]. The use of continuum models in solvation studies has become an attractive alternative to molecular simulation of aqueous electrolyte solutions, at a fraction of the computational effort. Continuum methods, however, are not without defects, starting with the intrinsic inability to describe the solvent structure and any solvation property associated with the restructuring of the solvent around the species in solution (including electrostriction and dielectric saturation).

If we consider the Born equation [222] as the quintessential continuum model for electrolyte systems, we could say that all contemporaneous continuum models have evolved from it, by introducing corrections to mimic the effects of the structuring of the solvent around species. For the particular case of near-critical and supercritical aqueous solutions, it was first recognized that the solvent compressibility must be accounted for in one way or another if the model is expected to show some realism in describing solvation at those conditions [40,287]. The failure of these incompressible continuum models to correctly describe the reaction path energetics in supercritical water was demonstrated by Tucker and Gibbons [333] for the hydrolysis of anisole, and resulted in the development of a compressible continuum solvation model for arbitrary charge distributions, which can predict electrostriction effects [334–336]. The first comparison between molecular simulation results and model predictions for the chloride plus methyl chloride reaction in supercritical water [337] has given encouraging results on the ability of the compressible continuum model to assess the solvent effect on the kinetic rate constants of reactions in supercritical water, even though the model contains no explicit solvent structural details. Thus, although more realistic than the noncompressible conterparts, the more accurate results from compressible models can be attributed in part to the fortuitous cancellation of other (neglected) molecular effects along the reaction path [219].

The dynamics of chemical reactions depend on the fluctuation of relevant quantities away from equilibrium, and the reaction kinetic rate is controlled by the corresponding fluctuation decay toward equilibrium. It is as yet unclear what the effects of the solvent might be on the dynamics of reactions at near criticality, i.e., when density and/or composition fluctuations become large. To gain some understanding on those effects for dynamic processes involving the crossing of an energy barrier, we invoke the TST [7], which involves only equilibrium and quasi-equilibrium quantities, i.e., it does not require any detailed description of the fluctuations themselves. Therefore, we can analyze the near-critical and supercritical solvent effects on the kinetic rate constants in terms of the equilibrium density distributions (and associated solvation quantities). To do so we apply the solvation formalism discussed earlier to the thermodynamic pressure effect on the kinetic reaction rate constant.

Let us consider the following simple reaction,

$$\nu_R R + \nu_S S \overset{\text{solvent}}{\rightleftharpoons} \mathfrak{R}^{\ddagger} \rightarrow \text{products} \tag{3.19}$$

where R is a reactive solute, S is a reactive cosolvent, $\mathfrak{R}^{\ddagger}$ is the activated complex in equilibrium with the reactants in solution, and the prefactors are the stoichiometry coefficients. According to the TST [7,338], the reaction rate constant is given by

$$k_{\mathrm{B}}^{\mathrm{TST}} = \frac{k_{\mathrm{B}} T K_o}{\hbar} K_c^{\ddagger} \tag{3.20}$$

where k_B and $\hbar$ are Boltzmann and Planck constants, respectively, T is the absolute temperature, K_o is a factor that gives the correct units for $k_{\mathrm{B}}^{\mathrm{TST}}$, and $K_c^{\ddagger}$ is the molar-based equilibrium constant for the activation process given by Eq. (3.19). To study the pressure effect on the kinetic rate constant we first rewrite $K_c^{\ddagger}$ in terms of the species fugacities, and then determine the pressure coefficient of $\ln k_{\mathrm{B}}^{\mathrm{TST}}$ [266]:

$$\left(\frac{\partial \ln k_{\mathrm{B}}^{\mathrm{TST}}}{\partial P}\right)_{T,x} = -\Delta\nu\kappa + (\nu_R \bar{v}_R^{\infty} + \nu_S \bar{v}_S^{\infty} - \nu_{\mathfrak{R}} \bar{v}_{\mathfrak{R}}^{\infty})/k_B T \tag{3.21}$$

where $\Delta v^{\ddagger} = \nu_R \bar{v}_R^{\infty} + \nu_S \bar{v}_S^{\infty} - \nu_{\mathfrak{R}} \bar{v}_{\mathfrak{R}}^{\infty}$ is the actual activation volume [339].

To introduce the solvation contributions into the pressure (or temperature) derivative of the TST kinetic rate constant at a microscopic level, the partial molar volumes involved in Eq. (3.21) are expressed in terms of the derivative

$(\partial P/\partial x_U)^{\infty}_{T,\rho,x_{k\neq U}}$. This derivative is an appealing quantity, because it measures the isothermal–isochoric finite pressure change induced by the solvent rearrangement in the solvation of an infinitely dilute solute and allows us to make an explicit separation between solvation and compressibility driven contributions to the species' partial molar volumes at infinite dilution. In fact, Eq. (3.20) can be rewritten as follows [266]:

$$\left(\frac{\partial \ln k_{\mathrm{B}}^{\mathrm{TST}}}{\partial P}\right)_{T,x} = \kappa\rho[\nu_R(\bar{v}_R^{\infty}(\mathrm{SR}) - \bar{v}_R^{\infty(IG)}(\mathrm{SR})) + \nu_S(\bar{v}_S^{\infty}(\mathrm{SR}) - \bar{v}_S^{\infty(IG)}(\mathrm{SR})) - \nu_{\Re}(\bar{v}_{\Re}^{\infty}(\mathrm{SR}) - \bar{v}_{\Re}^{\infty(IG)}(\mathrm{SR}))] \tag{3.22}$$

where $\rho = (1/\bar{v}_V^o)$, the superscript IG denotes ideal gas properties, and the solvation contribution to the partial molar volumes $\bar{v}_U^{\infty}(\mathrm{SR})$ can be interpreted microscopically in terms of either total or direct correlation function integrals, as discussed earlier. Consequently, the pressure coefficient of the TST rate constant can be factorized into a term involving the solvent's isothermal compressibility and another term containing short-range solvation contributions to the species in solutions relative to their ideal gas contributions (absence of solute–solvent interactions).

Note that, because the difference $(\bar{v}_U^{\infty}(\mathrm{SR}) - \bar{v}_U^{\infty(IG)}(\mathrm{SR}))$ subtracts the translational or ideal gas contribution from the $\bar{v}_U^{\infty}(\mathrm{SR})$, this term contains only contributions from intermolecular interactions. For reactions with $\Delta\nu = 0$ the ideal gas contributions cancel out, and the pressure effect results from the solvation contribution to the activation volume $[\nu_R\bar{v}_R^{\infty}(\mathrm{SR}) + \nu_S\bar{v}_S^{\infty}(\mathrm{SR}) - \nu_{\Re}\bar{v}_{\Re}^{\infty}(\mathrm{SR})]$, which is magnified by the solvent's isothermal compressibility.

According to Eq. (3.22), the isothermal thermodynamic pressure effect becomes predominant at near-critical conditions, unless the coefficient of the isothermal compressibility, the solvation factor, is zero (a truly unlikely situation). Because $\rho\bar{v}_U^{\infty}(\mathrm{SR})$ is a well-behaved finite quantity [50,235] and thermodynamic stability demands that the solvent isothermal compressibility be positive, the solvation factor determines the sign of the pressure effect at near-critical conditions. This prefactor is a function of the relative molecular asymmetry between the solute reactant and the activated complex (as well as the solvent, should this participate in the reaction).

The most critical aspect of the TST approach is to ascribe *a priori* some molecular asymmetry to the activated complex [340]. It is also clear that the activated complex is not the product of reaction; therefore, it is not appropriate to assign it the product's asymmetry, as is sometimes done in the literature [341,342]. In light of these difficulties, it appears more insightful to determine how the molecular asymmetry of the activated complex, and its solvation behavior, can affect the kinetic rate constant at near-critical conditions rather than trying to speculate on its molecular description.

For example, by assuming that the isothermal compressibility is a positive definite quantity, we can conclude that the sign of the pressure derivative on the left-hand side of Eq. (3.22) is determined by the linear combination of the short-range contributions to the species partial molar volumes $\bar{v}_U^{\infty}(\mathrm{SR})$ so that

$$\nu_{\mathfrak{R}}\bar{v}_{\mathfrak{R}}^{\infty}(\mathrm{SR}) > \nu_R\bar{v}_R^{\infty}(\mathrm{SR}) + \nu_S\bar{v}_S^{\infty}(\mathrm{SR}) + \Delta\nu\bar{v}_U^{\infty(IG)}(\mathrm{SR}) \tag{3.23}$$

will give rise to $(\partial \ln k_{\mathrm{B}}^{\mathrm{TST}}/\partial P)_{T,x} < 0$, where $\bar{v}_U^{\infty(IG)}(\mathrm{SR}) = \bar{v}_V^o(2 - (\kappa^{IG}/\kappa))$ is a positive quantity, and $\Delta\nu = \nu_R + \nu_S - \nu_{\mathfrak{R}}$ is typically positive or zero [266]. Eq. (3.23) clearly indicates that the pressure effect on the rate constant depends on a solvation balance between reactants and activated complex species, i.e., on the relative molecular asymmetries of all species in solution.

For the sake of simplicity (and for illustration purposes) let us assume that the transition state complex $\mathfrak{R}^{\ddagger}$ behaves as the reactant R, i.e., their molecular asymmetries are identical to each other regardless of their type. Thus a negative pressure effect on the kinetic rate constant, $(\partial \ln k_B^{\mathrm{TST}}/\partial P)_{T,x} < 0$, is always possible as long as $\bar{v}_S^{\infty}(\mathrm{SR}) < -(\Delta\nu/\nu_S)\bar{v}_U^{\infty(IG)}(\mathrm{SR}) < 0$, i.e., when the reactive cosolvent behaves as an attractive [241] species $[(\partial P/\partial x_S)_{T,\rho,x_{k\neq s}}^{\infty} < 0$ [236]], regardless of the type of solvation behavior of the other species. A similar situation occurs when $\mathfrak{R}^{\ddagger}$ and S exhibit the same solvation behavior, i.e., $(\partial \ln k^{\mathrm{TST}}/\partial P)_{T,x} < 0$ requires that $\bar{v}_R^{\infty}(\mathrm{SR}) < 0$.

Otherwise, $(\partial \ln k_B^{\mathrm{TST}}/\partial P)_{T,x} > 0$ will be observed at near-critical conditions as long as

$$\nu_{\mathfrak{R}}\bar{v}_{\mathfrak{R}}^{\infty}(\mathrm{SR}) < \nu_R\bar{v}_R^{\infty}(\mathrm{SR}) + \nu_S\bar{v}_S^{\infty}(\mathrm{SR}) + \Delta\nu\bar{v}_U^{\infty(IG)}(\mathrm{SR}) \tag{3.24}$$

Should the transition state complex $\mathfrak{R}^{\ddagger}$ behave similarly to the reactant R (or S) species, then the condition of Eq. (3.24), $\bar{v}_S^{\infty}(\mathrm{SR}) > -(\Delta\nu/\nu_S)\bar{v}_U^{\infty(IG)}(\mathrm{SR})$ [or $\bar{v}_R^{\infty}(\mathrm{SR}) > -(\Delta\nu/\nu_R)\bar{v}_U^{\infty(IG)}(\mathrm{SR})$], is satisfied by a species S (or R) behaving as a repulsive $(0 \lesseqgtr \bar{v}_S^{\infty}(\mathrm{SR}) \leq \rho^{-1})$ solute. Alternatively, note that when $\nu_{\mathfrak{R}}\bar{v}_{\mathfrak{R}}^{\infty}(\mathrm{SR}) \approx \nu_R\bar{v}_R^{\infty}(\mathrm{SR}) + \nu_S\bar{v}_S^{\infty}(\mathrm{SR}) + \Delta\nu\bar{v}_U^{\infty(IG)}(\mathrm{SR})$ in the compressible region of the solvent, $(\partial \ln k_B^{\mathrm{TST}}/\partial P)_{T,x}$ will exhibit a flat dependence with density (pressure), a behavior already observed by Knutson and co-workers [342] for Diels–Alder reactions in propane.

On the one hand, solvation phenomena by their very nature involve interactions occurring within nearest coordination shells. On the other hand, near-criticality is a handy condition that allows us to adjust the solvent's solvation power (large changes in density) by small perturbations of the system's pressure and temperature. Therefore, to observe the true solvent effect we must eliminate the compressibility contribution from Eq. (3.22) by determining the corresponding density coefficient, i.e.,

$$\left(\frac{\partial \ln k_{\mathrm{B}}^{\mathrm{TST}}}{\partial \rho}\right)_{T,x} = [\nu_R(\bar{v}_R^{\infty}(\mathrm{SR}) - \bar{v}_R^{\infty(IG)}(\mathrm{SR})) + \nu_S(\bar{v}_S^{\infty}(\mathrm{SR}) - \bar{v}_S^{\infty(IG)}(\mathrm{SR})) - \nu_{\Re}(\bar{v}_{\Re}^{\infty}(\mathrm{SR}) - \bar{v}_{\Re}^{\infty(IG)}(\mathrm{SR}))] \tag{3.25}$$

which, not surprisingly, indicates that the solvation effects are finite everywhere. In particular, at the solvent's criticality, the density coefficient becomes proportional to a linear combination of the species Krichevskii's parameters, i.e.,

$$\left(\frac{\partial \ln k_{\mathrm{B}}^{\mathrm{TST}}}{\partial \rho}\right)_{T,x}^{c} = \frac{[\nu_R(\partial P/\partial x_R)_{\rho_c T_c}^{\infty} + \nu_S(\partial P/\partial x_S)_{\rho_c T_c}^{\infty} - \nu_{\Re}(\partial P/\partial x_{\Re})_{\rho_c T_c}^{\infty}]}{k_{\mathrm{B}} T_c \rho_c^2} - \frac{\Delta \nu}{\rho_c} \tag{3.26}$$

Therefore, should any near-critical effects on the rate constant exist, Eq. (3.26) will show it as a change in the behavior of $(\partial \ln k_{\mathrm{B}}^{\mathrm{TST}}/\partial \rho)_{T,x}$ in the vicinity of the solvent's critical point. Eqs. (3.25) and (3.26) support earlier arguments on the advantages of isochoric rather than isobaric studies of the variation of $\ln k_B^{\mathrm{TST}}$ [339].

An alternative way to tune the solvation power of a near-critical solvent is by isobaric manipulation of the system's temperature. The factorization of the temperature coefficient between solvation and compressibility effects [266] results in the following expression:

$$\begin{aligned}\left(\frac{\partial \ln k_{\mathrm{B}}^{\mathrm{TST}}}{\partial T}\right)_{P,x} &= [\Delta H_o + k_{\mathrm{B}} T - \Delta \nu(\bar{h}_V^{ro} - (T/\rho)(\partial P/\partial T)_{\rho})/k_{\mathrm{B}} T^2] \\ &+ \int_0^{\rho(P)} \left(\frac{\partial(\nu_R \bar{v}_R^{\infty}(\mathrm{SR}) + \nu_S \bar{v}_S^{\infty}(\mathrm{SR}) - \nu_{\Re} \bar{v}_{\Re}^{\infty}(\mathrm{SR}))}{\partial T}\right)_{\rho} d\rho \\ &- \kappa\rho \left(\frac{\partial P}{\partial T}\right)_{\rho} [\nu_R(\bar{v}_R^{\infty}(\mathrm{SR}) - \bar{v}_R^{\infty(IG)}(\mathrm{SR})) + \nu_S(\bar{v}_S^{\infty}(\mathrm{SR}) - \bar{v}_S^{\infty(IG)}(\mathrm{SR})) \\ &- \nu_{\Re}(\bar{v}_{\Re}^{\infty}(\mathrm{SR}) - \bar{v}_{\Re}^{\infty(IG)}(\mathrm{SR}))]\end{aligned} \tag{3.27}$$

where ΔH_o is the standard heat of formation of the activated complex at the system temperature, and $\bar{h}_V^{or} = h_V^{or}(T, P)$ is the residual molar enthalpy of the pure solvent. Consequently, the solvation effect becomes

$$\left(\frac{\partial \ln k_{\mathrm{B}}^{\mathrm{TST}}}{\partial T}\right)_{\rho,x} = \frac{\Delta H_o + k_{\mathrm{B}}T - \Delta v(\bar{h}_V^{ro} - (T/\rho)(\partial P/\partial T)_\rho)}{k_{\mathrm{B}}T^2} + \int_0^{\rho(P)} \left(\frac{\partial(v_R\bar{v}_R^\infty(\mathrm{SR}) + v_S\bar{v}_S^\infty(\mathrm{SR}) - v_{\mathfrak{R}}\bar{v}_{\mathfrak{R}}^\infty(\mathrm{SR}))}{\partial T}\right)_\rho \partial\rho \tag{3.28}$$

In summary, this rigorous analysis tells us that the macroscopic pressure or temperature coefficient of the kinetic rate constant will reflect the behavior resulting from the actual (local) distributions of reactants and TS species in solution at a microscopic level. Within the framework of the TST, this analysis suggests two immediate consequences:

1. The solvation behavior and the pressure or temperature coefficient in the kinetic analysis are the microscopic and macroscopic manifestations of the same phenomenon.
2. The proper assessment of the solvation contribution to the pressure or temperature coefficient cannot be achieved by macroscopic correlations, such as equations of state, because of the lack of precise information about the molecular asymmetry of the transition state complex [343], its ambiguous connection to the adjustable parameters of the equations of state, and the consequent inherent inaccuracy of the species partial molar properties [344].

C. Discussion

Molecular-based modeling studies on supercritical fluids have most frequently focused on linking the solute mechanical partial molar properties at infinite dilution with the microstructure of the system; therefore, they eluded the crucial question regarding the discrimination between the coexisting solvation and compressibility driven phenomena. These phenomena involve rather different length scales; thus they require careful consideration to draw any definite link between the solute–solvent molecular asymmetry and the resulting solvation behavior.

Although solute-induced effects occur in all types of (nonideal) solutions, their manifestation in electrolyte systems deserve special attention. The presence of charged species in a dielectric solvent adds an important ingredient to the solvation phenomenon, i.e., the possible formation of neutral ion pairs. In fact, an outstanding property of water as a solvent at normal conditions is its intrinsic ability to solvate and thus dissolve ionic and polar species, owing this to its unusually large dielectric constant. This solvation process is typically described in terms of ion–solvent interactions, ion-induced solvent's microstructural changes, solvent's dielectric behavior, and their effects on the macroscopic properties of the solution.

Commonly, ion–ion interactions take place in either concentrated ionic solutions at normal conditions or dilute near-critical ionic solutions; consequently, ion speciation becomes a relevant solvation phenomenon (e.g., through electrostriction or dielectric saturation). Ion speciation involves a competition between the increasingly important ion-ion interactions, owing decreasing solvent's dielectric screening, and the ion–solvent interactions. In practical terms, the ability of water to attenuate the long-range electrostatic ion–ion interactions by dielectric screening becomes a key player in the ion solvation, because it allows us to tailor the water-solvating power by careful manipulation of its state conditions. It is important to understand that, even though we are tuning a macroscopic property (dielectric permittivity), the solvation process is microscopic and local in nature.

Regarding the solvent effects on the kinetic rate constants at near-critical and supercritical conditions, we have considered only equilibrium solvent effects, i.e., the potential breakdown of the TST hypotheses caused by the solvent dynamics is a real possibility and should not be forgotten [338]. For example, let us consider the kinetics of interconversion between CIP and SSHIP configurations in infinitely dilute NaCl aqueous solutions [Eq. (3.3)], for which we have some direct calculations of the corresponding transmission coefficient $\chi = k_B/k_B^{TST}$. At ambient conditions, the simulation results for the transmission coefficient indicates a value of $\chi \approx 0.2$ [269,275,345], already a large deviation from the TST value ($\chi = 1$). Preliminary results from molecular simulation at near-critical conditions indicate that this coefficient becomes small, $\chi \approx 0.05$ [310].

IV. FINAL REMARKS

Because an important goal behind these molecular-based studies is the development of successful correlations for engineering applications, the grand challenge we must deal with concerns the translation of any acquired microscopic understanding of the solvation phenomena into algebraically tractable macroscopic correlations. This translation also implies that we must be able to suggest, guided by the molecular formalism, the most adequate type of experimental data that are required (i.e., best-behaved properties) in the correlation process, in addition to having rigorous connections between these properties and the relevant solvation counterparts. Work toward this goal is already in progress [51,346,347].

In this context, molecular-based simulation has become a versatile and powerful tool for probing the microscopic behavior of high-temperature water and aqueous solutions, in that it provides essentially exact results for the structural and thermophysical properties of precisely defined model fluids, provided we bear in mind the limitations behind any simulation methodology [256]. Essentially, these limitations revolve around the combination of two ingredients: our incomplete knowledge of the phase diagram of the model solvent and the limits to the realism of the intermolecular models to mimic the real systems. Therefore, they suggest that an efficient way to advance our understanding of solvation phenomena in supercritical aqueous solutions is through the integration of experimental, theoretical (statistical mechanics), and molecular-based simulation approaches. This interplay has already proved to be an essential player in the determination of more accurate microstructural information of water at ambient and supercritical conditions [62].

Acknowledgments

This work was supported by the Division of Chemical Sciences, Office of Basic Energy Sciences, U. S. Department of Energy. The simulation work was facilitated by the generous allocation of computational resources on the IBM SP/2 at the University of Tennessee through the Shared University Resource (SUR) grant from IBM. The authors wish to express their gratitude to Dr. Markus Hoffmann (PNNL, Richland) and Dr. Andrey Kalinichev (IEM, Chernogolovka) for kindly providing tabulated data of their published figures on hydrogen bonding (Refs. [52,193], respectively) and to Dr. Istvan Borzsak and Dr. Thomas Driesner for the critical reading of an early version of the manuscript.

References

1. K. S. Davis and J. A. Day, *Water: The Mirror of Science.* Doubleday, Garden City, NY, 1961.
2. F. Franks, in *Water: A Comprehensive Treatise*, (F. Franks, ed.), Vol. 1, pp. 1–13. Plenum, New York, 1972.
3. D. Eisenberg and W. Kauzmann, *The Structure and Properties of Water.* Oxford University Press, New York, 1969.
4. L. Haar, J. S. Gallagher, and G. S. Kell, *Steam Tables.* Hemisphere Publishers, New York, 1984.
5. C. N. R. Rao, in *Water: A Comprehensive Treatise* (F. Franks, ed.), Vol. 1, pp. 93–113, Plenum, New York, 1972.
6. J. P. Hansen and I. R. McDonald, *Theory of Simple Liquids*, 2nd ed. Academic Press, Orlando, FL, 1986.
7. J. I. Steinfeld, J. S. Francisco, and W. L. Hase, *Chemical Kinetics and Dynamics.* Prentice-Hall, Englewood Cliffs, NJ, 1989.
8. E. S. Amis and J. F. Hinton, *Solvent Effects on Chemical Phenomena.* Academic Press, New York, 1973.
9. W. R. Shaw, T. B. Brill, A. A. Clifford, C. A. Eckert, and E. U. Frank, *Chem. Eng. News* **51**, 26–39 (1991).
10. E. Kiran and J. M. H. Levelt Sengers, *Supercritical Fluids: Fundamentals for Applications.* Kluwer Academic Publishers, Dordrecht, The Netherlands, 1994.
11. A. R. Katritzky, S. M. Allin, and M. Siskin, *Acc. Chem. Res.* **29**, 399–406 (1996).

12. R. Narayan and M. J. Antal, in *Supercritical Fluid Science and Technology*, (K. P. K. P. Johnston and J. M. L. Penninger, eds.), Vol. 406, pp. 226–241, American Chemical Society, Washington, DC, 1989.
13. M. Modell, R. C. Reid, and S. I. Amin, U.S. Pat. 1,113,446 (1978).
14. J. Yao and R. F. Evilia, *J. Am. Chem. Soc.* **116**, 11229–11233 (1994).
15. D. C. Elliott, *Ind. Eng. Chem. Res.* **33**, 558–565 (1994).
16. M. Modell, G. G. Gaudet, M. Simson, G. T. Hong, and K. Biemann, *Solid Wastes Manage.*, pp. 26–30 (1982).
17. T. B. Thomason, G. T. Hong, K. C. Swallow, and W. R. Killilea, *Therm. Process* **1**, 31–62 (1990).
18. T. B. Thomason and M. Modell, *Hazard. Waste* **1**, 453–467 (1984).
19. J. W. Tester, H. R. Holgate, F. J. Armellini, P. A. Webley, W. R. Killilea, G. T. Hong, and H. E. Barner, in *Emerging Technologies in Hazardous Waste Management III* (D. W. Tedder and F. G. Pohland, eds.), Vol. 518, pp. 35. American Chemical Society, Washington, DC, 1991.
20. W. R. Killilea, K. C. Swallow, and G. T. Hong, *J. Supercrit. Fluids* **5**, 72–78 (1992).
21. P. E. Savage, S. Golapan, T. I. Mizan, C. J. Martino, and E. E. Brock, *AICE J.* **41**, 1723–1778 (1995).
22. J. V. Sengers and J. T. F. Watson, *J. Phys. Chem. Ref. Data* **15**, 1291–1314 (1986).
23. W. J. Lamb, G. A. Hoffman, and J. Jonas, *J. Chem. Phys.* **74**, 6875–6880 (1981).
24. W. L. Marshall and E. U. Franck, *J. Phys. Chem. Ref. Data* **10**, 295–304 (1980).
25. T. W. de Loos, J. H. van Dorp, and R. N. Lichtenthaler, *Fluid Phase Equilib.* **10**, 279–287 (1983).
26. J. F. Connolly, *J. Chem. Eng. Data* **11**, 13–16 (1966).
27. H. A. Pray, C. E. Schweickert, and B. H. Minnich, *Ind. Eng. Chem.* **44**, 1146 (1952).
28. M. L. Japas and E. U. Franck, *Ber. Bunsenges. Phys. Chem.* **89**, 1268–1275 (1985).
29. M. L. Japas and E. U. Franck, *Ber. Bunsenges. Phys. Chem.* **89**, 793–800 (1985).
30. K. Tödheide and E. U. Franck, *Z. Phys. Chem.* [N.S.] **37**, 388 (1963).
31. T. M. Seward and E. U. Franck, *Ber. Busenges. Phys. Chem.* **85**, 2–7 (1981).
32. G. L. Huppert, B. C. Wu, S. H. Townsend, M. T. Klein, and S. C. Paspek, *Ind. Eng. Chem. Res.* **28**, 161–165 (1989).
33. J. M. L. Penninger and J. M. M. Kolmschate, in *Supercritical Fluid Science and Technology* (K. P. Johnston and J. M. L. Penninger, eds.), Vol. 406, pp. 242–258. American Chemical Society, Washington, DC, 1989.
34. J. R. Lawson and M. T. Klein, *Ind. Eng. Chem. Res.* **24**, 203–208 (1985).
35. P. A. Webley and J. W. Tester, in *Supercritical Fluid Science and Technology.* (K. P. Johnston and J. M. L. Penninger, eds.), Vol. 406, pp. 259–275. American Chemical Society, Washington, DC, 1989.
36. D. B. Mitton, J. C. Orzalli, and R. M. Latanision, in *Physical Chemistry of Aqueous Systems: Meeting the Needs of Industry* p. 638–643 (H. J. White, J. V. Sengers, D. B. Neumann, and J. C. Bellows, eds.). (Begell House, New York, 1995).
37. H. E. Barner, C. Y. Huang, T. Johnson, G. Jacobs, M. A. Martch, and W. R. Killilea, *J. Hazard. Mater.* **31**, 1–17 (1992).
38. M. Modell, in *Standard Handbook of Hazardous Waste Treatment and Disposal* p. 8.153–8.168 (H. M. Freeman, ed.). McGraw-Hill, New York, 1989.
39. L. A. Torry, R. Kaminsky, M. T. Klein, and M. R. Klotz, *J. Supercrit. Fluids* **5**, 163–168 (1992).

40. J. R. Quint and R. H. Wood, *J. Phys. Chem.* **89**, 380–384 (1985).
41. R. H. Wood, R. W. Carter, J. R. Quint, V. Majer, P. Thompson, and J. R. Boccio, *J. Chem. Thermodyn.* **26**, 225–249 (1994).
42. B. E. Conway, *Ionic Hydration in Chemistry and Biophysics*. Elsevier, Amsterdam, 1981.
43. W. J. Jones, J. A. Leigh, F. Mayer, C. R. Woese, and R. S. Wolfe, *Arch. Microbiol.* **136**, 254–261 (1983).
44. D. A. Cowan, in *Hydrothermal Vents and Processes* (G. Society, ed.), Publ. No. 87, pp. 351–364. Geol. Soc. Publ. House, London, 1995.
45. D. A. Cowan and J. A. Littlechild, in *Enzyme Technology for Industrial Applications* (L. M. Savage, ed.), pp. 197–237. IBC Biomedical Library, 1996.
46. K. S. Pitzer, ed., *Activity Coefficients in Electrolyte Solutions*, CRC Press, Boca Raton, FL, 1991.
47. K. S. Pitzer, *Thermodynamics*, 3rd ed. McGraw-Hill, New York, 1995.
48. R. E. Mesmer, D. A. Palmer, and J. M. Simonson, in *Activity Coefficients in Electrolyte Solutions* (K. S. Pitzer, ed.), 2nd ed., pp. 491–529. CRC Press, Boca Raton, RL, 1991.
49. D. A. Palmer and D. J. Wesolowski, eds., *Fifth International Symposium on Hydrothermal Reactions*, Gatlinburg, TN, 1997.
50. A. A. Chialvo and P. T. Cummings, *AICE J.* **40**, 1558–1573 (1994a).
51. A. A. Chialvo, P. T. Cummings, J. M. Simonson, and R. E. Mesmer, *J. Chem. Phys.* To appear in Jan. 8, 1999 issue.
52. M. M. Hoffmann and M. S. Conradi, *J. Am. Chem. Soc.* **119**, 3811–3817 (1997).
53. N. Matubayasi, C. Wakai, and M. Nakahara, *Phys. Rev. Lett.* **78**, 2573–2576 (1997).
54. M. Matubayasi, C. Wakai, and M. Nakahara, *J. Chem. Phys.* **107**, 9133–9140 (1997).
55. G. E. Walrafen, M. R. Fisher, M. S. Hokmabadi, and W.-H. Yang, *J. Chem. Phys.* **85**, 6970–6982 (1986).
56. K. Okada, Y. Imashuku, and M. Yao, *J. Chem. Phys.* **107**, 9302–9311 (1997).
57. D. L. Beveridge, M. Mezei, P. K. Mehrotra, F. T. Marchese, G. Ravi-Shanker, T. Vasu, and S. Swaminathan, in *Molecular-Based Study of Fluids* (J. M. Haile and G. A. Mansoori, eds.), Vol. 204, pp. 298–351. American Chemical Society, Washington, DC, 1983.
58. G. C. Lie, *J. Chem. Phys.* **85**, 7495–7497 (1986).
59. A. K. Soper and M. G. Phillips, *Chem. Phys.* **107**, 47–60 (1986).
60. P. Postorino, R. H. Tromp, M. A. Ricci, A. K. Soper, and G. W. Neilson, *Lett. to Nat.* **366**, 668–670 (1993).
61. No author, *Faraday Discuss.* **103**, 91–116 (1996).
62. A. A. Chialvo, P. T. Cummings, J. M. Simonson, R. E. Mesmer, and H. D. Cochran, *Ind. Eng. Chem. Res.* **37**, 3021–3025 (1998).
63. S. Scheiner, *Hydrogen Bonding. A Theoretical Perspective*. Oxford University Press, New York, 1997.
64. M. Mezei and D. L. Beveridge, *J. Chem. Phys.* **74**, 622–632 (1981).
65. M. G. Sceats and S. A. Rice, *J. Chem. Phys.* **72**, 3236–3247 (1980).
66. A. Rahman and F. H. Stillinger, *J. Chem. Phys.* **55**, 3336–3359 (1971).
67. F. H. Stillinger and A. Rahman, *J. Chem. Phys.* **57**, 1281–1292 (1972).
68. A. Luzar and D. Chandler, *J. Chem. Phys.* **98**, 8160–8173 (1993).
69. R. D. Mountain, *J. Chem. Phys.* **90**, 1866–1870 (1989).

70. W. L. Jorgensen, *J. Am. Chem. Soc.* **100**, 7824–7831 (1978).
71. T. I. Mizan, P. E. Savage, and R. M. Ziff, *J. Phys. Chem.* **100**, 403–408 (1996).
72. A. A. Chialvo and P. T. Cummings, *J. Phys. Chem.* **100**, 1309–1316 (1996).
73. M. Nakahara, T. Yamaguchi, and H. Ohtaki, *Recent Res. Devel. Phys. Chem.* **1**, 17–49 (1997).
74. A. G. Kalinichev and J. D. Bass, *Chem. Phys. Lett.* **231**, 301–307 (1994).
75. A. De Santis and D. Rocca, *J. Chem. Phys.* **107**, 10096–10101 (1997).
76. A. K. Soper, F. Bruni, and M. A. Ricci, *J. Chem. Phys.* **106**, 247–254 (1997).
77. M.-C. Bellissent-Funel, T. Tassaing, H. Zhao, D. Beysens, B. Guillot, and Y. Guissani, *J. Chem. Phys.* **107**, 2942–2949 (1997).
78. R. Schmidt and J. Brickmann, *Ber. Busenges. Phys. Chem.* **101**, 1816–1827 (1998).
79. A. G. Kalinichev and J. D. Bass, in *Physical Chemistry of Aqueous Solutions: Meeting the Needs of Industry* (H. J. White, J. V. Sengers, D. B. Neumann, and J. C. Bellows, eds.), pp. 245–252. Begell House, New York, 1995.
80. I. M. Svishchev and P. G. Kusalik, *J. Chem. Phys.* **99**, 3049–3058 (1993).
81. R. P. Murchi and H. J. Eyring, *J. Phys. Chem.* **68**, 221 (1964).
82. W. A. P. Luck, *Ber. Bunsenges Phys. Chem.* **69**, 627–637 (1965).
83. E. U. Franck and K. Roth, *Discuss. Faraday Soc.* **43**, 108–114 (1967).
84. G. V. Bondarenko and Y. E. Gorbaty, *Dokl. Phys. Chem. (Engl. Transl.)* **210**, 369 (1973).
85. A. A. Vetrov, O. I. Kondratov, and G. V. Yukhnevich, *Aust. J. Chem.* **28**, 2099–2107 (1975).
86. Y. E. Gorbaty and Y. N. Demyanets, *Chem. Phys. Lett.* **100**, 450–454 (1983).
87. Y. E. Gorbaty and Y. N. Demyanets, *J. Struct. Chem. (Engl. Transl.)* **24**, 385–392 (1983).
88. A. De Santis and D. Rocca, *J. Chem. Phys.* **107**, 9559–9568 (1997).
89. A. A. Chialvo and P. T. Cummings, *J. Chem. Phys.* **101**, 4466–4469 (1994).
90. A. A. Chialvo and P. T. Cummings, in *Encyclopedia of Computational Chemistry* (N. L. Allinger, ed.), Wiley, New York, in press.
91. J. E. Enderby and G. W. Nielsen, in *Water: A Comprehensive Treatise* (F. Franks, ed.), Vol. 6, pp. 1–46. Plenum, New York, 1979.
92. A. K. Soper, *Chem. Phys.* **107**, 61–74 (1986).
93. J. L. Finney and A. K. Soper, *Chem. Soc. Rev.* **23**, 1–10 (1994).
94. J. E. Enderby, *Chem. Soc. Rev.* **24**, 159–168 (1995).
95. B. Guillot and Y. Guissani, *Fluid Phase Equilib.* **151**, 19–32 (1998).
96. U. Buontempo, P. Postorino, M. A. Ricci, and A. K. Soper, *Europhys. Lett.* **19**, 385–389 (1992).
97. P. Postorino, M. A. Ricci, and A. K. Soper, *J. Chem. Phys.* **101**, 4123–4132 (1994).
98. R. H. Tromp, P. Postorino, G. W. Nielson, M. A. Ricci, and A. K. Soper, *J. Chem. Phys.* **101**, 6210–6215 (1994).
99. P. T. Cummings, H. D. Cochran, J. M. Simonson, R. E. Mesmer, and S. Karaborni, *J. Chem. Phys.* **94**, 5606–5621 (1991).
100. H. J. C. Berendsen, J. P. M. Postma, W. F. van Gunsteren, and J. Hermans, in *Intermolecular Forces: Proceedings of the Fourteenth Jerusalem Symposium on Quantum Chemistry and Biochemistry* (B. Pullman, ed.), pp. 331–342. Reidel Publ., Dordrecht, The Netherlands, 1981.
101. D. M. Carey and G. M. Korenowski, *J. Chem. Phys.* **108**, 2669–2675 (1998).
102. A. H. Narten and H. A. Levy, *J. Chem. Phys.* **55**, 2263–2269 (1971).

103. Y. E. Gorbaty and Y. N. Demyanets, *Mol. Phys.* **55**, 571–588 (1985).
104. G. Löffler, H. Schreiber, and O. Steinhauser, *Ber. Bunsenges. Phys. Chem.* **98**, 1575–1578 (1994).
105. P. Postorino, M. Nardone, M. A. Ricci, and M. Rovere, *J. Mol. Liq.* **64**, 221–240 (1995).
106. D. G. Montague, in *Hydrogen-Bonded Liquids* (J. C. Dore and J. Teixeira, eds.), Vol. C 329, pp. 129–137. Kluwer Academic Publishers, Dordrecht, The Netherlands, 1991.
107. E. S. Fois, M. Sprik, and M. Parrinello, *Chem. Phys. Lett.* **223**, 411–415 (1994).
108. R. Car and M. Parrinello, *Phys. Rev. Lett.* **55**, 2471–2474 (1985).
109. K. Yamanaka, T. Yamaguchi, and H. Wakita, *J. Chem. Phys.* **101**, 9830–9836 (1994).
110. Y. E. Gorbaty and Y. I. Demyanets, *J. Struct. Chem.* (*Engl. Transl.*) **24**, 716–722 (1983).
111. H. Ohtaki, T. Radnai, and T. Yamaguchi, *Chem. Soc. Rev.* **26**, 41–51 (1997).
112. T. Yamaguchi, *J. Mol. Liq.* **78**, 43–50 (1998).
113. Y. E. Gorbaty and A. G. Kalinichev, *J. Phys. Chem.* **99**, 5336–5340 (1995).
114. A. V. Iogansen and E. V. Brown, *Opt. Spectrosc.* **23**, 492–496 (1967).
115. T. Radnai and H. Ohtaki, *Mol. Phys.* **87**, 103–121 (1996).
116. P. Jedlovszky and R. Vallauri, *J. Chem. Phys.* **105**, 2391–2398 (1996).
117. R. L. McGreevy and L Pusztai, *Mol. Simul.* **1**, 359–367 (1988).
118. J. K. M. Sanders and B. K. Hunter, *Modern NMR Spectroscopy*, 2nd ed. Oxford University Press, New York, 1993.
119. H. J. C. Berendsen, J. R. Grigera, and T. P. Straatsma, *J. Phys. Chem.* **91**, 6269–6271 (1987).
120. A. Ben-Naim and F. H. Stillinger, in *Structure and Transport of Processes in Water and Aqueous Solutions* (R. A. Horne, ed.). Wiley-Interscience, New York, 1972.
121. W. L. Jorgensen, J. Chandrasekhar, J. D. Madura, R. W. Impey, and M. L. Klein, *J. Chem. Phys.* **79**, 926–935 (1983).
122. H. J. Strauch and P. T. Cummings, *Mol. Simul.* **2**, 89–104 (1989).
123. Y. Guissani and B. J. Guillot, *J. Chem. Phys.* **98**, 8221–8235 (1993).
124. F. H. Stillinger and A. Rahman, *J. Chem. Phys.* **60**, 1545–1557 (1974).
125. R. Mills, *J. Phys. Chem.* **77**, 685–688 (1973).
126. O. Teleman, B. Jönsson, and S. Engström, *Mol. Phys.* **60**, 193–203 (1987).
127. T. I. Mizan, P. E. Savage, and R. M. Ziff, *J. Phys. Chem.* **98**, 13067–13076 (1994).
128. C. D. Berweger, W. F. van Gunsteren, and F. Müller-Plathe, *Chem. Phys. Lett.* **232**, 429–436 (1995).
129. A. Z. Panagiotopoulos, *Mol. Phys.* **61**, 813–826 (1987).
130. D. A. Kofke, *Mol. Phys.* **78**, 1331–1336 (1993).
131. K. Kiyohara, K. E. Gubbins, and A. Z. Panagiotopoulos, *Mol. Phys.* **94**, 803–808 (1998).
132. J. R. Errington, K. Kiyohara, K. E. Gubbins, and A. Z. Panagiotopoulos, *Fluid Phase Equilibria* **151**, 33–40 (1998).
133. F. H. Stillinger, *Science* **209**, 451–457 (1980).
134. S. L. Carnie and G. N. Patey, *Mol. Phys.* **47**, 1129–1151 (1982).
135. A. A. Chialvo, *J. Chem. Phys.* **104**, 5240–5243 (1996).
136. G. C. Boulougouris, I. G. Economou, and D. N. Theodorou, *J. Phys. Chem. B* **102**, 1029–1035 (1998).
137. M. W. Evans, *J. Mol. Liq.* **32**, 173–181 (1986).

138. J. J. de Pablo, J. M. Prausnitz, H. J. Strauch, and P. T. Cummings, *J. Chem. Phys.* **93**, 7355–7359 (1991).
139. H. J. Strauch and P. T. Cummings, *J. Chem. Phys.* **96**, 864–865 (1992).
140. P. Bopp, G. Jancsó, and K. Heinziger, *Chem. Phys. Lett.* **98**, 129–133 (1983).
141. I. M. Svishchev, P. G. Kusalik, J. Wang, and R. J. Boyd, *J. Chem. Phys.* **105**, 4742–4750 (1996).
142. J. Brodholt, M. Sampoli, and R. Vallauri, *Mol. Phys.* **85**, 81–90 (1995).
143. L. X. Dang, *J. Chem. Phys.* **97**, 2659–2660 (1992).
144. A. A. Chialvo and P. T. Cummings, *J. Chem. Phys.* **105**, 8274–8281 (1996).
145. A. A. Chialvo and P. T. Cummings, *Fluid Phase Equilib.* **150–151**, 73–81 (1998).
146. W. L. Jorgensen, *J. Am. Chem. Soc.* **103**, 335–340 (1981).
147. F. H. Stillinger and C. W. David, *J. Chem. Phys.* **69**, 1473–1484 (1978).
148. M. Sprik and M. L. Klein, *J. Chem. Phys.* **89**, 7556–7560 (1988).
149. J. A. C. Rullmann and P. T. van Duijnen, *Mol. Phys.* **63**, 451–475 (1988).
150. P. Ahlström, A. Wallqvist, S. Engström, and B. Jönsson, *Mol. Phys.* **68**, 563–581 (1989).
151. S. Kuwajima and A. Warshel, *J. Phys. Chem.* **94**, 460–466 (1990).
152. S. W. Rick, S. J. Stuart, and B. J. Berne, *J. Chem. Phys.* **101**, 6141–6156 (1994).
153. K. Toukan and A. Rahman, *Phys. Rev. B* **31**, 2643–2648 (1985).
154. G. C. Lie and E. Clementi, *Phys. Rev. A* **33**, 2679–2693 (1986).
155. J. Anderson, J. J. Ullo, and S. Yip, *J. Chem. Phys.* **87**, 1726–1732 (1987).
156. S.-B. Zhu, S. Singh, and G. W. Robinson, *J. Chem. Phys.* **95**, 2791–2799 (1991).
157. U. Niesar, G. Corongiu, E. Clementi, G. R. Kneller, and D. K. Bhattacharya, *J. Phys. Chem.* **94**, 7949–7956 (1990).
158. A. Famulari, R. Specchio, M. Sironi, and M. Raimondi, *J. Chem. Phys.* **108**, 3296–3303 (1998).
159. M. Tuckerman, B. Berne, and G. Martyna, *J. Chem. Phys.* **97**, 1990–2001 (1992).
160. H. Lemberg and F. Stillinger, *J. Chem. Phys.* **62**, 1677–1690 (1975).
161. A. Wallqvist and O. Teleman, *Mol. Phys.* **74**, 515–533 (1991).
162. S.-B. Zhu and C. Wong, *J. Chem. Phys.* **98**, 8892–8899 (1993).
163. I. G. Tironi, R. M. Brunne, and W. F. van Gunsteren, *Chem. Phys. Lett.* **250**, 19–24 (1996).
164. L. X. Dang and B. M. Pettitt, *J. Phys. Chem.* **91**, 3349–3351 (1987).
165. J.-L. Barrat and I. R. McDonald, *Mol. Phys.* **70**, 535–539 (1990).
166. T. I. Mizan, P. E. Savage, and R. M. Ziff, *J. Comput. Chem.* **17**, 1757–1770 (1996).
167. H. J. C. Berendsen, J. P. M. Postma, W. F. van Gunsteren, A. DiNola, and J. R. Haak, *J. Chem. Phys.* **81**, 3684–3690 (1984).
168. T. I. Mizan, P. E. Savage, and R. M. Ziff, *J. Supercrit. Fluids* **10**, 119–125 (1997).
169. F. H. Stillinger and A. Rahman, *J. Chem. Phys.* **68**, 666–670 (1978).
170. G. D. Carney, L. A. Curtis, and S. R. Langhoff, *J. Mol. Spectrosc.* **61**, 371–379 (1976).
171. E. Clementi, G. Corongiu, and F. Sciortino, *J. Mol. Struct.* **296**, 205–213 (1993).
172. G. Corongiu and E. Clementi, *J. Chem. Phys.* **98**, 4984–4990 (1993).
173. J. Brodholt, M. Sampoli, and R. Vallauri, *Mol. Phys.* **86**, 149–158 (1995).
174. P. G. Kusalik and I. M. Svishchev, *Science* **265**, 1219–1221 (1994).
175. C. J. F. Böttcher, *Theory of Electric Polarization*. Elsevier, Amsterdan, 1973.

176. B. T. Thole, *Chem. Phys.* **59**, 341–350 (1981).
177. D. N. Bernardo, Y. Ding, and K. Krogh-Jespersen, *J. Phys. Chem.* **98**, 4180–4187 (1994).
178. D. N. Bernardo, Y. Ding, and K. Krogh-Jespersen, *J. Comp. Chem.* **16**, 1141–1152 (1995).
179. G. Ruocco and M. Sampoli, *Mol. Simul.* **15**, 281–300 (1995).
180. S. Nosé, *Mol. Phys.* **52**, 255–268 (1984).
181. H. C. Andersen, *J. Chem. Phys.* **72**, 2384–2393 (1980).
182. H. Goldstein, *Classical Mechanics*, 2nd ed. Addison-Wesley, Reading, MA, 1981.
183. L. R. Pratt, *Mol. Phys.* **40**, 347–360 (1980).
184. R. D. Mountain, *J. Chem. Phys.* **105**, 10496–10499 (1996).
185. J. E. Roberts, B. L. Woodman, and J. Schnitker, *Mol. Phys.* **88**, 1089–1108 (1996).
186. W. L. Jorgensen, *J. Chem. Phys.* **77**, 4156–4163 (1982).
187. A. G. Kalinichev, *Int. J. Thermophys.* **7**, 887–900 (1986).
188. H. D. Cochran, P. T. Cummings, and S. Karaborni, *Fluid Phase Equilib.* **71**, 1–16 (1992).
189. P. T. Cummings, A. A. Chialvo, and H. D. Cochran, *Chem. Eng. Sci.* **49**, 2735–2748 (1994).
190. P. T. Cummings, H. D. Cochran, and A. A. Chialvo, in *Physical Chemistry of Aqueous Systems* (H. J. White, Jr., J. V. Sengers, D. B. Neumann, and J. C. Bellows, eds.), pp. 253–260. Begell House, New York, 1995.
191. R. D. Mountain, *J. Chem. Phys.* **103**, 3084–3090 (1995).
192. T. I. Mizan, P. E. Savage, and R. M. Ziff, in *Innovations in Supercritical Fluids: Science and Technology* (K. W. Hutchenson and N. R. Foster, eds.), Vol. 608, pp. 47–64. American Chemical Society, Washington, DC, 1995.
193. A. G. Kalinichev and J. D. Bass, *J. Phys. Chem.* **101**, 9720–9727 (1997).
194. P. Jedlovszky, J. P. Brodholt, F. Bruni, M. A. Ricci, A. K. Soper, and R. Vallauri, *J. Chem. Phys.* **108**, 8528–8540 (1998).
195. A. Geiger, F. H. Stillinger, and A. Rahman, *J. Chem. Phys.* **70**, 4185–4193 (1979).
196. H. E. Stanley and J. Teixeira, *J. Chem. Phys.* **73**, 3404–3422 (1980).
197. A. Geiger and P. Mausbach, in *Hydrogen-Bonded Liquids* (J. C. Dore and J. Teixeira, eds.), pp. 171–183. Kluwer Academic Publishers, Dordrecht, The Netherlands, 1991.
198. K. Laasonen, M. Sprik, M. Parrinello, and R. Car, *J. Chem. Phys.* **99**, 9080–9089 (1993).
199. M. Sprik, J. Hutter, and M. Parrinello, *J. Chem. Phys.* **105**, 1142–1152 (1996).
200. M. Sprik, *J. Phys. Condens. Matter* **8**, 9405–9409 (1996).
201. G. Galli and M. Parrinello, in *Computer Simulation in Material Science* (M. Meyer and V. Pontikis, eds.), Vol. E 205, pp. 283–304. Kluwer Academic Press, Dordrecht, The Netherlands, 1991.
202. B. Johnson, P. M. W. Gill, and J. A. Pople, *J. Chem. Phys.* **98**, 5612–5626 (1993).
203. R. G. Parr and W. Yang, *Density-Functional Theory of Atoms and Molecules* Oxford University Press, Oxford, 1989.
204. A. D. Becke, *Phys. Rev. A* **38**, 3098–3100 (1988).
205. J. P. Perdew, *Phys. Rev. B* **33**, 8822–8824 (1986).
206. C. Lee, W. Yang, and R. C. Parr, *Phys. Rev. B* **37**, 785–789 (1988).
207. J. Ortega, J. P. Lewis, and O. F. Sankey, *Phys. Rev. B* **50**, 10516–10530 (1994).
208. F. J. García-Vidal, A. Martín-Rodero, F. Flores, J. Ortega, and R. Pérez, *Phys. Rev. B* **44**, 11412 (1991).

209. F. J. García-Vidal, J. Merino, R. Pérez, R. Rincón, J. Ortega, and F. Flores, *Phys. Rev. B* **50**, 10537–10547 (1994).

210. J. Ortega, J. P. Lewis, and O. F. Sankey, *J. Chem. Phys.* **106**, 3696–3702 (1997).

211. A. H. Narten, W. E. Thiessen, and L. Blum, *Science* **217**, 1033–1034 (1982).

212. G. Tóth and A. Baranyai, *J. Chem. Phys.* **107**, 7402–7408 (1997).

213. L. Pusztai and R. L. McGreevy, *Physica B* **234**, 357–358 (1997).

214. H. J. White, Jr., J.V. Sengers, D. B. Neumann, and J. C. Bellows, eds., *Physical Chemistry of Aqueous Systems: Meeting the Needs of Industry.* Begell House, New York, 1995.

215. L. L. Lee, P. G. Debenedetti, and H. D. Cochran, in *Supercritical Fluid Technology: Reviews in Modern Theory and Applications* (T. J. Bruno and J. F. Ely, eds.), pp. 193–226. CRC Press, Boca Raton, FL, 1991.

216. A. A. Chialvo, *J. Phys. Chem.* **97**, 2740–2744 (1993a).

217. S. T. Cui and J. G. Harris, *Chem. Eng. Sci.* **49**, 2749–2763 (1994).

218. P. M. Rodger, *Mol. Phys.* **89**, 1157–1172 (1996).

219. S. C. Tucker and M. W. Maddox, *J. Phys. Chem.* **102**, 2437–2453 (1998).

220. R. A. Robinson and R. H. Stokes, *Electrolyte Solutions*, 2nd. rev. ed. Butterworth, London, 1970.

221. S. H. Lee, P. T. Cummings, J. M. Simonson, and R. E. Mesmer, *Chem. Phys. Lett.* **293**, 289–294 (1998).

222. Y. Marcus, *Ion Solvation.* John Wiley, Chichester, 1985.

223. S. Petrucci, in *Ionic Interactions: From Dilute Solutions to Fused Salts* (S. Petrucci, ed.), Vol. 1, pp. 117–177. Academic Press, New York, 1971.

224. A. A. Chialvo, P. T. Cummings, J. M. Simonson, and R. E. Mesmer, *J. Chem. Phys.* **105**, 9248–9257 (1996).

225. B. E. Conway, *Chem. Soc. Rev.* **21**, 253–261 (1992).

226. J. T. Slusher and P. T. Cummings, *J. Phys. Chem. B* **101**, 3818–3826 (1997).

227. S. M. McDonald, J. W. Brady, and P. Clancy, *Biopolymers* **33**, 1481–1503 (1993).

228. M. Mathlouthi and A.-M. Seuvre, *J. Chem. Soc., Faraday Trans. 1* **84**, 2641–2650 (1988).

229. S. E. Kemp, J. M. Grigor, and G. G. Birch, *Experientia* **48**, 731–733 (1992).

230. G. G. Birch, S. Parke, R. Siertsema, and J. M. Westwell, *Pure Appl. Chem.* **69**, 685–692 (1997).

231. B. Guillot and Y. Guisani, in *Properties of Water and Steam: Physical Chemistry of Aqueous Systems Meeting the Needs of Industry* (H. J. White, J. V. Sengers, D. B. Neumann, and J. C. Bellows, eds.), pp. 269–277. Begell House, New York, 1995.

232. J. Walshaw and J. M. Goodfellow, *J. Mol. Biol.* **231**, 392–414 (1993).

233. A. Ben-Naim, in *Structure and Reactivity in Aqueous Solutions* (C. J. Cramer and D. G. Thrular, eds.), Vol. 568, pp. 371–380. American Chemical Society, Washington, DC, 1994.

234. M. Kinoshita, Y. Okamoto, and F. Hirata, *J. Chem. Phys.* **107**, 1586–1599 (1997).

235. A. A. Chialvo and P. T. Cummings, *Mol. Phys.* **84**, 41–48 (1995a).

236. J. M. H. Levelt Sengers, in *Supercritical Fluid Technology* (T. J. Bruno and J. F. Ely, eds.), pp. 1–56. CRC Press, Boca Raton, FL, 1991.

237. M. L. Japas and J. M. H. Levelt Sengers, *AICE J.* **35**, 705–713 (1989).

238. A. A. Chialvo, Y. V. Kalyuzhnyi, and P. T. Cummings, *AICE J.* **42**, 571–584 (1996).

239. A. H. Harvey, R. Crovetto, and J. M. H. Levelt Sengers, *AICE J.* **36**, 1901–1904 (1990).

240. J. M. H. Levelt Sengers, *J. Supercrit. Fluids* **4**, 215–222 (1991).
241. P. G. Debenedetti and R. S. Mohamed, *J. Chem. Phys.* **90**, 4528–4536 (1989).
242. J. G. Riess and M. L. Blanc, *Angew. Chem.* **90**, 654–668 (1978).
243. N. Schnoy, F. Pfannkuch, and H. Beisbarth, *Anaesthesist* **28**, 503–510 (1979).
244. T. Appenzeller, *Science* **252**, 1790 (1991).
245. R. Battino and H. L. Clever, *Chem. Rev.* **66**, 395 (1966).
246. E. Wilhelm, R. Battino, and R. J. Wilcock, *Chem. Rev.* **77**, 219 (1977).
247. D. Beutier and H. Renon, *AICE J.* **24**, 1122 (1978).
248. W. Schotte, *AICE J.* **31**, 154 (1985).
249. W. Wilhelm, in *Molecular Liquids: New Perspectives in Physics and Chemistry* (J. J. C. Teixeira-Dias, ed.), pp. 175–206. Kluwer Academic Publishers, Dordrecht, The Netherlands, 1992.
250. J. F. Galobardes, D. R. Van Hare, and L. B. Rogers, *J. Chem. Eng. Data* **26**, 363–366 (1981).
251. J. M. Simonson, H. F. Holmes, R. H. Busey, R. E. Mesmer, D. G. Archer, and R. H. Wood, *J. Phys. Chem.* **94**, 7675 (1990).
252. E. H. Oelkers and H. Helgeson, *Geochim. Cosmochim. Acta* **55**, 1235–1251 (1991).
253. J. M. H. Levelt Sengers, C. M. Everhart, G. Morrison, and K. S. Pitzer, *Chem. Eng. Commun.* **47**, 315–328 (1986).
254. K. S. Pitzer, *J. Phys. Chem.* **90**, 1502–1504 (1986).
255. P. T. Cummings, in *Supercritical Fluids: Fundamentals for Application* (E. Kiran and J. M. H. Levelt Sengers, eds.), Vol. E 273, pp. 287–312. Kluwer Academic Publishers, Dordrecht, The Netherlands, 1994.
256. P. T. Cummings, in *Supercritical Fluids. Fundamentals for Application* (E. Kiran and J. M. H. Levelt Sengers, eds.), Vol. E 273, pp. 387–410. Kluwer Academic Publishers, Dordrecht, The Netherlands, 1994.
257. B. Guillot and Y. Guissani, *Mol. Phys.* **79**, 53–75 (1993).
258. J. M. Rickman and S. R. Phillpot, *Phys. Rev. Lett.* **66**, 349–352 (1991).
259. S. R. Phillpot and J. M. Rickman, *J. Chem. Phys.* **94**, 1454–1464 (1991).
260. B. Guillot and Y. Guissani, *J. Chem. Phys.* **99**, 8075–8094 (1993).
261. E. Bycklyng, *Physica* **27**, 1030 (1961).
262. B. Widom, *J. Chem. Phys.* **39**, 2808–2812 (1963).
263. C.-L. Lin and R. H. Wood, *J. Phys. Chem.* **100**, 16399–16409 (1996).
264. A. Belonoshko and S. K. Saxena, *Geochim. Cosmochim. Acta* **55**, 3191–3626 (1991).
265. A. Belonoshko and S. K. Saxena, *Geochim. Cosmochim. Acta* **55**, 381–387 (1991).
266. A. A. Chialvo, P. T. Cummings, and Y. V. Kalyuzhnyi, *AICE J.* **44**, 667–680 (1998).
267. W. F. van Gunsteren and H. J. C. Berendsen, University of Groningen (1987).
268. T. Driesner, *Geochim. Cosmochim. Acta* (to be published) (1998).
269. E. Guàrdia, R. Rey, and J. A. Padró, *J. Chem. Phys.* **95**, 2823–2831 (1991).
270. E. Guàrdia, A. Robinson, and J. A. Padró, *J. Chem. Phys.* **99**, 4229–4230 (1993).
271. J. Gao, *J. Am. Chem. Soc.* **115**, 6893–6895 (1993).
272. J. Gao, *J. Phys. Chem.* **98**, 6049–6053 (1994).
273. A. A. Chialvo, P. T. Cummings, H. D. Cochran, J. M. Simonson, and R. E. Mesmer, *J. Chem. Phys.* **103**, 9379–9387 (1995).
274. P. T. Cummings and A. A. Chialvo, *J. Phys. Condens. Matter* **8**, 9281–9287 (1996).

275. D. E. Smith and A. D. J. Haymet, *J. Chem. Phys.* **96**, 8450–8459 (1992).
276. E. A. Carter, G. Ciccotti, J. T. Hynes, and R. Kapral, *Chem. Phys. Lett.* **156**, 472–477 (1989).
277. R. W. Zwanzig, *J. Chem. Phys.* **22**, 1420–1426 (1954).
278. S. Lüdermann, H. Schreiber, R. Abseher, and O. Steinhauser, *J. Chem. Phys.* **104**, 286–295 (1996).
279. M. Mezei and D. L. Beveridge, *Ann. N. Y. Acad. Sci.* **482**, 1–23 (1986).
280. P. B. Balbuena, K. P. Johnston, and P. J. Rossky, *J. Am. Chem. Soc.* **116**, 2689–2690 (1994).
281. P. B. Balbuena, K. P. Johnston, and P. J. Rossky, *J. Phys. Chem.* **99**, 1554–1565 (1995).
282. G. E. Bennett, P. J. Rossky, and K. P. Johnston, *J. Phys. Chem.* **99**, 16136–16143 (1995).
283. L.W. Flanagin, P. B. Balbuena, K. P. Johnston, and P. J. Rossky, *J. Phys. Chem.* **99**, 5196–5205 (1995).
284. K. A. Sharp, 3.0 ed. Columbia University, New York, 1988.
285. A. A. Rashin and B. Honig, *J. Phys. Chem.* **89**, 5588–5593 (1985).
286. M. Bucher and T. L. Porter, *J. Phys. Chem.* **90**, 3406–3411 (1986).
287. R. H. Wood, J. R. Quint, and J.-P. E. Grolier, *J. Phys. Chem.* **85**, 3944–3949 (1981).
288. K. P. Johnston, G. E. Bennett, P. B. Balbuena, and P. J. Rossky, *J. Am. Chem. Soc.* **118**, 6746–6752 (1996).
289. P. B. Balbuena, K. P. Johnston, and P. J. Rossky, *J. Phys. Chem.* **100**, 2716–2722 (1996).
290. P. B. Balbuena, K. P. Johnston, and P. J. Rossky, *J. Phys. Chem.* **100**, 2706–2715 (1996).
291. J. Chandrasekhar, D. C. Spellmeyer, and W. L. Jorgensen, *J. Am. Chem. Soc.* **106**, 903–910 (1984).
292. J. Aqvist, *J. Phys. Chem.* **94**, 8021–8024 (1990).
293. M. J. Field, P. A. Bash, and M. Karplus, *J. Comput. Chem.* **11**, 700–733 (1990).
294. A. S. Quist and W. L. Marshall, *J. Phys. Chem.* **72**, 684–703 (1968).
295. P. C. Ho, D. A. Palmer, and R. E. Mesmer, *J. Solut. Chem.* **23**, 997–1018 (1994).
296. G. H. Zimmerman, M. S. Gruszkiewicz, and R. H. Wood, *J. Phys. Chem.* **99**, 11612–11625 (1995).
297. J. M. Simonson and D. A. Palmer, *Geochim. Cosmochim. Acta* **57**, 1–8 (1993).
298. H. F. Holmes and R. E. Mesmer, *J. Chem. Thermodyn.* **15**, 709–719 (1983).
299. H. F. Holmes, R. H. Busey, J. M. Simonson, R. E. Mesmer, D. G. Archer, and R. H. Wood, *J. Chem. Thermodyn.* **19**, 863–890 (1987).
300. T. J. Bruno and J. F. Ely, *Supercritical Fluid Technology.* CRC Press, Boca Raton, FL, 1991.
301. B. M. Pettitt and P. J. Rossky, *J. Chem. Phys.* **84**, 5836–5844 (1986).
302. M. P. Tosi and F. G. Fumi, *J. Phys. Chem. Solids* **25**, 45–52 (1964).
303. F. G. Fumi and M. P. Tosi, *J. Phys. Chem. Solids* **25**, 31–43 (1964).
304. S. T. Cui and J. G. Harris, *J. Phys. Chem.* **99**, 2900–2906 (1995).
305. F. J. Armellini and J. W. Tester, *Fluid Phase Equilib.* **84**, 123–142 (1993).
306. E. H. Oelkers and H. C. Helgeson, *Science* **261**, 888–891 (1993).
307. M. Eigen and K. Tamm, *Z. Elektrochem.* **66**, 93–106 (1962).
308. E. Guàrdia, R. Rey, and J. A. Padró, *Chem. Phys.* **155**, 187–195 (1991).
309. G. Ciccotti, M. Ferrario, J. T. Hynes, and R. Kapral, *Chem. Phys.* **129**, 241–251 (1989).
310. A. A. Chialvo, P. T. Cummings, J. M. Simonson, and R. E. Mesmer, *J. Mol. Liq.* **73–74**, 361–372 (1997).

311. L. X. Dang, J. E. Rice, and P. A. Kollman, *J. Chem. Phys.* **93**, 7528–7529 (1990).

312. D. M. Pfund, J. G. Darab, J. L. Fulton, and Y. Ma, *J. Phys. Chem.* **98**, 13102–13107 (1994).

313. J. L. Fulton, D. M. Pfund, S. L. Wallen, M. Newville, E. A. Stern, and Y. Ma, *J. Chem. Phys.* **105**, 2161–2166 (1996).

314. S. L. Wallen, B. J. Palmer, and J. L. Fulton, *J. Chem. Phys.* **108**, 4039–4046 (1998).

315. H. Ohtaki and T. Radnai, *Chem. Rev.* **93**, 1157–1204 (1993).

316. J. E. Enderby and G. W. Neilson, *Rep. Prog. Phys.* **44**, 593–653 (1981).

317. B. J. Palmer, D. M. Pfund, and J. L. Fulton, *J. Phys. Chem.* **100**, 13393–13398 (1996).

318. A. A. Chialvo, P. T. Cummings, J. M. Simonson, and R. E. Mesmer, *J. Chem. Phys.* To appear in Jan. 8, 1999 issue.

319. J. O. M. Bockris and A. K. N. Reddy, *Modern Electrochemistry.* Plenum, New York, 1970.

320. Y. Marcus, *Introduction to Liquid State Chemistry.* Wiley, London, 1977.

321. J. Burgess, *Metal Ions in Solution.* Ellis Horwood, Chichester, 1978.

322. J. G. Kirkwood, *J. Chem. Phys.* **3**, 300–313 (1935).

323. A. A. Chialvo, in *Supercritical Fluids: Fundamental for Application*, NATO ASI. Kemer, Antalya, Turkey, 1993.

324. A. Ben-Naim, *Water and Aqueous Solutions.* Plenum, New York, 1974.

325. J. P. O'Connell, *Mol. Phys.* **20**, 27–33 (1971).

326. E. Matteoli and G. A. Mansoori, eds., *Fluctuation Theory of Mixtures*, Vol. 2. Taylor & Francis, New York, 1990.

327. K. E. Newman, *Chem. Soc. Rev.* **23**, 31–40 (1994).

328. P. G. Kusalik and G. N. Patey, *J. Chem. Phys.* **86**, 5110–5116 (1987).

329. J. P. O'Connell, in *Fluctuation Theory of Mixtures*, (E. Matteoli and G. A. Mansoori, eds.), Vol. 2, pp. 45–67. Taylor & Francis, New York, 1990.

330. B. E. Conway, *J. Solut. Chem.* **7**, 721–770 (1978).

331. A. A. Chialvo, *Fluid Phase Equilib.* **83**, 23–32 (1993).

332. J. G. Kirkwood and F. P. Buff, *J. Chem. Phys.* **19**, 774–777 (1951).

333. S. C. Tucker and E. M. Gibbons, in *Structure and Reactivity in Aqueous Solution* (D. G. Truhlar and C. J. Cramer, eds.), Vol. 568, pp. 196–211. American Chemical Society, Washington, DC, 1994.

334. H. Luo and S. C. Tucker, *J. Phys. Chem.* **100**, 11165–11174 (1996).

335. H. Luo and S. C. Tucker, *J. Am. Chem. Soc.* **117**, 11359–11360 (1995).

336. H. Luo and S. C. Tucker, *Theor. Chem. Acco.* **96**, 84–91 (1997).

337. H. Luo and S. C. Tucker, *J. Phys. Chem. B* **101**, 1063–1071 (1997).

338. J. T. Hynes, in *Theory of Chemical Reaction Dynamics* (M. Baer, ed.), Vol. 4, pp. 171–234. CRC Press, Boca Raton, FL, 1985.

339. C. A. Eckert, *Annu. Rev. Phys. Chem.* **23**, 239–264 (1972).

340. E. Kiran and J. F. Brennecke, eds., *Supercritical Engineering Science: Fundamentals, Studies and Applications*, Vol. 514. American Chemical Society, Washington, DC, 1993.

341. A. A. Clifford, in *Supercritical Fluids: Fundamentals for Application* (E. Kiran and J. M. H. Levelt Sengers, eds.), Vol. E-273, pp. 449–497. Kluwer Academic Publishers, Dordrecht, The Netherlands, 1994.

342. B. L. Knutson, A. K. Dillow, C. L. Liotta, and C. A. Eckert, in *Innovations in Supercritical Fluids: Science and Technology* (K. W. Hutchenson and N. R. Foster, eds.), Vol. 608, pp. 166–178. American Chemical Society, Washington, DC, 1995.

343. J. F. Brennecke, in *Supercritical Engineering Science: Fundamentals, Studies and Applications* (E. Kiran and J. F. Brennecke, eds.), Vol. 514, pp. 201–219. American Chemical Society, Washington, DC, 1993.

344. C. A. Eckert, D. H. Ziger, K. P. Johnston, and T. K. Ellison, *Fluid Phase Equilib.* **14**, 167–175 (1983).

345. R. Rey and E. Guàrdia, *J. Phys. Chem.* **96**, 4712–4718 (1992).

346. J. P. O'Connell, A.V. Sharygin, and R. H. Wood, *Ind. Eng. Chem. Res.* **35**, 2808–2812 (1996).

347. J. Sedlbauer, E. M. Yezdimer, and R. H. Wood, *J. Chem. Thermodyn.* **30**, 3–12 (1998).

POLAR AND NONPOLAR SOLVATION DYNAMICS, ION DIFFUSION, AND VIBRATIONAL RELAXATION: ROLE OF BIPHASIC SOLVENT RESPONSE IN CHEMICAL DYNAMICS

BIMAN BAGCHI AND RANJIT BISWAS

Solid State and Structural Chemistry Unit, Indian Institute of Science, Bangalore 560 012, India

CONTENTS

Advances in Chemical Physics, Volume 109, Edited by I. Prigogine and Stuart A. Rice
ISBN 0-471-32920-7

I. INTRODUCTION

Rate constants of chemical reactions in solutions are found to vary over a wide range. Although many chemical processes occur with extreme rapidity, with time constants in the tens of femtosecond range, there also exist many reactions that are quite slow, with time constants even in the seconds or minutes [1–23a]. Many factors control the rate of a given chemical reaction. These include the activation energy, the nature of the reaction potential energy surface, vibrational energy relaxation (VER) of a bond, the viscosity of the medium, and the polarity [1] of the solvent. The problem is complex because all these factors are often correlated with each other. For example, the activation energy may depend strongly on the solvent polarity, whereas the viscosity effects on chemical rates are determined, at least partly, by the shape of the reaction barrier [4,5]. The vibrational energy relaxation is profoundly affected by solvent forces that act as a sink of the energy dissipated. When the harmonic frequency of the bond is small (on the order of a few hundred cm^{-1}), then the frictional forces responsible for vibrational energy relaxation are essentially the same as those that couple to activated barrier crossing dynamics and to nonpolar solvation; all the three processes couple only to the high-frequency frictional response of the liquid. Recent progress in ultrafast laser spectroscopy has made study of these phenomena possible.

Much attention has recently been focused on the dynamics of dipolar liquids because there are many chemical reactions that involve separation or reorganization of charges. These reactions often occur in polar solvents because the polarity of the medium favors and stabilizes the charge separation. Because dipolar liquids consist of molecules having a permanent dipole moments, they exhibit strong intermolecular orientational correlations somewhat akin to the liquid crystals. At the same time, these liquids also sustain spatial density fluctuations that closely resemble the properties of the simple atomic fluids. Apparently, it is this unique combination of the spatial and orientational properties that renders the dipolar liquids so important among all the complex fluids. An additional factor favoring natural selection of dipolar liquids as chemical and biological solvents is that the stabilization energy of charged or polar molecules depend strongly on the nature of the charge distribution.

Solvation dynamics is not the only field that has witnessed intense activity in recent years. There has been a renewed interest in understanding the transport properties, such as the diffusion and viscosity of molecular liquids, the ionic mobility in electrolyte solution, and the vibrational phase and energy relaxations. Many of these processes occur on a femtosecond time scale. Aided by the recent advances in ultrafast laser spectroscopy, significant progress has been made in understanding many of these processes, and the correlations among these relaxation processes are beginning to emerge. It is fair to say that the whole field of classical physical chemistry has undergone a rejuvenation in recent times.

If there is one theme in solution phase chemical reaction dynamics that has now become recurrent, it is the importance of the biphasic solvent response in diverse relaxation processes. Before the 1980s, the emphasis was on understanding the slower aspects of the reaction dynamics. For example, what are the conditions in which the rate of a chemical reaction can be defined [4a]? In the case of slower relaxation processes, the use of a hydrodynamic description was acceptable. One representative example of such a use could be found in Kramers's theory of the activated barrier crossing reactions [4].

Perhaps the first break from this line of thinking appeared in the Grote–Hynes non-Markovian rate theory [5], which suggests that, for a vast majority of reactions, it is the high-frequency solvent response that couples to the barrier crossing dynamics. Interestingly, the zero frequency response has no direct relevance here. Vibrational energy relaxation is another example in which the high-frequency solvent response solely determines the rate, as embodied in the simple Landau–Teller rate expression [24]. It has been pointed out [23a] that even in vibrational phase relaxation, the ultrafast solvent response can give rise to a novel subquadratic quantum number dependence of the vibrational overtone dephasing [23,23a].

The objective of this review is to discuss the recent advances in understanding the role of this biphasic solvent response in various elementary relaxation processes, such as the ionic mobility, and vibrational energy and phase relaxations. Although the dominance of the ultrafast component came somewhat as a surprise, it was already well known in the dynamics of monoatomic liquids from computer simulation studies [25] and the mode-coupling theory [26,26a,26b]. It is the near universal presence of such an extremely fast component in molecular liquids that has created a lot of interest and motivated new progresses in various directions.

The organization of the rest of the following sections is as follows. We first introduce the polar solvation dynamics and present a general but brief discussion of earlier experimental and theoretical studies in this field. Then we turn to different but interconnected topics, such as the ionic mobility, nonpolar solvation dynamics, and vibrational energy relaxation. In between, we digress momentarily to examine solvation dynamics in supercritical water, which is rapidly becoming a topic of intense research because of its wide applications in chemical industry.

A. Polar Solvation Dynamics

1. *Solvation Time Correlation Function*

In experiments, the time-dependent progress of solvation of a newly created charge distribution is usually followed by measuring the time-dependent fluorescence Stokes shift (TDFSS) of the emission spectrum of the solute molecule [27–31]. A dilute solution of large dye molecules (such as coumarin, Nile red, or prodan) is used as solute probe. These molecules act as efficient probes for the following reasons. First, they undergo a large change in the dipole moment upon laser excitation, or may even photoionize. This creates a large polar stabilization energy in the excited state. Second, these dye molecules exhibit fluorescence in the excited state with a long life

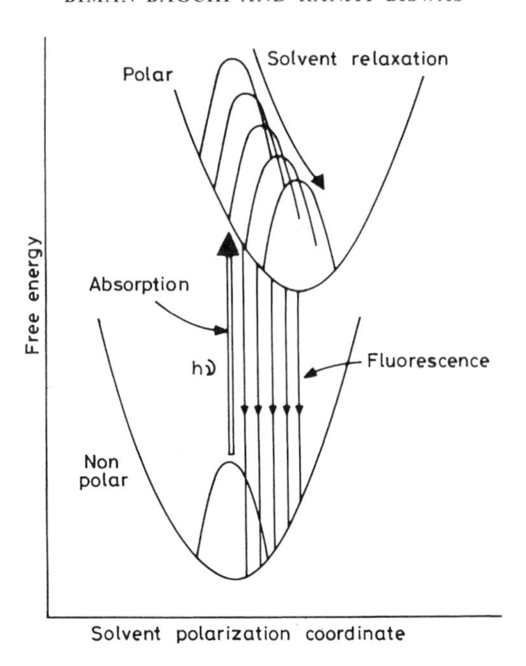

Figure 1. The physical processes involved in the experimental study of solvation dynamics. The initial state is a nonpolar ground state. Laser excitation prepares it instantaneously in a charge-transfer (CT) state, which is a polar state. This particular state is a highly nonequilibrium state, because at time $t = 0$, the solvent molecules are still in the Franck–Condon state of the ground state. Subsequent solvation of the CT state brings the energy of the system down to the final equilibrium state, which is detected by a red shift of the time-dependent fluorescence spectrum.

time (on the order of nanoseconds). More recently, higher-order nonlinear optical measurements such as three pulse photon echo peak shift (3PEPS) measurements have been carried out to study the solvation dynamics [32,33].

Physically, the process of solvation of a solute probe may be described as follows. Consider that a solute chromophore in its ground state is in equilibrium with the surrounding solvent molecules and the equilibrium charge distribution of the former is instantaneously altered by a radiation field. Ideally, when the solute–solvent system undergoes an optical Franck–Condon transition upon excitation, the equilibrium charge distribution of the solute is instantaneously altered. The solvent molecules still retain their previous spatial and orientational configuration. This is a highly nonequilibrium situation for the system. The energy of the Franck–Condon state is higher than the minimum of the potential energy in the excited state (Fig. 1). Subsequent to the excitation, the solvent molecules rearrange and reorient themselves to stabilize the new charge distribution in the excited state. The resultant energy is the solvation energy of the solute. The time dependence of the rearrangement of the solvent environment (solvation) is reflected in the continuous red shift of the emission spectrum. The temporal characteristics of solvation are then followed by monitoring the spectral response function [1,15,15a,29–31]

$$S(t) = \frac{\nu(t) - \nu(\infty)}{\nu(0) - \nu(\infty)} \tag{1.1}$$

where $\nu(t)$ denotes the time-dependent emission frequency of the solute chromophore. This is also termed the solvation time correlation function. This function is properly normalized and decays from unity at $t = 0$ to zero at $t = \infty$. The solvation time correlation function may also be written as [15,15a,34–36]

$$S(t) = \frac{E_{\text{sol}}(t) - E_{\text{sol}}(\infty)}{E_{\text{sol}}(0) - E_{\text{sol}}(\infty)} \tag{1.2}$$

where $E_{\text{sol}}(t)$ is the solvation energy of the solute at time t. The solvent response function $S(t)$ is a nonequilibrium correlation function. When the solvent response is linear to the external perturbation, $S(t)$ is equivalent to the equilibrium energy–energy time correlation function $S_E(t)$ [34–38], which is defined as [34–38a]

$$S_E(t) = \frac{< E_{\text{sol}}(0)E_{\text{sol}}(t) >}{< E_{\text{sol}}(0)E_{\text{sol}}(0) >} \ . \tag{1.3}$$

This assumption substantially simplifies the description of dynamics of solvation, because one now needs to consider only the equilibrium correlation function.

The time-dependent solvation energy $E_{\text{sol}}(t)$ may be decomposed (in a somewhat *ad hoc* but physically meaningful manner) into several components:

$$E_{\text{sol}}(t) = E_{\text{P}}(t) + E_{\text{NP}}(t) + E_{\text{intra}}(t) + E_{\text{solv}}(t) \tag{1.4}$$

where the relaxation of the potential energy derives contributions from the time-dependent electrostatic interaction energy between the polar solute and the dipolar solvent molecules, given by $E_{\text{P}}(t)$; the solute–solvent nonpolar interaction energy $E_{\text{NP}}(t)$; the relaxation of intramolecular vibrational modes of the solute, denoted by $E_{\text{intra}}(t)$; and the change in the interaction among the solvent molecules in the excited state $E_{\text{solv}}(t)$.

The observed spectral dynamics can be attributed to $E_{\text{P}}(t)$ only when the contributions of the other three terms are negligible. This is justified under the following conditions. First, the intramolecular vibrational energy relaxations occur on a much faster time scale. Second, the nonpolar solvent response is not probed, because there is little change upon excitation in shape and size of the solute probe. Third, the interactions among the solvent molecules remain the same in both the ground and the excited electronic states of the solute; therefore, the contribution of $E_{\text{solv}}(t)$ is negligible.

2. *Continuum Models*

Initial understanding of the time-dependent solvation process was offered by the continuum model [39–43]. This model is a generalization of the equilibrium solvation model of Born [44] and Onsager [45] to the time domain. In this model, the dipolar solvent is represented by a homogeneous dielectric continuum, characterized by the frequency-dependent dielectric function $\varepsilon(\omega)$, and the polar solute is replaced by a spherical molecular cavity [41]. The solvation energy is obtained by evaluating the reaction field of the polar solute molecule inside the cavity, which is obtained by a quasistationary boundary value calculation. For a dipolar solute, the continuum model predicts that the solvation energy time correlation function is a single exponential, with a time constant given by [41]

$$\tau_{\mathrm{L}}^{\mathrm{d}} = \left(\frac{2\varepsilon_{\infty} + \varepsilon_c}{2\varepsilon_0 + \varepsilon_c} \right) \tau_{\mathrm{D}} \tag{1.5}$$

where ε_0 and ε_{∞} are static and infinite frequency dielectric constants of the solvent, respectively. τ_{D} is the Debye relaxation time of the dipolar medium. These dielectric relaxation parameters (ε_0, ε_{∞} and τ_{D}) for a particular solvent are generally obtained by fitting to low-frequency experimental relaxation data and do not contain any high-frequency response. ε_c is the dielectric constant of the molecular cavity and is equal to unity for a spherical cavity. For ionic solvation, the continuum model prediction for the decay of $S(t)$ is also a single exponential; the time constant is given by [46]

$$\tau_{\mathrm{L}}^{\mathrm{ion}} = \left(\frac{\varepsilon_{\infty}}{\varepsilon_0} \right) \tau_D \tag{1.6}$$

For common dipolar solvents, $\varepsilon_0 >> \varepsilon_{\infty}$ and thus the difference between $\tau_{\mathrm{L}}^{\mathrm{d}}$ and $\tau_{\mathrm{L}}^{\mathrm{ion}}$ is not significant. Therefore, the homogeneous continuum model predicts that both for ionic and for dipolar solvation, the decay of the solvation energy time correlation function is exponential, with a time constant τ_{L}. Continuum models, of course, predict more complex dependence when the frequency-dependent dielectric relaxation $\varepsilon(\omega)$ is non-Debye.

3. *Early Experimental Investigations: Phase I*

Initial experimental studies on TDFSS subsequent to electronic excitation were performed in the late 1970s and early 1980s by using picosecond lasers. In one of the earliest studies [47], Stokes shift relaxation in 2-amino-7-nitrofluorene in isopropanol was investigated at various wavelengths and temperatures. A strong wavelength dependence of the relaxation was found, and the decay was distinctly slower in red tail of the spectrum than in the blue edge. The kinetics of solvation of aromatic dyes in polar solvents were studied, and the cooperative motion of solvent molecules was assigned an important role in solvation [48]. Okamura and co-workers [49] examined Stokes shift dynamics in 1-naphthylamine in 1-isopropanol. The relaxation times for both the decrease in intensity at blue end and the increase at red end were found to be equal to 52 ps, a value that is in somewhat good agreement with the continuum model prediction of 33 ps for this particular system. These studies were not unambiguous, however, as the solute probe could have formed exciplex with the solvent.

Subsequently, nonlinear spectroscopic techniques with subpicosecond laser pulses were developed. These techniques were characterized by a much improved signal : noise ratio and better time resolution. These develepements naturally spurred an upsurge in the study of TDFSS of several solute chromophores in different polar solvents. For example, the coumarin derivatives [34,50] 1-amino-naphthalene [5] and 4-amino-phthalimide [52,53] were used as probes. Molecules with more complicated photophysics — e.g., LDS-750 [54] bianthryl [53], 4-(9-anthryl)-*N,N*-dimethylaniline [53], bis(4-aminophenyl)sulfonate sulfone [55,56], and Nile red [57] — were also investigated. Protic, hydrogen-bonded solvents, such as alcohols and amides [51,52,54-58]; and aprotic solvents, such as *n*-nitriles [50], glycerol triacetate [53], dimethyl sulfoxide (DMSO) [54], dimethyl formamide (DMF), and propylene carbonate [34,50,59], were studied.

The main results of the above measurements may be summarized as follows.

1. The observed solvation times are largely insensitive to the probe size and depend primarily on the properties of the polar solvent studied.

2. The solvation time correlation function $S(t)$ was found to be nonexponential. This nonexponentiality prevailed not only at short times but also at long times ($t > \tau_L$), indicating the presence of a distribution of relaxation times. The $S(t)$ versus t curve could often be well represented by a stretched exponential form, such as the Kohlrausch–Williams–Watts function [60,61].

3. The average solvation time τ_s, defined as a time integral of $S(t)$, is generally larger than τ_L and usually lies between τ_L and τ_D. In some instances, solvation times were more than an order of magnitude longer than τ_L.
4. The solvation time for LDS-750 in methanol and butanol was found to be smaller than τ_L. Solvent translational modes were believed to be responsible for the faster relaxation observed in these solvents.
5. The observed deviation of τ_s from τ_L could be related to the static dielectric constant ε_0 (or $\varepsilon_0/\varepsilon_\infty$).
6. Strongly nonexponential relaxation and even exponential relaxation on a different time scale from τ_L^{ion} were observed [62].
7. Furthermore, the simulation studies of solvation dynamics in water by Maroncelli and Fleming [35] showed the importance of the solvent structure in the dynamics of solvation.

These observations indicate the breakdown of the simple continuum model. The failure of the continuum model could be attributed to neglect of the details of the solute–solvent interactions and the spatial and the orientational correlations that are present in dense liquid.

4. *Inhomogeneous Continuum Models*

Subsequently, the continuum model was extended in many directions to include the shape variation of the solute and the space dependence in the frequency-dependent dielectric function. Two inhomogeneous models were proposed [42–43] that considered the space dependence of the frequency-dependent dielectric function. The first model assumes a continuous variation of the dielectric function with distance (r) from the polar solute [42]. The second model treats the space inhomogeneity via a discrete shell representation of the position-dependent dielectric function [43]. In both models, $\varepsilon(\omega)$ is replaced by $\varepsilon(r, \omega)$. These models were found to be useful in studying the solvation dynamics in restricted environments, such as the γ-cyclodextrin cavity [63]; however, they are phenomenological and do not systematically incorporate the effects of molecular-level spatial and orientational orders. The effect of the specific interaction (such as hydrogen bonding) on solvation dynamics was also neglected.

5. *Experimental Discovery of Ultrafast Polar Solvation: Phase II*

The study of the ultrafast polar solvation dynamics using the subpicosecond laser pulses was first reported by Rosenthal and co-workers [27,27(a)], who investigated the solvation dynamics of the dye LDS – 750 in acetonitrile. The decay of the solvation energy time correlation function $S(t)$ was found to be dominated by an ultrafast, Gaussian component with a time constant of about 100 fs. This ultrafast component accounts for nearly 70% of the total decay, and the rest is carried out by an exponential relaxation. The solvation process is even more exciting and faster for water [29]. Here, the initial, ultrafast Gaussian component decays with a time constant as small as 40 fs, and the amplitude of the ultrafast component is nearly 80%. The long-time part of the decay is fitted to a biexponential, with time constants 240 fs and 860 fs. The presence of the ultrafast component in polar solvation dynamics has also been confirmed for many other liquids, such as methanol [64–66] and amides [38].

The dominance of the ultrafast polar component is not universal for all the dipolar liquids. For example, the amplitude of the ultrafast component in acetonitrile and water ranges between 60 and 80%, whereas that in methanol and formamide accounts for only 30–40%. This variation in the solvent response is clearly a manifestation of the inherent uniqueness of each of the dipolar liquids, for which the rapidity of the response is determined by the coupling of the static properties (e.g., ε_0, ε_∞, μ, and ρ) with the natural dynamics of the medium.

6. *Solvation Dynamics in Supercritical Water*

Supercritical water (SCW) has become a topic of intense research in recent years [67–70]. The critical point of water is located at $T_c = 647$ K and $P_c = 22.1$ MPa, and the critical density ρ_c is 0.32 g/cm^3. In the supercritical condition, water loses its three-dimensional hydrogen-bonded network and becomes completely miscible with organic compounds. The dielectric constant of SCW is close to six, which transforms water into an organic-like environment. This particular feature has made water a potential solvent for waste treatment in the chemical industry. The high compressibility of water above its critical point allows large variations of the bulk density through a minor change in the applied pressure. This characteristic can profoundly affect the solubility [68], transport properties [69], and the kinetics of the reactive processes in SCW [70]. Therefore, a clear understanding of the dynamical response of SCW is necessary and important for understanding and substantiating the dynamic solvent effects on the rate of the chemical reactions that occur in this medium. Recently, Re and Laria [71] performed molecular dynamic simulation studies on solvation dynamics in SCW and observed an ultrafast component in the solvation time correlation function. They also found that in SCW with $\rho = 1$ g/cm^3 the solvation process is nearly an order of magnitude faster at T=645 K than the simulated results for normal water at room temperature.

B. Solvation Dynamics in Mixed Solvents

Mixed dipolar solvent offers a good reaction medium, because one can tune the polarity of the solution by altering the composition of the mixture. This tunability may be the reason for many reactions occurring in mixed solvents inside the living cells. Mixed solvents are also commonly used in the chemical industry for solvent extraction of desirable compounds from the product mixture. Although a considerable number of theoretical [72] and simulation studies [73–75] on dielectric properties of dipolar mixtures has been carried out, solvation dynamics in mixtures has not received much attention — a melancholy tribute to the inherent difficulties in treating the complex interactions and their mutual effects on system dynamics in mixture. The preferential solvation by a species of the mixture can drastically alter the various relaxation processes of the medium; therefore, a change in the solute–solvent microdynamics in the mixture compared to that in the pure solvent may be observed [76].

The Stokes shift dynamics of 8-amino-1-naphthalenesulfonic acid in a water–ethanol mixture was investigated by Robinson and co-workers [77,78]. They observed a nonexponential decay of the fluorescence in the mixture. This is interesting, since the decay in either of the pure solvents was observed to be exponential. Robinson and co-workers [77,78] pointed out that in mixed solvents, the additional process of solvent exchange around the excited, highly polar solute would play a significant role in the Stokes shift dynamics.

C. Dynamics of Electron Solvation

When low-energy electrons are injected into polar fluids, a series of relaxation processes takes place, which evantually leads to the solvation of the electron. Many theoretical [79], experimental [80–85], and computer simulation studies [86–93] have already been focused on understanding the solvation process of a newly created electron. Rossky and co-workers [86–90] carried out extensive quantum molecular dynamic simulations to investigate in detail the electron solvation dynamics in water. Their simulations take full account of the quantum charge distribution of the solute coupled to the dielectric and mechanical response of the solvent. The solvent response function has been found to be bimodal with an initial Gaussian component of 25 fs, which carries ~40% of the total decay [88]. The rest is carried out by an exponential component with a time constant of about 250 fs. This study showed that the relaxation of the quantum energy gap between electronic eigenstates due to solvation plays a direct role in the nonradiative decay dynamics of the excited state electron. These studies have also indicated that the low-frequency translational motions of the solvent can affect both the inertial and diffusive dynamics rather significantly [88].

Barbara and co-workers [85] investigated the solvation dynamics of the hydrated electron in water by using femtosecond pump-probe spectroscopy and measuring in 35 fs resolution. These authors measured the solvent response function at different wavelengths for both normal water (H_2O) and heavy water (D_2O). This experimental study confirmed many of the predictions made by the simulation studies of Rossky and co-workers [88]. For example, the observed dynamics primarily reflected the predicted p-state adiabatic solvation, not the nonadiabatic $p \rightarrow s$ transition. The authors also experimentally confirmed that the hydrated electron solvation dynamics in water are predominantly biphasic, in which the initial ultrafast component arises from the librational motion of water molecules. The slow, long-time part is obviously the result of the diffusive motions of the solvent molecules. These machanisms are similar to those for ion solvation dynamics in polar liquids. The time constant of the initial ultrafast component measured in the visible region ranges between 50 and 70 fs. Unfortunately, this observation is not in accordance with the simulation results of Rossky and co-workers [88]. The reason for this discrepancy is not known and definitely deserves further study.

Rips [79] studied the electron solvation dynamics in polar liquids by using a hydrodynamic model [79]. In this model, the electron is assumed to be localized within a spherical cavity in an ideal incompressible liquid, in which the driving force for the solvation is the contraction of the cavity size to its equilibrium value. The competition between the gain in electron–solvent interaction energy and the increase in kinetic energy of the electron determines the contraction dynamics. The numerical results of this simple model study are in good agreement with the experimental results [80-81, 84]. Recently, a bubble model was proposed to study the adiabatic electron solvation dynamics in water, for which the molecular hydrodynamic theory was used to calculate the solvation time correlation function [94]. The results obtained in the bubble model calculation were also in good agreement with those from simulation studies of Rossky and co-workers [88].

D. Solvation Dynamics in Nonpolar Liquids

Nonpolar solvation dynamics usually refers to the relaxation of the instantaneously changed solute–solvent interaction energy owing to short-range (nonpolar) interactions. The distinction between polar and nonpolar is somewhat arbitrary because quadrupolar and octupolar interactions are certainly as short range as the Lennard–Jones interaction. Loosely, one attributes the nonpolar solvation dynamics to the energy relaxation (as manifested in the Stokes shift) observed when a nonpolar solute (e.g., dimethyl-s-tetrazene) is excited in nondipolar liquids, such as butyl benzene and 1,4-dioxane [95–104].

A considerable amount of understanding of nonpolar solvation dynamics has recently been achieved. This was possible only after the availability of sophisticated nonlinear spectroscopic techniques, such as 3PEPS [30,32,33] and transient hole burning measurements [96–104]. Theoretical [105–110] and instantaneous normal mode (INM) [111–118] studies have been performed to understand the basic mechanism of nonpolar solvation. The general observations are interesting and no less exotic than those found in polar solvation dynamics. For example, the solvation time correlation function constructed for nonpolar solvation dynamics has a pronounced biphasic character, with an ultrafast component on the subpicosecond time scale [95–103]. The scenario is even more interesting for supercooled liquid near the glass transition temperature, at which $S(t)$ exhibits highly nonexponential relaxation [105].

The driving force for nonpolar solvation is often the change in shape and size of the solute upon excitation [95–103]. The ultrafast component arises from the nearest neighbor solute–solvent cage dynamics. The degree of dominance of this component depends largely on the solute–solvent interacion strength. The long-time part of the solvent response is primarily governed by the structural relaxation of the solvent. The combination of these two relaxation processes is responsible for the bimodal frictional response of the solvent. Interestingly, for nonpolar solvation dynamics, unlike the case of polar solvation, the orientational relaxation is found to play an insignificant role.

E. Vibrational Relaxation

Vibrational relaxation in liquids is a subject of great importance in liquid phase chemistry [119–124], because the study of vibrational dynamics offers a window to probe directly the chemical bond's properties and motions. The advantage of this can be understood if one compares the study of vibrational mode with the other dynamic studies of the liquid state, such as dielectric relaxation, NMR, neutron scattering, and solvation dynamics. In latter experiments, one mainly observes the collective dynamics of the liquid, and information regarding the motion of a chemical bond and its interaction with the sorrounding solvent molecules is rarely present. In contrast, vibrational line-shape analyses and vibrational energy relaxation studies provide direct information about the interaction of a chemical bond with its surroundings.

The low-frequency ($<100\ \mathrm{cm}^{-1}$) vibrations of solvent molecules are usually described in terms of classical harmonic oscillators. In the liquid state, the molecules are colliding with each other incessantly. These collisions can be either elastic or inelastic. Elastic collisions give rise to a phase shift only because there is no energy transfer involved. This is called vibrational phase relaxation (VPR), or vibrational dephasing, and is an equilibrium phenomenon. On the other hand, inelastic collisions induce papulation redistribution in the vibrational levels, because it involves the transfer of momentum. This is known as vibrational energy relaxation (VER), or vibrational papulation relaxation. This is a nonequilibrium phenomenon, and is usually a much slower process than vibrational dephasing. For example, the typical time scale of VPR lies approximately between 0.1 ns and 1 ps, whereas that for VER ranges from tens of picoseconds to milliseconds [119–122] and may even be slower: VERs involving a time scale of seconds have been measured [125].

Vibrational dephasing has been traditionally studied by using Raman and IR line shapes. Recently developed time-resolved techniques, which complement earlier studies based on line shapes, have revealed interesting information about vibrational relaxation. For example, subquadratic quantum number dependence of the dephasing rate of the C–D stretching of $CDCl_3$ in neat liquid and in a $CHCl_3 + CD_3I$ solution has been observed [23,23a]. An unexpected fast VER of the first vibrationally excited state of the O–H mode for HOD in D_2O has also been observed [126].

F. Limiting Ionic Conductivity

1. Kohlrausch's Law and Walden's Rule

The discovery of the biphasic polar solvent response has generated renewed interest in the study of the transport properties in electrolyte solutions. One of the most important quantities here is the ionic conductance (Λ) of electrolyte solutions in various polar solvents such as water, acetonitrile, alcohols, and amides [127–131]. In the low concentration limit, the ionic conductance of strong electrolytes reveals the square root of the concentration (c) dependence. This is known as the Kohlrausch's law and is expressed as follows [127–130]:

$$\Lambda_m = \Lambda_0 - \kappa\sqrt{c} \tag{1.7}$$

where Λ_m represents the equivalent molar conductivity, Λ_0 is its limiting value at infinite dilution, κ is a coefficient that is found to depend more on the nature of the electrolyte (i.e., whether it is a uniunivalent or biunivalent) than on its specific identity. Λ_0 is generally obtained by extrapolating experimental Λ_m values at zero ionic concentration and is one of the most easily accessible transport quantities. Some earlier experimental results involving large ions indicated that the limiting ionic conductivity Λ_0 of an univalent ion in a particular solvent is proportional to the crystallographic radius of the ion and varies inversely as the solvent viscosity η_0. Thus, for a particular ion with a specific charge on it, the product of Λ_0 with solvent viscosity η_0 should be constant. This is known as Walden's rule and is expressed as follows [127–130]

$$\Lambda_0 \eta_0 = \text{constant} \tag{1.8}$$

This relation is purely an empirical one based only on experimental observations.

2. *Experimental Observations: Breakdown of Walden's Rule*

Eq. (1.8) can be rationalized in terms of the well-known Stokes law [127–130], which predicts that the friction on the ion is inversely proportional to the crystallographic radius of the ion (r_{ion}) and to the viscosity. The use of Einstein's relation between the friction and the diffusion coefficient (which is essentially Λ_0) then produces the Eq. (1.8). Experimental results, however, indicate that the ionic mobilities in polar solvents do not always decrease monotonically with increasing radius. Instead, there is often a maximum as $\Lambda_0 \eta_0$ is plotted against r_{ion}^{-1}. Figure 2 shows one set of these results depicting the breakdown of Walden's rule. Note that Stokes's law seems to work reasonably well for large ions, but for smaller ions it breaks down completely; and the breakdown starts showing up as the ion : solvent size ratio approaches unity. For smaller ions like Li^+ and Na^+, the breakdown is complete, and the deviations from Stokes's law are quite large.

G. Scope of the Review

The objective of this chapter is twofold. First, we present a review of the present status of the theory of solvation dynamics and its position with respect to recent experiments. The emphasis is admittedly on work performed by our own group; we present others' work as a brief review only. The second objective is to articulate the understanding that has emerged in several related problems of chemical dynamics in which solvation plays a role. In other words, we explore the interrelationship among several different problems of chemical reaction dynamics. The focus here is on solvation dynamics, ionic mobility, and the vibrational energy and phase relaxations.

Several reviews have appeared on related topics [1–3,13–15a,18,20]. Mention must be made of the elegant review by Maroncelli [31], which still provides an elaborate introduction to the field of solvation dynamics. The present review can be considered a continuation of Bagchi and Chandra's [15a]. Many of the earlier reviews were written before the full impact of the biphasic solvent response was realized. Thus the focus of this Chapter is somewhat different: The emphasis is not only on collective dynamics of pure

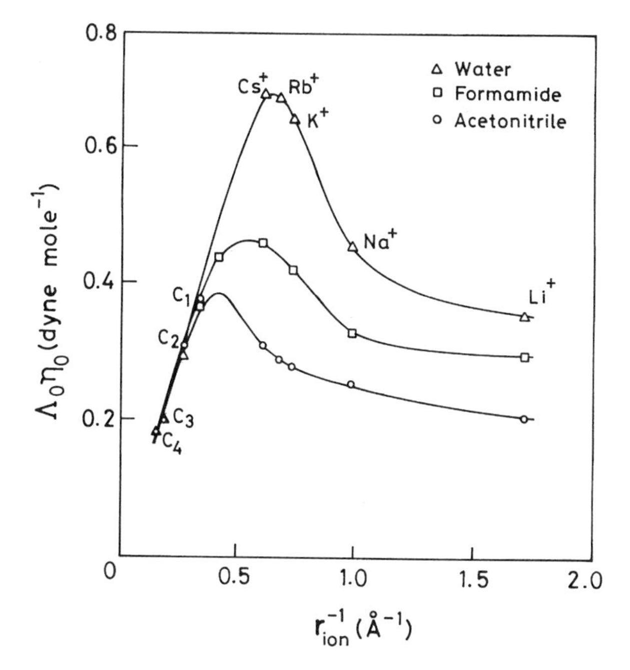

Figure 2. Experimental values of the Walden product ($\Lambda_0\eta_0$) of rigid, monopositive ions in water (*open triangles*), acetonitrile (*open circles*), and formamide (*open squares*) at 298 K are plotted as a function of the inverse of the crystallographic ionic radius r_{ion}^{-1}. The *solid line* in each case is simply a visual aid. Here, the tetra-alkyl ammonium ions are represented by C_1-C_4 where $C_n = (C_nH_{2n+1})_4N^+$; n is 1, 2, 3, or 4. The crystallographic radii of the ions are taken from Ref. [131a].

dipolar liquids but also on the role of the biphasic solvation dynamics in various elementary chemical processes.

Section II provides a brief review of theoretical and experimental work before 1993; the latter sections cover more recent work.

II. PROGRESS IN THE DEVELOPMENT OF MICROSCOPIC THEORIES

A. Introduction

Several molecular theories were developed for the polar solvation dynamics that took into account the microscopic structure of dense liquid to explain the experimental results. Early theoretical studies were also motivated by an interesting comment by Onsager [132]: "The polarization structure around an electron would form from outside in." This is the famous inverse snowball effect, which suggests that the relaxation of solvation energy is intrinsically nonexponential, because several length-dependent time scales are involved.

1. *Theory of Calef and Wolynes*

In an elegant approach, Onsager's inverse snowball model was tested by Calef and Wolynes [133], who used a new molecular hydrodynamic theory of solvation dynamics in dipolar liquids. In this approach the solvation was assumed to be carried out solely by the rotational diffusive motion of the solvent molecules. It was found that the inverse snowball picture was indeed valid in the absence of the translational modes. Calef and Wolynes's [133] calculations were based on the Smoluchowski–Vlasov equation. The physical picture behind this theory was later clarified by Chandra and Bagchi [15 (b)] (discussed later).

2. Dynamic Mean Spherical Approximation Model

Wolynes [134] further studied solvation dynamics by extending the linearized equilibrium theories of solvation [135–137] to the time domain. In this approach, the solute was treated as hard sphere and the solvent molecules as dipolar hard spheres. It was shown [134] that if the solvent response is linear to the time-dependent variation of the charge distribution on the solute, then the solvation energy relaxation could be described by a biexponential relaxation function, with one time constant being nearly equal to the longitudinal time constant of the solvent. The second time constant is comparatively larger and originates from the slow structural relaxation of the neighboring solvent molecules.

Although there are many appealing features in Wolynes's [134] theory, the expressions derived were not quantitatively accurate, because many approximations were involved and the solution was approximate. Rips and coworkers [138,139] presented an exact solution of Wolynes's dynamic mean spherical approximation (DMSA) model for both ion [138] and dipolar [139] solvation dynamics. The following expression of $S(t)$ for ion solvation was derived [138]:

$$S(t) = \mathcal{L}^{-1} \frac{[\chi(z) - \chi(0)]}{z[\chi(\infty) - \chi(0)]} \tag{2.1}$$

where $\chi(z)$ is the complex admittance of the solvent. This was related to $\varepsilon(z)$ as follows [138]:

$$\chi(z) = \frac{1}{2R_i[1 + \Delta(z)]} \left[1 - \frac{1}{\varepsilon(z)}\right] \tag{2.2}$$

where R_i is the radius of the ion, and $\Delta(z)$ is the dynamic correction to the ionic radius. This correction is equivalnt to the well-known Gurney–Frank correction in the equilibrium case and stands for the dynamic screening of the effects associated with the polarization of the solvent. $\Delta(z)$ was shown to be given by [138]

$$\Delta(z) = \frac{3r_s}{R_i} \frac{1}{(108)^{\frac{1}{3}}[\varepsilon(z)]^{\frac{1}{6}} - 2} \tag{2.3}$$

where r_s is the radius of a solvent molecule.

When Eqs. (2.1)–(2.3) were solved, Rips and co-workers [138] obtained a nonexponential decay of the solvation time correlation function. The energy relaxation was also found to be sensitive to the static dielectric constant of the solvent and the solute : solvent size ratio.

Wolynes's method was later extended to treat dipolar solvation [139]. The expressions are more involved. The resultant solvation time correlation function was found to be slower for dipolar solvation than that for the ionic solvation.

3. *Theory of Chandra and Bagchi: Importance of Solvent Translational Modes*

Chandra and Bagchi [15,15a,140–142] extended the molecular hydrodynamic treatment of Calef and Wolynes [133] to study the solvation dynamics in dense dipolar liquids. This approach led to a Markovian theory, as the memory effects in the frictional response were neglected. They used a generalized Smoluchowski–Vlasov equation to study the polarization density fluctuations. Both the rotational and the translational modes of the solvent were incorporated. This work was based on the idea that at the molecular length scale the relaxation of the number density would be slow; thus it could be treated as a conserved variable. Therefore, the hydrodynamic theory could be used to describe the dynamics of the medium. The only rectification needed was to modify suitably the hydrodynamic equations so that the effects of intermolecular correlations could be accounted for. This was successfully performed by using the density functional theory [26,26a,26b,143]. Chandra and Bagchi [15a] used the time correlation function formalism coupled with the basic principles of statistical mechanics to obtain the following expression for the longitudinal (i.e., $\ell = 1$, $m = 0$) component of the orientational polarization density relaxation of the solvent [15a]:

$$\langle P_{\mathrm{L}}(-\mathbf{q})P_{\mathrm{L}}(-\mathbf{q},t)\rangle = \langle P_{\mathrm{L}}(-\mathbf{q})P_{\mathrm{L}}(-\mathbf{q})\rangle e^{-t/\tau_{\mathrm{L}}(q)} \tag{2.4}$$

where the equilibrium polarization fluctuation correlation function $\langle P_{\mathrm{L}}(-\mathbf{q})P_{\mathrm{L}}(-\mathbf{q})\rangle$ was calculated from total correlation function $h(110; q)$

as follows [15a]:

$$\langle P_{\rm L}(-\mathbf{q})P_{\rm L}(-\mathbf{q})\rangle = \frac{N}{3}\left[1 + \frac{\rho_0}{4\pi}h(110; q)\right] \tag{2.5}$$

where N is the total number of solvent molecules present in the system. The longitudinal relaxation time constant $\tau_{\rm L}$ was also obtained from the pair correlation function [15a]:

$$\tau_{\rm L}^{-1}(q) = 2D_{\rm R}\{[1 + p'(q\sigma)^2][1 - (\rho_0/4\pi)c(110, q)]\} \tag{2.6}$$

where σ is the diameter of a solvent molecule and $p' = D_{\rm T}/2D_{\rm R}\sigma^2$. This theory was found to be useful in understanding various aspects of the collective orientational relaxation in dense dipolar liquids. The most important prediction of this theory was that a significant contribution from the translational mode could lead to the breakdown of Onsager's conjecture [132]. The importance of the solvent translational mode in the polarization relaxation is shown in Figure 3. This theory, however, was restricted to only the overdamped liquids with Markovian dissipative dynamics. Moreover, the dissipative kernels were approximated by the single particle friction [15a].

4. *Theory of Fried and Mukamel: Memory Function Approach*

Fried and Mukamel [144] studied the dynamics of solvation by using the memory function formalism. Their memory function theory was similar to the molecular hydrodynamic theory in spirit. The solvent studied was modeled as nonpolarizable polar liquid. The translational contribution to the solvent polarization relaxation was incorporated through a translational parameter [144]. The most important aspect of this study was the approximation of the rotational dissipative kernel by its long wavelength (i.e., $q = 0$) limit. This was then related to the frequency-dependent dielectric function $\varepsilon(\omega)$ of the solvent. The theory incorporated, for the first time, the effects of non-Debye dielectric dispersion on solvation dynamics.

5. *Theory of Wei and Patey: Use of the Kerr Approximation*

Wei and Patey [145] studied the ionic solvation dynamics by using the Kerr theory. They showed that the Smoluchowski–Vlasov approach implemented by Chandra and Bagchi [15a] could be regarded as a special case of a more generalized Kerr approximation for the time evolution of the two-particle density correlation function. Here the wavenumber and the frequency-dependent solvent polarization was obtained from the self part of the van Hove correlation function. The relevant static correlation functions were obtained from the equilibrium theories of liquid, such as the mean spherical approximation (MSA), the hypernetted chain (HNC), and the linearized

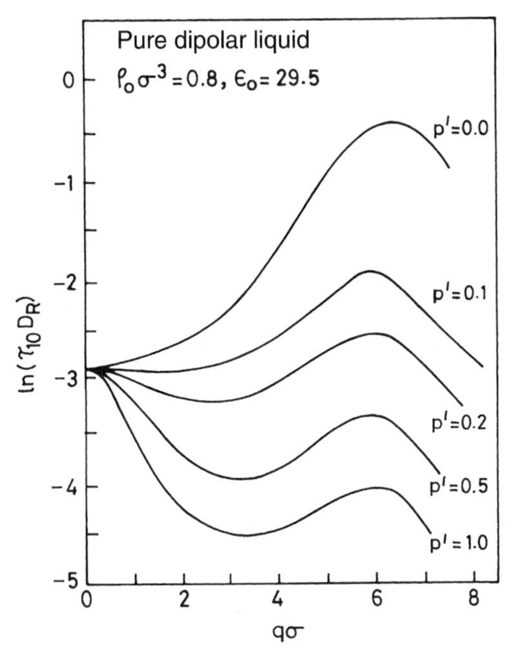

Figure 3. Effects of solvent translational modes on the wavenumber q dependent longitudinal polarization relaxation time $\tau_L(q)$. The values of $\tau_L(q)$ calculated by using Eq. (2.6) are plotted as a function of wavenumber for five different values of the translational parameter p'. D_T and D_R are the translational and rotational diffusion coefficients of a solvent molecule, respectively; $\varepsilon_0 = 29.5$; $\rho\sigma^3 = 0.8$. From Ref. [15a].

hypernetted chain (LHNC) theories. These authors found that for dipolar hard sphere solvents, the Kerr plus renormalized hypernetted chain (RHNC) and the Kerr plus MSA results for $S(t)$ were similar. For the water-like model, however, the Kerr plus RHNC results for $S(t)$ were significantly different from those for dipolar hard sphere solvents. This study also found that the translational modes of the solvent enhances the rate of the decay of $S(t)$.

These theoretical studies showed that the intermolecular correlations present in the dipolar liquids are responsible for the observed nonexponential decay of $S(t)$. The initial experimental studies [50,55,56,146–150], however, were carried out by using picosecond time resolution; thus the dynamics of solvation in ultrafast solvents like water and acetonitrile could not be performed. At this stage, computer simulation studies [35,151–152] provided the initial understanding of the solvation dynamics in ultrafast liquids. For example, Maroncelli [152] carried out molecular dynamic simulation studies of the solvation dynamics in acetonitrile and predicted the presence of a dominant ultrafast Gaussian component in the biphasic decay of $S(t)$. These predictions were later verified experimentally only when the ultrafast laser pulses became available, which can be considered the second stage in the study of the solvation dynamics. At this point, several new experimental techniques were designed to study the solvation dynamics in ultrafast solvents, such as water [29], acetonitrile [27,27a], and methanol [64–66]. The experimental results were exciting and fueled further interest in more sophisticated developments in the fields of experiment and theory. Next we describe a few representative examples of these theoretical studies.

B. Recent Theoretical Developments

1. *Surrogate Hamiltonian Theory of Friedman and Co-Workers*

Friedman and co-workers [153–158] developed a surrogate Hamiltonian method to study the solvation dynamics of a newly created charge distribution. In their theory, both the solute and the solvent molecules are described in terms of relatively realistic interaction site models. The solute–solvent coupling is expressed in terms of the solute–solvent site–site interaction. The authors derived an expression of a nonequilibrium distribution function for a surrogate Hamiltonian description of solvation dynamics in terms of renormalized solute–solvent interactions. The distribution function was then used to study the solvation dynamics. The theoretical results for solvation dynamics in water and acetonitrile obtained from the surrogate theory were in good agreement with the relevant nonequilibrium molecular dynamics simulation results. This is certainly the most detailed theory to date.

2. *Underdamped Non-Markovian Theory with Solvent Inertia: The Formulation of Roy and Bagchi*

The extended molecular hydrodynamic theory developed by our group [159–164] includes rather systematically the effects of inertial solvent response on solvation dynamics. In this approach, the extended molecular hydrodynamic theory was used to study the solvation dynamics in underdamped dipolar liquids for which the memory (non-Markovian) effect in the dissipative dynamics of the medium was considered explicitly. This theoretical formulation was successful for predicting the dynamics of solvation of an ion not only in Stockmayer liquid [163] but also in water [160], acetonitrile [159], and Brownian dipolar lattice [164]. The electrical part of the time-dependent solvation energy was assumed to be given by the following expression [15a]:

$$E_{\text{sol}}(t) = -\frac{1}{2}\int d\mathbf{r}\mathbf{E}_0(\mathbf{r}) \cdot \mathbf{P}(\mathbf{r}, t). \tag{2.7}$$

where $\mathbf{E}_0(\mathbf{r})$ is the bare electric field of the charged solute at a position $\mathbf{r}$, and $\mathbf{P}(\mathbf{r}, t)$ is the space- and time-dependent polarization of the solvent molecules owing to the presence of the ionic field. This polarization can be obtained from the molecular hydrodynamic theory (MHT) by solving the coupled conservation equations of number and momenta densities [15a] (see Section III). This approach [15a] led to the following approximate expression for $E_{\text{sol}}(t)$:

$$E_{\text{sol}}(t) = -\frac{2Q^2}{\pi\sigma}\int_0^\infty dq\left[\frac{\sin qr_c}{qr_c}\right]^2\left[1-\frac{1}{\varepsilon_{10}(q)}\right]\mathcal{L}^{-1}\left[\frac{1}{z+\Sigma_{10}(q,z)}\right] \tag{2.8}$$

where r_c is the distance of closest approach between the solute ion and a dipolar solvent molecule, Q is the charge of the ion, and $\varepsilon_{10}(q)$ is the 10 component of the wavenumber (q) dependent dielectric function. This function characterizes the space dependence of the static orientational correlation among the solvent molecules. $\mathcal{L}^{-1}$ stands for the Laplace inversion, z is the Laplace frequency conjugate to time, $\Sigma_{10}(q, z)$ is the 10 component of the wavenumber- and frequency-dependent generalized rate of solvent polarization, which is given by [159–164]

$$\Sigma_{10}(q,z) = \frac{2k_{\text{B}}Tf_{110}(q)}{I[z+\Gamma_{\text{R}}(q,z)]} + \frac{k_{\text{B}}Tq^2f_{110}(q)}{m\sigma^2[z+\Gamma_{\text{T}}(q,z)]} \tag{2.9}$$

where I is the average moment of inertia of a solvent molecule of diameter σ and mass m. $k_{\text{B}}T$ is the Boltzmann constant times the absolute temperature; $\Gamma_{\text{R}}(q, z)$ and $\Gamma_{\text{T}}(q, z)$ are the rotational and the translational dissipative kernels, respectively. The relevant dynamic response of the solvent is governed solely by these two kernels, and they play the most crucial role in accurately describing the dynamics of the medium. Note that Eq. (2.9) retains its form when the moment of inertia I is a tensorial quantity. Then the denominator of the first term is replaced by $I[z1 + \Gamma_{\text{R}}(q, z)]$ where 1 is the unit tensor and Γ_{R} is the dissipative tensor. The static structural correlations of the pure solvent are expressed by [15a]

$$f_{110}(q) = 1 - \left(\frac{\rho_0}{4\pi}\right)c(110; q) \tag{2.10}$$

where ρ_0 is the average number density of the pure liquid, $c(110; q)$ denotes the 110th component of the direct correlation function in the intermolecular frame with q parallel to the z axis. $f_{110}(q)$ is also related to the longitudinal (i.e., 10) component of the wavenumber-dependent dielectric function by the following relation [15a]:

$$1 - \frac{1}{\varepsilon_{10}(q)} = \frac{3Y}{f_{110}(q)} \tag{2.11}$$

where $3Y$ is the polarity parameter defined as $(4\pi/3)\beta\mu^2\rho_0$ for the solvent molecules, with the permanent dipole moment μ. $\beta = (k_B T)^{-1}$. Eq. (2.11) provides an accurate estimate of $f_{110}(q)$ in the small wave vector (i.e., $q \to 0$) limit.

A reliable scheme for the calculation of $\Gamma_R(q, z)$ and $\Gamma_T(q, z)$ was developed for underdamped liquids by Roy and Bagchi [159–164] (see below). The q dependence of the rotational memory function was neglected, and $\Gamma_R(q, z)$ was replaced by $\Gamma_R(q = 0, z)[\equiv \Gamma_R(z)]$. The latter was then obtained directly from the frequency-dependent dielectric function $\varepsilon(z)$ by using the same molecular hydrodynamic theory [159–164] that is used in calculating $P(\mathbf{q}, t)$, leading to the following expression for $\Gamma_R(z)$ [159–164]

$$\frac{k_B T}{I[z + \Gamma_R(z)]} = \frac{z}{2f_L(0)} \frac{\varepsilon_0[\varepsilon(z) - n^2]}{n^2[\varepsilon_0 - \varepsilon(z)]} \tag{2.12}$$

where ε_0 is the static dielectric constant of the medium, n^2 is the optical dielectric constant, $f_{110}(0)$ is the long wavelength limit of $f_{110}(q)$.

The translational memory kernel $\Gamma_T(q, z)$ was obtained directly from the dynamic structure factor of the liquid, using the following expression [165–166]:

$$\frac{k_B T}{m\sigma^2 \Gamma_T(q, z)} = \frac{S(q)[S(q) - zS(q, z)]}{q^2 S(q, z)} \tag{2.13}$$

where the dynamic structure factor was assumed to be given as $S(q, t) = S(q) \exp\left[\frac{-D_T q^2}{S(q)}\right]$; D_T is the translational diffusion coefficient of a solvent molecule. This can be calculated from the experimentally determined viscosity by using the Stokes–Einstein relation with the proper hydrodynamic (slip or stick) boundary condition.

In the description of Roy and Bagchi, $f_{110}(q)$ is determined by the two particle equilibrium orientational direct correlation functions between two dipolar solvent molecules. The ion–dipole interaction is assumed to be given by the interaction between the bare electric field of the ion with the solvent dipoles [Eq. (2.7)]. Thus the structural distortion of the solvent owing to the presence of the charge on the ion was neglected. Therefore, this description could be valid only in the $t \to 0$ limit, i.e., immediately after the charge is switched on a neutral solute molecule. This is because in that limit only, the solvent distortion is negligibly small. As the solvent orientational and spatial structures change because of the presence of the ion, there develops an orientational pair correlation between the solute and the solvent dipoles, leading to a breakdown of the above description in the limit $t \to \infty$. So a better description is needed to quantify this deviation, which can be done only by taking into account the proper ion–dipole correlation. In fact, a proper theory should include the time evolution of the ion–dipole orientational correlation function, which, however, requires a nonlinear theory. In the long time, one can obviously employ a linear theory that takes into account a proper ion–dipole pair correlation. This should certainly be a better approach than the earlier bare electric field description, as we shall demonstrate later. Another drawback of the earlier theory was that the self-motion of the ion was not taken into account consistently. It is known that the translational motion of the ion can significantly enhance its own rate of solvation [167]. This effect is quite large for the lighter ions, and a self-consistent description has now been developed.

C. Brownian Oscillator Model

The Brownian oscillator model is a well known model of vibrational energy relaxation in a classical oscillator. In the context of solvation dynamics, it was first used by van der Zwan and Hynes [40], who showed that the time dependent fluorescence (TDF) shift can be described in terms of a generalized Langevin equation for relaxation in a harmonic free-energy surface with frequency ω_L, where the dissipation is described by the time-dependent dielectric friction on a rotating dipole. Both the harmonic frequency and the dielectric friction were obtained by using the Onsager model [45]. Because the generalized Langevin equation was employed, both the inertial and the diffusive motions were naturally included. Zwan and Hynes then derived the following expression for the TDF shift [40]:

$$S(t) = \exp\left[-\frac{\omega_L^2 \tau_L^d t}{2}\right]\left[\cos(\omega_L \gamma t) + \left(\frac{\omega_L \tau_L^d}{2\gamma}\right)\sin(\omega_L \gamma t)\right] \tag{2.14}$$

where γ is the well-known quotient factor defined as follows [40]:

$$\gamma = \sqrt{[1 - (\omega_L \tau_L^d/2)^2]} \tag{2.15}$$

In Eqs. (2.14) and (2.15), τ_L^d and ω_L represent the longitudinal relaxation time and solvent frequency, respectively. The latter is a collective quantity, since it was obtained from the static and infinite frequency dielectric constants of the medium. The expression for ω_L is [40]

$$\omega_L = \frac{(2\varepsilon_0 + \varepsilon_\infty)(2\varepsilon_0 + 1)}{3\varepsilon_0(2\varepsilon_\infty + 1)} \times \omega_f^2 \tag{2.16}$$

where $\omega_f^2 = 2\frac{k_B T}{I}$ and represents the mean square thermal rotational frequency of a solvent dipole. Note that ω_L is the collective response.

Eq. (2.14) has two interesting conditional solutions. Although the condition $\omega_L \tau_L^d > 2$ represents the overdamped rotational diffusive relaxation of solvent dipoles, the condition $\omega_L \tau_L^d < 2$ brings the inertial (underdamped) motion into play. When the inertial motion becomes the dominant relaxation channel, the TDF shift exhibits damped oscillations with frequency ω_L, after an initial Gaussian decay.

Unfortunately, this simple and elegant model could explain the solvation experiments only qualitatively, and in many cases it offered inaccurate predictions. For example, the predicted pronounced oscillations (damped) in $S(t)$ for acetonitrile have not been observed in relevant experiments. In addition, $S(t)$ often becomes negative in this model, which has also not been observed in experiments.

Cho and co-workers [168,169] further developed the Brownian oscillator model to describe the ultrafast solvent dynamics and the heterodyne detected optical Kerr effect measurements. Here the nonequilibrium solvation correlation function was calculated from a time correlation function of solvation modes, in which the motion of the latter was again described by a generalized Langevin equation. Cho and co-workers thus obtained the following expression for the normalized correlation function [168]:

$$S(t) = \int_0^\infty d\omega \rho(\omega) R(\omega, t) \tag{2.17}$$

where $\rho(\omega)$ is the density of states and $R(\omega, t)$ represents the response function of the solvent collective mode with a certain frequency distribution. This is given by [168]

$$R(\omega, t) = \mathcal{L}^{-1}\left[\frac{s + \hat{\zeta}(s, \omega)}{s^2 + s\hat{\zeta}(s, \omega) + \omega^2}\right] \tag{2.18}$$

where s is the Laplace frequency conjugate to time t and $\hat{\zeta}(s, \omega)$ is the frequency-dependent damping term (friction).

In the work of Cho and co-workers [168], $\hat{\zeta}(s, \omega)$ was replaced by a constant quantity, and its mode dependence was neglected; i.e., a Markovian approximation was used for $\hat{\zeta}(t, \omega)$. Under this condition, a bimodal decay on short time scales (<2–3 ps) of $S(t)$ was observed. The important finding of this work was that the same density of states, $\rho(\omega)$ (with exponential distribution), could describe both the solvation dynamics and the Kerr relaxation. The main difficulty, however, was the identification of the solvent modes. This was partly solved by the instantaneous normal mode (INM) approach initiated by Stratt and co-workers [111–116] (discussed below).

D. Instantaneous Normal Mode Approach

In solids, the lattice dynamics is usually described in terms of the normal modes of vibration, which are essentially the phonons [170]. These normal modes are the linear combinations of displacements of molecules from the equilibrium lattice positions. On a short time scale, a dense liquid behaves like a solid, and the short-time solvent response is essentially elastic [171]. Therefore, like the case in solids, it may be possible to construct the phonon picture for classical liquids at short times. In liquids, these modes could be the harmonic oscillation around the equilibrium position of each atom or molecule. This idea of describing liquids at short times in terms of elementary excitations (phonon picture) has a long history and was elegantly described by Zwanzig [172,173]. Zwanzig showed that the variable $A(t)$ could be termed as elementary excitation only when it describes at least approximately the following periodic equation of motion [172,173]

$$\frac{dA(t)}{dt} \cong i\omega A(t) \tag{2.19}$$

where ω is some appropriate frequency. Zwanzig also pointed out that the elementary excitation can be a good collective mode if it is an approximate eigenfunction of the Liouville operator

$$LA(t) \cong \omega A(t) \tag{2.20}$$

where ω is the approximate eigenvalue.

Because ultrafast solvation occurs on a short time scale, it is natural that the process be described by a set of quasi-normal modes that, even with their short life time (because of mode-mixing), may suffice for this extremely fast relaxation process. A detailed study along this line has been initiated by Stratt [112], who coined the term instantaneous normal modes for these transient modes. Stratt and co-workers [111,113–116] performed pioneering work in establishing the mechanism of the ultrafast solvent response at short times. They used INM to show that it is the free inertial motion of the nearest neighbor solvent molecules that is responsible for this component. An INM study on solvation dynamics by Ladanyi and Stratt [114] revealed a nearly universal mechanism for the solvation of an excited solute in a solvent, irrespective of the nature of the solute–solvent interaction (nonpolar, dipolar, or ionic). According to Ladanyi and Stratt [114], the solvation is dominated by the simultaneous participation of the nearest-neighbor solvent molecules, for which the solvent libration is usually the most efficient route to the solvation.

III. POLARIZATION RELAXATION IN A DIPOLAR LIQUID: GENERALIZED MOLECULAR HYDRODYNAMIC THEORY

Not only the polar solvation dynamics but also many other chemical dynamical phenomena require an understanding of the position- and orientation-dependent density of a molecular liquid. A simple expression for solvation energy, as noted in Section II, is given by [15a]

$$E_{\mathrm{sol}}(t) = -\frac{1}{2}\int d\mathbf{r}\mathbf{E}_0(\mathbf{r}) \cdot \mathbf{P}(\mathbf{r}, t) \tag{3.1}$$

The fluctuating force on a tagged particle, which can be a neutral solute, an ion, or a dipole, depends on the density relaxation.

For atomic liquids, a detailed self-consistent microscopic theory of density relaxation was recently developed [26,26a,26b,174,175]. This is generally known as the mode coupling theory (MCT), although more than one version of it exists. An extension of the mode coupling theory to dipolar liquids requires the treatment of correlated orientational motion of dipolar molecules. This is a nontrivial problem. Below, we describe the advancement achieved in the recent years.

A. Theoretical Formulation

1. Molecular Hydrodynamic Theory and the Coupled Equations

In Section I, we noted that our group [15a] used the extended molecular hydrodynamic theory (EMHT) to calculate the space- and time-dependent polarization of the solvent $\mathbf{P}(\mathbf{r}, t)$ induced by the instantaneous creation of the ionic or dipolar field in the medium. An expression for the time-dependent fluctuation of the space- and orientation-dependent solvent number density $\delta\rho(\mathbf{r}, \mathbf{\Omega}, t)$ was obtained from the EMHT, which was then used to calculate $\mathbf{P}(\mathbf{r}, t)$ by using the following expression [15a]

$$\mathbf{P}(\mathbf{r}, t) = \int d\mathbf{\Omega}\, \mu(\mathbf{\Omega})\delta\rho(\mathbf{r}, \mathbf{\Omega}, t) \tag{3.2}$$

where $\mu(\mathbf{\Omega})$ is the orientation-dependent dipole moment of a solvent molecule.

For a molecular liquid, the conserved variables are the number density, the angular and spatial momenta densities, and the energy density. These conserved variables depend not only on the position $\mathbf{r}$ but also on the orientation $\mathbf{\Omega}$ of the molecules. For dense liquids, the relaxation of the temperature fluctuation is much faster than those of the momentum and density fluctuations at all length scales. Thus the effects of the temperature relaxation can be neglected. The intermolecular interactions among the solvent molecules generate a force field that, in turn, influences the density and momenta relaxations. This force can be obtained by using the well-known density functional theory (DFT) (discussed later). The extended hydrodynamic equations for the number and momenta densities are [15a]

$$\frac{\partial \rho(\mathbf{X}, t)}{\partial t} + \frac{1}{I} \nabla_{\mathrm{R}} \cdot \mathbf{g}_{\mathrm{R}}(\mathbf{X}, t) + \frac{1}{m} \nabla_{\mathrm{T}} . \mathbf{g}_{\mathrm{T}}(\mathbf{X}, t) = 0 \tag{3.3}$$

where $\mathbf{X} = (\mathbf{r}, \mathbf{\Omega})$; m is the mass of a solvent molecule; and ∇_{T} and ∇_{R} are the spatial translational and angular gradient operators, respectively. Eq. (3.3) is the continuity equation for number density. The relaxation of the number density is thus coupled to the relaxation of the spatial and angular momenta densities, denoted by $\mathbf{g}_{\mathrm{T}}(\mathbf{r}, \mathbf{\Omega}, t)$ and $\mathbf{g}_{\mathrm{R}}(\mathbf{r}, \mathbf{\Omega}, t)$, respectively.

In the ordinary hydrodynamics, the relaxation of momentum density (or the current mode) is described by the Navier–Stokes equation. In this description, the rate of change of momentum is related to the stress tensor on a volume element, the latter in turn is related to the pressure gradient and the spatial derivative of the velocity field with viscosity as the prefactor of the latter. This description becomes inadequate at molecular-length scales, at which the pressure gradient itself depends on density and the dissipative mechanism is coupled to friction rather than to viscosity [26,26a,26b]. In this regime, the density relxation is slow because of the presence of the pronounced short-range order, particularly at the molecular length scales. This slowing down of the density relaxation at intermediate wavenumbers gives rise to the well-known de Gennes's narrowing [176] of the dynamic structure factor of a dense liquid at intermediate wavenumbers. Thus one must modify the molecular hydrodynamic theory to describe the relaxation processes that are ocurring at molecular-length scales. This extension has already been carried out for atomic liquids [177]. For molecular liquids, we need also to consider the orientation of the molecules. This is particularly important for solvation dynamics, which probe primarily the orientational relaxation both at long- and at molecular-length scales.

The extended hydrodynamic equations for the relaxation of the momenta densities are given by [15a]

$$\frac{\partial \mathbf{g}_i(\mathbf{X}, t)}{\partial t} = -\rho(\mathbf{X}, t) \nabla_i \frac{\delta \mathcal{F}[\rho(t)]}{\delta \rho(\mathbf{X}, t)} - \int_0^t dt_1 \int d\mathbf{X}_1 \Gamma_i(\mathbf{X}, \mathbf{X}_1, t - t_1) \mathbf{g}_i(\mathbf{X}_1, t_1) + \mathbf{R}_i(\mathbf{X}, t) \tag{3.4}$$

The first term on the right hand side of Eq. (3.4) represents the systematic force field acting on a molecule at the molecular-length scale as a result of intermolecular interactions. The long wavelength limit of this term is equivalent to the pressure gradient term of the Navier–Stokes equation.

The second term on the right hand side of Eq. (3.4) corresponds to the dissipative part of the total force (or torque), as in the generalized Langevin equation. The dissipation is controlled by the corresponding memory kernels represented as $\Gamma_{\mathrm{T}}(\mathbf{X}_1, \mathbf{X}_2, t_1, t_2)$ and $\Gamma_{\mathrm{R}}(\mathbf{X}_1, \mathbf{X}_2, t_1, t_2)$, respectively. Additional contributions to the momentum relaxation come from the random force $\mathbf{F}(\mathbf{X}, t)$ and the random torque $\mathbf{N}(\mathbf{X}, t)$ acting on the particle at time t and is represented as $\mathbf{R}_i(\mathbf{X}, t)$ in Eq. (3.4). The memory kernels are related to the force–force and torque–torque autocorrelation functions by the second fluctuation-dissipation theorem [178]

$$\langle \mathbf{F}_{\mathrm{T}}(\mathbf{X}, 0).\mathbf{F}_{\mathrm{T}}(\mathbf{X}', t)\rangle = k_{\mathrm{B}}T\,\Gamma_{\mathrm{TT}}(\mathbf{X}, \mathbf{X}', t) \tag{3.5}$$

$$\langle \mathbf{N}(\mathbf{X}, 0).\mathbf{N}(\mathbf{X}', t)\rangle = k_{\mathrm{B}}T\,\Gamma_{\mathrm{RR}}(\mathbf{X}, \mathbf{X}', t) \tag{3.6}$$

$$\langle \mathbf{F}_{\mathrm{T}}(\mathbf{X}, 0).\mathbf{N}(\mathbf{X}', t)\rangle = k_{\mathrm{B}}T\,\Gamma_{\mathrm{TR}}(\mathbf{X}, \mathbf{X}', t) \tag{3.7}$$

where we have assumed $\Gamma_{\mathrm{TR}} = \Gamma_{\mathrm{RT}}$. For an underdamped solvent, the relaxations of the number density and the momenta densities may occur in the same time scale. For overdamped solvents, on the other hand, the momenta relaxations are much faster in the time scale of interest, and we may neglect the momentum relaxation. Clearly, the latter is not correct in underdamped solvents.

A full microscopic calculation of the dissipative kernels in complex molecular liquids is extremely difficult and has not yet been completely achieved. In the absence of a consistent description of the full position and orientation dependence of the memory kernel, it may be assumed to be local in space and orientation, but *non*local in time, using the following condition [15a]:

$$\Gamma_i(\mathbf{r}, \mathbf{r}', \boldsymbol{\Omega}, \boldsymbol{\Omega}', t, t') = \zeta_i(t - t')\delta(\mathbf{r} - \mathbf{r}')\delta(\boldsymbol{\Omega} - \boldsymbol{\Omega}') \tag{3.8}$$

where $\zeta_i(t)$ is the rotational or translational friction acting on the single particle. Eq. (3.8) then describes the dissipative dynamics in a non-Markovian system in which the forces and the torques acting on the solvent molecules are not correlated either in space or in orientation. But this assumption of the locality in space and orientation may not be correct for treating the collective motion such as the collective orientational relaxation of the solvent molecules. In fact, in strongly correlated systems, such as the dense dipolar liquids, the forces and torques are expected to be correlated both in space and in time. Then one must take into account the wavenumber dependence of the memory function. Furthermore, for the molecular liquids of interest, the forces (and the torques) may also depend on the orientations of the dipolar molecules. This would introduce a nontrivial rank dependence in the memory function. Note that the polar solvation dynamics probe only the orientational relaxation of rank $\ell = 1$. In the following discussions, we shall ignore the rank dependence of the memory function and use the following expression as an effective representation of the memory kernel [15a]:

$$\Gamma_i(\mathbf{r}, \mathbf{r}', \mathbf{\Omega}, \mathbf{\Omega}', t, t') = \Gamma_i(\mathbf{r} - \mathbf{r}', t - t')\delta(\mathbf{\Omega} - \mathbf{\Omega}') \tag{3.9}$$

Eqs. (3.5)–(3.9) complete the specification of the dissipative kernels (or memory functions). We shall return to the calculation of these functions after we discus the free energy functional.

2. *Free Energy Functional*

Now we must specify the interaction part of the free energy functional $\mathcal{F}[\rho(t)]$. It is obtained from the DFT, which provides an exact expression of the free energy as a functional of the density fluctuation. The DFT has been found to be enormously successful in recent years in explaining the dynamics of freezing, phase transitions, and transport properties of supercooled liquids [179–181]. Another remarkable success of the density functional theory was in prediction of the isotropic–nematic transition of hard ellipsoids [182]. The results obtained from this study was in agreement with the computer simulation results, showing the effectivenes of using the DFT in inhomogeneous systems. These studies are encouraging because we are primarily studying the macroscopic systems in terms of the microscopic direct correlation functions.

Let us consider a binary mixture of solvent and solute (ionic or dipolar). The presence of the solute reinforces the density inhomogeneity in the solute–solvent composite system. Inhomogeneity or nonuniformity can also arise either from the existence of an external force field or when the system is in a quasi-stationary nonequilibrium state, characterized by the local thermodynamic variable. Lebowitz and Percus [183] developed the density functional theory by reformulating the equilibrium statistical mechanics in terms of direct correlation function language. Apparently, the success of this formulation appears to originate from a rapid convergence of the Taylor expansion. The density functional theory expresses the free energy functional (interaction part) of the inhomogeneous system as a function of the position- ($\mathbf{r}$), orientation-, and time-dependent number density of the solute $n_s(\mathbf{r}, \mathbf{\Omega}, t)$ and solvent $\rho(\mathbf{r}, \mathbf{\Omega}, t)$. Then a Taylor expansion of the excess free energy of the inhomogeneous state over the homogeneous state is performed. The expansion is given by [15a]

$$\begin{aligned}\beta\mathcal{F}[n_s(\mathbf{r},\mathbf{\Omega},t),\rho(\mathbf{r},\mathbf{\Omega},t)] = &\int d\mathbf{r}\, d\mathbf{\Omega}\, \rho(\mathbf{r},\mathbf{\Omega},t)\left[\ln\frac{\rho(\mathbf{r},\mathbf{\Omega},t)}{\rho^0/4\pi} - 1\right] \\ &+ \int d\mathbf{r}\, d\mathbf{\Omega} n_s(\mathbf{r},\mathbf{\Omega},t)\left[\ln\frac{n_s(\mathbf{r},\mathbf{\Omega},t)}{n_s^0/4\pi} - 1\right] \\ &- \int d\mathbf{r}\, d\mathbf{r}'\, d\mathbf{\Omega}\, d\mathbf{\Omega}' c_{\mathrm{sd}}(\mathbf{r},\mathbf{\Omega};\mathbf{r}',\mathbf{\Omega}')\, \delta n_s(\mathbf{r},\mathbf{\Omega},t)\, \delta\rho(\mathbf{r}',\mathbf{\Omega}',t) \\ &- \frac{1}{2}\int d\mathbf{r}\, d\mathbf{r}'\, d\mathbf{\Omega}\, d\mathbf{\Omega}' c_{\mathrm{dd}}(\mathbf{r},\mathbf{\Omega};\mathbf{r}',\mathbf{\Omega}')\, \delta\rho(\mathbf{r},\mathbf{\Omega},t)\, \delta\rho(\mathbf{r}',\mathbf{\Omega}',t) \\ &+ \text{higher-order terms} \end{aligned} \tag{3.10}$$

where β is the inverse of the Boltzmann constant times the absolute temperature T; $\delta\rho(\mathbf{r},\mathbf{\Omega},t) = \rho(\mathbf{r},\mathbf{\Omega},t) - \frac{\rho_o}{4\pi}$ and $\delta n_s(\mathbf{r},\mathbf{\Omega},t) = n_s(\mathbf{r},\mathbf{\Omega},t) - \frac{n_s^o}{4\pi}$ and represent the time-dependent fluctuations in density; and ρ_o and n_s^o represent the average number density of the solvent and the solute, respectively.

In principle, Eq. (3.10) is exact. $c_{ij}(\mathbf{r},\mathbf{\Omega};\mathbf{r}',\mathbf{\Omega}')$ is two particle direct correlation function between a molecule of ith species and that of jth species at positions $\mathbf{r}$ and $\mathbf{r}'$ with orientations $\mathbf{\Omega}$ and $\mathbf{\Omega}'$, respectively. These direct correlation functions are the expansion coefficients in the density expansion of the free energy. i and j here are dummy indices and stand for both solute and solvent molecules. The direct correlation functions c_{ij} contain detailed microscopic information about the spatial and orientational correlations present in the molecular liquid. It is important to note that these two particle direct correlation functions are those of the reference homogeneous system. The solutions for the two particle direct correlation functions are available for several systems such as hard spheres [184,185], dipolar hard spheres [135] and multicomponent hard sphere mixtures [186,187]. In the present treatment, we ignore the effects of the higher-order density flutuations.

Note that the above free energy functional [given by Eq. (3.10)] has a minima with respect to density at equilibrium. Thus the free energy minimization with respect to solvent density will provide an expression for the space- and orientation-dependent inhomogeneous equilibrium density of the solvent $\rho^{\mathrm{eq}}(\mathbf{r},\mathbf{\Omega})$ in terms of the effective interaction potential $V_{\mathrm{eff}}(\mathbf{r},\mathbf{\Omega})$. In a binary mixture (solute plus solvent) this effective interaction potential derives contributions from both the solute–solvent and the solvent–solvent interactions, which are essentially the convolutions of respective density fluctuation and direct pair correlation function. The expression for the equilibrium density in turn gives the space-dependent equilibrium polarization $\mathbf{P}(\mathbf{r})$ [141].

We need, however, an equation of motion that will describe the time evolution of $\rho^{\mathrm{eq}}(\mathbf{r}, \boldsymbol{\Omega})$ and hence that of $\mathbf{P}(\mathbf{r})$. As shown in Ref. [15a], this was described by the internally coupled molecular hydrodynamic equations for the density and momenta relaxations [Eqs. (3.3) and (3.4)]. The solution of these equations at equilibrium should produce the same expression for the inhomogeneous density field as obtained from Eq. (3.10) via free energy minimization.

We have shown [15a] that the time-dependent general solution of Eqs. (3.3) and (3.4) contains detailed information regarding the complex dissipative dynamics of the medium. These dissipative dynamics are nothing but the different kinds of time-dependent frictional responses of the liquid, which are generally termed dissipative kernels or memory functions.

We next turn our attention to the solution of such a set equations of motion. We shall work in wavenumber space, since it offers considerable simplicity in the analytical work. The translational isotropicity of the homogeneous liquid renders this simplicity.

3. *The General Solution*

Let us first expand the fluctuating number density $\delta\rho(\mathbf{q}, \boldsymbol{\Omega}, t)$ in spherical harmonics [188]:

$$\delta\rho(\mathbf{q}, \boldsymbol{\Omega}, t) = \sum_{\ell m} a_{\ell m}(\mathbf{q}, t)\, Y_{\ell m}(\boldsymbol{\Omega}) \tag{3.11}$$

where $a_{\ell m}(\mathbf{q}, t) = \int d\boldsymbol{\Omega}\, Y^{\star}_{\ell m}(\boldsymbol{\Omega})\, \delta\rho(\mathbf{q}, \boldsymbol{\Omega}, t)$. The direct correlation function can also be expanded in terms of the spherical harmonics

$$c(\mathbf{q}, \boldsymbol{\Omega}, \boldsymbol{\Omega}') = \sum_{\ell_1 \ell_2 m_1} c(\ell_1 \ell_2 m_1; \mathbf{q})\, Y_{\ell_1 m_1}(\boldsymbol{\Omega})\, Y_{\ell_2 m_1}(\boldsymbol{\Omega}') \tag{3.12}$$

where $\mathbf{q}$ is chosen to be parallel to the z-axis. It is important to note that in the underdamped limit of momentum relaxation, $\partial \mathbf{g}_i/\partial t$ is nonzero so that the relaxation of the momenta densities occur on the same time scale as the number density relaxation. We then use the expansions in Eqs. (3.3) and (3.4). It is evident from Eq. (3.11) that the density relaxation essentially evolves from the time dependence of its Fourier coefficients $a_{\ell m}(\mathbf{q}, t)$. Therefore, the solution for the above set of equations is derived in terms of $a_{\ell m}(\mathbf{q}, t)$. The solution is obtained by differentiating both sides of Eq. (3.3) once more with respect to time, so that the momenta densities are easily eliminated using Eq. (3.4). Subsequently, a second-order partial differential equation in $\delta\rho(\mathbf{q}, \boldsymbol{\Omega}, t)$ is obtained, which reduces to a simple solvable algebraic equation in the Laplace plane with z as the Laplace frequency. This leads to the following expression [159–166]:

$$a_{\ell m}(\mathbf{q}, z) = \frac{a_{\ell m}(\mathbf{q}, t = 0)}{z + \Sigma_{\ell m}(\mathbf{q}, z)} \tag{3.13}$$

where $a_{\ell m}(\mathbf{q}, z)$ is the Laplace transform of $a_{\ell m}(\mathbf{q}, t)$ and $\Sigma_{\ell m}(\mathbf{q}, z)$ is the generalized rate of the orientational density relaxation, given by [159–166]

$$\Sigma_{\ell m}(\mathbf{q}, z) = \frac{k_B T \ell(\ell + 1) f(\ell\ell m; q)}{I[z + \Gamma_R(\mathbf{q}, z)]} + \frac{k_B T q^2 f(\ell\ell m; q)}{m\sigma^2[z + \Gamma_T(\mathbf{q}, z)]} \tag{3.14}$$

where $\Gamma_R(q, z)$ and $\Gamma_T(q, z)$ are the wavenumber- and frequency-dependent rotational and translational memory kernels, respectively. An important quantity that appears in Eq. (3.14) is the orientational caging parameter $f(\ell\ell m; q) = 1 - (-1)^m(\rho_0/4\pi)c(\ell\ell m; q)$. In the limit of the single particle dynamics (i.e., $\mathbf{q} \to \infty$), $f_{\ell\ell m}(q) = 1$; but it plays a nontrivial role in the collective density relaxation [159–166]. Note that the expansion coefficient $a_{\ell m}(\mathbf{q}, t)$ represents the time- and wavenumber-dependent relaxations of the solvent orientational polarization relaxation. Although the $\ell = 1$ and $m = 0$ component of $a_{\ell m}(\mathbf{q}, t)$ denotes the longitudinal component of the solvent polarization relaxation, the other two ($\ell = 1$, $m = \pm 1$) constitute the transverse component.

B. Calculation of the Dissipative Kernels

The dissipative kernels (which are essentially the memory functions or the space and time dependent frictions for rotational and translational motions) are defined by the fluctuation–dissipation theorems [178] and are given by Eqs. (3.5)–(3.9). Along with the static correlations, these dissipative kernels determine the rates of polarization relaxation and solvation dynamics.

As mentioned earlier, the calculation of the memory kernel for real liquids with complex intermolecular interactions is prohibitively difficult, and the full spectrum of the position, orientation, and frequency dependence of the memory kernel is still unknown. This often forms the bottleneck in any microscopic study of relaxation phenomena in condensed phases; however, significant progress has been made in recent years in calculating these functions.

1. Calculation of the Rotational Dissipative Kernel

In the underdamped limit of momentum relaxation of the solvent molecules, the friction on a molecule is expected to be rather small. In the short time, the binary interaction should determine the friction, and the collective effects are not important. In this limit, the wavenumber dependence of the rotational memory kernel is not significant. Consequently, $\Gamma_R(q, z)$

may be replaced by either its single particle value $\hat{\zeta}_R(z) = \Gamma_R(q \to \infty, z)$ or its collective limit $\Gamma_R(z) = \Gamma_R(q = 0, z)$. The first approximation turns out to be extremely useful in simple model liquids, such as the Stockmayer liquid, for which $\hat{\zeta}_R(z)$ can be obtained analytically. The second approximation, on the other hand, forms the basis of an inversion procedure to calculate the rotational kernel in complex liquids, such as water. The details of the steps involved in the calculation of the rotational kernel in these limiting cases are discussed next.

a. Single Particle Limit of the Rotational Kernel. The determination of $\hat{\zeta}_R(z)$ is rather simple, because we are interested in only the ultrashort time scales. The starting point is the following exact relation between the frequency-dependent friction and the angular velocity correlation function [174–177]

$$\hat{\Psi}_R(z) = \frac{\Psi_R(t = 0)}{z + \hat{\zeta}_R(z)} \tag{3.15}$$

where $\hat{\Psi}_R(z)$ is the Laplace transform of the angular velocity correlation function $\Psi_R(t)$. At short times, $\Psi_R(t)$ can be well approximated by the following Gaussian form [174–177]

$$\Psi_R(t) = \Psi_R(t = 0) \exp\left[-\frac{1}{2}\Omega_R^2 t^2\right] \tag{3.16}$$

where Ω_R is the orientational Einstein frequency of the solvent [15a]. This is essentially the mean square torque acting on the solvent molecules and can be considered as the rotational analogue of the well-known Einstein frequency Ω_0. If one assumes that the dipolar interaction is the only angle-dependent part in the intermolecular potential, then Ω_R can be expressed exactly in terms of the orientational pair correlation function of the dipolar liquid as [189]

$$\Omega_R^2 = -\frac{\mu^2 \rho_0}{3I\pi^2\sigma^3} \int_0^\infty dq\, q j_1(q)[h(110; q) + h(111; q)] \tag{3.17}$$

where $j_1(q)$ is the spherical Bessel function of order unity and $h(\ell\ell m; q)$ is the harmonic coefficient of the Fourier transformed orientational pair correlation function, defined as [188]

$$h(\mathbf{q}, \mathbf{\Omega}, \mathbf{\Omega}') = \sum_{\ell_1 \ell_2 m} h(\ell_1 \ell_2 m; q) Y_{\ell_1 m}(\mathbf{\Omega}) Y_{\ell_2 m}(\mathbf{\Omega}') \tag{3.18}$$

In simple model systems, such as a Stockmayer liquid, an accurate estimation of the orientational pair correlation function can be obtained analytically from the mean spherical approximation (MSA) for not too high polarity of the solvent. This makes the short-time evaluation of the orientational Einstein frequency, and hence the determination of $\hat{\zeta}_R(z)$, feasible in this model system.

It needs to be pointed out that, although Eq. (3.15) is exact, Eq. (3.16) is valid only at the short times. There exist better short time representations of the velocity correlation functions [177], but they tend to get more involved and somewhat approximate in nature. In this chapter, we shall limit ourselves to Eq. (3.16), which is reliable at short times and simple and amenable to totally microscopic calculation without invoking any adjustable parameter.

b. Collective Limit of the Rotational Kernel: The Inversion Procedure. The frictional effects become really important in the long time when the collective effects are also experienced the dynamics. In this limit, therefore, one may approximate the friction by its low wavenumber limit. In the formulation of the inversion procedure, one assumes that the full wavenumber dependence of the rotational memory kernel may be replaced by its $q = 0$ limiting value so that $\Gamma_R(q, z) = \Gamma_R(q = 0, z)$. As we shall demonstrate later, this approximation works rather well, even in the underdamped limit of relaxation. The main merit of the inversion procedure is that the memory kernel is obtained directly from experiments. This may be elaborated as follows.

Of the several known experimental methods of studying the short-time orientational response of dipolar liquids, the ones that are of interest here are the far-IR spectroscopy and the measurement of Kerr relaxation of the neat liquid. The observed relaxations in these experiments may be explained in terms of the collective orientational correlation function, $C_{\ell m}(q = 0, t)$, which is defined as [15a,190,191]

$$C_{\ell m}(q = 0, t) = \frac{4\pi}{3}\mu^2 \langle a_{\ell m}(q = 0, t = 0) a_{\ell m}(q = 0, t) \rangle \qquad (3.19)$$

It is important to note that it is the collective (i.e., $q = 0$ limit of) orientational relaxation that is measured experimentally. In the far-IR absorption studies, the line-shape function $\alpha(\omega)$ is proportional to $\omega\varepsilon''(\omega)$, where ω is the Fourier frequency and $\varepsilon''(\omega)$ is the imaginary part of the frequency-dependent dielectric function $\varepsilon(\omega)[\varepsilon(\omega) = \varepsilon'(\omega) - i\varepsilon''(\omega)]$. $\varepsilon(\omega)$, in turn, is given by $C_{11}(q = 0, \omega)$ by the following relation [15a]

$$\varepsilon(\omega) = \varepsilon_\infty + \frac{4\pi\beta}{V}[C_{11}(q = 0, t = 0) + i\omega C_{11}(q = 0, \omega)] \qquad (3.20)$$

where ε_∞ is the infinite frequency dielectric constant of the medium and V the volume of the system. Note that for the present theoretical formulation, we need to work in the Laplace plane where the Laplace frequency is related to the experimental frequency ω by the simple expression: $z = i\omega$. In a dynamic optical Kerr effect measurement, the off-resonant birefringence signal studied essentially probes the response function of the total polarizability of the medium in all possible polarization directions [168–169]. In a previous study on acetonitrile, Cho [168] assumed that the Kerr relaxation of acetonitrile is dominated by the $C_{20}(t) = \langle P_2[\cos\{\theta(0)\}]P_2[\cos\{\theta(t)\}]\rangle$, as the main contribution to the relaxation seems to come from the single particle orientational motion.

In the definition of the collective orientational correlation function, as before, the $a_{\ell m}(q, t)$ terms are the coefficients of spherical harmonic expansion of $\delta\rho(\mathbf{q}, \mathbf{\Omega}, t)$. Therefore, the molecular hydrodynamic equations provide the following expression for $C_{\ell m}(q = 0, t)$ in the frequency plane [159–164]

$$C_{\ell m}(q = 0, z) = \frac{C_{\ell m}(q = 0, t = 0)}{z + \Sigma_{\ell m}(q = 0, z)} \tag{3.21}$$

Eq. (3.21) is then inverted to obtain $\Gamma_R(z)$ in terms of $C_{\ell m}(q = 0, z)$ as [159–164]

$$\frac{k_B T}{I\left[z + \Gamma_R(q = 0, z)\right]} = \frac{1 - zC_{\ell m}(q = 0, z)}{\ell(\ell + 1)f(\ell\ell m; q)C_{\ell m}(q = 0, t)} \tag{3.22}$$

Using this equation, we can now calculate $\Gamma_R(q = 0, z)$ in terms of the experimental quantities like $C_{11}(q = 0, t)$ (as measured by dielectric relaxation/FIR) or $C_{2m}(k = 0, t)$ (which is measured in Kerr relaxation). A particularly useful form of Eq. (3.22) is obtained when $\Gamma_R(z)$ is related to the frequency-dependent dielectric relaxation $\varepsilon(z)$ of the solvent [159–163]:

$$\frac{k_B T}{I\left[z + \Gamma_R(q = 0, z)\right]} = \frac{z}{2f(110; q = 0)} \frac{\varepsilon_0[\varepsilon(z) - n^2]}{n^2[\varepsilon_0 - \varepsilon(z)]} \tag{3.23}$$

where n and ε_0 are the refractive index and the static dielectric constant of the solvent, respectively. This expression was presented earlier as Eq. (2.12). Note that all the quantities appearing on the right hand side of Eq. (3.23) may be obtained from known experimental results. In particular, $f(110; q = 0) = 3Y/(1 - 1/\varepsilon_0)$ may be evaluated for any system in which the polarity parameter $3Y$ and the static dielectric constant ε_0 are known. The procedure outlined above is an inversion technique because a fundamental microscopic quantity is derived from an experimentally measurable property of the system. The validity and limitation of Eq. (3.23) are discussed at length later in this section. For complex real liquids, such as water, acetonitrile, and methanol, we shall use this approximate representation of the rotational memory kernel to calculate the relaxation of the solvation energy time correlation function $S_E(t)$.

2. *Calculation of the Translational Dissipative Kernel*

The same considerations as discussed in the case of the rotational kernel apply to the determination of the translational kernel. First, for simple model systems $\Gamma_T(q, z)$ can be replaced by its single particle limit $\hat{\zeta}_T(z)$, which is related to the Laplace transform of the velocity time autocorrelation function $\hat{C}_v(z)$ by the following exact relation [166]

$$\hat{C}_v(z) = \frac{C_v(t = 0)}{z + \hat{\zeta}_T(z)} \tag{3.24}$$

In this case, calculation of the short-time approximation of the velocity autocorrelation function $C_v(t)$ requires the Einstein frequency, which is given by the following well-known statistical mechanical expression [174]

$$\Omega_0^2 = \frac{\rho_0}{3m} \int d\mathbf{r}\, g(r) \nabla^2 u(r) \tag{3.25}$$

where $g(r)$ and $u(r)$ are the radial distribution and the interaction potential of the solvent molecules, respectively; and Ω_0 can be determined analytically for model Stockmayer liquids.

The second method of determining the translational kernel $\Gamma_T(q, z)$ involves calculation of the quantity directly from the dynamic structure factor $S(q, z)$. The relation was given in Eq. (2.13).

IV. POLAR SOLVATION DYNAMICS: MICROSCOPIC APPROACH

In Section III, we noted that the progress in the study of polar solvation dynamics took place in several distinct steps [1,12–13a,27–39,46–66]. The initial studies, restricted mainly to slow liquids, found that the decay of the solvation time correlation function is markedly nonexponential, because of intermolecular correlations, and that the continuum model predictions are generally inadequate [1,15,15a,27–30]. The translational modes, however, were found to provide an indirect (hidden) route to the continuum limit. In the second stage, attention was focused on understanding solvation in fast solvents, such as water and acetonitrile [27,27a,30,35,192–195]. The latter study revealed several startling results, such as the bimodality of solvent response with an initial, ultrafast component that decays in <100 fs with 50–70% of the total amplitude. This is followed by an exponential-like slow component, which decays with a time constant of about 1 ps. Although many questions remain unanswered and some facts unexplained, theoretical studies [40,114,153–164] have been largely successful in explaining the observed dynamics.

Recently, attention seems to have returned to the problem of solvation in slow liquids. The emphasis this time, however, is on understanding the dynamics of slow liquids themselves by using solvation dynamics as a reliable probe [38,38a,194]. Furthermore, the questions being asked now are somewhat different. For example, the role of the solute itself in the solvation dynamics remains to be quantified. Here two different aspects need to be understood. First and the most discussed part is the distortion of the dipolar solvent by the polar solute. Thus both static and dynamic correlations of the solvent molecules near the solute are different from those far away. Although the change in dynamic correlations has not been treated properly yet, the effects of differing static correlations owing to an ion have been included in theoretical descriptions. It has been shown that structural distortions can lead to slower decay at the long times and, in fact, must be included to describe relaxation in slow solvents [195].

As discussed in Section I, both experiments and computer simulations study the time-dependent progress of the solvation in terms of the following normalized time correlation function:

$$S(t) = \frac{E_{\text{sol}}(t) - E_{\text{sol}}(\infty)}{E_{\text{sol}}(0) - E_{\text{sol}}(\infty)} \tag{4.1}$$

where $E_{\text{sol}}(t)$ is the time-dependent solvation energy due to the interaction of the instantaneously changed solute field (ionic or dipolar) with the fluctuating polarization of the surrounding solvent molecules. $E_{\text{sol}}(\infty)$ represents the average energy after the attainment of the new equilibrium state. A large number of theoretical studies have been carried out to calculate the $S(t)$ from first principles [15,15a,40,114,159–164]. However, most of the earlier studies neglect the effects of the self-motion of the solute on its own rate of solvation—the solvation has been assumed to proceed through the rotational and translational motions of the solvent molecules only. In many cases, such as for a light solute ion or for a fast-rotating solute dipole in a sluggish environment, the self-motion of the solvated solute may play an important role. In the case of dipolar solvation, the continuum model [41,196] predicts that the rotational motion of the probe solute accelerates the solvation through a simple addition of $2D_{\text{R}}$ (where D_{R} is the rotational diffusion coefficient of the solute) to the solvation rate. This discussion, however, did not address the effect of the translational motion of the dipolar solute on its own solvation. In the case of ionic solvation, Roy and Bagchi [197] explored the role of solute motion by using a semiphenomenological approach. The study was partly motivated by the simulation studies of Neria and Nitzan [167]. They concluded that for small light ions the self-motion can have significant effects. However, the limitation of Roy–Bagchi approach was that only the average effects of solute motion were included via an effective diffusion coefficient—no detailed treatment of solute motion was considered.

In this section, a simple but self-consistent microscopic expression for the polar solvation energy relaxation is derived. The effects of self-motion of the polar solute on its own solvation is systematically incorporated. As will be seen, the theory includes the equilibrium structural distortion of the solvent near the polar solute probe. Another interesting feature of this work is that the main equations for the solvation time correlation function $S(t)$ have the same structure as the mode-coupling theory for transport properties (such as the friction coefficient and the viscosity) in a supercooled liquid near its glass transition temperature [26,26a,26b,175].

A. Molecular Expression for Multipolar Solvation Energy

We now discuss the molecular expression for the time-depenedent solvation energy of a newly created multipole inside a dipolar liquid. The energy that we are concerned with in this section is the polar solvation energy only; the nonpolar solvation will be discussed later. The polar solvation energy arises from the interaction of the ionic or dipolar field with the polarization mode of the solvent molecules. Because this energy is a result of many-body interaction, this is an average energy that can be time dependent. This energy is collective; therefore, a time-dependent mean field approach is appropriate.

We take the derivative of the free energy functional [given by Eq. (3.10)] with respect to the density of the solute

$$\frac{\partial \beta \mathcal{F}[n_s(\mathbf{r}, \mathbf{\Omega}, t), \rho(\mathbf{r}, \mathbf{\Omega}, t)]}{\partial n_s(\mathbf{r}, \mathbf{\Omega}, t)} = \ln\left[\frac{n_s(\mathbf{r}, \mathbf{\Omega}, t)}{n_s^0/4\pi}\right] - \int d\mathbf{r}'\, d\mathbf{\Omega}'\, c_{\mathrm{sd}}(\mathbf{r}, \mathbf{\Omega}; \mathbf{r}', \mathbf{\Omega}')\, \delta\rho(\mathbf{r}', \mathbf{\Omega}', t) \tag{4.2}$$

At equilibrium, the derivative of the free energy with respect to density is zero. Therefore, we obtain the following expression for the equilibrium density of the solute:

$$n_s^{\mathrm{eq}}(\mathbf{r}, \mathbf{\Omega}) = \frac{n_s^0}{4\pi} \exp[-\beta V_{\mathrm{eff}}^{\mathrm{s}}(\mathbf{r}, \mathbf{\Omega})] \tag{4.3}$$

where $V_{\mathrm{eff}}^{\mathrm{s}}(\mathbf{r}, \mathbf{\Omega})$ is the effective potential energy on a solute at position $\mathbf{r}$ with orientation $\mathbf{\Omega}$; this energy is the result of the interaction with the surrounding solvent molecules. This is expressed as follows:

$$\beta V_{\text{eff}}^{\text{s}}(\mathbf{r}, \boldsymbol{\Omega}) = -\int d\mathbf{r}' \, d\boldsymbol{\Omega}' \, c_{\text{sd}}(\mathbf{r}, \boldsymbol{\Omega}; \mathbf{r}', \boldsymbol{\Omega}') \, \delta\rho(\mathbf{r}', \boldsymbol{\Omega}') \tag{4.4}$$

Similarly, we obtain the following expression for the equlibrium inhomogeneous density of the solvent owing to the presence of the solute

$$\rho_{\text{d}}^{\text{eq}}(\mathbf{r}, \boldsymbol{\Omega}) = \frac{\rho_{\text{d}}^{0}}{4\pi} \exp[-\beta V_{\text{eff}}^{\text{d}}(\mathbf{r}, \boldsymbol{\Omega})] \tag{4.5}$$

where the effective potntial energy of a solvent molecule is expressed as follows:

$$\beta V_{\text{eff}}^{\text{d}}(\mathbf{r}, \boldsymbol{\Omega}) = -\int d\mathbf{r}' \, d\boldsymbol{\Omega}' \left[\delta n_s(\mathbf{r}', \boldsymbol{\Omega}') c_{\text{sd}}(\mathbf{r}, \boldsymbol{\Omega}; \mathbf{r}', \boldsymbol{\Omega}') + \frac{1}{2} \delta\rho(\mathbf{r}', \boldsymbol{\Omega}') c_{\text{dd}}(\mathbf{r}, \boldsymbol{\Omega}; \mathbf{r}', \boldsymbol{\Omega}') \right] \tag{4.6}$$

Note that the average of $\beta V_{\text{eff}}^{\text{s}}(\mathbf{r}, \boldsymbol{\Omega})$ [given by Eq. (4.4)] is zero, since $\delta\rho$ is the fluctuation from the equilibrium. Thus the effective energy is temporally modulated by the time-dependent solvent density fluctuation $\delta\rho(\mathbf{r}, \Omega, t)$. One can consider the fluctuation either from the initial ($t = 0$) state or from the final ($t = \infty$) state. Within the linear response approximation, these two dynamics are not different. That is, the net perturbation owing to the solute is assumed to be small. Thus Eq. (4.4) can be generalized in time domain to obtain the following expression for the time-dependent potential energy:

$$\beta V_{\text{eff}}^{\text{s}}(\mathbf{r}, \boldsymbol{\Omega}, t) = -\int d\mathbf{r}' \, d\boldsymbol{\Omega}' \, c_{\text{sd}}(\mathbf{r}, \boldsymbol{\Omega}; \mathbf{r}', \boldsymbol{\Omega}') \, \delta\rho(\mathbf{r}', \boldsymbol{\Omega}', t) \tag{4.7}$$

In writing Eq. (4.7) we have assumed that the solute–solvent two-particle direct correlation function c_{sd} remains unchanged for all the time although the microscopic density $\delta\rho(\mathbf{r}, \boldsymbol{\Omega}, t)$ evolves from the nonpolarized state at $t = 0$ to the final polarized state at $t = \infty$. This can be justified if the presence of the foreign solute induces an infinitesimal perturbation on the correlations of the pure solvent [133]. That is, for small perturbations, the solvent response is linear, implying that both the statics and the dynamics follow those of the pure liquid. One can easily generalize Eq. (4.7) by making c_{sd} also time dependent by introducing an implicit time-dependent density. This, in turn, introduces an additional term on the right hand side of Eq. (4.7), which is second-order in density, with a triplet direct correlation function as the coefficient [198,199]. Thus Eq. (4.7) is exact (within mean field treatment) to the first order; which was first noted by Calef and Wolynes [133]. The systematic inclusion of these higher-order terms will, of course, introduce nonlinearity in solvation energy relaxation [200].

Eq. (4.7) relates the fluctuation in the effective potential energy to the fluctuation in the solvent number density. This is certainly a valid representation in dense liquid. Eq. (4.7), however, represents the time dependent solvation energy of an immobile solute. The motion (translational and/or rotational) of the solute is expected to accelerate the rate of its own solvation. When this particular feature is considered, the expression for the time-dependent solvation energy of the mobile solute becomes

$$E_{\mathrm{sol}}(\mathbf{r}, t) = -k_{\mathrm{B}} T n_s(\mathbf{r}, \Omega, t) \int d\mathbf{r}' \, d\boldsymbol{\Omega}' \, c_{\mathrm{sd}}(\mathbf{r}, \boldsymbol{\Omega}; \mathbf{r}', \boldsymbol{\Omega}') \, \delta\rho(\mathbf{r}', \boldsymbol{\Omega}', t) \tag{4.8}$$

Eq. (4.8) is physically reasonable. The convolution of c_{sd} and $\delta\rho$ is simply the excess chemical potential μ_{excess} owing to the solute–solvent interaction, and thus $n_s\mu_{\mathrm{excess}}$ is the total energy density on the solute.

We need the expression for the solvation energy time autocorrelation function, since in experiments this quantity is monitored as a function of time. Ion solvation dynamics are discussed next, followed by an examination of dipolar solutes.

B. Ion Solvation Dynamics

Eq. (4.8) leads to the following expression for the time-dependent solvation energy of a moblie ion:

$$E_{\mathrm{sol}}(\mathbf{r}, t) = -k_{\mathrm{B}} T n_{\mathrm{ion}}(\mathbf{r}, t) \int d\mathbf{r}' \, d\boldsymbol{\Omega}' c_{\mathrm{id}}(\mathbf{r}, \mathbf{r}', \boldsymbol{\Omega}') \, \delta\rho(\mathbf{r}, \boldsymbol{\Omega}', t) \tag{4.9}$$

$c_{\mathrm{id}}(\mathbf{r}, \mathbf{r}', \boldsymbol{\Omega}')$ is the ion–dipole direct correlation function (DCF) [188,209], which gives the equilibrium structure of the solvent around the ion. In many earlier theories [15,15a,159–164], the ion–dipole interaction was assumed to be given by the effective interaction between the bare electric field of the ion and the solvent dipoles. This assumption fails to take into account properly the solvent structural distortion (both orientational and spatial) around the ion, and this distortion may be important in the long-time limit. The ion–dipole direct correlation function c_{id} used here includes this effect correctly.

Eq. (4.9) leads to a microscopic expression of solvation energy that includes the effects of self-motion of the ion. After transforming Eq. (4.9) from real-space to wavenumber space, the usual spherical harmonic expansions of $c_{\mathrm{id}}(q, \Omega')$ and $\delta\rho(\mathbf{q}, \Omega', t)$ [195] and subsequent integration over Ω' leads to the following expression of the time- and wavenumber-dependent solvation energy [202]:

$$E_{\text{sol}}(\mathbf{k}, t) = -k_{\text{B}}T \int d\mathbf{q}\ c_{\text{id}}^{\ell m}(q) n_{\text{ion}}(\mathbf{k} - \mathbf{q}, t) a_{lm}(\mathbf{k}, t) \tag{4.10}$$

where $c_{\text{id}}^{\ell m}$ is the (ℓm)-th component of the ion–dipole direct correlation function when expanded in the intermolecular frame and $a_{\ell m}(\mathbf{q}, t)$ is the (ℓm)-th component of the solvent polarization fluctuation [15,15a]. For the ionic field, $\ell = 1$ and $m = 0$ component survives only. Consequently, the solvation energy time autocorrelation function (SETCF) $C_{EE}(\mathbf{k}, t)$ takes the following form [202]:

$$C_{EE}(\mathbf{k}, t) = (k_{\text{B}}T)^2 \rho_0 \int d\mathbf{q}\ S_{\text{ion}}(\mathbf{k} - \mathbf{q}, t) |c_{\text{id}}^{10}(q)|^2 S_{\text{sol}}^{10}(q, t) \tag{4.11}$$

where ρ_0 is the average number density of the pure liquid and the S terms are defined as follows

$$S_{\text{ion}}(\mathbf{k} - \mathbf{q}, t) = \langle n_{\text{ion}}(\mathbf{k} - \mathbf{q}, t) n_{\text{ion}}(\mathbf{q} - \mathbf{k}, t) \rangle \tag{4.12}$$

and

$$S_{\text{solv}}^{10}(q, t) = \frac{\langle a_{10}(-q) a_{10}(q, t) \rangle}{N} \tag{4.13}$$

where N is the total number of the solvent molecules that are present in the volume V.

Note that Eq. (4.11) has the structure of a mode-coupling equation [26,175], where for relaxation at a particular wavenumber $\mathbf{k}$, all other modes (i.e., $\mathbf{q}$ modes) contribute. In fact, to derive Eq. (4.11) we have used the Gaussian decoupling approximation [178] common in mode-coupling theory.

The $\mathbf{k} = 0$ limit of this equation is the total solvation energy that is measured in solvation experiments. This is because in experiments solvation energy relaxation derives contribution from all the solute molecules. Thus the experimentally observed correlation function is given by the following expression [202]:

$$C_{EE}(t) = 4\pi (k_{\text{B}}T)^2 \rho_0 \int_0^\infty dq\ q^2 S_{\text{ion}}(q, t) |c_{\text{id}}^{10}(q)|^2 S_{\text{solv}}^{10}(q, t). \tag{4.14}$$

Eq. (4.14) is an important result. It includes the effects of self-motion of the ion through $S_{\text{ion}}(q, t)$, the self-dynamic structure factor. One expects this term on physical grounds also. The ion–dipole correlation enters through c_{id}, which couples the solvation dynamics and ion motion with the collective solvent dynamics, given by $S_{\text{solv}}^{10}(q, t)$.

The normalized solvation energy–energy time correlation function $S_E(t)$. is therefore given as follows [202]

$$S_E(t) = \frac{\int_0^\infty dq\, q^2 S_{\rm ion}(q,t)|c_{\rm id}^{10}(q)|^2[1-\frac{1}{\varepsilon_{\rm L}(q)}]\mathcal{L}^{-1}[\frac{1}{z+\Sigma_{10}(q,z)}]}{\int_0^\infty dq\, q^2 |c_{\rm id}^{10}(q)|^2[1-\frac{1}{\varepsilon_{\rm L}(q)}]}. \tag{4.15}$$

The $S_E(t)$ provides information that is equivalent to the TDFSS of a probe solute. This expression reduces correctly to the previously used expression of solvation energy derived by Roy and Bagchi [159–163] if the asymptotic form of $c_{\rm id}(q)$ is used. Note that Eq. (4.15) is a fully microscopic expression that does not involve any cavity or other continuum model concepts. Detailed numerical predictions of Eq. (4.15) and comparisons with relevant experiments will be discussed when we present the numerical results.

Below we discuss some of the terms that appear in Eq. (4.15).

The self dynamic structure factor $S_{\rm ion}(q,t)$ has been assumed to be given by the well-known following expression [203]:

$$S_{\rm ion}(q,t) = \exp\left\{\frac{-q^2 k_{\rm B}T}{m\zeta_{\rm ion}}\left[t + \frac{1}{\zeta_{\rm ion}}(e^{-t\zeta_{\rm ion}}-1)\right]\right\} \tag{4.16}$$

where $\zeta_{\rm ion}$ is the total translational friction on the ion. This is calculated as an algebraic sum of two components [204]. They are the Stokes friction owing to the zero frequency shear viscosity (η) of the solvent ζ_0 and the translational dielectric friction $\zeta_{\rm T}^{\rm DF}$. ζ_0 has been calculated from the relation $\zeta_0 = 4\pi\eta r_{\rm ion}$ (using the slip boundary condition), where $r_{\rm ion}$ is the radius of the ion. The translational dielectric friction, which arises solely because of the coupling of the ionic field to the fluctuating solvent polarization mode has been computed self-consistently from the force–force correlation function by using the well-known Kirkwood formula [205]. The details in this regard are discussed when we discuss the ionic mobility.

The wavenumber-dependent ion–dipole direct correlation function $c_{\rm id}^{10}(q)$ couples the dynamic structure factor of the ion with that of the solvent. This was obtained from the MSA solution of Chan and co-workers [206] (discussed below).

The solvent orientational dynamic structure factor $S_{\rm solv}^{10}(q,z)$ is assumed to be given by the well-known memory function approximation [174,177]:

$$S_{\rm solv}^{10}(q,z) = S_{\rm solv}^{10}(q)\left[\frac{1}{z+\Sigma_{10}(q,z)}\right] \tag{4.17}$$

where $S(q)$ is the static structure factor of the solvent and has been calculated using the following statistical mechanical relation [15,15a]:

$$S^{10}_{\text{solv}}(q) = \frac{1}{4\pi 3Y}\left[1 - \frac{1}{\varepsilon_L(q)}\right] \tag{4.18}$$

Here $3Y$ is the polarity parameter, defined as usual by $3Y = (4\pi/3)\beta\mu^2\rho_0$ for the solvent molecules with a permanent dipole moment μ; $\beta = (k_B T)^{-1}$; $\varepsilon_L(q)$ is the 110 component of the wavenumber-dependent dielectric function, characterizing the space dependence of the static correlation between the solvent molecules. This is related to the polarity parameter and the orientational direct correlation functions of the solvent via Eq. (2.10). $\Sigma_{10}(q, z)$ is the wavenumber- and frequency-dependent 10-component of the generalized rate of solvent polarization relaxation that contains both the rotational and the translational dissipative kernels. The calculation of these quantities was described above.

Let us next discuss the dipolar solvation dynamics.

C. Dipolar Solvation Dynamics

The solvation dynamics of a dipole can be rather different from that of an ion. The former is not only much slower but also depends more strongly on the local short-range correlations present in the dipolar solvent. Earlier theories based on MHT and the DMSA model [139] led to rather similar results. Both these studies, however, neglected the self-motion (rotation and translation) of the dipolar solute. As mentioned, the continuum model predicts that the self-rotation should accelerate solvation, and the modified rate should be equal to $\tau_S^{-1} = \tau_R^{-1} + \tau_{S0}^{-1}$, where τ_R is the total rotation time of the tagged dipoar solute, and τ_S and τ_{S0} are the solvation times in the presence and absence of the self-rotation. Below, a self-consistent microscopic theory is developed that, for the first time, includes the effects of both the translational and the rotational self motion in dipolar solvation dynamics.

For a mobile (both rotationally and translationally) dipolar solute, one can use the density functional theory to write down the expression for the time-dependent solvation energy, given by

$$E_{\text{sol}}(\mathbf{r}, \boldsymbol{\Omega}, t) = -k_B T n_s(\mathbf{r}, \boldsymbol{\Omega}, t) \int d\mathbf{r}'\, d\boldsymbol{\Omega}'\, c_{dd}(\mathbf{r}, \boldsymbol{\Omega}; \mathbf{r}', \boldsymbol{\Omega}')\, \delta\rho(\mathbf{r}', \Omega', t), \tag{4.19}$$

where $E_{\text{sol}}(\mathbf{r}, \boldsymbol{\Omega}, t)$ the solvation energy of a mobile dipolar solute located at position ($\mathbf{r}$) with orientation ($\boldsymbol{\Omega}$) at a particular time t. $n_s(\mathbf{r}, \boldsymbol{\Omega}, t)$ is the position-, orientation- and time-dependent number density of the dipolar solute. $c_{dd}(\mathbf{r}, \boldsymbol{\Omega}; \mathbf{r}', \boldsymbol{\Omega}')$ represents the direct correlation function between the solute dipole and a solvent dipole at positions **r** and **r**′ with orientations Ω and Ω′, respectively. The direct correlation functions c_{dd} contains detailed microscopic information about the orientational structure of the liquid mixture. The density of the solute dipole is assumed to be small so that the interaction among them can be neglected. Then tedious but straightforward algebric calculations (see Appendix A) lead to the following expression for normalized SETCF of mobile dipolar solute

$$S_E(t) = \frac{[F_{10}(t) + 2F_{11}(t)]}{[F_{10}(t=0) + 2F_{11}(t=0)]} \tag{4.20}$$

where $F_{10}(t)$ and $F_{11}(t)$ are the longitudinal and the transverse components of the time-dependent solvation energy of the dipolar solute, respectively. The expressions of these two quantities are given below.

$$F_{10}(t) = \int_0^\infty dq\, q^2 S_{10}^{\text{self}}(q,t)|c_{dd}(110;q)|^2 S_{10}^{\text{solv}}(q,t) \tag{4.21}$$

and

$$F_{11}(t) = \int_0^\infty dq\, q^2 S_{11}^{\text{self}}(q,t)|c_{dd}(111;q)|^2 S_{11}^{\text{solv}}(q,t) \tag{4.22}$$

The derivation of the above two expressions (Eqs. 4.21 and 4.22) is rather involved (see Appendix A).

So, the primary inputs required to follow the rate of solvation using Eqs. (4.15) and (4.20) are the ion–dipole and dipole–dipole direct correlation functions and the dynamic structure factors, $S_{\text{solv}}^{\ell m}(q,t)$ and $S_{\text{self}}^{\ell m}(q,t)$. The calculational details of these quantities are discussed briefly in the next section. As mentioned, if the self-motion is neglected, the present theory leads to results that are quite similar to the predictions of the dipolar dynamical mean spherical approximation model [139].

D. Details of the Method of Calculation

1. Calculation of the Wavenumber-Dependent Direct Correlation Functions

We obtain the ion–dipole direct correlation function $c_{id}(q)$ by Fourier transforming the expression of microscopic polarization $P_{\text{mic}}(r)$ given by Chan and co-workers [206]. An interesting aspect of this expression is that it predicts a region of negative values of $P_{\text{mic}}(r)$, which indicates the alignment of the solvent dipole just outside the first solvation shell in a direction opposite to that of those as the nearest neighbors [201]. The expression for the ratio of the microscopic polarization $P_{\text{mic}}(r)$ to the macroscopic polarization

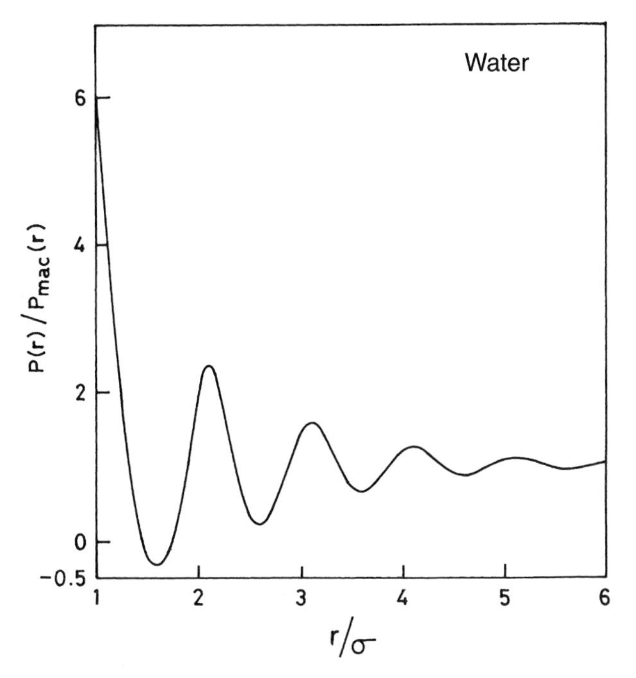

Figure 4. The solvent microscopic polarization: macroscopic polarization ratio $P_{\text{mic}}(r) : P_{\text{mac}}(r)$ is plotted as a function of r for water at 298 K. The solute : solvent size ratio is 1. Note that r is scaled by the solvent diameter σ.

$P_{\text{mac}}(r)$ is given as follows [201]:

$$\frac{P_{\text{mic}}(r)}{P_{\text{mac}}(r)} = 1 - r \times \frac{dF(r)}{dr} + F(r), \tag{4.23}$$

where $F(r)$ is a dimensionless quantity and appears as a correction factor to the macroscopic polarization; the latter is given by the following expression [206]:

$$P_{\text{mac}}(r) = \frac{(\varepsilon_0 - 1)zQ}{4\pi\varepsilon_0 r^2} \tag{4.24}$$

where Q denotes the electronic charge and z the valency of the ion. We have calculated $\frac{P_{\text{mic}}(r)}{P_{\text{mac}}(r)}$ using the two mutually exclusive conditions given in [206]. Figure 4, shows the calculated variation of the polarization ratio with r (scaled by solvent diameter) in water for a solute : solvent size ratio of 1. It is clear from Figure 4 that the polarization density around an ion is oscillatory in nature, which is maximum at the contact. This is in contrast to the dielectric saturation theory, which predicts a divergence in the local static dielectric constant and thus in the local polarization near the ion.

For ion solvation dynamics, we need to calculate $S^{10}_{\text{solv}}(q, t)$ [Eq. (4.17)]. For water, the longitudinal component of the wavenumber-dependent dielectric function $\varepsilon_L(q)$ is obtained from MSA after correcting it at both the limits of $q \to 0$ and $q \to \infty$ by using the XRISM results of Raineri and co-workers [157]. The above procedure gives nearly perfect agreement between theory and experiment for solvation dynamics in acetonitrile [159], water [163,207], and methanol [208]. For these liquids, it was found that the solvation was dominated mainly by the zero wavenumber modes. But for slow liquids like propanol, the solvent structure at small wavelengths (i.e., $q\sigma \geq 2\pi$) may also be important, especially for the proper description of the slower dynamics inherent to these solvents. In the absence of any microscopic calculation, we have used the following systematic scheme to find $f_L(q)$ and hence $\left[1 - \frac{1}{\varepsilon_L(q)}\right]$. At the intermediate wavenumbers, we have taken the orientational correlations from the MSA model [188]. The relation between $f_L(q)$ and the orientational direct correlation function is given by Eq. (2.10). In the $q \to 0$ limit, we have used Eq. (2.10) to obtain the correct low wavenumber behavior of these static correlations. The corresponding wavenumber-dependent dielectric function $\varepsilon_L(q)$ is obtained by using Eq. (2.11). In the $q \to \infty$ limit, we used a Gaussian function, which begins at the second peak height of $\left[1 - \frac{1}{\varepsilon_L(q)}\right]$, to describe the behavior at large wavenumbers. This Gaussian function eliminates the wrong large wavenumber behavior of the MSA. We have verified that the results are insensitive to the details of the Gaussian function [195]. In Figure 5 the function $\left[1 - \frac{1}{\varepsilon_L(q)}\right]$ is plotted against q to show the static correlations used in our calculations.

2. *Calculation of the Generalized Rate of the Polarization Relaxation*

The calculation of the longitudinal component of the generalized rate of the solvent polarization relaxation $\Sigma_{10}(q, z)$ involves the calculations of the rotational $\Gamma_R(q, z)$ and translational $\Gamma_T(q, z)$ kernels. $\Gamma_R(q, z)$ is obtained from the frequency-dependent dielectric function $\varepsilon(z)$ of the solvent by using a novel inversion procedure. The relation between $\Gamma_R(q, z)$ and $\varepsilon(z)$ is expressed in Eq. (2.12). Fortunately, the experimental dielectric relaxation data of almost all the common dipolar liquids are available.

The translational dissipative kernel can be obtained by using Eq. (2.13). Here the translational diffusion coeffeicient of a solvent molecule is taken from experiments.

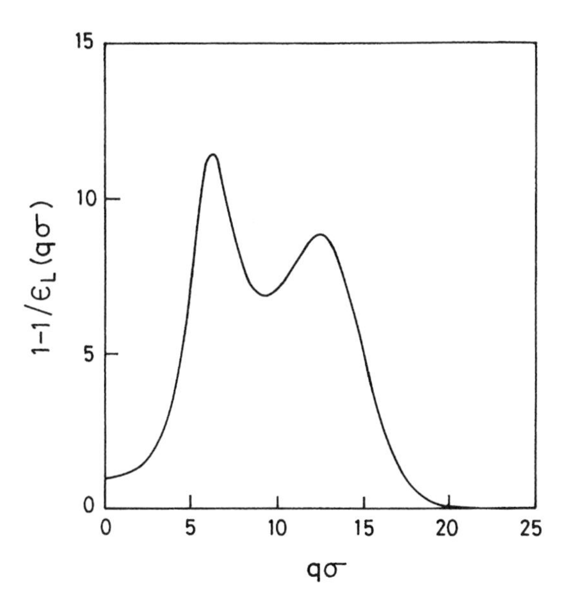

Figure 5. The wavenumber dependence of $[1 - 1/\varepsilon_L(q)]$ in *N*-methyl propionamide (NMP). The longitudinal dielectric function $\varepsilon_L(q)$ has been obtained from the MSA with proper corrections at both the $k \to 0$ and the $q \to \infty$ limits. The wavenumber is scaled by σ (see Section VII).

3. Calculation of the Solvent Dynamic Structure Factor

The transverse component of the solvent dynamic structure factor, $S_{\text{solv}}^{11}(q, t)$ can be expressed as follows [15,15a]:

$$S_{\text{solv}}^{11}(q, t) = S_{111}(q)\mathcal{L}^{-1}\left[\frac{1}{z + \Sigma_{11}(q, z)}\right] \tag{4.25}$$

where $S_{111}(q)$ is the transverse component of the wavenumber-dependent static structure factor and $\Sigma_{11}(q, z)$ is the transeverse component of the generalized polarization relaxation rate of the pure solvent. The transverse component of the static structure factor is related to orientational correlation function $f_{111}(q)$ by the following expression [15,15a]:

$$S_{111}(q) = \frac{1}{4\pi} f_{111}(q) \tag{4.26}$$

where $f_{111}(q)$ can be related to the $c(111;q)$ component of the direct correlation functions as follows [15]:

$$f_{111}(q) = 1 + \left(\frac{\rho_0}{4\pi}\right) c(111; q) \tag{4.27}$$

The transverse component of the generalized rate is given by [15]

$$\Sigma_{11}(q, z) = \frac{2k_B T f_{111}(q)}{I[z + \Gamma_R(q, z)]} + \frac{k_B T q^2 f_{111}(q)}{m\sigma^2[z + \Gamma_T(q, z)]} \tag{4.28}$$

In our calculation $c(111;q)$ was obtained from the MSA. Because the large wavenumber processes are more important in dipolar solvation, the MSA can give fairly accurate values for this quantity. We also compared this quantity thus obtained with that of Raineri and co-workers [153]; good agreement was observed.

We should comment here about the presence of the 111 component of the static orientational structure factor to determine the overall polarization relaxation rate of the solvent. The introduction of this component in describing the orientational dynamics of dipolar solvation is still a matter of considerable debate. If one totally ignores the transverse component and allows only the longitudinal (i.e., 110) component to govern the related dynamics of multipolar solvation, then the rate of solvation becomes much faster than what it could be if both of the components were simultaneously incorporated. In some cases, the rate may even become comparable to that of the solvation of an ion. The neglect of the transverse component in dipolar solvation has sometimes been supported by the notion that it is the artifact of the point dipole approximation. But even for real dipoles or quadrupoles, the presence of this component cannot be ruled out in a microscopic theory. Our present theory systematically includes both these components and predicts a slower rate of dipolar solvation. The same trend has also been found from the DMSA model [139].

4. *Calculation of the Wavenumber-Dependent Orientational Self-Dynamic Structure Factor*

When the solute has both rotational and translational motion, the orientational self-dynamic structure factor $S_{\text{self}}^{\ell m}(q, t)$ is assumed to be given by the following expression:

$$S_{\text{self}}^{\ell m}(q, t) = \frac{1}{4\pi} \times \mathcal{L}^{-1}[z + \Sigma_{\text{sol}}(q, z)]^{-1} \tag{4.29}$$

Again, $\Sigma_{sol}(q, z)$ can be calculated in terms of the frequency-dependent single-particle total rotational friction $\zeta_R(z)$ as follows

$$\Sigma_{sol}(q, z) = \frac{k_B T\ell(\ell+1)}{I[z + \zeta_R(z)]} + D_T^{sol} q^2 \tag{4.30}$$

where D_T^{sol} is the translational diffusion coefficient of the solute dipole. The limit of $D_T^{sol} = 0$ for Eq. (4.30) naturally provides the expression of orientational self-dynamic structure factor for rotationally mobile dipolar solute. Then this can be given as [165,209]:

$$S_{self}^{\ell m}(t) = \frac{1}{4\pi} \times \mathcal{L}^{-1}\left[z + \frac{\ell(\ell+1)k_B T}{z + \zeta_R(z)}\right]^{-1} \tag{4.31}$$

where the prefactor $\frac{1}{4\pi}$ comes from the normalization. The calculation of $\zeta_R(z)$ for a rotating dipolar solute is a nontrivial exercise. For dipolar liquids, the total single particle friction $\zeta_R(z)$ may be resolved into a short-range part, denoted by ζ_R^0 and a long-range dipolar part, usually termed the rotational dielectric friction $\zeta_R^{DF}(z)$. The bare rotational friction, ζ_R^0 may also include the friction from all the angle-dependent, nondipolar interactions. If there are no significant viscoelastic effects in the solvent, ζ_R^0 is independent of frequency and can be obtained using the Debye–Stokes–Einstein (DSE) relation. This quantity depends critically on the geometry of the tagged rotating dipole [13,201,210,211]. The rotational dielectric friction, can be calculated from the torque–torque correlation function by using the well-known Kirkwood's formula [205]. But this leads to a complex four-dimensional integration over the torque–torque autocorrelation function.

Traditionally, it has been assumed that the tagged dipole is immobile. This leads, after some tedious algebra, to the following expression for the dielectric friction [165,166]

$$\begin{aligned}\beta\zeta_R^{DF}(z) = \frac{\rho_0}{2(2\pi)^4}\int_0^\infty dq\, q^2 \Bigg[& \frac{c_{dd}^2(110; q)[1 + (\rho_0/4\pi)h(110; q)]}{z + \Sigma_{10}(q, z)} \\ & + 2\frac{c_{dd}^2(111; q)[1 - (\rho_0/4\pi)h(111; q)]}{z + \Sigma_{11}(q, z)}\Bigg]\end{aligned} \tag{4.32}$$

The most nontrivial part of this calculation is the calculation of the direct correlation functions c_{dd} in a binary mixture for unequal sizes [187]. In numerical calculations, we have used the MSA model to evaluate the direct correlation functions. For this model, analytic expressions for binary dipolar mixture are available [201]. One has to solve three simultaneous nonlinear equations to calculate the usual kappa parameters. We have solved these equations numerically by using a nonlinear root-search technique (finite difference Levenberg–Marquardt algorithm).

We shall use the present formulation first to study the solvation dynamics in the model Stockmayer liquid to demonstrate the robustness of the scheme. The basic idea is to show that with a reasonable short-time representation of the memory kernels, the formulation developed above provides an accurate description of the initial Gaussian solvation dynamics in an underdamped solvent. We then proceed to implement the same scheme to the study of the solvation dynamics in real liquids, which will be reported in Section V.

E. Numerical Results for a Model Solvent: Ion Solvation Dynamics in a Stockmayer Liquid

In this section, we present results of the detailed numerical calculations carried out in a model Stockmayer liquid. We compare the theoretical predictions with those from simulations of Neria and Nitzan [167]. We also present comparisons between the predictions of the DMSA and the MHT theory for solvation in the underdamped Stockmayer liquid. The advantage of studying solvation in this liquid is that all the necessary quantities can be calculated from the first principles, without using any *ad hoc* or adjustable parameters.

To obtain the necessary orientational correlation function, we mapped the Lennard–Jones system into the hard sphere fluid by finding the density and temperature-dependent effective hard sphere diameter σ_{HS}. A dipolar hard sphere fluid is characterized by only two dimensionless quantities: the reduced dipole moment μ_{HS} and the reduced density $\rho^{\star}_{\mathrm{HS}}$, defined respectively as

$$\mu^{\star 2}_{\mathrm{HS}} = \frac{\mu^2}{k_{\mathrm{B}} T \sigma^3_{\mathrm{HS}}} \tag{4.33}$$

$$\rho^{\star}_{\mathrm{HS}} = \rho_0 \sigma^3_{\mathrm{HS}} \tag{4.34}$$

where μ is the magnitude of the dipole moment of a Lennard–Jones (LJ) molecule and ρ_{HS} corresponds to the number density of the hard sphere fluid. We subsequently use the Verlet–Weiss [212] scheme of implementing the Weeks–Chandler–Andersen (WCA) perturbation theory [213]. This procedure is expected to be reasonable at high densities. For the systems simulated by Neria and Nitzan [167], it is found that the corresponding hard sphere diameter is $1.01\sigma_{\mathrm{LJ}}$. The Stockmayer fluid in the simulation performed by Neria and Nitzan is characterized by the following parameters:

$$\rho^\star = \rho_0 \sigma_{\mathrm{LJ}}^3 = 0.81 \tag{4.35}$$

$$T^\star = \frac{k_{\mathrm{B}} T}{\varepsilon} = 1.23 \tag{4.36}$$

$$\mu^\star = \left(\frac{\mu^2}{\varepsilon \sigma_{\mathrm{LJ}}^3} \right)^{\frac{1}{2}} = 1.32 \tag{4.37}$$

where ε is the Lennard–Jones energy parameter. It is now straightforward to calculate the values of the reduced parameters for the corresponding hard sphere fluid. The reduced parameters thus obtained are $\rho^\star = 0.83$ and $\mu^\star = 1.17$. These values are used throughout our calculations. The static orientational correlation functions are then obtained from the MSA for this dipolar hard sphere liquid. The MSA is expected to be fairly reliable for the low polarity liquid considered here.

In the subsequent discussion, the time is scaled by τ_I, the time constant of free inertial decay of the solvent molecules. $\tau_I = \sqrt{\frac{I}{k_{\mathrm{B}} T}}$. This appears to be the natural choice to study the guiding role of the intermolecular correlations on the rate of solvation at molecular lengths. The time constant τ_{G} of the initial Gaussian solvation is obtained by fitting the decay of the solvation time correlation function $S(t)$ from $S(t) = 1.0$ to $S(t) \approx 0.30$. The inverse Laplace transformation [Eq. (4.15)] was carried out numerically using the Stehfest algorithm [214].

As discussed earlier, we need the frequency-dependent dielectric function $\varepsilon(z)$ of the solvent to obtain the solvation time correlation function. This quantity is not easily available *a priori*, and this prevented researchers in the past from making detailed comparisons among different theories. Fortunately, in the underdamped limit, we can calculate this function accurately by using the same molecular hydrodynamic description as employed here.

Neria and Nitzan [167] extensively studied the effects of the solute translational motion on the rate of its own solvation in their simulation of Stockmayer liquid. The solute motion in their system was included through its contribution to the Lagrangian. The results of these calculations (with the effective hydrodynamic description) are shown in Figures 6 and 7, in which the dimensionless parameter is varied keeping everything else fixed. In particular, the solute : solvent mass ratio is kept fixed at 0.5. As expected, the decay of the solvation energy is Gaussian in the early time. These figures show the almost perfect agreement between the theory and simulation. Here also DMSA fails by not taking into account the translational motion of either the solute or the solvent. A significant decrease in the value of

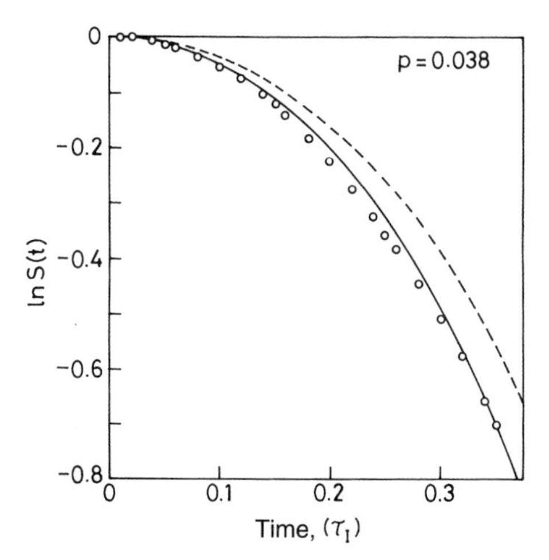

Figure 6. The effects of solute's translational motion on its own rate of solvation. The logarithmic (natural) values of the solvation time correlation function $S(t)$ is plotted as a function of time (t) for the case in which both the solute ion and the solvent (Stockmayer liquid) molecules are translationally mobile. The theoretical predictions are shown by the *solid line* and those from the DMSA by the *dashed line*. The *open circles* denote the simulation of Neria and Nitzan. The following parameters are used: $p = 0.038$; the solute : solvent mass ratio $= 0.5$; and size ratio $= 0.875$. Time is scaled by τ_I where $\tau_I = (\sqrt{I/k_BT})p = I/m\sigma^2$. From Ref. [197].

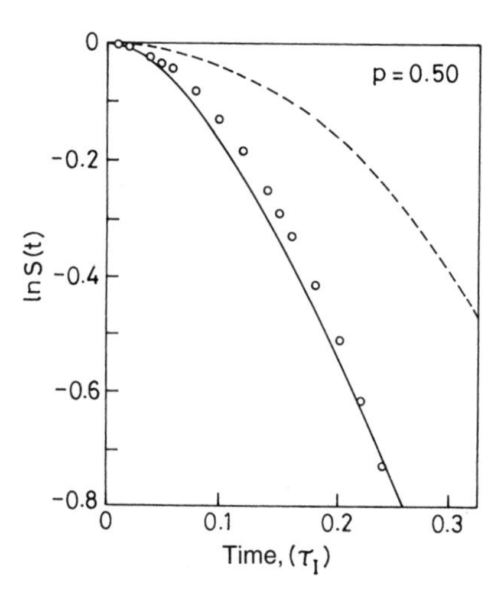

Figure 7. Comparison of different theoretical predictions and computer simulation results for the decay of $S(t)$ when both the solute and the solvent molecules are translationally mobile. The results are for $p = 0.5$. All other representations and parameters remain the same as in Figure 6. From Ref. [197].

TABLE I
Calculated Values of the Time Constant of the Initial Gaussian Decay of the Solvation Time Correlation Function for the Solvation of an Ion in Model Stockmayer Liquid[a]

Solute	Solvent	p	τ_I, ps	τ_G/τ_I	
				Simulation	Theory
Immobile	Immobile	0.000	0.41	0.50	0.484
Immobile	Mobile	0.038	0.41	—	0.46
		0.250	1.05	—	0.39
		0.500	1.49	—	0.34
Mobile	Mobile	0.038	0.41	0.42	0.44
		0.250	1.05	0.33	0.32
		0.500	1.49	0.28	0.28

[a]When available, the calculated values are compared to the simulation results of Neria and Nitzan [197].

τ_G shows important contribution resulting from the the motion of light solute to its own solvation.

Neria and Nitzan also studied the solvation dynamics in a Stockmayer liquid in the other two conditions: both the solute and the solvent are translationally immobile and the solute is kept fixed but the solvent molecules are allowed to rotate and translate. The principal results of these studies are presented in Table I along with the theoretical predictions of the molecular theory. The comparison shows that the theoretical value of the Gaussian time constant τ_G is in good agreement with the simulation results.

V. SOLVATION DYNAMICS IN WATER: POLARIZABILITY AND SOLVENT ISOTOPE EFFECTS

In this section we address the solvation of a nascent ion in water. The development of our understanding of molecular dynamics in water has been rather slow compared to that in simple, nonassociated liquids, such as acetonitrile. This is primarily because of the additional complications that arise because of the presence of an extended hydrogen bond network in water. The well-known study by Migus and co-workers [80] on the dynamics of electron solvation indicated that the solvation in liquid water can be complete within 250 fs or less. Barbara and co-workers [12] reported a study of solvation of Coumarin in which the solvation was found to be biexponential with time constants approximately equal to 250 fs and 1.2 ps. This study could not resolve the initial part of solvation dynamics. Computer simulations, on the other hand, provided more detailed information. The important simulation studies of Maroncelli and Fleming [35] showed three notable features. First, the solvation is dominated by a Gaussian component that decays within a few tens of femtoseconds and that carries 70–80% of the solvation energy. This is followed by a marked oscillation in the solvation time correlation function. The last phase of the decay is slow and exponential-like, with a time constant on the order of a picosecond or so. But the question was whether solvation in water in real systems could really proceed as fast as revealed by this simulation. Recent experiments on solvation of coumarin-152 in water by Jimenez and co-workers [29] showed that the dynamics of solvation in water is indeed ultrafast and biphasic in nature. The Gaussian component (taken as $\exp[-(t/\tau_G)^2]$) was found to dominate the relaxation, with a time constant of about 40 fs, whereas the slower decay at longer times can be described by a sum of two exponentials, with time constants equal to 126 fs and 880 fs, respectively. Several interesting key questions still remain unresolved regarding the dynamics of polar solvation in water; we'll list a few of them below.

First, we must understand the reason for the great separation of time scales between the initial Gaussian decay and the subsequent slow, exponential-like decay. It has been suggested [27,27a,168] that the Gaussian decay is the result of the librational modes of water [215–217], whereas the long time decay is owing to the diffusive dynamics involving primarily the nearest-neighbor molecules. This interpretation, in turn, raises several questions: Why is the relative contribution of the Gaussian component so large when the librational modes themselves contribute only a small amount to the total dielectric relaxation? Can we explain quantitatively the slow, long-time decay? Its description in terms of nearest-neighbor molecules is not fully understood yet. In addition, much controversy still persists regarding the differences in the prediction of simulation and the experimental results. The molecular dynamic simulations [35] predicted a much faster Gaussian relaxation, with a time constant that is two to three times smaller than the experimental value. In the simulations, the Gaussian decay was found to be followed by a marked oscillation in $S(t)$, which was not detected later in the experiments [29]. Therefore, we need to understand the molecular origin of the ultrafast decay to answer any of these detailed questions.

In this section, the extended molecular hydrodynamic theory developed in Section IV is used to calculate the solvation time correlation function in water. Some of the results presented here are new. In this calculation, the recent experimental dielectric relaxation data for water [218] have been used. The results thus obtained are indeed good. The present treatment enables us to study the effects of the intermolecular vibration and librational motions that are well known in the dielectric relaxation of water. It is found that, if the 193 cm^{-1} intermolecular vibrational mode is moderately damped, then a good agreement with the experimental result is obtained without any other adjustable parameter. This study indicates that the mode arising out of the interaction induced effects [219–222] plays a significant role in determining the initial inertial response of the solvent molecules. The good agreement observed here and further analysis presented later also reinforce the conclusion that the initial Gaussian decay is dominated by the macroscopic, long wavelength polarization modes of water.

Another important aspect discussed in this section is the theoretical prediction of the presence of an isotope effect in the aqueous solvation. The static dielectric constant and the refractive indices of H_2O and D_2O are almost the same [223,224]; but the parameters, such as the Debye relaxation time [225] and the librational frequencies, are different. Deuterated water (heavy water) is also a more structured and ordered liquid, exhibiting a stronger hydrogen bond, compared to normal water [226]. Zero point energies of the vibrational modes [227] are also different for H_2O and D_2O. The effect of isotope substitution on the dynamics of aqueous electron solvation is well studied [82,228,229]. It is thus worth examining the solvation dynamics in heavy water and analyzing the relative importance of the effects of isotope substitution in aqueous solvation dynamics. Such an analysis is presented here. We find that there is a noticeable difference in the solvation rate of water and deuterated water in the long time.

Eq. (4.15) is used to calculate $S(t)$ in water. The calculational details of the relevant functions and parameters are described below.

A. Method of Calculation

1. *Calculation of the Wavenumber-Dependent Ion–Dipole Direct Correlation Function*

The wavenumber-dependent ion–dipole direct correlation function $c_{id}(q)$ is an important quantity, because it couples the solute dynamics with those of the solvent. This has been obtained from the MSA solution of Chan and co-workers [206]. The calculation details regarding this function were described in Section IV.D.

2. *Calculation of the Static Orientational Correlations*

The wavenumber-dependent longitudinal dielectric relaxation $\varepsilon_L(q)$ was obtained from the recent XRISM calculations of Raineri and co-workers [157]. The variation of $1/\varepsilon_L(q)$ as a function of $q\sigma$ shows features qualitatively similar to the universal characteristics discussed earlier. In particular, the amplitude of negative excursion of $1/\varepsilon_L(q)$ is found to be quite large on account of its high polarity. This, in turn, produces a pronounced softening of $f(110; q)$ at the intermediate wavenumbers. Therefore, the polarization relaxation at the intermediate wavenumbers is expected to be substantially slower than the macroscopic relaxation.

3. *Calculation of the Rotational Dissipative Kernel*

The rotational dissipative kernel in water has been calculated using the inversion scheme described in Section III. In this scheme, the dielectric relaxation of the solvent is needed to calculate $\Gamma_R(q, z)$. One thus requires a reliable description of the frequency-dependent dielectric function $\varepsilon(z)$.

The frequency dependence of the dielectric function of water is rather complex and derives important contributions from both slow Debye dispersion and high-frequency inertial modes of the solvent [221,222,230,231]. The low-frequency behavior can be fitted to a sum of two Debye relaxations [230,231] that cause the dielectric constant to decrease from the static dielectric constant $\varepsilon_0 = 78.36$ to a value of 4.93, which is conventionally termed the infinite frequency dielectric constant ε_∞. It is important to note that ε_∞ is different from n^2. The high-frequency behavior of the dielectric function is extracted from two far-IR peaks of water centered roughly at 193 and 685 cm^{-1}, respectively [230,231]. The high-frequency relaxation is dominated by the 193 cm^{-1} peak, which arises from translational intermolecular vibrations. The peak at 685 cm^{-1}, on the other hand, is assigned to the librational motions of the hydrogen-bonded network and carries only a small fraction of the relaxation. In the calculations reported here, the recent dielectric relaxation data for water by Kindt and Schmuttenmaer [218] have been used. These are tabulated in Table II. The high-frequency dielectric dispersions are summarized in Table III.

A statistical mechanical expression for the frequency-dependent dielectric function can be obtained from the total dipole moment correlation function $\phi(t)$ [15,15a] by using the linear response theory. This is a normalized correlation function and is defined as follows [15,15a]

$$\phi(t) = \frac{\langle \mathbf{M}(0)\mathbf{M}(t) \rangle}{\langle \mathbf{M}(0)\mathbf{M}(0) \rangle} \tag{5.1}$$

where $\mathbf{M}(t)$ is the time-dependent total moment. This is defined as:

$$\mathbf{M}(t) = \frac{1}{\sqrt{N}} \sum_{n=1}^{N} \mu_n(t) \tag{5.2}$$

where N denotes the total number of solvent molecule present in the system and $\mu(t)$ is the time-dependent dipole moment vector of a single isolated molecule. Note that in defining $\mathbf{M(t)}$, the cross-terms are neglected.

TABLE II
The Dielectric Relaxation Parameters of H_2O and D_2O[a]

Solvent	ε_0	τ_1, ps	ε_1	τ_2, ps	ε_2
H_2O	78.36	8.24	4.93	0.18	3.48
D_2O	78.3	10.37	4.8	—	—

[a] See also Ref. [163].

TABLE III
The High-Frequency Contributions to the Dielectric Relaxation Data of H_2O and D_2O[a]

Solvent	n_1^2	Ω_1, (cm^{-1})	n_2^2	Ω_2, (cm^{-1})	n_3^2	Ω_3, (cm^{-1})	n_4^2
H_2O	3.48	193	2.1	685	1.77	—	—
D_2O	4.8	64.0	4.2	184	2.1	505	1.77

[a] See also Ref. [163].

Bertolini and Tani [222] performed a detailed molecular dynamics simulation study of the dielectric relaxation of water in which they simulated 343 water molecules interacting through transferable intermolecular potential-4 point (TIP4P) potential. Their simulated $\phi(t)$ values provide the following general expression for the dielectric relaxation in the frequency plane [222]:

$$\varepsilon(z) - n^2 = \frac{[\varepsilon_0 - \varepsilon_1]}{[1 + z\tau_1]} + \frac{[\varepsilon_1 - \varepsilon_2]}{[1 + z\tau_2]} + \sum_{j=1}^{3}(n_j^2 - n_{j+1}^2)[1 - z\Phi_j^{\mathrm{lib}}(z)] \tag{5.3}$$

where τ_i is the time constant of the ith Debye dispersion, which reduces the dielectric constant of the solvent from ε_i to ε_{i+1}. On the other hand, the n_j^2 values are the intermediate dielectric constants related to the high-frequency librational (or vibrational) modes, such that the jth libration is responsible for the reduction in the value of the dielectric constant from n_j^2 to n_{j+1}^2. $\Phi_j^{\mathrm{lib}}(z)$ is the Laplace transform of the librational moment correlation function corresponding to the jth librational mode and n^2 is the optical dielectric constant.

We used Eq. (5.3) with known experimental results to calculate the rotational kernel from Eq. (2.12).

The expression for the jth mode librational moment correlation function, $\phi_j^{\mathrm{lib}}(z)$ can be derived as follows. Let us assume that the librational motion takes place in a harmonic potential well and the oscillator executes a marginally damped motion with damping constant γ. This damped harmonic motion can then be described by a generalized Langevin equation as follows:

$$I\ddot{\mu}(t) = -\Omega_0^2\mu(t) - \gamma\dot{\mu}(t) + N \tag{5.4}$$

where Ω_0 is the frequency of oscillation, I is the moment of inertia, $\mu(t)$ represents the time-dependent angular displacement, and N is the random torque. The double dots stand for the second-order time derivative, and the single dot for the first order. When this equation is Laplace transformed, the following expression for the normalized librational moment correlation function is obtained [232]

$$\phi_j^{\text{lib}}(z) = \frac{z + \gamma_j}{\Omega_{0,j}^2 + z^2 + z\gamma_j} \tag{5.5}$$

Recently, Kindt and Schmuttenmaer [218] studied the dielectric relaxation of water using femtosecond tera-Hertz (fs-THz) pulse spectroscopy. Their results on dielectric relaxation of water provide a good fit with two Debye relaxations and $\varepsilon_\infty = 4.93$. For water, however, $n^2 = 1.77$, which indicates that the high-frequency librational (or vibrational) motion of the hydrogen bond network may be responsible for the calculated $\varepsilon_\infty - n^2$ dispersion. We have attributed this dispersion to two damped librational modes with different frequencies $\Omega_{0,j}$ (Table II). The damping constant γ_j is taken as twice the value of that of the respective frequencies $\Omega_{0,j}$ of the jth mode.

4. *Calculation of the Translational Dissipative Kernel*

The translational dissipative kernel $\Gamma(q, z)$ is calculated directly from the translational dynamic structure factor of water by using Eq. (2.13). The details are given in Section II.

B. Calculation of Other Parameters Necessary in Calculations for Water

The following parameters have been used to characterize water in our calculations.

1. The polarity parameter $3Y$ has been calculated using $\mu = 1.8$ D [233] and a density $\rho_0 = 0.03334$ Å^{-3}, which leads to a value of $3Y = 11.5$.
2. The ionic solute is the same size as coumarin-343 (C-343). In our calculation, C-343 is approximated by a spherical ionic solute with the charge situated at the center of the sphere. An important parameter is thus the solute : solvent size ratio, which is equal to 3.0.

C. Parameters Required in the Calculation of the Isotope Effect

To study the influence of deuterium isotope substitution in solvent, the following changes are necessary.

1. For D_2O, we calculated the polarity parameter using the proper dipole moment ($\mu = 1.8545$ D [234]) and number density ($\rho_0 = 0.0301$ Å^{-3}), which results in a value of 10.5671 for $3Y$ in D_2O.
2. The solvent mass m changes from 18 to 20 amu, and the other mass-dependent quantities change accordingly.
3. The rotational dissipative kernel is calculated from Eq. 2.12 with the respective values of the related parameters for D_2O.

The dielectric relaxation in deuterated water is rather different from that of water. Far-IR and Raman studies are available for D_2O [223–224,235–236], and in the present investigation we used these results; but the dielectric relaxation data with a broad frequency coverage are not available [225]. For the sake of comparison, we studied the solvation dynamics in water with a single Debye relaxation process. As the far-IR band positions are not much different in H_2O and D_2O, it is expected that the n_2^2, n_3^2, and n_4^2 should also not be much different in these solvents. We have, therefore, used the same value of these quantities in both the solvents. Tables II and III, summarize the dielectric relaxation data used to calculate and compare the dynamics of ion solvation in H_2O and D_2O. The rest of the parameters remain the same as in water.

D. Results

1. *Ion Solvation Dynamics in Water: Theory Meets Experiment*

Figures 8–10 depict the theoretically calculated $S(t)$ for water obtained after using three different sets of dielectric relaxation data. For comparison, the experimental results of Jimenez and co-workers [29] are also shown in these figures. The theoretical results shown in Figure 8 were obtained by using only the two Debye dispersions. The contributions from the high-frequency intermolecular vibration (IMV; hydrogen bond excitation) and librational modes are not included. As seen in Figure 8, the decay of the theoretically predicted $S(t)$ is much slower than the experimental results. This indicates the importance of the high-frequency modes in determining the solvation energy relaxation in water.

Figure 9 describes the comparison between the theory and experiments when the theoretical calculations are conducted with two Debye contributions plus one high-frequency IMV mode contribution. The frequency of the IMV band used here is taken as 193 cm^{-1} which accounts for the $\varepsilon_\infty = 4.93$ to $n_1^2 = 2.1$ dispersion. The agreement is now better. To facilitate the comparison, the calculated $S(t)$ is fitted to the following form:

$$S(t) = A_G \exp[-\omega_G^2 t^2/2] + A_1 \exp[-t/\tau_1] + A_2 \exp[-t/\tau_2] \qquad (5.6)$$

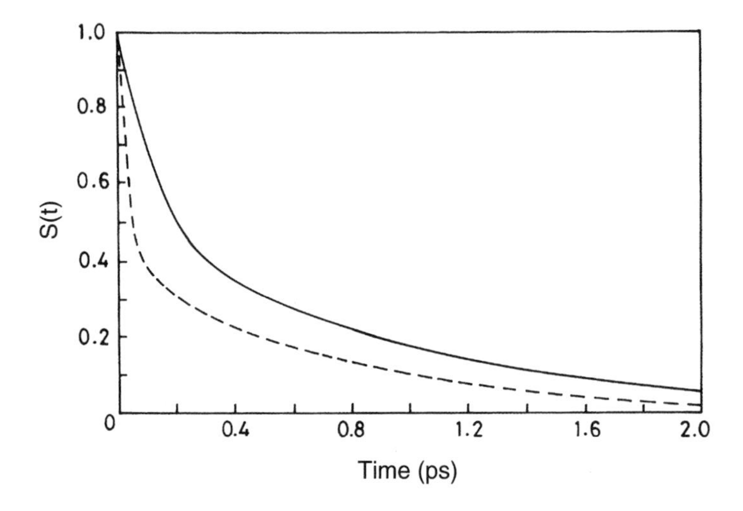

Figure 8. The prediction of the MHT (*solid line*) with only two Debye dispersions [218] in the dielectric relaxation compared to experimental results (*dashed line*) for the solvation time correlation function $S(t)$. Eq. (4.15) was used to calculate the ionic solvation dynamics of excited coumarin-350 in water. Note that no high-frequency contribution to $\varepsilon(z)$ has been included.

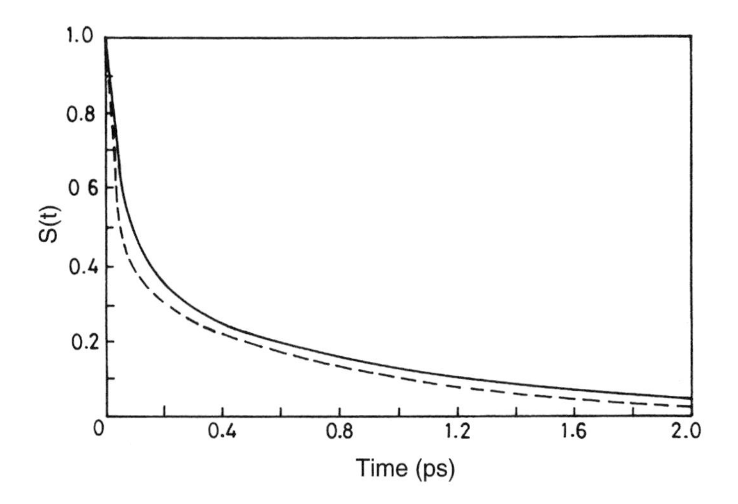

Figure 9. The same comparison as that shown in Figure 8 but with the addition of one high-frequency contribution. The frequency of the high-frequency mode is taken as 193 cm^{-1}, which is the intermolecular vibration (hydrogen bond excitation) and which is responsible for the decrease of the high-frequency dielectric constant $\varepsilon_\infty = 3.48$ to $n_1^2 = 2.1$. The representations remain the same as those in Figure 8.

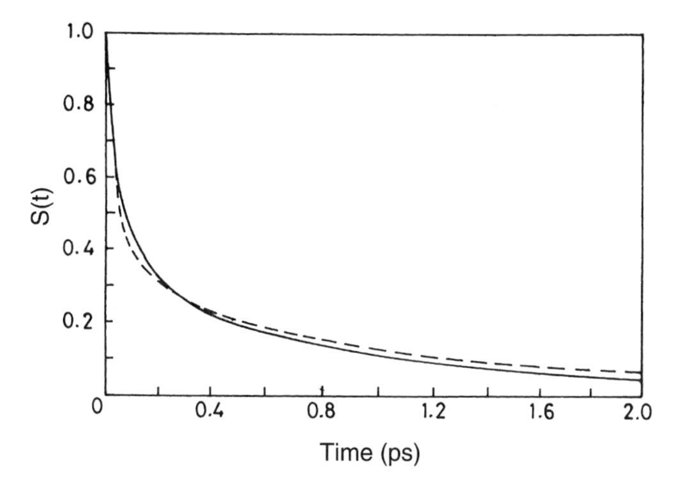

Figure 10. The same comparison as that shown in Figure 9 but with the addition of one librational contribution. The intermolecular vibration mode frequency remains the same as that in Figure 9. The frequency of the librational mode is 685 cm^{-1}, which is responsible for the decrease of the high-frequency dielectric constant $n_1^2 = 2.1$ to $n_2^2 = 1.77$. The representations remain the same as those in Figures 8 and 9.

TABLE IV
Parameters Obtained by Fitting the Experimental [29] and Theoretical Curves to Eq. (5.6)

		Theoretical	
Fit Parameters	Experiment	2 Debye + 1 IMV	2 Debye, 1 IMV, + 1 librational
A_G	0.48	0.23	0.255
A_1	0.20	0.415	0.420
A_2	0.32	0.355	0.325
A_3	0.283	0.365	0.4
ω_G (ps^{-1})	38.5	37.75	56.036
τ_1,(fs)	126	111.3	100.7
τ_2, (fs)	880	1051	1042

where A_i and τ_i stand for the amplitude and their time constants, respectively, and ω_G denotes the frequency associated with the ultrafast Gaussian decay. The fit parameters thus obtained (by using the Levenbarg–Marquardt nonlinear root search algorithm) are shown in Table IV along with the experimental data. It is clear from the table that all three calculated time constants (after using the two Debye dispersions plus the 193 cm^{-1} IMV band) are in good agreement with those from the experiment [29]; however, the respective amplitudes are somewhat different.

Figure 10 shows the theoretical results with two Debye dispersions, one IMV dispersion, plus one librational dispersion. The frequency of the IMV band remains the same as that in Figure 9. The extra libration added here is 685 cm^{-1}, which reduces the high-frequency dielectric constant $n_1^2 = 2.1$ to $n_2^2 = 1.77$. This extra channel makes the decay of $S(t)$ faster than that observed in Figure 9 and even faster than that from experiment [29]. The fit parameters for this case are shown in Table IV. This result appears to be in good agreement with the recent theoretical studies of Marcus and co-workers [237].

2. *Role of Intermolecular Vibrations in the Solvation Dynamics of Water: Effects of Polarizability*

The IR peak at the 193 cm^{-1} band is related to the interaction-induced effects [219,220]. This also appears in Raman [223,226] and inelastic neutron scattering [223,224] and was assigned to the O...O stretching mode of the O–H...O unit. This band is located approximately at the same frequency in the Raman and the far-IR spectrums of H_2O and D_2O [236]. Because the frequency shift is rather small in D_2O, 193 cm^{-1} was assigned to the hydrogen-bonded O...O stretching. This band is thus related to the intermolecular oscillation of the hydrogen bond network and is believed to be governed by the dipole-induced dipole mechanism [219,220]. As a result, this band may be weak or absent in molecular dynamic simulations [238,239] if the polarizability effects are neglected. Note that simulations do locate a weak Raman peak around 193 cm^{-1}, even when polarizability of the solvent is not taken into account.

If the water molecules were not polarizable, the 193 cm^{-1} IMV would not affect the dielectric relaxation. And, as a result, would not affect the polar solvation dynamics. The IMV and the extended hydrogen bond network would continue to exist even in the absence of the polarizability. Such a situation indeed arises in computer simulations [35].

One immediate consequence of the neglect of polarizability is that the optical dielectric constant is equal to unity: $n^2 = 1$. Another important consequence is that the dielectric relaxation from $\varepsilon_\infty = 4.93$ to $n^2 = 1$ now must proceed via the libration and higher frequency modes. What would be the consequence of these rather artificial changes? We now turn our attention to this problem.

Maroncelli and Fleming performed a detailed simulation study of solvation dynamics in nonpolarizable water [35]. The neglect of the polarizability means setting n^2 equal to unity. Roy and Bagchi [160] performed a separate set of calculations for $S(t)$ in which two Debye dispersions and a single libration (underdamped) are included. The frequency of this libration is taken to be 685 cm^{-1}, which is responsible for the dispersion $\varepsilon_\infty = 4.92$

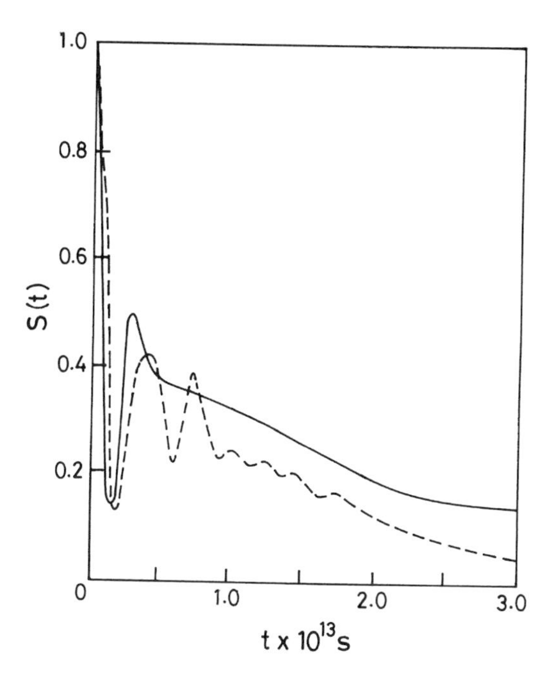

Figure 11. Comparison of the calculated solvation time correlation function $S(t)$ (*solid line*) and the simulation results of Maroncelli and Fleming [35] (*dashed line*). To facilitate the comparison, calculations were done here by replacing the 193 cm^{-1} IMV band with a 600 cm^{-1} band. See the text for further details.

to $n^2 = 1$ (Fig. 11). However, the relaxation rate becomes faster than that observed in experiments [29]. As can be clearly seen from Figure 11, the initial rates agree rather well with those from the simulation [35]. The present study, therefore, suggests that the classical computer simulations without polarizability effects may not be reliable for initial solvation, because they fail to describe the high-frequency dielectric response of water correctly. The marked oscillation in $S(t)$ may be entirely due to the assumed underdamping of the used mode, and thus is not present in the experimental situation.

Recently Lang and co-workers [239a] performed a photon-echo experiment on solvation dynamics of eosin dye in water. The solvation response function thus obtained consists of a Gaussian time constant of 30 femtosecond with an amplitude of ~60%. Thus the time constant is in the same range as reported in their earlier paper published in *Nature* [29] although the amplitude of it is somewhat less. The time constant of the long time biexponential decay remain virtually the same. It is worthwhile to note that the earlier experiments of Jimenez and co-workers [29] and the most recent

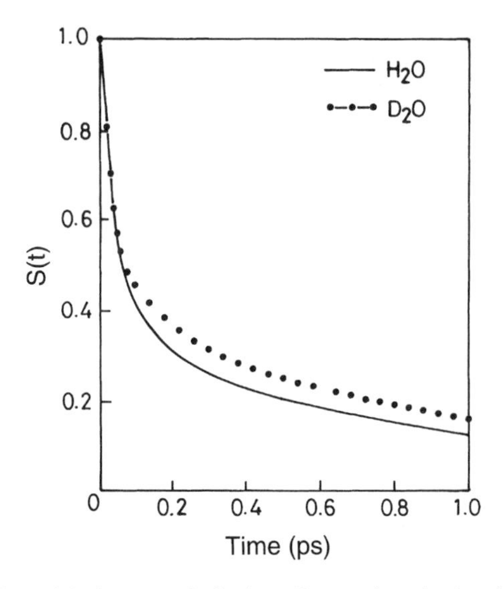

Figure 12. Prediction of the isotope substitution effect on the solvation dynamics of an ion in water from MHT. The *solid line* represents the calculated solvation time correlation function as a function of time in water; the line connecting the *solid circles* corresponds to that in heavy water. The solute : solvent size ratio is 3.0; the dielectric relaxation data of these two liquids are summaried in Tables II and III.

work [239a] employ two completely different time domain spectroscopic techniques. Therefore, the agreement between these two experiments is impressive. The main difference between these experiments is that there are some oscillations observed immediately after the initial Gaussian decay in the photon-echo measurements.

Lang and co-workers [239a] attributed the initial Gaussian decay to 193 cm^{-1} intermolecular vibration (IMV) band as was proposed originally by Roy and Bagchi [160]. Subsequent theoretical studies [163a] have shown that the oscillations can be attributed to the 193 cm^{-1} IMV and 685 cm^{-1} libration bands.

3. *Solvation Dynamics in Heavy Water: Solvent Isotope Effect*

Figure 12 represents the comparison of predicted $S(t)$ from MHT in H_2O and D_2O, respectively. Our studies clearly indicate that an isotope effect may be observed in solvation dynamics in the long-time tail of $S(t)$. As pointed out by Nandi and co-workers [63], this is because there is a marked difference between the frequency-dependent rotational dissipative kernel of H_2O and D_2O at small frequencies. An isotope effect has been observed [82,85,88,240] in the case of electron solvation in water in which the initial short-time relaxation in D_2O is found to be slower than that in H_2O. Studies of the isotope substitution effect on electron transfer rates in aniline and deuterated aniline [229] reveal that this effect is related to a process that is faster than the solvent diffusive motions. It is, however, interesting to note that the intensity of the fluorescence decay of coumarin-152 and coumarin-153 in deuterated aniline shows a significant difference in the long-time decay [229]. Thus a study of isotope effects on solvation in aniline may provide useful insight on electron transfer reaction. Yoshihara and co-workers [241] investigated the dynamics of liquid aniline and its N deuterated and N methylated derivatives by heterodyne-detected optical Kerr effect measurements. Their experimental studies revealed a small isotope effect that marginally increases the frequency of the librational motion [241].

At this point we would like to discuss the relation of the present work with the important work of Bader and Chandler [151], who simulated the solvation dynamics after a photoinduced electron transfer of an aqueous Fe^{+2}–Fe^{+3} system. In this case the solvation energy contains contributions both from an inner shell (the ligands) and the bulk. The theory presented here is, obviously, applicable only to the latter part. In the simulation of Bader and Chandler the minimum after the Gaussian decay occurs around 20 fs, which is in good agreement with the calculations shown in Figure 11.

E. Conclusions

In this section, we presented a detailed analysis of ion solvation in water and heavy water, using an extended molecular hydrodynamic theory. A good agreement is obtained between the theory and the available experimental results. The present theoretical formulation suggests that the initial ultrafast relaxation is dominated by the macroscopic polarization modes of the solvent. In water, the initial ultrafast component seems to derive major contributions from the intermolecular vibrational and librational modes. The static properties of the solvent are found to critically determine the biphasic nature of the solvation energy relaxation. In highly polar liquids, large dipole moments and static dielectric constants lead to large values of the force constant $f(110; q = 0)$. This implies that the underdamped polarization relaxation takes place in a steep free energy surface. The large value of $f(110; q = 0)$, in turn, amplifies the contribution of the high-frequency inertial motions of the solvent to the generalized rate of polarization relaxation. This may be the reason why the intermolecular vibrations and librations are so important in the ultrafast relaxation in water, whereas their relative contribution to dielectric relaxation is rather small. A moderate isotope effect is observed in the long time part of aqueous solvation. It follows from the present study that the vibrational mode at 193 cm^{-1} in water, which is the result of the O...O stretching of O–H...O units, may not be as underdamped as revealed in earlier experiments [221]. But the detailed nature of the dynamics of this is not fully understood and requires further investigation.

The main limitation of the present study is the assumption of the linear response of the dipolar solvent to the sudden creation of charge distribution. However, the good agreement between theory and experiment, especially at the short times, suggests that linear response may hold for solvation of a small, rigid solute ion. The validity of the linear response theory has also been confirmed for electron solvation dynamics in water by the quantum molecular dynamics simulation studies of Rossky and co-workers [87–88]. This is because the early part of solvation dynamics is primarily dominated by the long wavelength polarization mode of the solvent molecules. The long wavelength modes are, by definition, least affected by the distortion of the solvent structure owing to the presence of the ion. But these distortions are certainly important for the intermediate wavenumber response that arises from the molecular-length scale processes. The latter, however, contributes mainly to the long-time decay, when the bulk of the relaxation is over. Therefore, it is expected that the assumption of linear response may not significantly alter the ion solvation dynamics.

The second limitation of the present study is the neglect of the wavenumber dependence of the dissipative kernel. As discussed, neither of these may turn out to be a serious limitation in the present context.

Let us finally address the much-debated question of nonlinearity in solvation dynamics. Conventionally, in the linear response treatment for a weak solute–solvent coupling, the solvent relaxation in the presence of the solute is replaced by the dynamics of the unperturbed solvent. The solute thus provides only the driving force for the solvent relaxation and does not at all affect its dynamics. By nonlinearity one essentially means the effect of the solute on the dynamics of the solvent, and this may arise in several ways. First, in a wide variety of systems, specific solute–solvent interactions are nonnegligible, and one needs to consider explicitly the solvent dynamics in the presence of the solute, and not that of the unperturbed solvent. This is a situation often encountered with dye molecules, which can form complex structures (because of the hydrogen bonds) with water and methanol; and the whole scenario may change completely. Recent simulation studies [242] have shown that the solvation of the Na^+ ion in a series of ethers belongs to the nonlinear response regime. The nonlinear behavior is associated with the specific binding of the cation to the negative oxygen sites [242].

Second, the assumed step function change in solute charge on excitation may not accurately represent the experimental situation [31]. Here, one essentially assumes that the electronic states of the probe molecules are of fixed properties, negligibly perturbed by the state of the surrounding solvent molecules. But the general validity of such assumption is not obvious, given that the strength of solute–solvent interactions in the systems of interest amounts to many thousands of wavenumbers [31]. For example, one may imagine that the solute's excited state dipole moment would increase slightly as its solvent surroundings become more polarized. Such solvation-dependent changes, if present, would introduce substantial nonlinearity in the dynamics. Furthermore, participation of the intramolecular vibrational modes of the solutes may lead to significant nonlinearity effects.

The formulation of any theory beyond the assumption of linear response is indeed a formidable task that has not yet been achieved. For this purpose, the solvent distortion in the presence of the solute needs to be taken into account. This, in turn, invalidates the wavenumber space description of the present MHT. The problem must be formulated in the real space, as was done initially by Calef and Wolynes [133]. This, however, requires extensive numerical work, even if we can describe the distortion by methods of equilibrium liquid state theories. It should also be pointed out here that to account for the nonlinear response it is necessary not only to include the static nonlinearity (which is because of the difference in molecular arrangement near the solute from that in the bulk) but also to take properly into account the dynamic nonlinearity. The latter can be quite significant, because the transport properties of the nearest-neighbor molecules (such as the translational and rotational diffusion coefficients) can be markedly different from those in the bulk because of the hydrogen bonding effects. Although incorporation of the static nonlinearity can be achieved by equilibrium statistical mechanics [155–158], the dynamic nonlinearity still remains a difficult and open problem.

VI. IONIC AND DIPOLAR SOLVATION DYNAMICS IN MONOHYDROXY ALCOHOLS

Monohydroxy alcohols constitute an important class of solvents for chemical and biological reactions. It is known that the structure and the polarity of these alcohols largely control the rate and even the course of many chemical reactions. An in-depth understanding of the dynamics of these solvents, however, has eluded chemists for a long time. There are many reasons for this lacuna, the foremost among them is the nonavailability of sufficient time resolution to resolve the fast components that are critically important in the dynamics of these hydrogen-bonded liquids. This has been changed in the last 5 years with the application of femtosecond spectroscopy to the study of these solvents. Notable among these studies is the study of polar solvation dynamics.

As discussed in Section I, the polar solvation dynamics of a newly created ion or dipole has been studied extensively [27–66]. Experimental studies

TABLE V
Comparison of Experimental Time Constants and Amplitudes

	Methanol		Butanol	
Study	τ_1, ps	A_1, %	τ_1, ps	A_1, %
Joo and co-workers [243]	0.065	52.6	0.0585	38.2
Horng and co-workers [66]	0.03[a]	10.1	0.243	15.9
Bingemann and Ernsting [65][b]	0.07	30.0	—	—

[a] Arbitrary, because the response was too fast to be detected by the experiment performed, which used subpicosecond time resolution.
[b] Data for butanol have not yet been reported by this group.

[27–34], theoretical studies [153–164,202,204,208,243], and computer simulations [35,151] suggested that in many common solvents, like water and acetonitrile, the time-dependent solvent response to an instantaneously created charge distribution is biphasic, with an ultrafast component that carries 60–70% of the total solvation energy and a time constant of 50–100 fs. The other component of the biphasic response is rather slow with a time constant in the picosecond range. The ultrafast component in these solvents has been shown to be rather generic in nature, which originates from the fast modes of the solvent, such as librational and intermolecular vibrational modes.

The scenario appears to be rather different and controversial (at this stage) for the homologous series of the straight-chain monohydroxy alcohols. An earlier experimental study [148] on solvation dynamics in methanol showed that it exhibits much slower dynamics, in which the solvent response to an instantaneously created charge distribution could be given by a biexponential response function with two different time constants in the picosecond range. Subsequent theoretical investigations [162,197] and computer simulation studies [27a,36] in methanol, however, indicated the presence of an ultrafast component. This has been confirmed by a series of recent experiments [64–66,243] that shows there is indeed an ultrafast component in methanol, with a time constant of about 70 fs and that carries nearly 50% of the total solvation. Although the experimental and computer simulation studies are in agreement with the existing theoretical results for water and acetonitrile, the situation has not been the same for methanol.

Recently solvation dynamics in ethanol, propanol, and butanol have also been investigated experimentally by using nonlinear spectroscopic techniques groups [64–66,243]. The results obtained are rather different for different experiments. Table V shows time constants associated with the ultrafast component of the total solvent response function for the four alcohols, as obtained by three different experiments, to emphasize this point. The obvious disagreement between different experiments for the same alcohols immediately raises the following questions:

1. Why do these experiments reveal such diverse time scales?
2. Which of these experiments correctly probes the dynamical processes that are relevant in determining the polar solvent response to a sudden change in the charge distribution on a dye molecule?
3. Is there any relation between apparently diverse results of these three experiments?

Answers to these questions are crucially important in determining the dynamic solvent effects on the kinetics of electron transfer reactions and other chemical processes. Note that no theoretical study exists for the solvation dynamics in these three alcohols.

An important feature of some of these experiments is the use of a large dye molecule as a probe. As mentioned, the techniques applied in these experiments [65,66,243], which followed the time-dependent progress of solvation, differ from each other. In the experiment of Joo and co-workers [243], which uses a dye molecule (IR144 or HITCI) as a probe, the time-dependent solvent response was measured by three pulse stimulated photon echo peak shift (3PEPS). Here the response function contains contribution from two potentially different sources. One is the relaxation of the solvation energy and the other is the equilibrium energy–energy time correlation function. In the experiment of Horng and co-workers [66], the time-dependent progress of solvation of excited C-153 was followed directly by measuring the usual time-dependent Stokes shift of fluorescence emission spectrum of the probe. On the other hand, Bingemann and Ernsting [65] studied the solvation dynamics of the styryl dye DASPI in methanol. In their experiment, the stimulated emission spectrum was analyzed by broad-band transient absorption. These authors have not yet reported the experimental results for other higher alcohols.

Clearly, the difference in the results observed may arise at least partly because of the sensitivity of the different experimental techniques to the vari ous aspects of solute–solvent dynamics. A theoretical study of the polar s vation dynamics of an excited solute in these alcohols may answer some of the questions like the presence and amplitude of the ultrafast component in these solvents.

In this section we present such a theoretical study of polar solvation dynamics in methanol, ethanol, propanol, and butanol and compare the theoretical predictions with all the available experimental results [65,66,243]. The study reported here has been carried out by using the most recent experimental results of dielectric relaxation in monohydroxy alcohols [218], in the simple theoretical approach developed in Section IV. The present study has produced several interesting results. For methanol, the theoretical results are in almost quantitative agreement with the experimental results of Bingemann and Ernsting [65] and are also in satisfactory agreement with results of Horng and co-workers [66]. For ethanol, propanol, and butanol, the agreement among the theoretical predictions and the experimental results of Horng and co-workers [66] are also excellent; however, the theoretical results are in total disagreement with the experimental results of Joo and co-workers [243]. The reason for this is not certain. One possibility is that the results of Joo and co-workers might be sensitive to the nonpolar part of the solvation, the dynamics of which are still poorly understood. We shall come back to this point later.

The solute probes used in experimental studies can be either ionic or dipolar in nature. For example, if a well-separated charge-transfer complex is formed in the excited state, then the subsequent solvation dynamics can be described as ionic. On the other hand, if no charge separation but only charge redistribution takes place on excitation, then the dynamics of solvation of the solute are regarded as that of a dipole. However, even in the case of Coumarin-153, the charge distribution is an extended one, with variable charges at different atomic sites. Thus moments higher than dipole can be involved [244]. This aspect is rather difficult to include in a molecular theory. The calculations for the solvation of a neutral dipole are possible, and the expressions involved are somewhat more complicated than those for the ion. To understand the results of Horng and co-workers [66], we also calculated the solvation energy time correlation function (STCF) for the dipolar solute. The theoretical predictions appear to be in good agreement with the experimental results of Horng and co-workers [66] for ethanol and propanol. For methanol, the long-time decay of the theoretically calculated STCF for the dipole is also in good agreement with that of Horng and co-workers [66].

The theory used here is rather simple and easy to implement. Earlier, we showed that this theory is successful in explaining the experimentally observed solvation dynamics not only in some simple systems like water [160,161,163] and acetonitrile [159,164] but also in amide systems [195]. The reason for the success can be attributed to several factors. First, an accurate calculation of the relevant memory function was carried out by using the full dielectric relaxation data. Second, the polar solvation dynamics are found to be dominated by the macroscopic polarization mode of the solvent. This long wavelength polarization fluctuation was properly treated by our theory, which also systematically includes the effects of short-range local correlations on the generalized rate of polarization relaxation of the solvent.

Note that even the moderate success of the continuum model description [138] in predicting the qualitative features of the observed solvation dynamics in these polar solvents is entirely the result of the dominance of this macroscopic polarization mode. This aspect will be discussed later.

The organization of the rest of this section is as follows. In Section VI.A we briefly describe the theoretical formulation; Section VI.B contains the details of the calculational procedure. Numerical results on ionic and dipolar solvation dynamics of coumarin-153 in alcohols are given in the next two sections, and in Section VI.E, we give a tentative explanation on the origin of the ultrafast component in ethanol and butanol observed by Joo and co-workers [243].

A. Theoretical Formulation

As discussed in Section I, the solvation time correlation function is defined by the following expression:

$$S(t) = \frac{E_{\text{solv}}(t) - E_{\text{solv}}(\infty)}{E_{\text{solv}}(0) - E_{\text{solv}}(\infty)} \tag{6.1}$$

where $E_{\text{solv}}(t)$ is the time-dependent solvation energy of the probe at time t and $E_{\text{solv}}(\infty)$ is the solvation energy at equilibrium. This energy can be obtained by following the time-dependent Stokes shift of the emission spectrum after an initial excitation, as in the experiments of Horng and co-workers [66], who used C-153 to obtain the $S(t)$. The time-dependent progress of solvation of a laser-excited dye was also studied by various highly nonlinear spectroscopic techniques especially suited to study the ultrafast component [243]. These techniques are, however, often sensitive to the total energy–energy time correlation function $M(t)$, defined by the following expression [243]:

$$M(t) = \frac{\langle \Delta E(t) \Delta E(0) \rangle}{\langle |\Delta E(0)|^2 \rangle} \tag{6.2}$$

where $\Delta E(t)$ is the fluctuation in the energy difference between two levels. It is usually stated that $S(t)$ and $M(t)$ are the same within the linear response of the liquid. However, different experiments may be sensitive to different aspects of $S(t)$ and $M(t)$. Thus, although the experiments of Horng and co-workers [66] probe the polar solvation dynamics [i.e., $S(t)$] those of Joo and co-workers [243] are sensitive to the full $M(t)$. For a large dye molecule, this might make a significant difference, as discussed below.

To study the time-dependent progress of solvation in these alcohols, we need the full expression of either $M(t)$ or $S(t)$. For ionic solvation dynamics, this is given by Eq. (4.15), and we use Eq. (4.20) to study the dipolar solvation dynamics. As discussed earlier, the important ingredients for calculating $S(t)$ from Eqs. (4.15) and (4.20) are the ion–dipole and dipole–dipole direct correlation functions [$c_{id}(q)$ and $c_{dd}(q)$], the longitudinal (i.e., 10) and the transverse (i.e., 11) components of the static orientational correlation functions [$\varepsilon_{10}(q)$ and $\varepsilon_{11}(q)$], and the memory functions. We shall discuss briefly the calculation procedures of these quantities.

B. Calculational Procedure

1. *Calculation of the Static Correlation Functions*

For methanol, the 10 component of the static correlations has been taken from the XRISM calculation of Raineri and co-workers [157]. For higher alcohols, these correlations are obtained from the MSA model. We have, of course, used them after properly correcting at both the $q = 0$ and the $q \to \infty$ limits. The calculation procedure was described in Section IV. The transverse 11 component is calculated directly from the MSA model.

The calculation of the ion–dipole DCF is taken directly from the solution of Chan and co-workers [206] of the mean spherical model of an electrolyte solution. We have, of course, used it in the limit of zero ionic concentration. Note that the MSA model for the ion–dipole correlation is known to be fairly accurate [135]. We would like to emphasize here that because of the properties of $\varepsilon_{10}(q)$ and $c_{id}(q)$, the most important contribution to the ionic solvation dynamics comes from the $q = 0$ limit, at which these functions are determined by the macroscopic, well-defined parameters. As already discussed, we require radii of the different solute and solvent molecules for the calculation of $c_{id}(q)$, which have been calculated from their respective van der Waal's molar volumes [245]. Adjustable parameters do not need to be used at any stage of the calculations.

To further highlight the role of the long wavelength polarization modes in ionic solvation dynamics, Figure 13 shows the integrand of the denominator of Eq. (4.15) $I(q\sigma)$ as a function of $q\sigma$, where

$$I(q\sigma) = (q\sigma)^2[c_{id}^{10}(q\sigma)]^2[1 - 1/\varepsilon_{10}(q\sigma)] \tag{6.3}$$

Here the solvent is ethanol. The other static parameters such as dipole moment, density, and static dielectric constant for ethanol are given in Table VI. Figure 13 clearly shows that the polar solvation dynamics derives a major contribution from the long wavelength ($q\sigma \to 0$) region. This is because of the long-range nature of ion–dipole interactions. We shall come back to this point again.

2. *Calculation of the Memory Functions*

We calculated the rotational memory function $\Gamma_R(k, z)$ from the frequency-dependent dielectric function $\varepsilon(z)$ by using Eq. (2.12). The $\varepsilon(z)$ values for

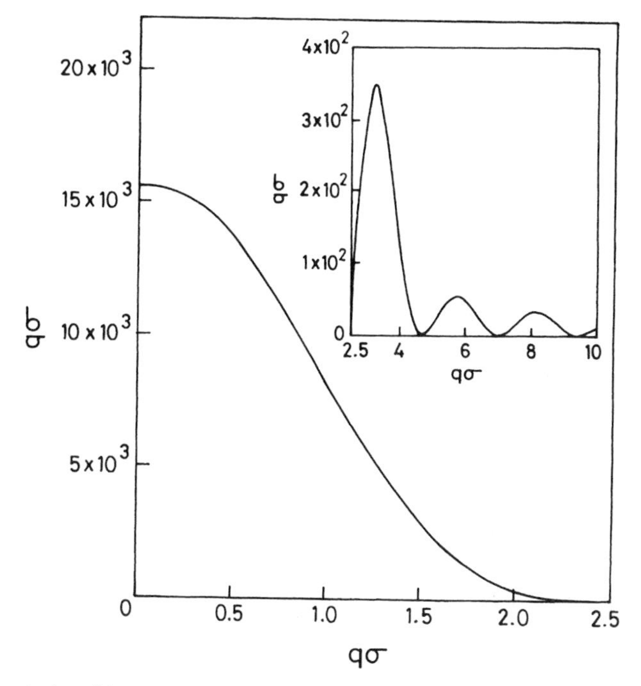

Figure 13. A plot of the wavenumber dependence of the integrand $I(q\sigma)$ for ethanol showing the dominance of small wavenumbers in determining the ionic solvation energy. *Inset*: The wavenumber dependence of the above integrand at $q\sigma > 2.5$. Note the scale difference along the ordinates of the two graphs. The solvent parameters needed for this calculation are given in Tables VI and VII. See the text for further details.

TABLE VI
Solvent Parameters Needed for the Theoretical Calculation[a]

Solvent	Diameter, Å	$\mathcal{R}$[b]	μ, D	ρ, g/mL	η, cP
Methanol	4.10	1.9	1.7	0.78	0.5428
Ethanol	4.67	1.67	1.69	0.79	1.075
Propanol	5.2	1.52	1.7	0.8	2.2275
Butanol	5.532	1.41	1.66	0.806	2.271

[a] Data taken at room temperature.
[b] The solute : solvent size ratio for C-153.

these monohydroxy alcohols are often described by the following rather general expression [218]:

$$\varepsilon(z) = \varepsilon_\infty + \sum_{i=1}^{m} \frac{(\varepsilon_i - \varepsilon_{i+1})}{[1 + z\tau_i]^\beta} \tag{6.4}$$

TABLE VII
Dielectric Relaxation Parameters[a] for Four Alcohols

Solvent	Processes	ε_0	τ_1, ps	ε_1	τ_2, ps	ε_2	τ_3, ps	ε_3	n_D
Methanol	3 D	32.63	48	5.35	1.25	3.37	0.16	2.10	1.327
Ethanol	3 D	24.35	161	4.15	3.3	2.72	0.22	1.93	1.36
Propanol	3 D	20.44	316	3.43	2.9	2.37	0.20	1.97	1.384
Butanol	1 D-C	18.38	528.4	—	—	—	—	2.72	1.40

[a] Measured at room temperature.

where z is the Laplace frequency; ε_1 is the static dielectric constant; the ε_i values are intermediate steps in the dielectric constant, with $\varepsilon_{m+1} = \varepsilon_\infty$ as is its limiting value at high frequency; m is the number of distinct relaxation processes; the τ_i values are their relaxation time constants; and β is the fitting parameter. For the first three alcohols in the series studied here, an accurate fit is obtained with $\beta = 1$ (i.e., Debye processes) and $m = 3$. For butanol, $\beta = 0.924$ [246], which means a Cole–Davidson model is necessary to describe the relaxation process in this solvent.

For methanol, ethanol, and propanol, we use the most recent dielectric relaxation data measured by Kindt and Schmuttenmaer (KS) [218], who used the femtosecond teraHertz pulse transmission spectroscopic technique to characterize the multiple Debye behavior of these alcohols. The details of the dielectric relaxation data are summarized in Table VII. The notable aspect of this femtosecond pulse transmission technique is that for the three alcohols studied, the measured ε_∞ approaches closely to the square of the optical refractive index of the medium, n_D^2. This suggests that the dielectric relaxation processes observed and characterized by them are complete and reliable. Another interesting aspect of this study is that this technique can detect precisely even a fast process with a time constant of 160 fs (for methanol).

Unfortunately, KS [218] did not report the $\varepsilon(z)$ for butanol. Mashimo and co-workers [246] found that for this liquid the relaxation process is a Cole–Davidson one and is responsible for reducing the dielectric constant from 18.32 to 2.72, with a relaxation time constant of 528.4 ps. But the value of the square of the refractive index n_D^2 for butanol is 1.96. On the other hand, Garg and Smyth [247] reported a fit by three exponentials with time constants that do not appear to be reliable [246]. This indicates that although the first two Debye processes have been successfully fitted to a Cole–Davidson function with the fitting parameter $\beta = 0.924$ by Mashimo and co-workers [246], the latter authors missed the last fast process, which is responsible for the decrease of the dielectric constant from 2.72 to 1.96. It is not difficult, however, to estimate the third time constant by using the results of KS. These authors noted that this time constant is rather universal for alcohols. Hence we attribute this missing region of the dielectric dispersion from 2.72 to 2.22 to a relatively faster Debye relaxation process with a time constant of 2.5 ps. The choice of this range for the τ_D is also supported by the work of Barthel and co-workers [230] in propanol, for which the fastest time constant measured is 2.4 ps. Thus we have supplemented the dielectric relaxation data of Mashimo and co-workers [246] with a fast process with time constant of 2.5 ps.

The translational kernel $\Gamma_T(q, z)$ is obtained from the dynamic structure factor of the liquid by using Eq. (2.13). Once these two dissipative kernels are obtained, we can calculate both the longitudinal 10 and the transverse 11 components of the generalized rate of polarization relaxation by using Eqs. (2.12) and (4.28), respectively. Subsequently, the normalized $M(t)$ is obtained by using Eqs. (4.15) and (4.20).

C. Numerical Results: Ionic Solvation

1. *Methanol*

Let us first discuss the theoretical results on solvation dynamics in methanol. Figure 14 shows the decay of the calculated, normalized SCTF $S(t)$ of an ion with time t. The solute probe is C-153, which was used in the experiment of Horng and co-workers [66]. The solute : solvent size ratio is 1.9. For comparison, Figure 14 also plots the experimental results available from three different experiments [65,66,243]. It is obvious that in methanol, the solvation dynamics is dominated primarily by an ultrafast Gaussian component. The component with a theoretically calculated ultrafast time constant of about 70 fs accounts for 40–50% of the total solvation. This is in excellent agreement with the experimental results of Bingemann and Ernsting [65]. The decay rate observed by Joo and co-workers [243] is considerably faster than the theoretical prediction. The results of Horng and co-workers [66], on the other hand, are only slightly slower than the theoretical results, with an overall good agreement. The exact reason for the deviation from the experimental results of Joo and co-workers [243] is not known (see Section VI.D).

In solvation dynamics, the high-frequency modes make a contribution that is much larger than their role in dielectric relaxation. Figure 15 shows the progressive acceleration of the solvation time correlation function for metanol as the high-frequency components of dielectric relaxation are sequentially added. The theoretical predictions are compared to those of Horng and co-workers and Bingemann and Ernsting. Note the important role played by the last Debye relaxation time [$\tau_D(3) = 160$ fs].

Several comments on this almost quantitative agreement among the present theory and the experimental results [65,66] are in order. First, this is

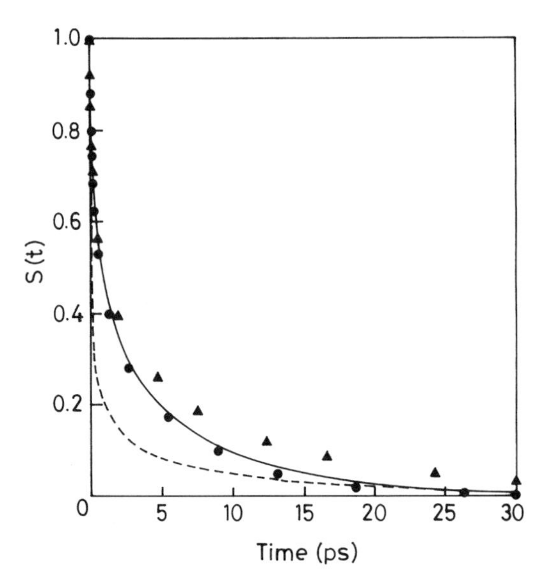

Figure 14. Comparison of the theoretical predictions and three different experimental results for the ionic solvation dynamics of excited C-153 in methanol at room temperature. The *solid line* represents the calculated normalized solvation energy–energy time correlation function $S(t)$ obtained using the Eq. (4.15); the experimental results are shown as *filled triangles* (Horng and co-workers), *filled circles* (Bingemann and Ernsting) and the *small dashed line* (Joo and co-workers). Note the agreement between the theoretical predictions and the experimental results of Bingemann and Ernsting and Horng and co-workers. The solute : solvent size ratio is 1.9; the dielectric relaxation data and other static parameters are given in Tables VI and VII.

obtained by using the most recent dielectric relaxation data made available by KS [218]. These authors found that in methanol, three consecutive Debye processes bring $\varepsilon(z)$ down all the way to 2.1, which is close to n_{D}^2. Thus no librational mode seems to be significantly coupled to the dielectric relaxation in methanol. If this is indeed correct, then the results obtained here may be the fastest polar solvation dynamics present in methanol. Second, it is interesting to contrast the ultrafast solvation dynamics in methanol with those in water and acetonitrile. In water, a high static dielectric constant ($\varepsilon_0 = 78$) and the 193 cm^{-1} librational band are responsible for the ultrafast solvation; whereas in acetonitrile, the same is carried out by the extremely fast single particle rotational motion of the solvent molecules, as probed by the Kerr relaxation [248]. In methanol, the situation is rather close to acetonitrile in the sense that the ultrafast component arises from the fast dielectric response of the solvent. Third, no adjustable parameter was used in the present calculation. Finally, we have also checked that the calculated $S(t)$ is rather insensitive to the probe size.

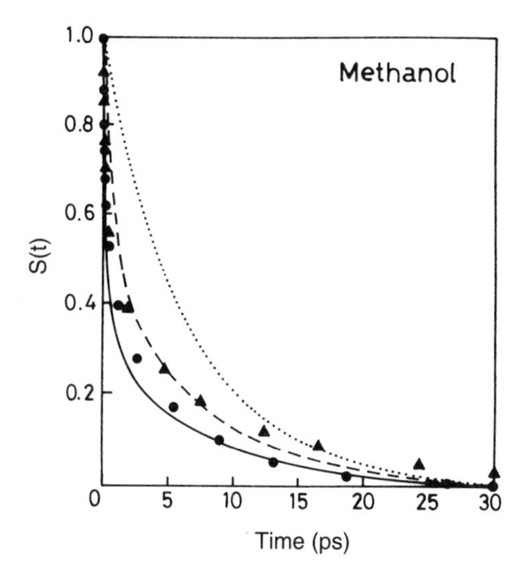

Figure 15. The effect of the sequential addition of the ultrafast components of dielectric relaxation data [218] on the solvation dynamics in methanol. The calculated $S(t)$ values are plotted for the following cases: only the first Debye relaxation is considered (*small dashed line*), only the first two Debye relaxations are considered (*large dashed line*), and all three Debye relaxations are considered (*solid line*). The experimental results of Horng and co-workers (*filled triangles*) and Bingemann and Ernsting (*filled circles*) are also shown. Note that the agreement is quantitative only when the full dielectric relaxation data are systematically incorporated.

2. *Ethanol*

Figure 16 presents the theoretical results of solvation dynamics in ethanol. The solute is the same, but now the solute : solvent size ratio is 1.67 owing to the larger molecular size of ethanol. Figure 16 also plots the experimental results of Joo and co-workers [243] and Horng and co-workers [66]. The solvation is much slower than that observed in methanol. The most recent dielectric relaxation data of KS [218] were used to obtain the dynamic polar response function of ethanol (Table VII). The solvation, however, is still biphasic, possessing a fast component that is not the ultrafast mode in the range of 50–100 fs. The theoretical results are in good agreement with the experimental results of Horng and co-workers [66].

A few comments are in order. First, the excellent agreement observed between the theory and experimental results [66] was obtained by using the recent dielectric relaxation data [218]. These data provide dielectric relaxation down to 1.93, which differs from n_D^2 by only 0.08. Given the small value of $\varepsilon_\infty - n_D^2$, there seems to be little scope of any significant ultrafast component in ethanol, although there can still be a presence of 1–5% of

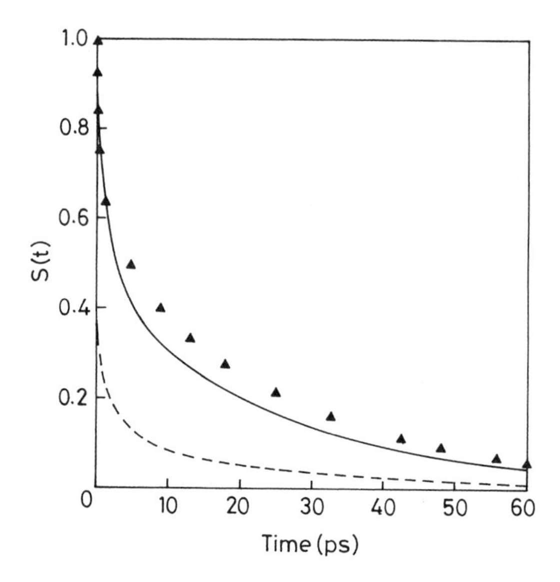

Figure 16. Ion solvation dynamics of excited C-153 in ethanol at room temperature. The calculated normalized solvation energy–energy time correlation function $S(t)$ is shown by the *solid line*. The experimental results of Horng and co-workers (*filled triangles*) and Joo and co-workers (*small dashed lines*) are also shown. The solute : solvent size ratio is 1.67.

such a component if the remaining dispersion ($\varepsilon_\infty - n_D^2$) comes from a high-frequency solvent mode. No adjustable parameter was used in the theoretical calculation.

3. *Propanol*

Next we present the results on ionic solvation dynamics of C-153 in propanol (Fig. 17). The solute : solvent size ratio is 1.52. Note that here the solvation becomes even slower, which could be attributed to the higher value of the first Debye relaxation time, which manifests itself in the slower rotation of a solvent molecule in the hydrogen-bonded pseudocrystalline-type network [247]. We also plotted the experimental results of Horng and co-workers [66], and the agreement is excellent. Here we do not have the results of Joo and co-workers [243]. Again, no adjustable parameter was used in the calculation of $S(t)$.

4. *Butanol*

Figure 18 shows the results for butanol; the solute : solvent size ratio is 1.41. For comparison, we also plotted the experimental results of Joo and co-workers [243] and Horng and co-workers [66]. The theoretical results are in good

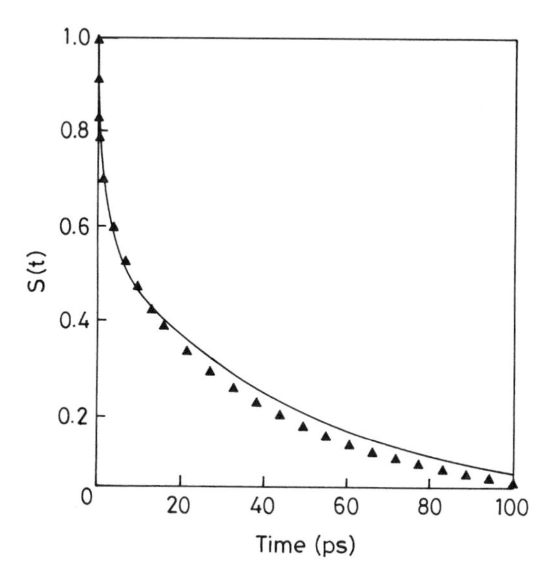

Figure 17. Ion solvation dynamics of excited C-153 in propanol at room temperature. The theoretical results are shown by the *solid line*; those of Horng and co-workers [32] are represented by the *filled triangles*. The solute : solvent size ratio is 1.52.

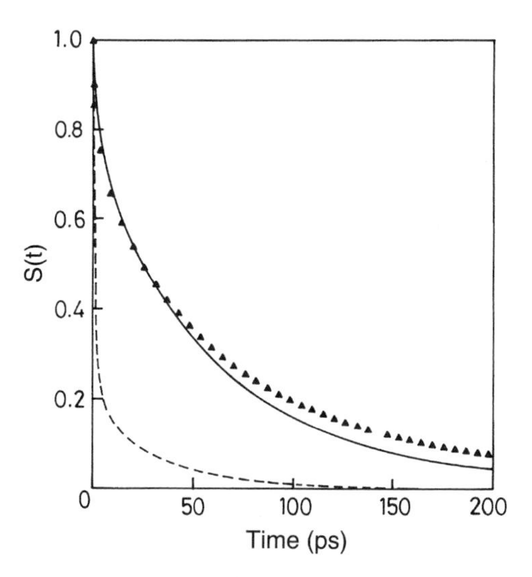

Figure 18. Ion solvation dynamics of excited C-153 in butanol at room temperature. The *solid line* represents the theoretical results, the *filled triangles* those of Horng and co-workers, and the *small dashed lines* those of Joo and co-workers. The solute : solvent size ratio is 1.41.

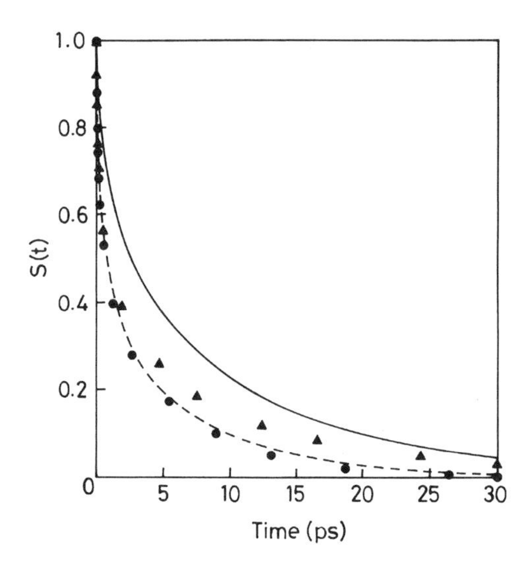

Figure 19. Dipolar solvation dynamics of excited C-153 in methanol at room temperature. The *solid line* and the *dashed line* represent the theoretical results on dipolar and ionic solvation dynamics, respectively. The *filled circles* and the *filled triangles* represent the experimental results of Bingemann and Ersting and Horng and co-workers, respectively.

agreement with one set of experimental results [66]. However, the disagreement with the results of Joo and co-workers [243] is most dramatic in this case.

Unlike for methanol, ethanol, and propanol, for butanol dielectric relaxation data to the degree of desired accuracy are not available. Thus we approximated the fastest solvent response by a time constant of 2.5 ps, taken by extrapolating the data of Barthel and co-workers [230] (discussed below).

D. Numerical Results: Dipolar Solvation

In this section we present the theoretical results on dipolar solvation dynamics of C-153 in monohydroxy alcohols. The necessary theoretical formulation is given in Section IV. We used Eq. (4.20) to calculate the solvation energy time correlation function for dipolar solvation dynamics. The solvent static parameters and experimental dielectric relaxation data needed for theoretical calculation are given in Tables VI and VII.

1. *Methanol*

We first present the results on dipolar solvation dynamics of C-153 in methanol. Figure 19 shows the decay of the calculated normalized STCF values.

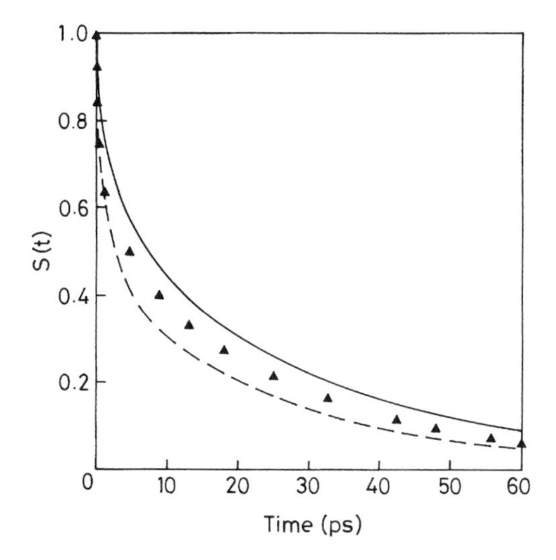

Figure 20. Dipolar solvation dynamics of excited C-153 in ethanol at room temperature. See Figure 19 for identification of the symbols.

For comparison, experimental results [65,66] are also shown. The decay rate of the theoretically predicted $S(t)$ values for the dipolar solute C-153 in methanol, is somewhat slower than that of the experimental results. In fact, the agreement between the theoretical predictions and experimental results becomes better, if the solute is assumed to acquire predominantly an ionic character on excitation (see Fig. 14). An interesting point to note is that the long-time decay of calculated $S(t)$ values for the dipolar solute is now nearly identical to that of Horng and co-workers [66].

2. *Ethanol*

Figure 20 presents the theoretical results of the dipolar solvation dynamics of C-153 in ethanol. It also shows experimental results [66]. The agreement is good. The theoretical results on ionic solvation dynamics of C-153 in ethanol are also shown in Figure 20. Note that there is no significant difference between the theoretically predicted ionic and dipolar solvation dynamics of C-153 in ethanol.

3. *Propanol*

Next we present the results for propanol. Figure 21 shows the decay of the theoretically predicted normalized STCF values for propanol along with experimental results [66]. The agreement between the theoretical predictions and experimental results is also good here. The theoretical results on

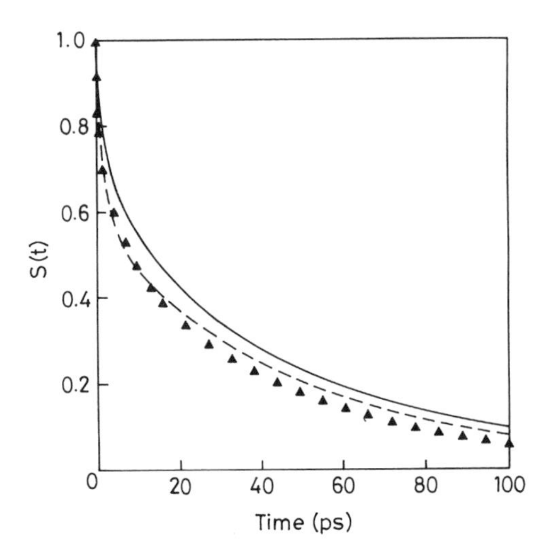

Figure 21. Dipolar solvation dynamics of excited C-153 in propanol at room temperature. See Figure 19 for identification of the symbols.

ionic solvation dynamics of C-153 in propanol are also presented in Figure 21. In propanol, there is no appreciable difference between the polar solvent response either to an ionic field or to a dipolar field.

It is important to note that there is no significant difference between the ionic and dipolar solvation dynamics of C-153 in ethanol and propanol. This is in contrast to the results obtained for water [163]. The difference between the two types (ionic and dipolar) of solvation dynamics of a solute probe in a particular solvent strongly depend on the polarity of the medium (quantified by ε_0 and $3Y$). In weakly polar liquids, the intermolecular correlations and orientational caging [15,15a] are not important. However, the translational modes of the solvent can still play a role in the long times.

E. Can Nonpolar Solvation Dynamics be Responsible for the Ultrafast Component Observed by Joo and Co-Workers?

As mentioned earlier, the large difference observed among the three experimental results is surprising. The theoretical studies reported here seem to suggest that, except for methanol, there is no ultrafast component in polar solvation dynamics with a time constant <100 fs in normal alcohols. This and the other aspects of theoretical predictions are in good agreement with experimental results of Horng [66]. This, in turn, implies that the latter work might have explored only the polar solvation dynamics in these liquids. This suggestion raises two questions: First, what really is the origin of the ultrafast component observed in the experiment of Joo and co-workers [243] in ethanol and butanol? Second, why does methanol display an ultrafast component whereas the others do not?

It is easy to answer the second question because methanol is quite mobile and possesses a fast component in the dielectric relaxation data that couples to polarization relaxation through the frequency-dependent dielectric function $\varepsilon(z)$. It also has a considerably high dielectric constant. It is, therefore, not surprising that polar solvation dynamics in methanol would exhibit an ultrafast component.

The answer to the first question on the origin of the ultrafast component observed by Joo and co-workers [243] in ethanol and butanol is much more complex. We have the following tentative explanation in terms of nondipolar solvation dynamics.

The dye molecules (HITCI or IR144) used by Joo and co-workers [243] are massive, with large surface areas, and both contain several aromatic rings. It is known experimentally that these molecules can make an appreciable contribution to the total solvation energy from the nondipolar interactions. One can then divide the total time-dependent solvation energy $E_{\text{total}}(t)$ into two parts:

$$E_{\text{total}}(t) = E_{\text{NP}}(t) + E_{\text{P}}(t) \tag{6.5}$$

where, the subscripts P and NP represent the polar and the nonpolar contribution to the total solvation energy, respectively. Then the corresponding solvation time correlation functions (unnormalized) are defined as follows:

$$S(t) = S_{\text{NP}}(t) + S_{\text{P}}(t) \tag{6.6}$$

where $S_{\text{NP}}(t) = \langle E_{\text{NP}}(t)E_{\text{NP}}(0)\rangle$ and $S_{\text{P}}(t) = \langle E_{\text{P}}(t)E_{\text{P}}(0)\rangle$. Here the angle brackets represent the canonical average, and we have assumed that there is no coupling between the polar and nonpolar parts of the solvation energies. Among the three techniques employed in the nonlinear spectroscopic experiment of Joo and co-workers [243], we shall discuss only the 3PEPS. Here the time dependence of the peak shift is related to the line broadening function $g(t)$ [18]. The line broadening function is determined both by the magnitude of the Stokes shift and by the STCF as follows [243]:

$$g(t) = i\lambda_{\mathrm{P}} \int_0^t dt_1 \, S_{\mathrm{P}}(t_1) + i\lambda_{\mathrm{NP}} \int_0^t dt_1 \, S_{\mathrm{NP}}(t_1) + \langle |\Delta E_{\mathrm{P}}|^2 \rangle \int_0^t dt_1 \int_0^{t_1} dt_2 \, S_{\mathrm{P}}(t_2) + \langle |\Delta E_{\mathrm{NP}}|^2 \rangle \int_0^t dt_1 \int_0^{t_1} dt_2 \, S_{\mathrm{NP}}(t_2) \quad (6.7)$$

where λ_{P} and λ_{NP} are the reorganization energies that are together responsible for the time-dependent shift of the center of the absorbing frequency of the solute probe. ΔE is the fluctuation in the absorption energy caused by the fluctuating solvent environment surrounding the chromophore (the probe molecule).

Determination of $g(t)$ requires the values of λ_X ($X = P, NP$) and ΔE_X for both the polar and nonpolar solvation dynamics. Our reason for presenting Eq. (6.7) is to emphasize the point that in 3PEPS and other nonlinear optical techniques, it is not immediately possible to separate the observed response into polar and nonpolar parts. Some recent experimental studies by Berg and co-workers [95–103] and Maroncelli and co-workers [104] revealed that for nondipolar solvation, the frequency shift ranges from 300 to 1000 cm^{-1}. The larger value is obtained when a large probe molecule with aromatic rings is dissolved in polar solvents. The polar solvation, on the other hand, gives a shift ranging from 1000 to 2500 cm^{-1}. Because a big dye molecule was used in the experiments of Joo and co-workers [243], a contribution of 30–60% may easily come from the nonpolar part [95–104,107]. This is significant.

We next need an estimate of the fastest time scale possible in the nonpolar solvation dynamics. As discussed by several authors [105,107], the ultrafast component of the nonpolar solvation dynamics may arise from a mechanical (i.e., viscoelastic) response of the liquid. This can indeed be very fast. This may also be related to the fact that many high-frequency intermolecular vibrational modes that are present in a liquid may couple not to the polar solvation dynamics but to the nonpolar solvation dynamics. We next present an analysis of the time constant of the viscoelastic response of the medium.

The theoretical calculations of the nonpolar solvation time correlation function $S_{\mathrm{NP}}(t)$ have been discussed by several authors [95–108]. In contrast to the ionic solvation in polar solvents, the nonpolar solvation dynamics is controlled essentially by the dynamic structure factor of the liquid; the orientational relaxation is seen to play a less important role. It has been shown that $S_{\mathrm{NP}}(t)$ is directly proportional to the dynamic solvent structure factor, and the final expression is given by the following relation [105]:

$$S_{\mathrm{NP}}(t) \propto \int_0^\infty dq\, q^2 c_{21}^2(q) S(q, t) \tag{6.8}$$

where c_{21} is the wavenumber-dependent direct correlation function between the probe solute (labeled 2) and the solvent molecules (labeled 1). We calculated the normalized dynamic structure factor $S(q, t)$ of the solvent at two large wavenumbers—at $q\sigma = 2\pi$ and $q\sigma = 4\pi$, respectively, for a Lennard–Jones liquid, at density $\rho\sigma^3 = 0.8$ and temperature, $T^* = 2.0$ (where $T^* = k_B T/\varepsilon$ and ε is the Lennard–Jones energy parameter). We used the following dynamic structure factor, given by the well-known Mori continued fraction, which is expressed as follows [175,177,249–252]:

$$S(q, z) = \frac{S(q)}{z + \langle\omega_q^2\rangle[z + \Delta_q(z + \tau_q^{-1})^{-1}]^{-1}} \tag{6.9}$$

where ω_q is a wavenumber-dependent frequency of the solvent cage $\Delta_q = \omega_l^2(q) - \langle\omega_q^2\rangle$ and τ_q is the wavenumber-dependent relaxation time, defined as $\tau_q = 2(\Delta_q/\pi)^{\frac{1}{2}}$. The details of these definitions are given in Appendix B. The results for two wavenumbers are shown in Figure 22. It is obvious from this figure that the ultrafast time scale in nonpolar solvation dynamics can indeed originate from the large wavenumber processes. The $S(q, t)$ values at intermediate to large wavenumbers decay with a large rate, which in turn can give solvation times of 150–200 fs. This is also the finding of the INM analysis of nonpolar solvation dynamics [111–116] and in the experimental studies of nonpolar solvation dynamics in aprotic solvents [104]. However, this is still longer than the time scale observed by Joo and co-workers [243], and it is unlikely that this mechanism can explain the large amplitude (60–70%) of the ultrafast component, but this nonpolar mechanism might be present in some other cases for which solvation is slow.

Until now we seem to have eliminated both the polar solvation dynamics and the ordinary viscoelastic response of the medium as the probable candidates. We now need to look for a different source for explaining the ultrafast component. There are two other possible candidates. One is the suggestion by Ernsting and co-workers that when the time scale of solvation falls below 100 fs, one needs to consider the involvement of vibrational redistribution in the excited state, as they can interfere with solvation dynamics in a way that can be indistinguishable from the desired solvent relaxation [65]. Unfortunately, we cannot comment on this interesting scenario. The next choice is the one suggested by Cho and co-workers [253] and by Ladanyi and Stratt [114]. For a large dye molecule like HITCI or IR144, a simple calculation shows that 30–40 nearest-neighbor solvent molecules may interact

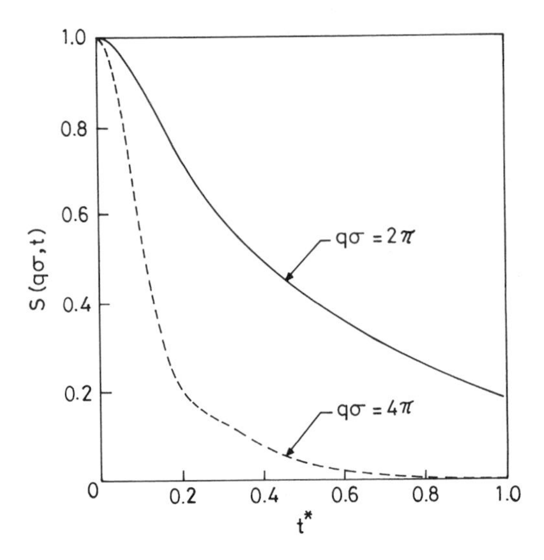

Figure 22. Normalized solvent dynamic structure factors in a time plane for two wavenumbers. The normalized solvent dynamic structure factor is defined as $S(q, t) = \mathcal{L}^{-1}[S(q, z)/S(q)]$ and calculated from the Eq. (6.9) for the Lennard–Jones potential. The wavenumber is scaled by the solvent diameter and the time is scaled by the quantity $[m\sigma^2/k_B T]^{\frac{1}{2}}$. The mass m and diameter used are those of an ethanol molecule. Note that the decay of $S(q, t)$ at the larger wavenumber is faster than that at the smaller one. In nonpolar solvation dynamics, the short-time solvent response originates from this type of mechanical and elastic response of the nearest-neighbor solvent molecules.

directly with the surface of the solute molecule. This may give rise to a contribution of 0.3–0.5 eV in the dispersion energy alone. This can indeed make a large contribution to the equilibrium solvation time correlation function $\langle \Delta E(t)\Delta E(0)\rangle$. Because alcohols are structured liquids, the molecules next to the probe solute can experience high-frequency motions from the bulk. These motions can easily have frequencies in the 100–1000 cm^{-1} range. Therefore, it is possible to obtain a $M(t)$ that decays very fast, and the experiments of Joo and co-workers [243] are sensitive to this dynamics. This mechanism is quite similar to the nonpolar solvation dynamics discussed above, except that it is more sensitive to specific short-range solute–solvent interactions.

Clearly, our present understanding is still far from complete. The way the solvation time correlation function was constructed by Horng and co-workers [66] and the time resolution available to them suggest that they left out such ultrafast nonpolar solvations, as discussed above. This certainly explains the agreement with the theory—but we do not yet have a quantitative explanation of the ultrafast component observed from 3PEPS measurements in ethanol, propanol, and butanol.

If the tentative explanation proposed here in terms of nondipolar, nearest-neighbor, solute–solvent interactions turns out to be true, then the experiments of Joo and co-workers [243] might provide us with the much-desired information about dynamics of solute–solvent interactions. This raises the following interesting question: What role would their results play in determining the kinetics of electron-transfer reactions occuring in these solvents? If they are not related to dipolar solvation, then one may need to reinvestigate the dynamic solvent effects on the kinetics of these reactions by appropriately generalizing the Marcus model [254,255]. This ultrafast solvation may well play the role of nonpolar vibrational modes (the Q-modes) envisaged by Sumi and Marcus [11] in their theory of electron-transfer reaction.

F. Discussion

Let us summarize the main results of this section. We used a simple molecular theory to study the time-dependent progress of solvation of excited C-153 in the first four member of the homologous series of the straight-chain monohydroxy alcohols. The use of this theory was encouraged by its earlier success in studying the solvation dynamics not only of water and acetonitrile but also of amides and substituted amides at lower temperature. The theoretical predictions are in excellent agreement with the experimental results of Bingemann and Ernsting [65] for methanol and of Horng and co-workers [66] for all four alcohols.

We find that the solvation dynamics in methanol is primarily governed by the short-time inertial modes of the solvent. The ultrafast, Gaussian component with a time constant of about 70 fs accounts for 40–50% of the total solvation in this solvent. This is also in good agreement with the experimental results of Bingemann and Ernsting [65]. For the other three alcohols, we do not find any such ultrafast component, which is also the observation of Horng and co-workers [66]. We also found that there is no appreciable difference between the dipolar and ionic solvation dynamics of excited C-153 in ethanol and propanol.

The most intriguing part of the present study is the absence of any ultrafast component for polar solvation in ethanol, propanol, and butanol. On the basis of the work reported here, we suggest that the universal ultrafast component observed by Joo and co-workers [243] might be the result of nonpolar solvation originating from high-frequency motions of the hydrogen-bonded, nearly polycrystalline liquids.

Let us now comment on the validity of the present hydrodynamic approach to describe the ultrafast polar solvation. Our first point concerns the controversy regarding collective versus single particle motion being the crucial step in ultrafast solvation. Here we note that the relevant response function (or the memory function) of the dipolar liquid, probed at the ultrafast times, is essentially single particle in nature. Because the theory also suggests that at short times translational modes are not important, the relevant motion of a dipolar molecule is then given by the following generalized Langevin equation:

$$I\frac{d^2\rho(\mathbf{\Omega})}{dt^2} = N(\Omega) - \int dt'\, \Gamma(t - t')\omega(t') + R(t) \tag{6.10}$$

where the memory function $\Gamma(\mathbf{t})$ is approximated by its short-time limit, $N(\mathbf{\Omega})$ is the systematic torque acting on the molecule in question from all other molecules, $R(t)$ is the random force, and ω is the angular velocity. This is the equation of motion used in our calculations, with the torque term given by density functional theory. Note that $\rho(\Omega)$ is the single particle orientational density. Thus, if we freeze the motion of all other molecules (the Rigid cage version of Maroncelli [152]), then the equation of motion becomes that of a free particle (no friction) but in the force field of all other molecules of the system. We have shown elsewhere [197] that this leads to a simple and transparent derivation of the relation between the single particle orientational correlation function and the solvation time correlation function at the short times, recently proposed by Maroncelli and co-workers [36]. Thus, there is really no difference between the single particle and the collective picture.

There is another important issue: As the probe solute is perturbed from its initial state at time $t = 0$, the response of the liquid is largely linear at short times. This response involves the motion of each molecule in the force field of others. In our theory, each molecule behaves as a free particle (i.e., experiences no friction) at short times. The small rotation of the solvent molecules leaves the force field on any given molecule essentially time invariant during the motion of any molecule. This force can be Fourier decomposed. For dipolar liquids, the long wavelength part of this force (i.e., the one involving all the molecules) has the largest force constant. Thus the long wavelength polarization relaxation is the fastest and gives rise to the ultrafast solvation. Single particle motion alone (without this collective force field) will never give rise to the observed amplitude for the ultrafast solvation.

Second, we consider the relative importance of the bulk polarization versus the nearest-neighbor contribution. The bulk polarization mode of our theory is essentially the same mode that appears in the continuum model discussions. It is certainly correct that the long wavelength longitudinal polarization modes in water, acetonitrile, and methanol decay in the same ultrafast time scale as found in the experiments. This can be proved rigorously by using expressions that involve only macroscopic variables and so does not involve any assumption or use of any microscopic argument. So the question then comes down to the relative importance of the long wavelength modes in solvation dynamics. In present and related theories, the long wavelength modes together make a large contribution to the total solvation energy, because of the long-range nature of the ion–dipole interaction.

It should be pointed out that for ionic solvation, computer simulation studies of a small system can run into difficulties. In the language of the present theory, the simulation studies of a small system mean the truncation of all the modes with wavenumbers between zero and a number, which can be rather high. Note that in simulations, the situation is further complicated because the probe solute is often placed in the center of the system, further curtailing the length of the allowed solvent polarization by half. Thus there is always the danger that the computer simulation studies of polar solvation dynamics—partcularly of an ion—may lead to the conclusion that only the nearest-neighbor molecules are important and the contribution of the bulk is negligible. The situation is different for dipolar solvation, in which the system size dependence could be weak.

The above discussion is not meant to suggest that the nearest-neighbor molecules do not contribute significantly and that their dynamics are not in the ultrafast time scale. Rather, our aim is to point out the possible error that may easily arise in the simulations of polar liquids, especially in the study of solvation dynamics of a foreign solute. In the present theory, the relaxation at the lowest wavenumber (around $q\sigma = 1$) accessible in the solvation simulation of 255 molecules (plus the probe solute) is almost of the same time constant as that of $q = 0$. For example, the short-time rate of the wavenumber-dependent solvent polarization relaxation $\Sigma(q\sigma, z)$ at $q\sigma = 1$ is only $\sim$10% smaller than that at $q\sigma = 0$. Actually, this weak wavenumber dependence of the relaxation rate is the reason for the dominance by the long wavenumber modes and is also the reason (in the theory) why the ultrafast component dominates in water, acetonitrile, and methanol. In fact, in the simulation studies of ionic solvation dynamics in acetonitrile [152], Maroncelli noted the dominance by the longest wavelength mode.

This discussion can be made more quantitative by plotting the generalized rate of solvent polarization relaxation $\Sigma_{10}(q\sigma, z)$ [Eq. (4.15)] against the wavenumber $q\sigma$. This is shown in Figure 23 for two limits of frequency for ethanol. Note that both for low and high frequencies, the wavenumber dependence is rather weak, as noted earlier. The pronounced flatness at low frequency is due to the contribution of translational modes, which becomes relevant as the orientational relaxation becomes slow. This is because in ethanol the single particle orientation is slow, as evident from the large value of τ_1 (Table VII). The rate $\Sigma_{10}(q\sigma, z)$ at intermediate wavenumbers is slower

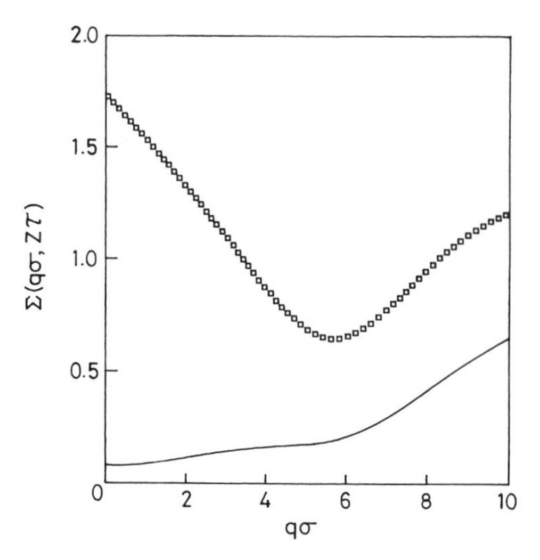

Figure 23. A plot of the calculated wavenumber- and frequency-dependent generalized rate of solvent polarization relaxation for two limits of frequency. The results are obtained for ethanol using Eq. (2.9). The *diamonds* represent the wavenumber dependence of the generalized rate at a very high-frequency, whereas the *solid line* describes the same at a very low frequency. Note that z is scaled by τ, which is equal to 1×10^{-12} s.

than at all other wavenumbers. So computer simulation studies in a finite system may probe this slower decay. However, this may not be greatly significant, as the wavenumber dependence itself is weak.

As already pointed out, the solutes that are routinely used as probes are of complicated shapes. In addition, they undergo complex changes in their charge redistribution upon excitation. The solvation dynamics of such a complex charge distribution can be rather different from that of an ion or a dipole, particularly at short times, in which the nearest-neighbor molecules would react to the extended charge distribution in a way that could be entirely different from what is expected from a simple model calculation. The field from a complex charge distribution can be entirely represented by a multipolar expansion, the convergence of which is, however, not clear. In the light of this ambiguous situation, the theoretical results obtained here that, at least for alcohols, the solvation dynamics of an ion are not significantly different from those of a dipole is important and reassuring from theoretical point of view.

Last, the results presented were obtained with no adjustable parameter. The near-perfect agreement obtained between theory and experiments for a large number of systems studied so far—water, acetonitrile, amides, and the

Brownian dipolar lattice — suggest that the present theory is capable of accurately describing polar solvation dynamics in dipolar liquids.

The present work suggests the following future problems for experimental study. It will be useful to perform both the dielectric relaxation and TDFSS experiments over a wide range of temperatures in alcohols. Second, dielectric relaxation using the teraHertz technique for butanol and higher alcohols are needed. In addition, the effect of ultrafast nonpolar solvation dynamics on electron transfer reactions is an interesting theoretical problem.

VII. ION SOLVATION DYNAMICS IN SLOW, VISCOUS LIQUIDS: ROLE OF SOLVENT STRUCTURE AND DYNAMICS

When the viscosity of a liquid becomes large near its glass transition temperature, several interesting anomalies are observed in its transport properties [26a,256–257]. In this slow regime, local short-range correlations (both spatial and orientational) can play an important role in determining the nature of the response of the liquid to a given perturbation. The latter is often provided by instantaneously changing the nature of a suitable probe placed inside the liquid. The response of the liquid may now derive significant contributions from a collection of a small number (say 10–100) of molecules that are spatially close to the solute probe. However, understanding the nature of this response is often difficult. Fortunately, the contribution of the translational modes of the liquid can be unambiguously studied on the molecular-length scale by using the inelastic neutron-scattering experiments. On the other hand, there does not seem to exist any such efficient experimental technique at this point that probes the collective orientational dynamics at a local level, as inelastic neutron-scattering does for the translational dynamics. This has severely limited our understanding of molecular relaxation and hence that of chemical dynamics in slow liquids. Thus we may need to study chemical dynamics in slow liquids with the twin goal of understanding both the natural dynamics of these liquids and their effects on chemical relaxation processes.

Recently, Angell [256] proposed an elegant classification of the large number of seemingly unrelated results on relaxation in glassy liquids. He showed that most of the liquids fall between the two extreme limits of liquid behavior, termed fragile and strong. Liquids made of simple rigid molecules, like chlorobenzene and neopentane, are in the fragile limit, whereas network liquids, like silica, correspond to the strong limit. The fragile liquids show anomalous behavior, such as adherence to the Vogel–Fulcher temperature dependence of relaxation time [256] and nonexponential dynamics. Strong liquids, on the other hand, show Arrhenius temperature dependence and nearly exponential kinetics. Most of the chemically important solvents fall in the fragile limit. Thus the study of chemical relaxation in these liquids may provide a valuable tool to understand and further characterize the relaxation in the slow, viscous liquids. Note that from a theoretical point of view, the study of slow liquids has remained a difficult and a challenging problem for several decades, and the progress itself has been slow.

Of the various chemical relaxation processes studied in supercooled liquids, orientational relaxation, diffusion-limited chemical reactions, solvation dynamics, and ionic mobility are known to exhibit interesting dynamics [256]. In this chapter, we present theoretical studies of orientational relaxation and solvation dynamics. Recent advances in the study of solvation dynamics in the normal regime [159–164] make one optimistic that these techniques can also be used to obtain valuable information regarding the dynamics in supercooled liquids. We have chosen a specific system, namely the amides, for which experimental results are available.

Since the nature of relaxation in slow liquids is quite different from that in fast solvents, the theoretical formulation that is successful in the ultrafast limit [15,159–164] may not be adequate in the opposite limit of slow liquids. Fortunately, we find that the molecular hydrodynamic theory developed in Section IV can be useful to treat the slow regime as well. The extended theory properly includes the short-range correlations between the polar solute and the dipolar solvent molecules. Friedman and co-workers [153–158] have made important contributions in this aspect.

In this section we present the theoretical results of the ionic solvation dynamics of excited Coumarin in *N*-methyl formamide (NMF), *N*-methyl acetamide (NMA), and *N*-methyl propionamide (NMP). The extended theory has been found to be remarkably successful in explaining the solvation dynamics of Coumarin in these viscous solvents. This agreement is over many decades of time evolution and is particularly pleasing because it does not involve any adjustable parameter. Furthermore, our study of orientational relaxation reveals the following interesting fact. We find that even though the relaxation is highly non-Markovian and the decay of orientational relaxation markedly nonexponential, the average relaxation times for the first- and second-rank correlation functions still follow closely the Debye $\ell(\ell+1)$ law. The reason for this has been discussed.

The organization of the rest of this section is as follows. The next part contains the theoretical formulation of ion solvation dynamics and orientational relaxation and section VII.B contains the calculational details. We present the numerical results and the comparison with experiments in Section VII.C.

A. Theoretical Formulation

1. Ion Solvation Dynamics

The time-dependent progress of the solvation of excited Coumarin in amides is followed by calculating the normalized STCF $S_E(t)$ from Eq. (4.15). As discussed, Eq. (4.15) involves calculations of the ion–dipole direct correlation function, the static correlation functions among the solvent molecules, the solute dynamic structure factor, and the generalized rate of the solvent orientational polarization relaxation. The calculational details of these quantities are described briefly below.

2. Orientational Relaxation

The orientational relaxation in dipolar liquids exhibits rich dynamics, much of which is still ill-understood [258,259]. For example, we do not yet fully understand the reason for the difference between the single particle and the collective dynamics in dense dipolar liquids [165,166]. Another example is the orientational relaxation in supercooled liquids, which shows an interesting dynamical phenomenon known as the $\alpha - \beta$ bifurcation. These problems have been addressed and discussed at length [165,166,260]. Although some aspects have been clarified [165,166], much remains unclear, because a detailed theoretical calculation of the orientational correlation functions has not yet been possible owing to the complexity of the systems involved. Another problem has been the lack of experimental techniques to measure the first rank (i.e., $\ell = 1$), single particle orientational correlation function $C_1(t)$ directly in the time domain. In the absence of this information, the relaxation time of $C_1(t)$ is often equated to the time constant of the dielectric relaxation. But these two may be quite different, because the collective memory function contains the contribution from only the $q = 0$ mode, whereas the single particle friction is more susceptible to the short-range orientational correlations. In principle, we need the full wave vector and frequency-dependent rotational dissipative kernel $\Gamma_R(q, z)$, which, unfortunately, is not available yet.

To understand orientational relaxation in the slow liquids, we have carried out the following calculation to gain insight into the single particle orientational relaxation. The orientational correlation functions $C_{\ell m}(t)$ are defined as [15,15a]

$$C_{\ell m}(t) = \langle Y^*_{\ell m}[\Omega(0)]\, Y_{\ell m}[\Omega(t)] \rangle \tag{7.1}$$

where $Y_{\ell m}(\Omega)$ is the spherical harmonic of rank ℓ and projection m. $Y^\star_{\ell m}(\Omega)$ is the complex conjugate to $Y_{\ell m}(\Omega)$ and $\langle\ \rangle$ denotes the average over the initial orientations. For isotropic systems, the m dependence can be ignored, and we refer to the orientational correlation functions as $C_\ell(t)$. On rather general grounds, this correlation function can be represented in terms of the memory function as follows

$$C_\ell(t) = \mathcal{L}^{-1}\left[z + \frac{\ell(\ell+1)k_{\mathrm{B}}T}{I[z + \Gamma_{\mathrm{R}}(z)]}\right]^{-1} \tag{7.2}$$

where the orientational memory function $\Gamma_{\mathrm{R}}(z)$ is related to $\Gamma_{\mathrm{R}}(q, z)$. In our work, this has been obtained from experimental dielectric relaxation data by using Eq. (2.12).

B. Calculation Procedure

1. *Calculation of the Ion–Dipole Direct Correlation Function*

We have obtained the wavenumber-dependent ion–dipole direct correlation function, $c_{\mathrm{id}}(q)$ from Chan and co-workers [206] in the limit of zero ionic concentration. The calculation procedure of this quantity was discussed in Section IV. The evaluation of $c_{\mathrm{id}}(q)$ requires the knowledge of the solute : solvent size ratio, which is equal to 1.73, 1.6, and 1.5 for the solvation of Coumarin in its excited state in NMF, NMA and NMP, respectively.

2. *Calculation of the Solvent Static Correlation Functions*

The most important solvent static correlation function is the longitudinal component of the wavenumber-dependent dielectric function $\varepsilon_{\mathrm{L}}(q)$. An accurate determination of this quantity is important, because it can have significant effects on solvation dynamics in slow liquids, such as the three amide systems studied here. The most interesting feature of these liquids is that they possess high static dielectric constants and large Debye relaxation times. Actually, the high values of static dielectric constants are indicative of extensive intermolecular hydrogen-bonded network that provides a suitable geometry to align the molecular dipoles, producing a large dipole moment. The large values of ε_0 and τ_{D} make these solvents considerably slower in responding to the external perturbation. For these amides, $\varepsilon_{\mathrm{L}}(q)$ is obtained from the MSA model after correcting it both at $q = 0$ and $q \to \infty$ limits. The calculation procedure of this quantity was described in Section IV.

3. *Calculation of the Generalized Rate of Solvent Orientational Polarization Relaxation*

The longitudinal component of the frequency- and wavenumber-dependent generalized rate of solvent polarization relaxation $\Sigma_{10}(q, z)$ is calculated from Eq. (2.9). The rotational memory kernel is obtained from the frequency-dependent dielectric function $\varepsilon(z)$ of these solvents by using

TABLE VIII
Solvent Parameters Needed for the Theoretical Calculations

Solvent	Molar volume, Å^3	μ, D	T, K	ε_0	ε_∞	τ_D, ns	η, cp
NMF	60.2	3.82	234	304	10	1.50	6.7
NMA	76.0	3.71	302	182	10	0.39	4.0
NMP	93.0	3.59	273	216	6	4.02	10.8
	93.0	3.59	253	270	6	11.1	20.7
	93.0	3.59	244	300	6	18.6	28.7

Eq. (2.12). For slow liquids, the latter has been given by the Cole–Davidson formula [258–259] as follows:

$$\varepsilon(z) = \varepsilon_\infty + \frac{\varepsilon_0 - \varepsilon_\infty}{\left[1 + z\tau_{\mathrm{D}}\right]^\beta} \tag{7.3}$$

where τ_{D} is the Debye relaxation time and β is the Cole–Davidson fitting parameter. Fortunately, for all these solvents, the dielectric relaxation data are available experimentally [230,231]. These experimental data have found the best fit with $\beta = 0.91$ [231] for all the three amides. The radii of the different solvent molecules have been calculated from their respective van der Waal's molar volumes. The other parameters needed for the calculations are given in Table VIII.

The translational memory kernel $\Gamma_{\mathrm{T}}(q, z)$ is obtained by using Eq. (2.13), and the solute dynamic structure factor is calculated by using Eq. (4.16).

C. Numerical Results and Discussion

To understand the dynamic response of the liquid, Figure 24 shows the collective rotational memory function plotted as a function of the frequency for NMP. It shows a biphasic dependence that is typical of slow liquids, for which non-Debye relaxation assumes a greater significance. Note the large value in the $z \to 0$ limit and the relatively smaller friction at higher frequencies.

Figures 25–27 contain the results of our calculations for the solvation dynamics of excited Coumarin in NMF, NMA, and NMP, respectively. To study the solvation dynamics in these systems, we used Eq. (4.15), in which the Laplace inversion was performed numerically by using Stehfest algorithm [214]. In Figures 25–27 the experimental results [261] and the prediction of the DMSA model [134, 138] shown. From these figures, it is clear that the agreement between the present extended molecular theory and the experimental results is quite good. The agreement for NMF is par-

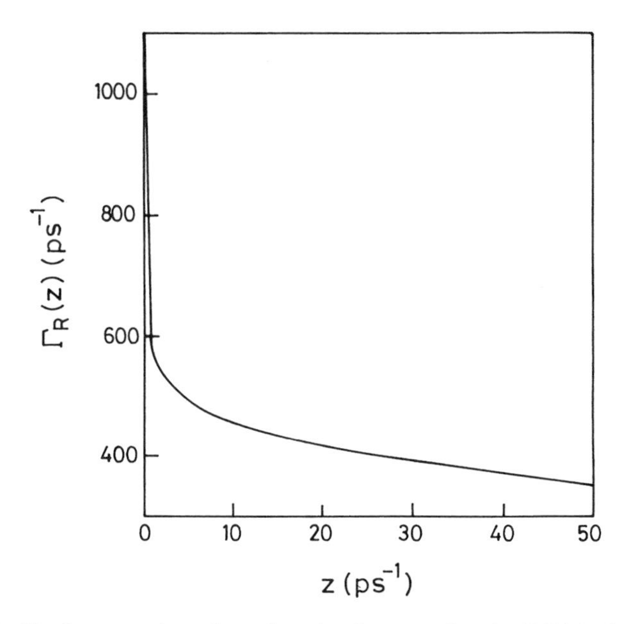

Figure 24. The frequency-dependent orientational memory function $\Gamma_R^*(z)$ is plotted as a function of the Laplace frequency z for NMP at 253 K. Eq. (2.12) was used to calculate the $\Gamma_R(z)$ values. The necessary dielectric relaxation data needed for the calculation are given in Table VIII. Note the biphasic character of $\Gamma_R(z)$, which is typical of slow liquids, like NMP, at lower temperatures.

ticularly good, for reasons not yet clear. It is obvious that our theory predicts somewhat faster solvation at short times than has been observed experimentally. This may arise from the fact that the experiments missed an initial part in the relaxation of the total solvation energy [261]. This early response may come from an ultrafast component of the solvent response, as revealed by the recent studies of Chang and co-workers [262]. We shall come back to this point later.

It is also clear from Figures 25–27 that the DMSA model completely fails to describe the solvation dynamics. The reason for this is not clear but appears to arise from two factors. First, the mean spherical approximation gives wrong correlations both at the long and at the short wavelengths. The error in the short wavelength is that it predicts large correlations to persist even when $q \to \infty$. This will certainly slow down the decay. Second, the way the dynamics are introduced in the DMSA model is rather *ad hoc*.

To understand the effects of the ion–dipole correlation function, we compared $C_{EE}(t)$ with the earlier linear theory, in which the solvation energy $E_{\rm sol}(t)$ was calculated by Eq. (4.14) (Fig. 28). The linear theory seems to

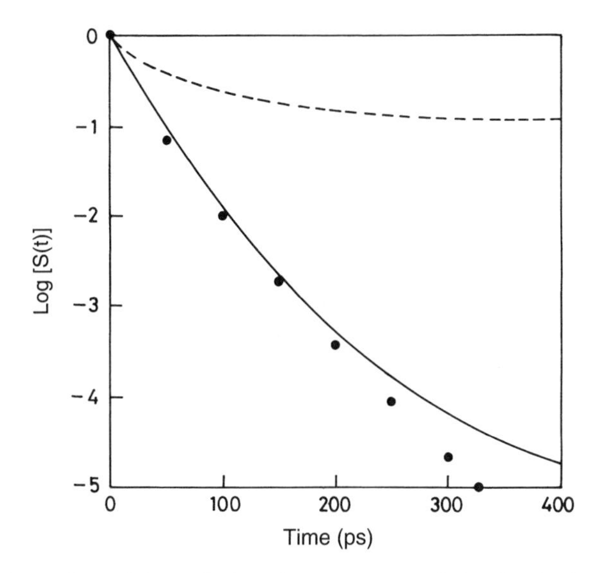

Figure 25. Comparison of the prediction of the present extended molecular theory (*solid line*) and the experimental results (*filled circles*). Here the normalized solvation time correlation function $S(t)$ is plotted against time (t) for the solvation of the excited Coumarin in NMF at 234 K. The prediction of the same by the DMSA model (*dashed line*) is also shown. The theoretical calculations (ion solvation dynamics) were carried out with a solute : solvent size ratio of 1.73. For the calculation of $S(t)$, we made use of Eq. (4.15). The other parameters needed for the computation are given in Table VIII. Note that the ordinate is in the logarithmic scale.

break down in the longer times. This is expected, because the theory ignored the solvent structure around the ion, which is incorporated in this extended molecular theory via $c_{\mathrm{id}}(q)$.

We also studied the temperature dependence of the solvation dynamics in NMP. In this regard, the required values of the necessary parameters are available in Table VIII [230,231,261]. As the temperature is reduced, the rate of solvation is expected to be progressively slower as the relaxation times of the medium become larger. In fact, there can be an interesting competition between ε_0 and τ_{D}—the former tends to make the solvation faster whereas the latter does the reverse. Figure 29 compares the solvation dynamics at three different temperatures: 244, 253, and 273 K. The relaxation slows down considerably at $T = 244$ K.

In view of the good agreement obtained between the theory and experiment, one wonders about the sensitivity of the theoretical calculations on the experimental parameters used. In particular, one must have an estimate of the effects of small deviations in the dielectric function on the calculated STCF. It is important to note here that there are considerable uncertainties

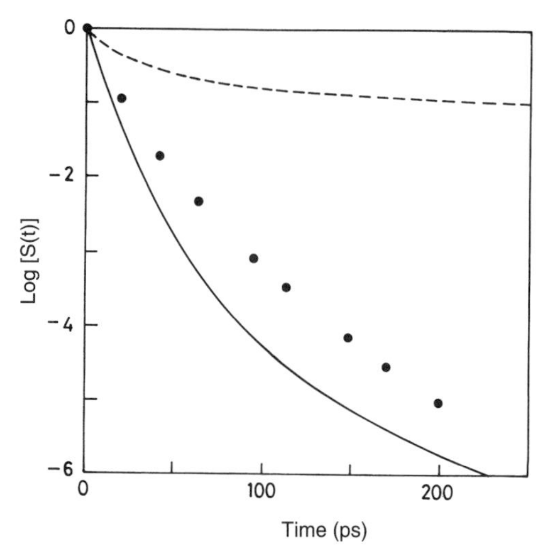

Figure 26. Comparison of the prediction of the present extended molecular theory (*solid line*) and the experimental results (*filled circles*). Here the normalized solvation time correlation function $S_E(t)$ is plotted against time (t) for the solvation of the excited Coumarin in NMA at 302 K. The *dashed line* represents the prediction of the DMSA model of the same liquid at the same temperature. For the theoretical calculation, the solute : solvent size ratio was taken to be 1.6. The other parameters needed to characterize the solvent are given in the Table VIII. Note that the ordinate is in the logarithmic scale.

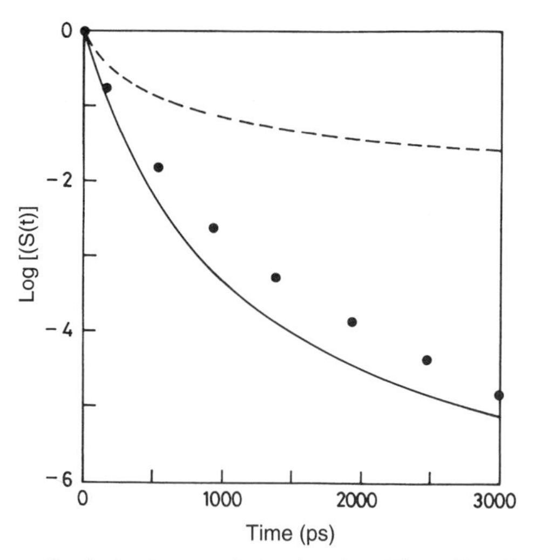

Figure 27. The normalized solvation correlation function $S(t)$ predicted by the present extended molecular theory for NMP at 253 K (*solid line*) compared to that of the experiment (*filled circles*). The solute is the excited Coumarin molecule. The prediction of the DMSA model is also shown (*dashed line*). In this case, the solute : solvent size ratio is 1.5. For other parameters, see Table VIII.

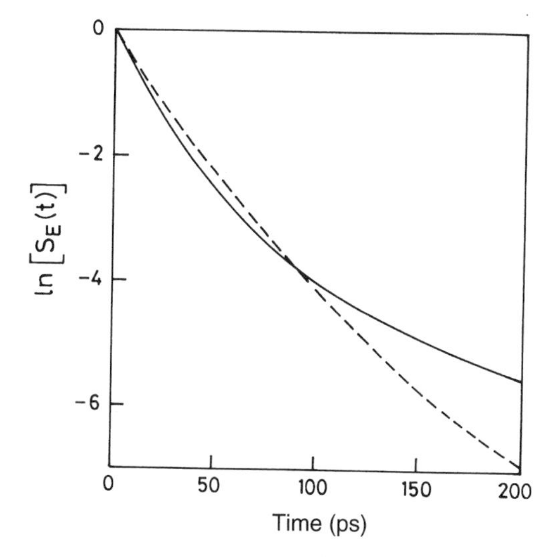

Figure 28. Comparison of the predictions of the earlier linear theory [Eq. (2.8)] (*dashed line*) and the present extended molecular theory (*solid line*) for the solvation time correlation function is plotted against time t for the solvation of excited Coumarin in NMP at 253 K. Note the deviation of the linear theory from the present extended theory in the long time. The solute : solvent size ratio is 1.5. Other parameters used in the theoretical calculations are given in Table VIII.

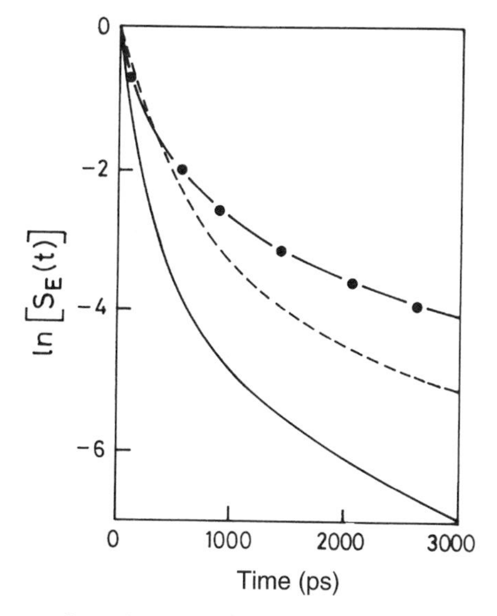

Figure 29. Temperature dependence of the rate of solvation energy relaxation in NMP. Here, logarithmic values of the normalized STCF $S(t)$ at temperatures $T = 244$ K (*dashed and dotted line*), $T = 253$ K (*dashed line*), and $T = 273$ K (*solid line*) have been plotted as a function of time t. Note that the dynamics are considerably slower at 244 K.

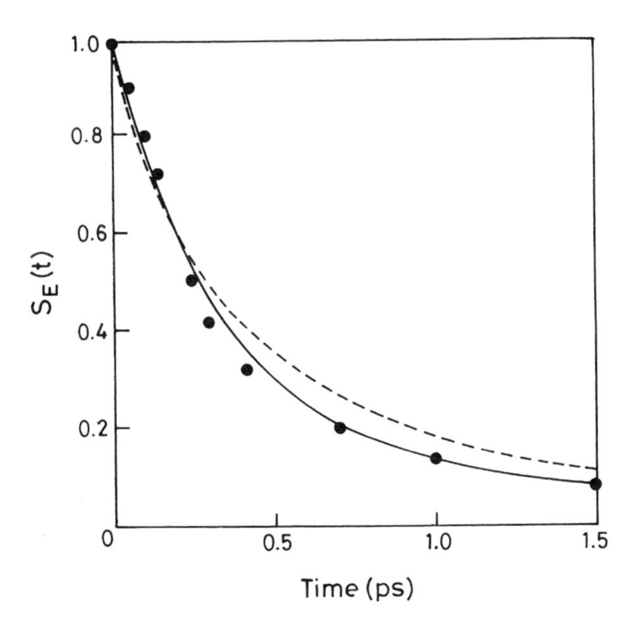

Figure 30. The decay of the solvation time correlation functions $S(t)$ calculated by using the fits to the Davidson–Cole (*dashed line*) and the biexponential (*solid line*) form of dielectric function $\varepsilon(\omega)$ are compared to the experimental results (*filled circles*). The liquid is dimethylformamide (DMF) at 298 K. The solute : solvent size ratio is 1.70.

in the values of the Davidson–Cole parameters employed here. Also note that the Davidson–Cole form of dielectric function tends to overemphasize the slower decay in the long time. All these points suggest that we should use an alternative form of $\varepsilon(z)$, such as a biexponential (known as Budo formula [263]) to check the robustness of the present theoretical scheme. Unfortunately, the necessary parameters to be used in such a form are not available for any of the amide systems at the temperatures studied here. We could find only one amide, namely dimethylformamide (DMF), for which both the fits to the Davidson–Cole and the biexponential forms are available [231]. This is, however, at room temperature (298 K) only. We have, therefore, carried out a calculation for this amide system, using both of these fits to study the robustness of the present scheme. In Figure 30, we compare the predicted STCFs thus obtained with the available experimental results [262]. It is clear from Figure 30 that, although the Davidson–Cole form is successful in predicting the decay of STCF at shorter times, the biexponential form of $\varepsilon(z)$ provides a better description in the long time. Thus we suggest that in dielectric relaxation experiments fits to the

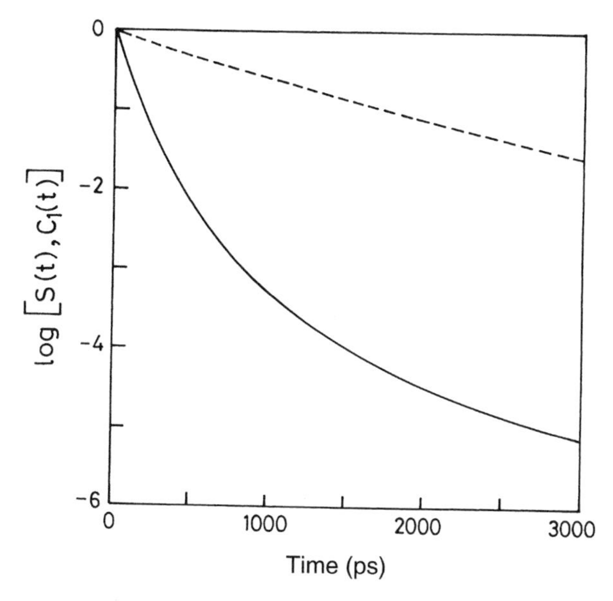

Figure 31. The relaxation of the single particle orientational correlation function $C_1(t)$ (*dashed line*) in NMP at $T = 253$ K plotted against time t and compared to the solvation time correlation function $S(t)$ (*solid line*) in NMP. The $C_1(t)$ values were obtained by Eq. (7.2) and the $S(t)$ values were calculated from Eq. (4.15). For other solvent parameters, see Table VIII.

Budo formula [263] should also be performed. This will perhaps allow for a better description of the solvation dynamics in these otherwise slow liquids.

The above analyses seem to suggest that the present theoretical scheme can be regarded at least semiquantitatively successful for a long range of time. They also suggest that a more reliable form of the dielectric relaxation function for these liquids is needed to describe the dynamics more accurately.

In Figure 31, the first-rank (i.e., $\ell = 1$) single particle orientational relaxation rate $C_1(t)$ is compared to the normalized SETCF $S_E(t)$. In the short time, $S_E(t)$ decays faster than $C_1(t)$, whereas the rates of the decays are comparable in the long time. This is because in the short times, $S_E(t)$ derives contribution from the long wavelength (i.e., $q \simeq 0$) modes, which relaxes much faster than the single particle orientation. In the long time, on the other hand, the intermediate wave vectors (the molecular-length scale processes) control the dynamics; thus the rate of the decay becomes comparable to $C_1(t)$.

To understand the orientational relaxation in the amide system, we calculated $C_1(t)$ and $C_2(t)$ by using the memory function $\Gamma_R(z)$ in Eq. (7.2). The results of these calculation are shown in Figure 32. Note that both

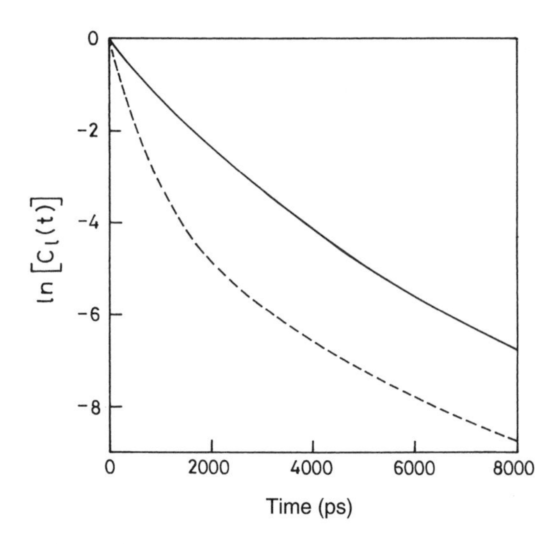

Figure 32. Comparison of the rates of relaxation of the orientational correlation functions $C_\ell(t)$, (where ℓ denotes the rank of the correlation function) for $\ell = 1$ (*solid line*) and $\ell = 2$ (*dashed line*). The solvent is NMP at 273 K. We used Eq. (7.2) to calculate both the correlations. Note the strong nonexponentiality of $C_2(t)$.

TABLE IX
Comparison of the Calculated Average Relaxation Times of the Orientational Correlation Functions

Solvent	T, K	τ_D, ps	$\langle\tau_1\rangle$, ps	$\langle\tau_2\rangle$, ps	$\langle\tau_1\rangle/\langle\tau_2\rangle$, ps
NMF	234	1500	756.24	251.903	3.00
NMA	302	390	215.98	71.95	3.00
NMP	273	4020	1927.6	635.241	3.03

functions are nonexponential. The nonexponentiality in $C_2(t)$ is more pronounced than that in $C_1(t)$. This nonexponentiality indicates the marked non-Markovian effects and can be understood from Figure 24, which shows the frequency dependence of the memory function. Since, the decay of $C_1(t)$ is slower, it probes mostly the low-frequency values. The $C_2(t)$, on the other hand, may probe a substantial range of the frequency, along which the $\Gamma_R(z)$ changes considerably. Thus the $C_1(t)$ is expected to be strongly nonexponential. What is most surprising, however, is that the average relaxation time $\langle\tau_\ell\rangle$ [where $\langle\tau_\ell\rangle = \int_0^\infty dt\ C_\ell(t)$] closely follows the $\ell(\ell+1)$ law prescribed by the rotational diffusion model (Table IX). This, we think, is rather interesting and should be studied in detail.

Now we shall turn our attention to the missing component of the inertial response of the solvation dynamics. As mentioned, the initial experimental study might have missed up to 40% of the total response because of the limited temporal resolution used in those experiments [261]. Subsequent experiments by Chang and Castner [262] revealed that both in formamide (FA) and dimethylformamide (DMF), there is a dominant ultrafast component that may carry up to 60–70% of the total strength. Note that although formamide is a strongly hydrogen-bonded system no such bonding should be present in DMF. Even this system, however, shows ultrafast solvation. From theoretical studies for water, acetonitrile, and methanol, the origin of ultrafast component and the biphasic decay is now fairly well understood [208,232,237,264,265]. In the amide systems, the reason is somewhat less clear. We have shown in this work that the extended molecular hydrodynamic theory (EMHT) can explain the long-time decay satisfactorily. Therefore, the problem that must be addressed is the importance of the ultrafast component in these systems. We now turn our attention to this problem.

For the ultrafast component to make a significant contribution, the following criteria should be satisfied. First and, most important, the difference between ε_∞ and n^2 (n is the refractive index of the medium) must be large. For the amide systems, $\varepsilon_\infty \simeq 10$ and $n^2 \simeq 2.1$ so that the difference between them is quite significant. Second, the most of the fast relaxation responsible for the decrease of the dielectric constant from ε_∞ to n^2 should come from the high-frequency librational modes of the system. The experiments carried out by Chang and Castner [262] reveal that these amide systems may contain more than one librational mode, with frequencies between 50 and 100 cm^{-1}. We have, therefore, carried out the following theoretical investigation, which is motivated purely by the desire to explore the possibility of the ultrafast component in otherwise slow amides. We have taken only one librational mode [266] at 110 cm^{-1}, and we have assumed that this is responsible for the decrease of the dielectric constant from ε_∞ to n^2. The rest of the parameter values remain the same. In addition, we assumed that this librational mode is overdamped. The results shown in Figure 33, suggest that the ultrafast component is indeed responsible for about 70% of the solvation dynamics. This estimate seems to be in surprisingly good agreement with that of Chang and Castner (see Fig. 9 of Ref. [262]), although their approach was completely different from ours. We should also caution the reader that our estimate is rather crude, because of the approximations involved.

D. Conclusions

Let us first summarize the main results of this section. We considered solvation dynamics and orientational relaxation in the slow amide liquids at low temperatures. To describe the local, short-range correlations that are probed by the polar solute in a slow liquid, we extended the MHT to include the

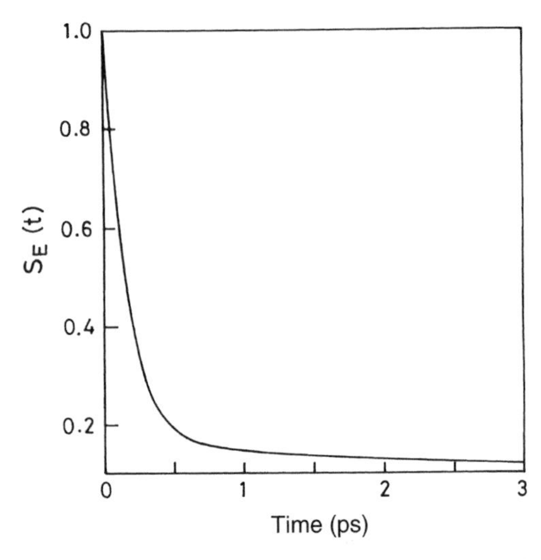

Figure 33. The normalized solvation time correlation function $S(t)$ of Coumarin in NMF at $T = 253$ K plotted as a function of time t. Note the presence of the ultrafast component, which was missed by Chapman and co-workers [*J. Phys. Chem.* **94**, 4929 (1990)] in their experimental studies. An important contribution to the ultrafast polar response of the solvent comes from the high-frequency librational modes of the solvent, assumed to be centered at 110 cm^{-1}. See the text for further details.

solute–solvent spatial and orientational correlations. The extended theory has been remarkably successful in describing the solvation dynamics over many decades of temporal evolution. We find that the memory function for these slow liquids has an interesting biphasic structure, which reflects the marked nonexponentiality of the orientational correlation functions. Another interesting finding is that although the Debye model of rotational diffusion is totally inadequate to describe the decay of the orientational correlation functions, the average relaxation time still approximately obeys the $\ell(\ell + 1)$ Debye relation for rotational diffusion. We also explored the possible magnitude of the ultrafast component in the solvation dynamics in these liquids. A simple calculation indicates that the ultrafast component may contribute up to 70% of the total energy, and this decay may be over within 500 fs. This estimate seems to be in agreement with the estimate of Chang and Castner [262] for DMF. Although the time scale and the magnitude of the ultrafast component are comparable to those of acetonitrile, the slow long-time component is entirely different. For amides, the long-time component is slower by many orders of magnitudes than the fast component. This makes the solvation here truly biphasic.

In the present work, we used the ion–dipole DCF to describe the solvent structure around the dipolar solute. This approximation is superior to our earlier studies [15,15a,159–164] in which an electric field with a spherical cutoff (to include the size of the ion) was employed. A realistic molecule, however, often has an extended charge distribution. In such a case, the solute–solvent interaction is more complex. The formalism employed here can be recast in terms of solute–solvent DCFs; however, the calculation of such correlation functions may be difficult. Some progress toward this goal has been made by Raineri and co-workers [153–158]. However, the satisfactory agreement among the theoretical investigations and the existing experimental results obtained here and elsewhere [159–164,195,202,208] seems to suggest the existence of a significant ionic character of the probe molecule in its excited state. But the story may be quite complicated in actual practice, and certainly a thorough and proper investigation is needed to understand this aspect in detail. Another limitation of the present work is that the specific solute–solvent interaction effects, such as hydrogen bonding, have not been included. This implies that the long-time part of the solvation dynamics may depend on the nature of the specific solute–solvent interaction. Again, the satisfactory agreement obtained in all the four liquid systems studied here indicates that even if these specific effects are important, their dynamics may occur at the same time scale as generated by the present theory.

Our investigation seems to indicate that the study of solvation dynamics can be a powerful tool for exploring the relaxation spectrum of fragile molecular liquids. More important, the solvation dynamics can directly reveal the collective orientational motion of a small number of molecules. This information is not easily available by other techniques. One possibility is to use both quenched and instantaneous normal mode analyses, which have been found useful for revealing many interesting dynamic properties of complex liquids.

VIII. IONIC SOLVATION DYNAMICS IN NONASSOCIATED POLAR SOLVENTS

Solvation dynamics in nonassociated polar solvents (e.g., acetonitrile, acetone, dimethyl sulfoxide) have recently been studied with considerable interest [27,27a,66,151,152,159]. These solvents are important because many of them have potential use in chemical industry. For example, both acetone and dimethyl sulfoxide are commonly used as reaction media for many important organic reactions that occur in solution. Acetonitrile, when mixed with water at desired proportions, acts as a good solvent in reversed-phase liquid chromatography [267].

The dynamic response of these nonassociated polar solvents are distinctly different from those of the associated liquids, such as water, methanol, and formamide. The solvation dynamics in the nonassociated liquids have been found to be rather slow compared to those in water, methanol, and formamide. This is expected, because the nonassociated solvents cannot sustain high-frequency librations or vibrations since there is no hydrogen bond network. Some of them, however, may differ from this trend and can indeed exhibit ultrafast dynamics by virtue of rapid single particle orientation. One such example is acetonitrile.

As discussed in Section I the first measurements on solvation dynamics using ultrafast nonlinear laser spectroscopy were carried out for acetonitrile. These experimental studies were performed by Rosenthal and co-workers [27,27a], who used a big dye molecule (LDS-750) as a probe. Their observations were quite fascinating, since they reported for the first time the biphasic nature of the solvent response. The most interesting outcome of this experiment was the discovery of the presence of an ultrafast component with a time constant <100 fs.

Subsequently, theoretical investigations [159] and computer simulation studies [151,152] were carried out to understand the origin of the biphasic solvent response and the mechanism that drives the ultrafast relaxation in these nonhydrogen-bonded solvents. Computer simulation studies of Maroncelli [152] revealed some important aspects of ion solvation dynamics in acetonitrile. Bagchi and co-workers [159] carried out a theoretical study on ionic solvation dynamics in this solvent and used Kerr relaxation data to calculate the polarization relaxation. The theoretical results thus obtained [159] were found to be in good agreement with the experimental observations [27,27a] and simulation studies [152]. However, the use of Kerr relaxation data in studying solvation dynamics may not be correct, because although the solvation dynamics probes the solvent response of rank one (i.e., $\ell = 1$) the Kerr relaxation [248] measures the collective response of rank two ($\ell = 2$). In addition, the latter often includes contributions from the intermolecular vibrations (hindered translations); these motions influence solvation dynamics in a different manner. Nevertheless, this theoretical study [159] pointed out that the ultrafast solvation in acetonitrile originates from the fast single particle orientation of a solvent molecule around its principal axis. The authors also studied the solvation dynamics in acetonitrile by using dielectric relaxation data [159, 232]. The agreement was found to be poor, because the accuracy of the dielectric relaxation data available then was not reliable [159,232]. In particular, the high-frequency dispersion data were not available.

Recently, Venables and Schmuttenmaer [268] studied the dielectric relaxation of acetonitrile using the femtosecond teraHertz pulse spectroscopy.

Their dielectric relaxation data are expected to be reliable. We used these data when calculating $S(t)$ by Eq. (4.15). The resulting theoretical predictions are again found to be in good agreement with the experimental results of Rosenthal and co-workers [27,27a].

The organization of the rest of this section is as follows. The next part contains a brief discussion of the calculation details. Section VIII.B presents the results of solvation dynamics in acetonitrile. Acetone and dimethyl sulfoxide are reviewed in Section VIII.C.

A. Calculation Procedure

1. Calculation of the Ion–Dipole Direct Correlation Function

The ion–dipole DCF was calculated from the MSA solution of Chan and co-workers for strong electrolytes [201]. We, of course, used it in the limit of zero ion concentration. The details in this regard are given in Section IV.D.1.

2. Calculation of the Static Orientational Correlations

For acetonitrile, we calculated the static orientational correlations from the MSA model by approximating acetonitrile as a sphere with a radius of 2.24 Å. We corrected the wrongly large k limit of the MSA model by using the results of Raineri and co-workers [157]; this gives us an accurate estimation of $\varepsilon_L(k)$.

3. Calculation of the Rotational Memory Kernel

The rotational memory kernel $\Gamma_R(z)$ was calculated from $\varepsilon(z)$ by using Eq. (2.12). The recent frequency-dependent dielectric relaxation data of acetonitrile describe the $\varepsilon(z)$ values as a sum of two Debye relaxations [268] of the following form

$$\varepsilon(z) = \varepsilon_\infty + \frac{\varepsilon_0 - \varepsilon_1}{1 + z\tau_{D1}} + \frac{\varepsilon_1 - \varepsilon_2}{1 + z\tau_{D2}} \tag{8.1}$$

The dielectric relaxation data for acetonitrile are summarized in Ref. [268].

4. Calculation of the Translational Memory Kernel

The translational memory kernel was calculated by using the translational dynamic structure factor of the liquid. The calculational details regarding the calculation of $\Gamma_T(q, z)$ were described in Section II and Eq. (2.13).

5. Calculation of Other Parameters

Let us finally summarize the other parameters used to characterize acetonitrile that are important for our calculation of $S(t)$.

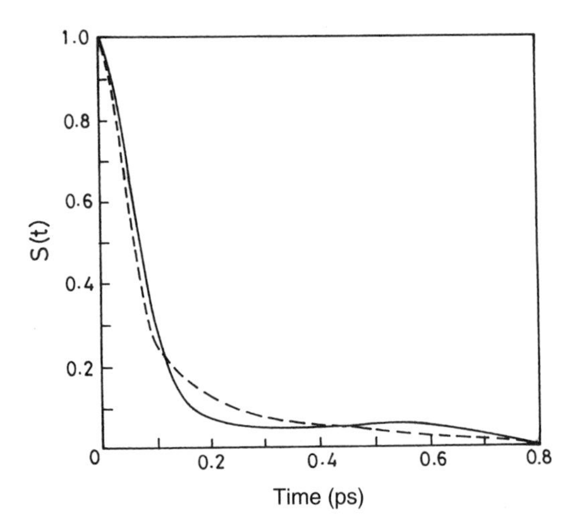

Figure 34. The calculated solvation time correlation function $S(t)$ is compared to the experimental one for the solvation of LDS-750 in liquid acetonitrile. The theoretical results are represented by the *solid line*, and the experimental observations by the *dashed line*. The solvation time correlation function is calculated here by using the Kerr relaxation data [159]. The solute : solvent size ratio is 3.

1. Experimental measurements of the dipole moment of acetonitrile provides a value of $\mu = 3.5$ D [233]. The density is 0.7857 g/cm^3.
2. The solute is chosen to be a sphere with a radius three times that of the solvent. This is expected to mimic the solute : solvent size ratio corresponding to the experimental solute LDS-750. The solvent diameter, σ_{MeCN} is known to be equal to 4.48 Å.
3. The coefficient of viscosity η of acetonitrile has a value of 0.341 cP at $T = 298$ K.

B. Results

In this section we review the earlier results of our group [159] on solvation dynamics in acetonitrile and then present the recent numerical calculations. We also present the theoretical predictions of solvation dynamics in acetone and DMSO.

1. *Ion Solvation Dynamics in Acetonitrile*

Figure 34 displays the comparison of the earlier theoretical results of our group [159] (obtained by using the Kerr relaxation data) and those from the experiments of Rosenthal and co-workers [27,27a]. The relevant calculational scheme that uses the Kerr relaxation data for solvation dynamics

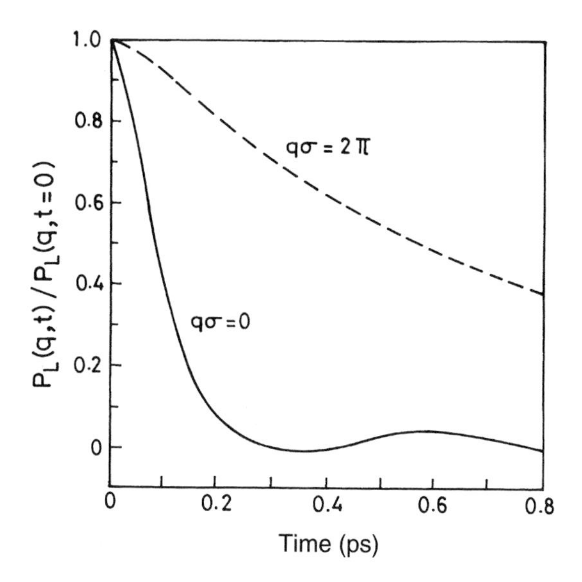

Figure 35. Comparison of the relative contributions to ion solvation dynamics in acetonitrile from the collective ($q\sigma \simeq 0$) and the microscopic ($q\sigma \simeq 2\pi$) components of the polarization fluctuations. The rotational kernel is calculated from the Kerr relaxation data [159].

are described in Ref. [159]. As mentioned, the agreement is good. One interesting aspect to note here is that there is a small oscillation in the theoretical curve at long time, which has not been observed experimentally. The oscillation is just around the experimental data. The theory also produces the slow long-time decay fairly well.

In Figure 35, the decomposition of the dynamics into the two main contributors of solvation—macroscopic ($q = 0$) region and molecular ($q\sigma = 2\pi$) regions—are shown. The results are obtained using the Kerr relaxation data. The slow long-time decay in the theory comes from the molecular-length scale processes in the liquid state, whereas the fast initial decay comes from the long wavelength processes. Needless to say, the same trend is obtained when the dielectric relaxation data are used.

Figure 36 displays the theoretical results obtained by using the most recent dielectric relaxation data of Venables and Schmuttenmaer [268]. The agreement with the experiment is good and surprisingly similar to that when Kerr data were used. The similar agreement reflects and reinforces the fact that in acetonitrile the ultrafast solvation is carried out by the rapid single particle orientation of acetonitrile molecule, which is, of course, substantially dressed by the collective polarization fluctuation.

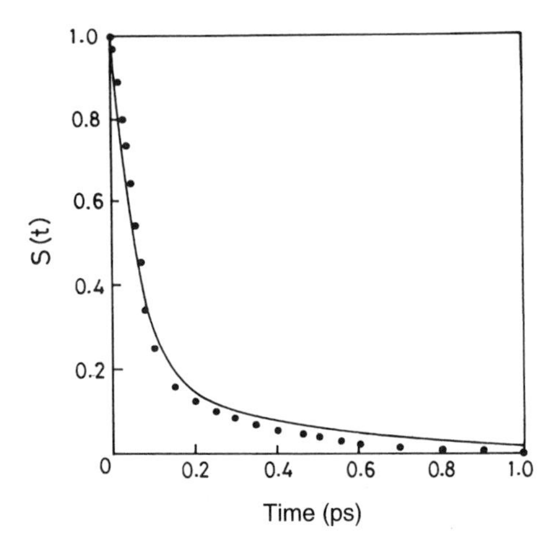

Figure 36. The same comparison as shown in Figure 34 except here the theoretical calculations (*solid line*) are performed by using the data of Venbles and Schmuttenmaer [268]. Note the agreement with the experimental data (*solid circles*) is surprisingly similar to that obtained by using Kerr data.

The dominance of the collective solvent polarization fluctuation on solvation dynamics in acetonitrile is shown in Figure 37. The theoretical results are obtained after including only the $q\sigma \simeq 0$ polarization mode. The agreement with the relevant experimental results is almost the same as that when contributions from all wavenumbers are included (Figure 36). This particular feature again highlights the fact that the long wavelength polarization mode governs the ionic solvation dynamics in ultrafast liquids.

Recently, Ernsting and co-workers [269a] investigated the solvation dynamics of aminonitrofluorene (ANF) in acetonitrile by using transient hole-burning spectroscopy [269a]. The dynamics were found to be biphasic as before, but with one extra feature: a weak oscillation in the $S(t)$ versus time curve at intermediate times. The latter may arise from the underdamped solvent libration mode. This is interesting. However, this oscillation has neither been detected by the experiments of Rosenthal and co-workers [27,27a] nor been reproduced by the present extended MHT.

2. *Ion Solvation Dynamics in Acetone*

The time-dependent progress of solvation of Coumarin-152 in acetone is shown in Figure 38. The solute : solvent size ratio is 1.56; the dielectric relaxation data are given in Ref. 66. We attributed the calculated $\varepsilon_\infty - n^2$ dispersion to a libration with frequncy 60 cm^{-1} [270]. This librational mode of acetone at that frequency was detected in the far-IR studies of Gadzhiev [270]. The

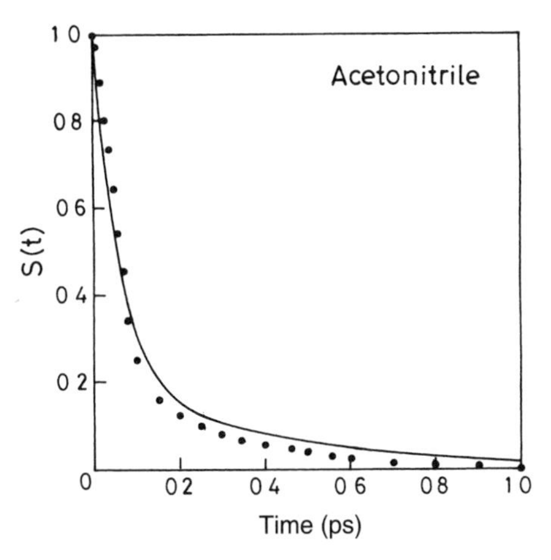

Figure 37. The dominance of the $q\sigma = 0$ polarization mode in ionic solvation dynamics. The theoretical results (*solid line*) are calculated from Eq. (4.15) by considering only the $q\sigma \simeq 0$ solvent polarization mode. Note the agreement with the experimental data (*solid circles*). The other static parameters are defined in Figure 35.

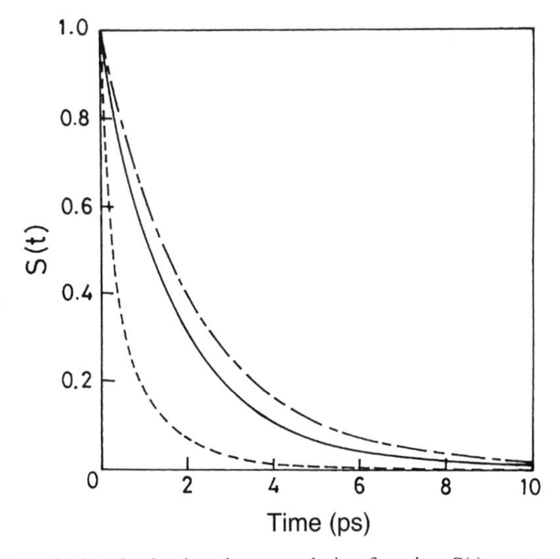

Figure 38. The calculated solvation time correlation function $S(t)$ compared to the experimental one for ionic solvation of excited Coumarin-152 in acetone. The theoretical results (*solid line*) were obtained by using Eq. (4.15) after adding a libration of 60 cm^{-1} in $\varepsilon(z)$ of acetone. The *large dashed line* denotes the theoretical result without any librational contribution. Note the agreement with the experiemental data (*small dashed line*) is poor. The reason for this may be the solute–solvent specific interaction (Section VI). The static parameters used for the calcualtion are as follows: ρ = 0.79 g/cm^3, $\mu = 2.7$ D, $\eta_0 = 0.337$ cP, and $T = 300$ K. The solute : solvent size ratio is 1.56; The diameter of Coumarin-152 is taken to be 7.8 Å.

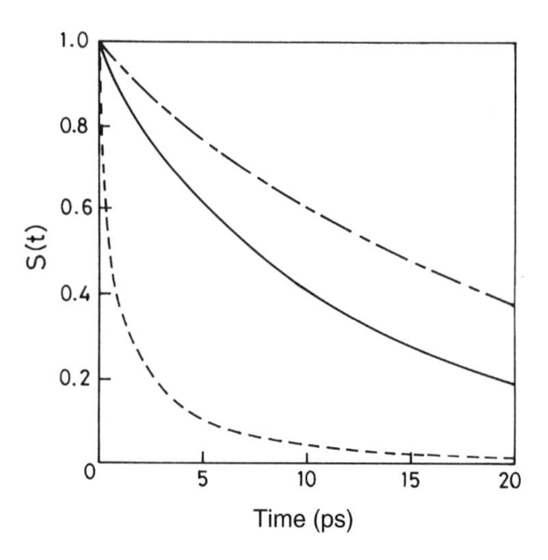

Figure 39. Ion solvation dynamics of Coumarin-152 in DMSO. The theoretical results are shown by the *solid line*, obtained after adding a libration of 40 cm^{-1} in $\varepsilon(z)$ of DMSO. The *large dashed line* represents the calculation without any libration. The *small dashed line* denotes the experimental observations. The poor agreement may be attributed to the specific solute–solvent interaction. The following static parameters were used in the calculation: ρ = 1.014 g/cm^3, $\mu = 4.1$ D, $\eta_0 = 1.1$ cP. and $T = 300$ K. The solute : solvent size ratio is 1.48.

experimental results of Horng and co-workers are also presented in Figure 38. The comparison shows that the theoretically predicted decay of $S(t)$ is slower than what has been observed in experiments [66]. As we pointed out in Section VI, the faster decay in the experimental data may arise from the solute–solvent specific interaction. This specific interaction coupled with solute–solvent binary cage dynamics may provide an extra channel for faster energy relaxation.

The effects of the specific solute–solvent interaction are more pronounced if the solvent is reactive and tends to coordinate with the probe solute. Such might be the case for DMSO. The high reactivity of DMSO to a large variety of organic chromophores may be the reason why this solvent did not attract as much attention as acetonitrile did. We have carried out a theoretical study of the solvation dynamics in this solvent, the results of which are presented next.

3. *Ion Solvation Dynamics in Dimethyl Sulfoxide*

Figure 39 shows the comparison of the theoretical predictions and experimental results of ion solvation dynamics in DMSO. The solute : solvent size ratio is 1.5. The dielectric relaxation data are summarized in [66]. Here the difference between ε_∞ and n^2 is 1.9755. This dispersion is attributed to a libration of 40 cm^{-1}. Castner and Maroncelli [271] observed a peak at this frequency in DMSO in their studies of the polarizability anisotropy response using optical heterodyne-detected Raman-induced Kerr effect spectroscopy (OHD-RIKES). It is clear from Figure 39 that the disparity between the theory and the experiment is strong in this case. The origin of this pronounced disagreement may again be the enhanced solute–solvent specific interaction.

Recently, Tembe and co-workers [269b] carried out molecular dynamic simulation studies on solvation dynamics and barrier crossing dynamics of DMSO. These simulation studies clarified several aspects of various elementary relaxation processes occurring in this solvent.

C. Conclusions

Let us first summarize the main results of this section. The ionic solvation dynamics of acetonitrile, acetone, and DMSO have been studied. These solvents represent a particular class, the nonassociated polar solvents. For acetonitrile, a good agreement between the theory and the experiments has been observed. The ultrafast component observed in this solvent seems to arise from the fast single particle orientation. This fast orientation is further intensified by the large value of the collective force constant. However, for acetone and DMSO, the theoretical predictions are wildly off from the experimental observations. The specific solute–solvent interaction may play a significant role in determining the initial part of the solvent response in these liquids. A systematic inclusion of this particular aspect in a simple theory is a nontrivial task that needs further attention.

The role of the specific chromophore–solvent interaction on solvation dynamics can be assessed semiquantitatively by designing some ingenious experiments. One such suggestion is to study the solvation dynamics in halogen-substituted DMSO. The presence of the halo atom (*F* or *Cl*) in the methyl carbon will reduce the coordinating tendency of the carbonyl oxygen. This should in turn reduce the degree of specific interaction. Second, it would be very interesting to study the ultrafast solvation dynamics in binary mixtures. Here one can tune the magnitude of the specific interaction. Third, one requires accurate dielectric relaxation data both for acetone and DMSO — the existing data are rather old. Thus, several interesting problems remain for future study.

IX. ORIGIN OF THE ULTRAFAST COMPONENT IN SOLVENT RESPONSE AND THE VALIDITY OF THE CONTINUUM MODEL

Recently, much attention has been focused on the origin of ultrafast solvation, which has led to several interesting discussions and has deepened our understanding of both polar and nonpolar solvation dynamics. There appear to be two broad classes of explanations. The first one originally comes from the continuum model, later put to more microscopic formulation by the MHT in the work of Roy and Bagchi [159–164,232]. The second explanation comes from the work of Stratt and co-workers [111–116] via instantaneous normal mode (INM) analysis. While there is agreement on some issues between these two approaches, some differences do remain. In the following, a brief summary of the two explanations is presented. We also touch on the general validity of the continuum approach, which seems to have found new applicants in recent times. Then we discuss several plausible explanations put forward by various authors for the observed ultrafast decay of $S(t)$.

A. Plausible Explanations

1. *Extended Molecular Hydrodynamic Theory: Role of the Collective Solvent Polarization Mode*

The extended MHT developed by Bagchi and co-workers offers the following picture for the ultrafast response [15,15a,159–164,232,265]. First, the dipolar solvent must contain ultrafast dynamic components that can couple to the polarization. They may be librations, H-bond excitations, or fast single particle orientations. Second, the force constant for polarization fluctuation must also be large [15,15a,159–264]. As discussed below, this is large if the static dielectric constant ε_0 of the solvent is large. For small fluctuations, the free energy can be regarded harmonic in polarization fluctuations at equilibrium. Under this condition, the following microscopic expression for the free energy of the system was derived [15,15a]:

$$F(\{\mathbf{P}_{\mathrm{L}}(\mathbf{q})\}) = \frac{(2\pi)^3}{2V}\int d\mathbf{q}\; K_{\mathrm{L}}(q)\mathbf{P}_{\mathrm{L}}^2(\mathbf{q}) \tag{9.1}$$

where $K_{\mathrm{L}}(q)$ represents the wavenumber dependent force constant of the longitudinal (i.e., $\ell = 1$ and $m = 0$) component of the wavenumber-dependent polarization fluctuation $P_{\mathrm{L}}(q)$, and V is the volume of the system. The force constant is related to the wavenumber-dependent longitudinal dielectric function $\varepsilon_{\mathrm{L}}(q)$ as follows [15,15a]

$$K_{\mathrm{L}}(q) = \frac{2}{(2\pi)^2}\frac{\varepsilon_{\mathrm{L}}(q)}{\varepsilon_{\mathrm{L}}(q) - 1} \tag{9.2}$$

The $\varepsilon_{\mathrm{L}}(q)$ of a dipolar liquid has the following interesting wavenumber dependence. At small q, $\varepsilon_{\mathrm{L}}(q)$ is large for polar liquids. $K_{\mathrm{L}}(q)$ is, therefore, nearly equal to $\frac{2}{(2\pi)^2}$. At intermediate q (i.e., $q\sigma \simeq 2\pi$), $\varepsilon_{\mathrm{L}}(q)$ is negative, with a small absolute value, much less than unity. Thus $K_{\mathrm{L}}(q)$ exhibits a pronounced

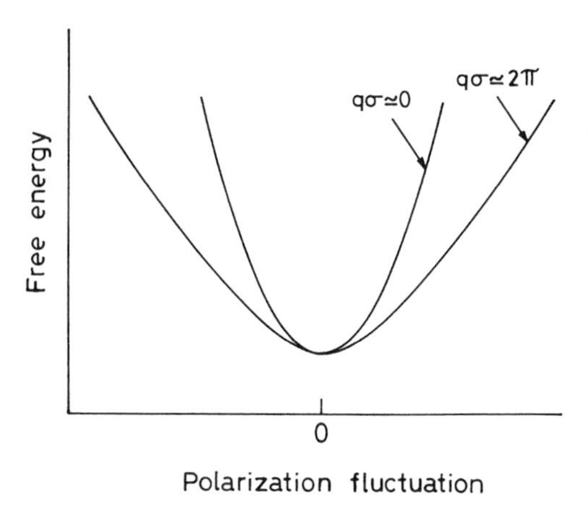

Figure 40. Representation of the polarization potential surfaces for two wavenumbers at which solvent orientational polarization relaxation takes place at different rates. The large value of the force constant at $q \simeq 0$ makes the surface steep and hence the relaxation becomes fast. At intermediate wavenumbers ($q \simeq 2\pi/\sigma$), the value of the force constant becomes small. This induces a flattening of the surface; therefore, the relaxation becomes slow.

softening in this region; $\varepsilon_L(q)$ becomes equal to unity at large q, where $K_L(q)$ diverges. However, the contribution to solvation energy from such large q is negligible. Most of the contribution comes from small q, with a significant amount also coming from the intermediate q. Therfore, for the practical purpose, the force constant of polarization relaxation is the largest at small q.

As first stressed by van der Zwan and Hynes [40], solvation dynamics of an ion can be viewed as a relaxation in a harmonic polarization potential surface where the curvature of the potential well is determined by the force constant $K_L(q)$. At small wavenumbers, the curvature is steep, since $K_L(q)$ is large and thus the relaxation is fast. At intermediate wavenumbers, on the other hand, the softening of the force constant makes the relaxation slow (Fig. 40).

The effects of the force constant and the natural dynamics on the ultrafast response can be understood more convincingly if one considers the following generalized Langevin equation

$$\ddot{\mathbf{P}}_L(\mathbf{q}) = K_L(q)\mathbf{P}_L(\mathbf{q}) - \int dt' \, \Gamma_R(\mathbf{q}, t-t')\mathbf{P}_L(\mathbf{q}, t') + \mathbf{R}(t) \qquad (9.3)$$

which describes the relaxation of the longitudinal component of the solvent orientational polarization density. The double dot denotes the second-order time derivative. Γ_R is the same rotational dissipative kernel as that enters into the solvation dynamics. The effects of the fast orientations and high-frequency librations are embodied in this kernel. Eq. (9.3) makes it abundantly clear the effects of the force constant and the natural fast dynamics of the medium on the ultrafast polarization relaxation. Thus the thermodynamic driving force for ultrafast ion solvation exists in all strongly polar liquids. Note that the translational memory kernel is not included in Eq. (9.3). A systematic incorporation of this quantity will need a more sophisticated and rigorous treatment.

Therefore, it is obvious that a large value of $K_L(q)$ is not sufficient to switch on the ultrafast relaxation. It has to couple with the high-frequency, natural dynamics of the medium. In case of water, this dynamics is provided by the intermolecular vibrations (the hydrogen bond excitation at 193 cm^{-1}) and the libration modes. For acetonitrile, the very fast single particle orientation is partly responsible. For methanol, the ultrafast component (relatively smaller in amplitude) derives contributions from both the fast single particle orientation and the libration.

The point of this explanation is that the observed ultrafast polar solvation is mainly collective in nature. As this relaxation is driven by the small wavenumber fluctuations, it is the same as the one predicted by the continuum model–based theories [40,43,237,264]. The predominance of the collective (i.e., $q = 0$) modes in ionic solvation dynamics in polar solvents is displayed in Figure 41. The theoretical results are calculated by using Eq. (4.15) for methanol. The experimental results of Bingemann and co-workers are also shown in Figure 41. The quantitative agreement with experimental results supports the notion that the ionic solvation dynamics in polar liquids is largely governed by the long wavelength polarization fluctuations. Another interesting feature of this comparison is that the long-time decay of $S(t)$ is surprisingly insensitive to the molecular-length scale processes. One reason behind such insensitivity could be the ultrafast nature of the solvent, in which the solvation becomes complete well before the molecular-length scale processes take over.

Note that this interpretation is applicable only for polar solvation dynamics. For nonpolar solvation, the MHT also attributes the ultrafast solvation to the binary solute–solvent dynamics.

2. *Instantaneous Normal Mode Approach: Nonpolar, Nearest-Neighbor Solute–Solvent Binary Dynamics*

Recently, an alternative interpretation of the ultrafast component was provided by the INM analysis, which finds that it is the free inertial motion of the nearest-neighbor solvent molecules that is responsible for this component [114]. This interpretation comes from the detailed numerical analysis of the modes that couple to ultrafast solvation. Another important aspect

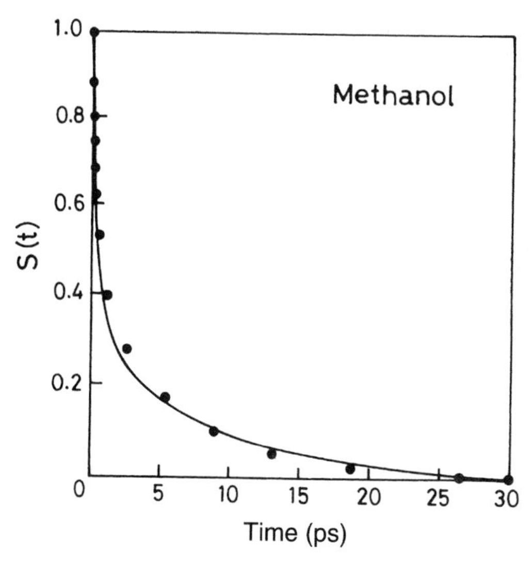

Figure 41. The predominance of $q = 0$ mode in ionic solvation dynamics of excited Coumarin-152 in methanol. The theoretical results are calculated after considering only the $q = 0$ mode solvent response. The *solid line* denotes the calculated $S(t)$ values from Eq. (4.15). The experimental results of Bingemann and co-workers are represented by the *solid circles*. This is the only mode that is included in the continuum model–based calculations of Marcus and co-workers. (See Refs. [237] and [264].)

of this explanation is that the microscopic origin of ultrafast polar and nonpolar solvation is the same. It is the simultaneous binary dynamics of the solute–solvent system that is responsible for the ultrafast response. Actually, for nonpolar solvation, the nature of the response is easy to understand. It is the same cage dynamics that is responsible for the ultrafast frictional or viscoelastic response in a dense liquid [107,108,272].

As mentioned earlier, our analysis also finds that it is the solute–solvent binary dynamics that are responsible for the ultrafast nonpolar solvation. The interpretation obviously varies for the polar solvation. The only way the MHT could predict a nearest-neighbor-mediated polar solvation that is faster than the collective (q=0) solvation is a large contribution from the translational modes. There is, of course, the other possibility that the computer simulations have missed some of the collective response, because the sizes of the systems simulated are far too small. Note that although one can use some approximate scheme (like the reaction boundary method or Ewald summation) to capture the total energy, the dynamic partitioning of this energy among various modes will surely be affected by the small size of the system. This certainly remains a problem to be debated.

3. *Competition between the Polar and the Nonpolar Solvent Responses*

Because the initial decay of the collective solvent polarization mode for ionic solvation can occur on a time scale comparable to that of the nonpolar binary part, an interesting competition between these two contributing components might be present. However, one is still not sure about the relative contributions of the binary and the collective modes in determining the initial course of the ionic solvation and how far and to what extent the domination of any one of the components vary with solvents, temperature, and density. This is certainly a problem that deserves further study. In conclusion, we note that for nonpolar solvation dynamics, the ultrafast component arises from the solute–solvent cage dynamics in which the rapidity of the process is largely determined by the strength of the attractive Lennard–Jones interaction. This is contrasted with the case of the polar solvation dynamics, in which the ultrafast component originates exclusively from the relaxation of the long wavelength polarization mode [15,15a,159–164,208,273].

B. The Validity of the Continuum Model Description: Recent Works of Marcus and Co-Workers

Recently, Marcus and co-workers [237,264] carried out detailed studies of solvation dynamics in various solvents with results that appear to be in good agreement with the experimental results. This work is based entirely on the continuum model. The main emphasis of this work is the use of an accurate dielectric dispersion dataset. The success of this approach again brings back the question of the validity of the continuum model.

In Section I, we noted that the limitation of the continuum model is the total neglect of the molecularity of the solute–solvent intercations and the microscopic structure around the solute. There is, however, a good physical reason why the continuum model is expected to capture at least some of the dynamics correctly, as detailed below.

The MHT shows that the initial, ultrafast part of ionic solvation is dominated by the long wavelength modes. Now the dynamics of these modes are obviously well described by the continuum model. There are several other reasons why the continuum model can be successful. First, in many liquids, the solvent translational modes accelerate the decay of the short-range correlations and bring their relaxation times to par with the long wavelength modes. This fact is particularly important for dipolar solvation. Second, many of the solute probes are rather large compared to the solvent molecules. This is particularly true for water, acetonitrile, and methanol. In such cases, the response that is probed is essentially the long wavelength response, as the large wavenumber correlations hardly make any contribution.

The continuum model approach, however, can break down for small ions. This can easily be verified by considering small ions such as Li^+. Although it is not possible to experimentally study the solvation dynamics of such ions, the effects of solvation do enter in ionic mobility via dielectric friction. This subject will be discussed later.

X. ION SOLVATION DYNAMICS IN SUPERCRITICAL WATER

Supercritical water has become a topic of intense research recently [67–71]. The critical point of water is located at $T_c = 647$ K, $P_c = 22.1$ MPa and the critical density d_c is 0.32 g/cm^3. In supercritical conditions, water loses its three-dimensional hydrogen-bonded network [274,275] and becomes completely miscible with organic compounds. The dielectric constant of SCW is close to six, which makes water behave like an organic solvent [276]. This particular feature has made water a potential solvent for the waste treatment in the chemical industries [68–70]. The high compressibility of water above its critical point allows large variations of the bulk density through a minor change in the applied pressure. This characteristic can profoundly affect the solubility [68], transport properties [69], and the kinetics of the reactive processes in SCW [70]. Therefore, a clear understanding of the dynamic response of SCW is necessary and important for understanding and substantiating the dynamic solvent effects on the rate of the chemical reactions in this medium.

The objective here is to present a theoretical study of the solvation dynamics of an ion in supercritical water. We use the theoretical scheme developed in Section II. The calculated solvation time correlation function is compared with the results of the recent molecular dynamics simulation studies of Re and Laria [71]. A good agreement is observed in the initial (up to 70% of the total solvent response) decay. What is even more interesting is the similarity in the solvation time correlation function of SCW with that in ambient water, although the dynamics of the two systems are entirely different. The molecular theory here shows that, while the intermolecular vibration at 193 cm^{-1} is partly responsible for ultrafast solvation in ambient water [163], it is the very fast single particle rotation that makes $S(t)$ decay so fast in SCW. Agreement between the molecular theory and simulation worsens at long times. This may be attributed to cluster formation around the ion and/or to the contribution of the nonpolar solvation dynamics. Approximate estimates of the time scales of both the processes are also presented.

The organization of the rest of this section is as follows. The next part contains the method of calculation. We present the numerical results in Section X.B and in Section X.C we present two plausible explanations for the slow decay of the simulated solvation time correlation function at long times.

A. Calculation Procedure

1. *Calculation of the Static Correlation Functions*

To calculate the normalized solvation energy time correlation function $S_E(t)$ using Eq. (4.15), we need the 10 component of the wavenumber-dependent dielectric function $\varepsilon_{10}(q)$ and the orienational structure factor $f_{110}(q)$. Since a method of accurate calculation of these quantities for SCW is not available, we obtained the correlation functions from the MSA model. We corrected the MSA static correlations at both the $q \to 0$ and the $q \to \infty$ limits by using the known results (see Section VII).

We obtain the ion–dipole direct correlation function $c_{\mathrm{id}}^{10}(q)$ directly from the solution of Chan and co-workers [206] of the mean spherical model of an electrolyte solution. The calculation is performed in the limit of zero ionic strength.

2. *Calculation of the Memory Kernels*

The rotational memory kernel $\Gamma_{\mathrm{R}}(q, z)$ is obtained from the frequency-dependent dielectric function $\varepsilon(z)$ by using Eq. (2.12).

The frequency-dependent dielectric function $\varepsilon(z)$ for SCW is described by the following expression [276]:

$$\varepsilon(z) = \varepsilon_{\infty} + \frac{(\varepsilon_0 - \varepsilon_{\infty})}{[1 + z\tau_{\mathrm{D}}]} \tag{10.1}$$

The translational kernel $\Gamma_{\mathrm{T}}(k, z)$ was obtained by using Eq. (2.13).

We next present the theoretical results on ionic solvation dynamics in supercritical water. The thermodynamic state of SCW at which the calculations are performed is characterized by $P = 21.7$ MPa, $T = 647$ K and $\rho = 0.32$ g/cm^3. At this state, the frequency-dependent dielectric function of SCW is described by a single Debye process with $\tau_{\mathrm{D}} = 0.98$ ps and $\varepsilon_0 = 7.5$ [276]. Note that in Ref. [276], ε_{∞} is taken equal to unity in the fitting of $\varepsilon(z)$, which, strictly speaking, should be about 1.8. However, computer simulation studies use nonpolarizable water, which means in their case $\varepsilon_{\infty} = 1$. We have, therefore, used the experimental results without any modification.

B. Numerical Results

Figure 42 presents the theoretical results on polar solvation dynamics in SCW for Cl^-. The solute : solvent size size ratio is 1.3. For comparison, we also plotted the simulated equilibrium solvation time correlation function [71] in

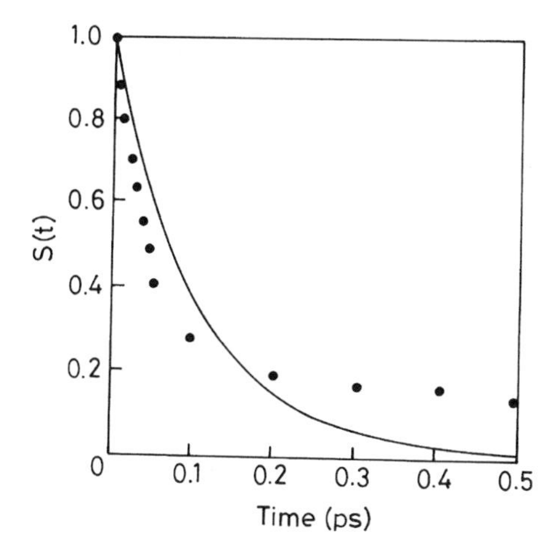

Figure 42. Comparison of the theoretical predictions and the computer simulation results for the ionic solvation dynamics in SCW of Cl^- in water. The *solid line* represents the calculated normalized solvation time correlation function $S(t)$; *solid circles* represent the simulated equilibrium solvation time correlation function of Re and Laria [71].

Figure 42. It is clear from the comparison that the short time decay of the calculated STFC is in good agreement with the simulated results. The predicted long-time decay rate is, however, faster than what is observed in simulation studies [71]. We shall discuss this point below. We investigated the effects of the self-motion of the ion on its own solvation dynamics. The effect was found to be insignificant, since the time scale of solute diffusion is much slower than that of the solvent orientational polarization relaxation.

As discussed by Re and Laria [71], the average rate of ionic solvation in SCW is faster than the experimental rate in ambient water [29]. However, one should compare the simulation results of Re and Laria [71] with those of Maroncelli and Fleming [35] as both neglect the molecular polarizability of water. Both these simulations predict a fast decay at short times. The simulated short-time decay of the solvation time correlation function in ambient water [35] was found to be much faster than what was later observed in experiments [29]. Subsequently, it was suggested [161, 163] that this is because of the neglect of polarizability in simulations, which in turn removes the contribution to the solvation dynamics of the IMV at 193 cm^{-1}. In the absence of polarizability, it is the hindered rotation of the O–H...O moiety at 685 cm^{-1} that gives rise to the very fast decay observed in simulations of Maroncelli and Fleming [35]. In supercritical water, the extensive hydrogen bond network is largely destroyed; therefore, the contribution of IMV at around 193 cm^{-1} is expected to be negligible.

The hindered rotation is also absent. As a result, the experimental studies of ionic solvation dynamics in SCW may indeed reveal the same rate as has been found in the simulation studies of Re and Laria [71]. This is a major prediction of this work that can be tested experimentally.

C. Origin of the Slow Long-Time Decay Rate of the Simulated Solvation Time Correlation Function

We next turn our attention to explore the origin of the slow long-time decay of the STCF observed in the simulation studies [71]. The time constant of the long-time decay was found to be about 1 ps [71], for which no experimental confirmation is yet available. We offer here two plausible explanations. In the first we calculated the average rate of cluster formation by following Becker–Doring–Zeldovich (B–D–Z) scheme [277]. Here one water molecule at a time is added or evaporated from a cluster around the ion. This can be represented as

$$A_{n-1} \rightleftharpoons A_n \rightleftharpoons A_{n+1} \tag{10.2}$$

The average rate of formation of cluster is given by [277]

$$\langle J \rangle = A\beta' c(1)_0 \left[\int_1^m dn\, n^{-\frac{2}{3}} e^{\Delta G / k_B T} \right]^{-1} \tag{10.3}$$

where β' is the impringing parameter, defined as $\beta' = \frac{P}{\sqrt{2m\pi kT}}$; P is the pressure; m is the mass of a molecule; and ΔG is the change in total Gibb's free energy for the formation of a cluster of n molecules from the embryo of concentration $c(1)_0$. We assume $c(1)_0 = 1$ and $m = 6$. A is the surface area of the cluster and can be calculated from its effective volume ($v^\star$), given by $A = 4\pi[\frac{3v^\star}{4\pi}]^{\frac{2}{3}}$. Note that here no nucleation barrier is involved, as cluster size is too small. If one now assumes that ΔG can be given by the following relation [277]:

$$\Delta G = -n\Delta g + \gamma n^{\frac{2}{3}} \tag{10.4}$$

where Δg is the change in Gibb's free energy per molecule and γ is the surface tension of the cluster, then one can calculate the average rate from the above B–D–Z expression. For water, Δg is experimentally known and is approximately equal to 10 Kcal/mol [278]. Eqs. (10.3) and (10.4) give a value of 2.4 ps for the time constant, which is in the correct range.

In the second scenario, we assume that the slow decay arises from the nonpolar solvation dynamics, which is neglected in the study described above.

Here the long-time part of $S(t)$ is given by [105]

$$S_{\mathrm{NP}}(t) \propto \int_0^{\infty} dq\, q^2 F_s(q, t) c_{12}^2(q) F(q, t) \tag{10.5}$$

where $F_s(q, t)$ and $F(q, t)$ are the self- and the solvent dynamic structure factors, respectively, and C_{12} is the solute–solvent isotropic direct correlation function. The slow decay can come from the nearest neighbors, in which $F(q, t)$ decays as [105]

$$F(q, t) \propto \exp\left[-\frac{Dq^2 t}{S(q)}\right] \tag{10.6}$$

It has been shown [105] that the nonpolar solvation dynamics derive maximum contribution from the solvent dynamic modes at $q\sigma \simeq 5$. When this particular feature is considered, we obtain a time constant (given by $S(q)/Dq^2$) of about 1.2 ps, which is also in the correct range.

It thus appears that both the cluster formation and the nonpolar solvation dynamics can explain the slow long-time tail. Unfortunately, we are unable to estimate the relative contributions of each of these different processes analytically. Note that these two contributions can be significant in SCW because the magnitude of polar solvation energy here is rather small, in contrast to that in ambient water where the latter dominates.

D. Conclusions

In conclusion, we note that several aspects of ion solvation dynamics can be explained from a molecular theory. The theoretical predictions are in agreement with the recent computer simulation studies [71]. The slow long-time decay observed in simulations may originate from a combined effect of a nonpolar solvent response and cluster formation at the critical solvent density. The ultrafast solvation in SCW is somewhat similar to what has been observed in normal water at ambient conditions; however, the molecular mechanisms behind the ultrafast solvent response in water at these two thermodynamic conditions are dramatically different. Although the intermolecular vibration of the O–H...O moiety at 193 cm^{-1} is primarily responsible for the observed ultrafast dynamics in ambient water, it is the very fast single particle rotation of water molecule that makes the solvent response ultrafast in the supercritical state. In addition, we predict that for solvation in SCW, the effects of molecular polarizability of water may be negligible. Thus the simulation results here can indeed be close to experiments, unlike in the case of ambient water, in which simulation predicted a faster decay than what was experimentally observed.

There is another interesting point left untouched here. Re and Laria [71] observed a significant difference between the nonequilibrium and the equilibrium solvation time correlation function — the former was markedly slower in the long time. Although such a nonlinear response is expected if cluster formation is involved in the late stage of solvation, a quantitative understanding of this interesting result is yet to be developed.

XI. NONPOLAR SOLVATION DYNAMICS: ROLE OF BINARY INTERACTION IN THE ULTRAFAST RESPONSE OF A DENSE LIQUID

In the preceding sections, we discussed the polar solvation dynamics in detail, with emphasis on several startling results that have been discovered in the last decade. Although theoretical discussions centered largely around the polar component, experimentally it is impossible to separate the polar contributions from the nonpolar ones. The nonpolar component, however, can be studied systematically by studying the solvation of a nonpolar solute in nonpolar solvent; and such studies have been carried out in the recent past. As it turned out, the study of nonpolar solvation paid rich dividends, because it threw light on some aspects not recognized previously.

As mentioned in Section I, the nonpolar solvation can be of various kinds. It can be the result of the dispersive force included in the Lennard–Jones-type short-range interactions or of quadrupolar or higher multipolar interactions. Nonpolar solvation energy can be significant if there is a change in the solute's size and shape of the probe molecule upon electronic excitation. The identification of this component as nonpolar is somewhat arbitrary. This is because the quadrupolar and the octupolar interactions are also as short range as the Lennard–Jones. One should note that although the contribution of the nonpolar component could be small compared to the polar component when there is a significant charge redistribution on excitation, the former can also be rather large when the probe molecule is large, as is often the case with dye molecules. Thus the nonpolar component can be important, even dominant, when the equilibrium fluctuations determine the relevant time correlation function, as in the line shape analysis. Recent experiments have shown that the nonpolar component can also decay with the time constant in the 100 fs range [95–103].

Berg and co-workers [95–103] have employed transient hole-burning spectroscopy to study the nonpolar solvation dynamics of nonpolar molecules, such as tetrazene in butyl benzene. Recently, he also studied the nonpolar solvation dynamics of DNA fragments and proteins in various organic solvents [279]. Here, also, the solvation is marked by the presence of an ultrafast component that originates from the nearest-neighbor cage dynamics around the solute. Berg [110] developed an elegant continuum model for nonpolar solvation dynamics by which the solvent response function is calculated from the time-dependent shear and longitudinal moduli of the solvent. The interesting feature of this study is that the latter may be called as an analog of the continuum model of polar solvation dynamics. This analogy appears from the fact that in the continuum model for polar solvation dynamics one needs $\varepsilon(\omega)$ as an input whereas in Berg's model the frequency-dependent shear and longitudinal moduli are necessary to obtain the nonpolar solvent response function.

The motivation to study nonpolar solvation dynamics is manyfold. It has been known for a long time that the rate of VER is dominated by binary interactions [119,280]. Stratt and co-workers [114] pointed out that both VER and nonpolar solvation dynamics (NPSD) are governed largely by the same type of binary solvent frictional responses. Although the important role of binary collisions in VER has been discussed [119,281–283], the same for NPSD has not been anticipated before. The objective of this section is to demonstrate that such similarity can be understood from an entirely different theoretical framework, i.e., the mode-coupling theory [26,26a, 26b,249–252], which provides a reliable description of the frequency- (or time-) dependent response of the liquid.

There is, however, a more fundamental aspect of this problem. As mentioned, recent studies on solvation dynamics have found a nearly universal ultrafast component in the 50–100 fs range in many solvents. The presence of such components in water and acetonitrile has been rationalized by various theories. It is the presence of such an ultrafast component in higher normal alcohols (ethanol, butanol), reported recently by Joo and co-workers [243], that deserves special attention. These authors studied the solvation dynamics of a large dye molecule in alcohols by using 3PEPS techniques and found that 30–60% of the decay of the total solvation energy correlation function is carried out by an ultrafast Gaussian component with a time constant τ_G less than 100 fs. Joo and co-workers [243] also found that the presence of such an ultrafast component in alcohols (methanol, butanol) is rather generic in nature. These observations were different from the experimental results of Horng and co-workers [66], who studied the polar response of the solvent to an instantaneously created charge distribution and found no such ultrafast component. The scenario has become even more interesting by the recent experimental studies of Ernsting and co-workers [284] on the solvation dynamics of LDS-750 dye in solvents such as chloroform and acetonitrile. In this work, the authors argued that the experimentally observed fluorescence Stokes shift for the dye molecule at times <70 fs may be attributed to various intramolecular relaxation processes, such as isomerization and/ or vibrational energy relaxation [284]. Recently, a theoretical investigation on the polar solvation dynamics in these monohydroxy alcohols [208] found that there is no ultrafast component in the polar solvent response when $\tau_G \leq 100$ fs in these solvents, except in methanol. The same work also showed that a microscopic theory based on a realistic model for solvent dynamic response can give rise to a time constant of 150–200 fs for nonpolar solvation, which is, again, higher than that reported by Joo and co-workers [243].

A possible interpretation of this ultrafast component was provided by the instantaneous normal mode analysis that finds free inertial motion of the solvent molecules responsible for this component. The INM studies on solvation dynamics by Ladanyi and Stratt [114] have revealed a nearly universal mechanism for the solvation of an excited solute in a solvent, irrespective of the nature of the solute–solvent interaction (nonpolar, dipolar, or ionic). According to the picture provided by Ladanyi and Stratt [114], the solvation is dominated by the simultaneous participation of the nearest-neighbor solvent molecules in which the solvent libration is usually the most efficient route to the solvation. Recently, Skinner and co-workers [108] presented an interesting analysis of the nonpolar solvation dynamics in which energy relaxation in the range of 100 fs was observed. In this work, the ultrafast component arises from the solute–solvent two-particle translational dynamics in the cage of the other solvent molecules. Thus here, too, the binary part plays the dominant role.

Earlier theoretical studies of nonpolar solvation dynamics [105,208] did not consider the contribution of the binary part to the solvation energy explicitly; the binary part was included only in the frictional response. The situation for nonpolar solvation could be quite different from polar solvation. In the latter, energy relaxation may be dominated by the coupling of the polar solute's electric field to the solvent polarization whose dynamics is controlled by the collective density relaxation of the solvent. This density relaxation, of course, contains the binary component. Note that the above separation is somewhat arbitrary, because the total energy of a polar solute may also contain a significant nonpolar contribution. The separation into binary and the rest — the correlated contribution — to the energy, becomes necessary when the short-range part of the intermolecular interaction makes a dominant contribution to the total energy; and this is certainly the case for nonpolar solvation. It is important to note that in time-dependent fluorescence Stokes shift experiments on polar solutes in polar solvents, the nonpolar part may not make a large contribution, because it may not change significantly upon excitation.

In this section an analytical study of the importance of the binary part to the nonpolar solvation dynamics is discussed. The approach is based on ideas borrowed from the mode-coupling theory. A particular advantage of this approach is that it explicitly separates the energy and the force into a binary part from short-range interactions and a collective part from the correlated dynamics involving many particles. This approach has been rather successful in dealing with the dynamics of dense liquids. We report the results of a theoretical analysis of NPSD, VER, and frequency-dependent friction. We found that 60% of the total decay of the nonpolar solvation energy correlation function with a time constant ~100 fs is carried out by the binary interaction part only. The important point here is that both of them are controlled almost entirely by the binary part of the frequency-dependent friction.

The organization of the rest of this section is as follows. Section XI.A discusses the theory for the calculation of the frequency-dependent friction. The next part describes the theoretical details of the nonpolar solvation dynamics of a Lennard–Jones particle in a Lennard–Jones fluid; the results are presented in Section XI.C.

A. Theoretical Details

In this section we develop a theory that describes the time-dependent solvent response at a very short time. The analysis presented here was motivated partly by the recent work of Skinner and co-workers [108]. Our main concern is the normalized solvation energy–energy time correlation function $S_{NP}(t)$ of the solute, with the nonpolar interaction as the only source of energy. The expression for $S_{NP}(t)$ is given by

$$S_{NP}(t) = \frac{C_{EE}(t)}{C_{EE}(t=0)} \tag{11.1}$$

where $C_{EE}(t)$ is the STCF. If $v_{12}(r)$ denotes the interaction energy between the solute and a solvent molecule, then the instantaneous energy of the solute can be written as

$$E_{1,\text{solv}}(t) = \sum_j v_{1j}[r_{1j}(t)] \tag{11.2}$$

where the summation runs over all the j solvent molecules. The calculation of the corresponding energy–energy time correlation function is rather complicated. The energy–energy correlation function of the probe is now defined by the following expression

$$C_{EE}(t) = \left\langle \sum_j v_{1j}[r_{1j}(0)] \sum_k v_{1k}[r_{1k}(t)] \right\rangle \tag{11.3}$$

Thus $C_{EE}(t)$ contains contributions both from the binary and the three-particle dynamics. An accurate treatment of the latter is difficult.

Skinner and co-workers [108] carried out an elegant analysis of the correlation functions. A notable feature of this analysis is a treatement of the three-particle term by use of the Kirkwood superposition approximations [285]. The study revealed the presence of an ultrafast component with a time constant on the order of a few hundred femtoseconds. An important aspect of this study is the demonstration that the three-particle term slows down the decay of the binary term. Such a role of the three-particle term is known from other studies. This effect can be as large as 30% in favorable cases.

In this section an alternative analysis of the nonpolar solvation dynamics is presented. The treatment is entirely analytical and is accurate both at short and long times. The results obtained are quite similar to those of Skinner and co-workers [108]. The limitation of the work at present is the neglect of the triplet term, which can, however, be included.

In mode-coupling theory, the short-time force autocorrelation function can be decoupled from the collective ones. The details in this regard are given in Section XII. We assume here that the energy–energy time correlation function can also be separated into a short-time binary part and a long-time collective part. The binary part relaxes on a fast time scale, whereas the slow collective part, which is coupled to the solvent density fluctuation, relaxes on a much longer time scale. Thus the expression of the energy time correlation function is assumed to be given by

$$C_{EE}(t) = C^{B}_{EE}(t=0)\exp[-(t/\tau^{B}_{E})^2] + C_{\rho\rho}(t) \tag{11.4}$$

where we have already assumed that the binary part is Gaussian. The expression for the collective density fluctuation part $C_{\rho\rho}(t)$ is obtained from the density functional theory and is given by [105]

$$C\rho\rho(t) = \frac{(k_{\mathrm{B}}T)^2\rho}{2\pi^2}\int_0^{\infty} dq\, q^2 c_{12}^2(q) F^s(q,t) F(q,t) \tag{11.5}$$

where $c_{12}(q)$ is the wavenumber-dependent solute–solvent two-particle direct correlation function, which was obtained by using the well-known Weeks–Chandler–Andersen scheme [213]. $F(q,t)$ and $F^s(q,t)$ represent the solvent dynamic structure factor and the self-dynamic structure factor, respectively. To calculate $C_{\rho\rho}(t)$ by using Eq. (11.5) we need the expressions for $F(q,t)$ and $F^s(q,t)$.

We use the following expression of the solvent dynamic structure factor [174,175,249–252]:

$$F(q, t) = \mathcal{L}^{-1}\left[\frac{S(q)}{z + \frac{\langle \omega_q{}^2 \rangle}{z + \frac{\Delta_q}{z + \tau_q{}^{-1}}}}\right] \tag{11.6}$$

where the solvent dynamic structure factor in the frequency z plane is obtained from a Mori continued fraction expansion truncated at the second order. The Laplace inversion is performed numerically by using the Stehfest algorithm. $S(q)$ is the static structure factor of the solvent and is calculated from the solution of the Percus–Yevick equation for pure liquids: $\langle \omega_q{}^2 \rangle = \frac{k_B T q^2}{m S(q)}$ and $\tau_q^{-1} = 2\sqrt{\frac{\Delta_q}{\pi}}$, $\Delta_q = \omega_l^2(q) - \langle \omega_q{}^2 \rangle$, where $\omega_l^2(q)$ is the second moment of the longitudinal current correlation function. The other equations and parameters necessary for calculating $F(q, t)$ are given in Appendix B.

Next we describe the expression for the self-dynamic structure factor $F^s(q, t)$. The expression is given by [203]

$$F^s(q, t) = \exp\left\{\frac{-q^2 k_B T}{m\zeta_0}\left[t + \frac{1}{\zeta_0}(e^{-t\zeta_0} - 1)\right]\right\} \tag{11.7}$$

where $\zeta_0 = \zeta(z = 0)$, the latter has been calculated self-consistently from Eq. (12.3) (presented below). When expressing $F^s(q, t)$, the zero frequency value of the friction is clearly an approximation. It is the time dependent friction which should be used to obtain the correct result. But doing an infinite-loop calculation with the time-dependent friction is highly nontrivial. Note that the form of the self-part is accurate both in the short- and in the long-time, but it precludes any nonexponential behavior of $F^s(q, t)$ that it may exhibit in the intermediate time. This, however, is only a minor limitation of the present analyses [249–251], because the nonexponentiality at the intermediate times is expected to be small for tracers with sizes smaller than the solvent molecules [249–251]. We assume that the zero frequency limit does not introduce much error, as we know that the major contribution of $R_{\rho\rho}(t)$ to the total friction is in the long-time limit.

Eq. (11.4) has a simple physical meaning. The binary part arises from the relaxation owing to the direct interaction of two particles, whereas the cage around the solute (whose energy is being considered) remains fixed. The second term, on the other hand, involves the relaxation of the cage itself.

We next derive a microscopic expression of the time constant associated with the binary part of the solvation energy, τ_E^B. We follow the same steps as followed in the calculation of the force–force time correlation function.

The expression for the time constant is given by

$$\tau_E^B = \sqrt{\frac{-2C_{EE}^B(t=0)}{\ddot{C}_{EE}^B(t=0)}} \tag{11.8}$$

where

$$C_{EE}^B(t=0) = 4\pi\rho \int_0^\infty dr\, r^2[v_{12}(r)]^2 g_{12}(r) \tag{11.9}$$

and

$$\ddot{C}_{EE}^B(t=0) = \frac{4\pi\rho}{m} \int_0^\infty dr\, r^2[v_{12}(r)]^2 \nabla_r g_{12}(r) \nabla_r v_{12}(r) \tag{11.10}$$

Note that the $t = 0$ limit of Eq. (11.4) gives the mean square energy fluctuation of the nonpolar solvation energy. This is an approximate expression. The exact expression involves three-particle ($g_3(r)$) and higher-order correlation functions, as discussed by Skinner and co-workers [108]. The contribution of these higher-order correlation terms, however, can be absorbed in the collective density fluctuation part $C_{\rho\rho}(t)$; hence the $t = 0$ limit of Eq. (11.4) is expected to provide the mean square fluctuation of the nonpolar solvation energy without any significant error.

We discuss the theoretical results obtained by using Eqs. (11.4), (11.8), and (11.10) in the next section.

B. Numerical Results: Significance of the Solute–Solvent Two-Particle Binary Dynamics

In this section we discuss the theoretical results on nonpolar solvation dynamics calculated by using Eq. (11.4). Figure 43 shows the plot of the normalized energy time correlation function $S_{NP}(t)$ obtained for a Lennard–Jones system at $\rho\sigma^3 = 0.844$ and $T^*(\equiv k_B T/\varepsilon) = 0.728$ with $\sigma = 3.41$ Å, $m = 40$ amu and $\varepsilon = 120\, k_B$. The calculation is carried out for the size and mass ratios 1. The Gaussian time constant (τ_E^B) obtained at this state point is 110 fs. Figure 43 also shows the decomposition of the total energy into binary and collective parts. It is clear from the figure that the normalized binary relaxation part accounts for ~62% of the decay of the total nonpolar solvation energy time correlation function. The remaining part is carried by the collective density fluctuation. The domination of the binary part in nonpolar solvation dynamics indicates that the nearest-neighbor participation makes the solvation ultrafast for nonpolar solvation dynamics.

We now comment on the time constant associated with the binary part of the solvation energy correlation function. For a Lennard–Jones system,

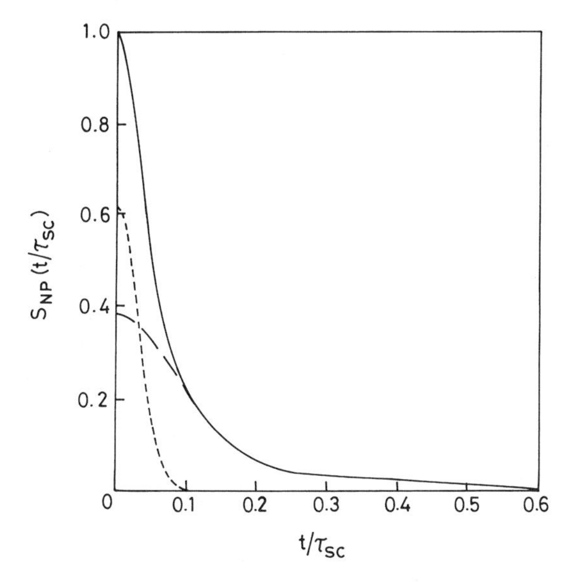

Figure 43. Comparison of the binary and collective components of the nonpolar solvation energy time correlation functions. The normalized nonpolar solvation energy time correlation function $S_{NP}(t)$ is plotted as a function of time t for a solute : solvent size ratio of 1. The *solid line* represents the decay of the total nonpolar solvation energy time correlation function, which has been calculated by using Eq. (11.1). The *small dashed line* shows the decay of the binary component and the *long dashed line*, the decay of the collective component of the total $S_{NP}(t)$. The solvent considered is argon at $T^\star = 0.728$ with $\rho^\star = 0.844$. The mass of the solute is taken as that of an argon atom. Note that the time is scaled by the quantity $\tau_{sc} = \sqrt{\frac{m\sigma^2}{k_B T}}$, which is equal to 2.527×10^{-12} s.

the present theory gives a value of 110 fs for the binary time constant at $\rho^\star = 0.844$ and $T^\star = 0.728$. This is in agreement with the results of Skinner and co-workers [108]. Another important observation is that it is close to the Gaussian time constant of the frictional response, which is close to 100 fs. The time constant of energy relaxation is, however, still larger than the fastest time scale observed experimentally by Joo and co-workers [243].

The reason for the discrepancy may be the following. Joo and co-workers [243] used a huge dye molecule, which may lead to a specific solute–solvent interaction in both the ground and the excited states. This type of specific interaction is stronger than the usual solute–solvent binary interaction. This, in turn, means the movement of the solute particle in a much deeper potential well created by the nearest-neighbor solvent molecules, which would certainly reduce the value for the binary time constant. This trend

TABLE X
The Time Constant Associated with the Binary Energy Relaxation Calculated at Different Solute : Solvent Ratios[a]

$\varepsilon_{sol} : \varepsilon_{solv}$	τ_E^B, fs
1	110
2	99
3	90
4	80
5	72

[a] The time constants were calculated by using Eq. (11.8). The solute : solvent size and mass ratios are taken as unity. The solvent is argon at 120 K.

is shown in Table X in which we have calculated the time constants by varying the solute : solvent ε ratio at size and mass ratios unity. The same trend was also found by Ohmine [286], although in a different context.

C. Conclusions

Let us first summarize the main results presented in this section. We investigated the importance of the solute–solvent binary interaction in determining the initial ultrafast response in nonpolar solvation dynamics. The present investigation is completely different from the earlier in that the present approach is based on the mode-coupling theory, which provides an accurate description of both short-time and long-time dynamics. We used simple expressions for the initial decay of the force–force and energy–energy time correlation functions. The application of these expressions was partly motivated by the mode-coupling theory and partly by the work of Skinner and co-workers. The nonpolar solvation dynamics were found to be determined by the binary friction at the initial times. The slow long-time decay observed from the nonpolar STCF still remains slaved to the hydrodynamic motion of the solvent molecules. We further found that the time constant and the decay of the initial nonpolar solvation energy time correlation function is determined to a large extent by the strength of the solute–solvent interaction. As shown in Table X, the Gaussian time constant of the initial decay can easily be in the order of 50–100 fs, if the solute–solvent attractive interaction (parametrized by ε) is sufficiently large. This seems to substantiate our earlier argument that the nonpolar solvation dynamics can be responsible for the experimental observations of Joo and co-workers, for whom 30–60% of the decay of the total STCF was carried out by an ultrafast component with a time constant <100 fs.

As discussed above, the previous theories of nonpolar solvation dynamics neglected the binary part completely, which led to the conclusion that the initial part of the nonpolar STCF decays with a time constant of 150–200 fs. It was shown in this work that the binary part can lead to the initial decay with Gaussian time constant $\tau_G \sim 50$ fs, which is in agreement with the INM analysis of Stratt and co-workers and the simulation results of Skinner and co-workers.

We shall see in Sections XII and XIII that both the vibrational energy and the phase relaxations are intimately related to the dynamics of nonpolar solvation. This close interrelationship seems to originate from the fact that the same binary component largely determines the vibrational dynamics of a solute immersed in a nonpolar solvent.

XII. VIBRATIONAL ENERGY RELAXATION

The study of vibrational energy relaxation has a long history [119,121,280–283]. Theoretical studies on VER have mostly been carried out by invoking two basic models: the isolated binary collision (IBC) model, in which the collision frequency is modified by the liquid structure, and the weak coupling model, in which the vibrational motion of the molecule weakly couples to the translational and rotational degrees of freedom so that a perturbative technique can be employed. The IBC model was developed by Herzfeld and co-workers [119], and the VER rate for two-level systems is assumed to be given by $\tau_{ij}^{-1} = P_{ij}\tau_c^{-1}$, where τ_c^{-1} is the collision frequency and P_{ij} is the probability per collision that a transition from level i to level j will take place. Note that P_{ij} is independent of density but does depend on temperature, whereas τ_c depends on both state parameters.

The application of the IBC model (proposed originally for the gas phase) to study the vibrational energy relaxation in the liquid phase was criticized on several grounds by many contemporary authors. Fixman [281] argued that the transition probability should be density dependent and the dynamics of the relaxation be modified by three-body and higher-body interactions. He considered the total force acting on the vibrating molecule at any time as a sum of hard binary collision and a Brownian random force. This treatment led him to suggest that the interactions higher than the two-body collision interactions are important for determining the VER rate in liquids. Zwanzig [282] criticized the fundamental assumption of the IBC model that the rate can be given by the product of the collision frequency and the transition probability. He used the weak-coupling perturbative technique to obtain the VER rate, and showed that the two- and three-body interactions are equally crucial in determining the rate of energy relaxation.

In a subsequent work [287], Herzfeld pointed out that the the models applied by Fixman and Zwanzig were internally inconsistent. He showed that the IBC model could be used to obtain the VER rate in dense liquids as well. The only condition here is that one should not use the zero-frequency Enskog friction to calculate the rate, since it overestimates the high-frequency spectrum of the collisional frequency. Therefore, a systematic approach is needed to obtain the proper high-frequency spectrum of the collisional friction. This can be significantly different for molecules interacting with continuous potential, like Lennard–Jones, from those interacting with hard sphere potentials. A detailed analytic treatment of the dynamic friction based on the generalized Langevin equation (GLE) was first provided by Berne and co-workers [288], who tested the validity of IBC approach by molecular dynamic simulation studies [288]. Their simulation studies on diatomic dissolved in Lennard–Jones argon (with rigid bond approximation) indicated that the IBC model is accurate in describing the vibrational energy relaxation at moderately high frequency.

The second approach considers the VER as a classical process in which energy is dissipated to the medium by the usual frictional process; then one can adopt a stochastic approach. For VER involving low frequency, it is reasonable to assume that the vibration is harmonic. Under this condition, the rate of the VER of a classical oscillator is given by the simple Landau–Teller expression [24]:

$$\frac{1}{T_1} = \frac{\zeta_{\text{real}}^{\text{bond}}(\omega_v)}{\mu} \tag{12.1}$$

where μ is the reduced mass of the diatomic making up the vibrating bond, ω_v is the harmonic vibrational frequency of the bond, and $\zeta_{\text{real}}^{\text{bond}}(\omega_v)$ is the real part of the friction acting on the vibrational coordinate. The molecular dynamic simulation studies of Berne and co-workers [288] showed that, if the cross-correlations between the solvent forces on each atom of the diatomic are neglected and the bond is held rigid, then $\zeta_{\text{real}}^{\text{bond}}(\omega_v)$ could be approximated as

$$\zeta_{\text{real}}^{\text{bond}}(\omega_v) = \frac{\zeta_{\text{real}}(\omega_v)}{2} \tag{12.2}$$

where $\zeta_{\text{real}}(\omega_v)$ is the friction experienced by one of the atoms of the vibrating homonuclear diatomic. Eq. (12.2) was also used by Oxtoby in his theory of vibrational dephasing.

The friction at the bond frequency is the cosine integral of force–force time corelation function (FFTCF) acting on the bond. This friction is responsible for papulation redistribution in vibrational levels since energy dissipates through friction.

Recent INM [111–116] and theoretical studies have shown that both VER and nonpolar solvation dynamics are dominated by the same frictional response that are governed primarily by the binary interactions (modes). The objective of this section is to discuss why the solvent forces responsible for VER are essentially the same as those that determine the initial part of the nonpolar solvation dynamics. The study reported here differs from earlier studies on this problem in that the mode-coupling theory was used to calculate the time-dependent friction, which was then used to obtain the VER rate in dense liquids. The agreement between the theoretical predictions and computer simulation results on VER of low-frequency modes indicates that the VER rate is crucially governed by only a few high-frequency modes but not by all the binary modes [288,288a]. This is in corroboration with the recent INM studies of Goodyear and Stratt [113], who found that a particular set of high-frequency instantaneous normal modes is responsible for carrying out the vibrational energy relaxation. All these results seem to suggest that the IBC model is useful for calculating the VER rate in dense liquids.

A. Calculation of the Frequency-Dependent Friction

The calculation of the frequency-dependent friction is rather involved. For the present purpose, the main interest is the high-frequency limit of this friction, which can be obtained two different ways. Both approaches generate numerically similar results, although the semiempirical approach is much simpler. Both approaches are outlined below.

In both the modern kinetic theory and the sophisticated mode coupling theory [26,26a,26b] the frequency-dependent total friction $\zeta(z)$ of a dense liquid is assumed to be given by the combination of three terms: the binary collision term $\zeta_{\mathrm{B}}(z)$, the density fluctuation term $R_{\rho\rho}(z)$, and the transverse current term $R_{tt}(z)$. The contribution of these terms, however, are not additive. The final expression is given by the following relation [26,26a,26b]:

$$\zeta^{-1}(z) = [\zeta_{\mathrm{B}}(z) + R_{\rho\rho}(z)]^{-1} + R_{tt}(z) \tag{12.3}$$

Note that the binary friction is determined by the short-range interactions between the solute and the solvent molecules, whereas the $R_{\rho\rho}(z)$ and $R_{tt}(z)$ are governed by the relatively long-range interactions. The calculation of $\zeta_{\mathrm{B}}(z)$ is nontrivial, and the details regarding its calculation is described in the next section.

We first describe the calculation of $R_{\rho\rho}(t)$. This contribution originates from the coupling of the solute motion to the collective density fluctutation of the solvent through the solute–solvent two-particle direct correlation function. The expression for $R_{\rho\rho}(t)$ is as follows [26,249–252]:

$$R_{\rho\rho}(t) = \frac{\rho k_B T}{[m(2\pi)^3]} \int d\mathbf{q}' \, (\hat{q}\hat{q}')^2 q'^2 [F^s(q', t) - F^o(q', t)][c_{12}(q')]^2 F(q', t) \quad (12.4)$$

where $F^s(q, t)$ is the dynamic structure factor of the solute, $F^o(q, t)$ represents the inertial part of the dynamic structure factor of the solute, $c_{12}(q)$ denotes the wavenumber-dependent two-particle direct correlation function, and $F(q, t)$ is the dynamic structure factor of the solvent. The expressions for $F(q, t)$ and $F^s(q, t)$ are given by Eqs. (11.6) and (11.7), respectively. An interesting aspect of Eq. (12.4) is that it accounts for the microscopic distortion of the solvent around the solute through the product $c_{12}(q)F(q, t)$. For particles interacting via Lennard–Jones interaction, $c_{12}(q)$ can be obtained by employing the well-known Weeks–Chandler–Andersen (WCA) theory, which requires the solution of the Percus–Yevic equation for the binary mixture. The latter is obtained in the limit of zero solute concentration. The calculation procedures of the other quantities have been described in detail elsewhere [249–252]; thus we discuss them only briefly here.

The inertial part of the self-dynamic structure factor $F^o(q, t)$ is given by [249–252]:

$$F^o(q, t) = \exp\left(-\frac{k_B T}{m} \frac{q^2 t^2}{2}\right) \quad (12.5)$$

The third contribution in Eq. (12.3) comes from $R_{tt}(z)$, which originates from the coupling of the solute's motion with the transverse mode of the collective density fluctuation. The calculation of $R_{tt}(z)$ is highly nontrivial. However, we do not need $R_{tt}(z)$ in the present calculation, since it was shown explicitly in Ref. [249] that the R_{tt} contribution becomes negligible in a solvent at high density with a solute of comparable size; but for the sake of completeness, we give the expression for $R_{tt}(z)$ in Appendix C.

1. *Microscopic Expression for Binary Friction*

The quantity that is particularly relevent in the present work is the binary part of the friction $\zeta_B(z)$. This contribution originates from the instantaneous two-body collision between the solute and solvent particles. Since these collisional events are always associated with the ultrashort time ($t \leq 100$ fs) dynamics of the medium, the time dependence of the binary friction could be well approximated by the following Gaussian decay function [249–252]:

$$\zeta_B(t) = \Omega_0^2 \exp[-(t/\tau_B)^2] \quad (12.6)$$

where τ_B is the time constant associated with the above binary Gaussian relaxation and Ω_0 is the Einstein frequency of the solute in the solvent cage. This quantity can be evaluated from the solute–solvent radial distribution function $g_{12}(r)$ and the interaction potential $v_{12}(r)$ between them as follows [249–252]

$$\Omega_0^2 = \frac{4\pi\rho}{3m} \int_0^\infty dr\, r^2 g_{12}(r) \nabla_r^2 v_{12}(r) \tag{12.7}$$

We need the expression for the time constant τ_B to calculate $\zeta_B(t)$ from Eq. (12.6). The expression for τ_B can be derived from the definition of $\zeta(t)$ by using the short-time expansion, which is given by [249–252]

$$\begin{aligned} \Omega_0^2/\tau_B^2 = {} & (\rho/3m)^2 \int d\mathbf{r}\, [\nabla^\alpha \nabla^\beta v_{12}(\mathbf{r})] g_{12}(\mathbf{r}) [\nabla^\alpha \nabla^\beta v_{12}(\mathbf{r})] \\ & + (\tfrac{1}{6}\rho) \int [d\mathbf{q}/(2\pi)^3] \gamma_{d12}^{\alpha\beta}(\mathbf{q}) [S_{12}(q) - 1](q) \gamma_{d12}^{\alpha\beta}(\mathbf{q}) \end{aligned} \tag{12.8}$$

where summation over repeated indices is implied. The other quantities in Eq. (12.8) are described in Refs. [249–252]. It is clear from Eq. (12.8) that although it provides an accurate estimation of τ_B, its implementation requires extensive numerical work.

In the following we derive a simpler (although approximate) expression for $\zeta_B(t)$. First, note that the expression for the same FFCTF, is given exactly by

$$C_{FF}(t=0) = (4\pi\rho k_B T) \cdot \int_0^\infty dr\, r^2 g_{12}(r) \nabla_r^2 v_{12}(r) \tag{12.9}$$

This is determined by the binary interaction terms only and is proportional to the Einstein frequency. Thus a straightforward generalization to the time dependence of the binary part of the force–force time correlation function would be [249–252]

$$C_{FF}^B(t) = (4\pi\rho k_B T) \cdot \int_0^\infty dr\, r^2 G_{12}(r, t) \nabla_r^2 v_{12}(r) \tag{12.10}$$

where $G_{12}(r, t)$ is the distinct part of the van Hove time correlation function, defined by $G_{12}(r, t) = \langle \delta[r - r(t)] \rangle$, where $r(t)$ is the time-dependent separation between the solute and a solvent molecule [174]. Note that Eq. (12.10) is approximate but sensible in the short time. If one now assumes that the initial decay of the correlation function is given by $C_{FF}^B(t) = C_{FF}^B(t=0) \exp[-(t/\tau_F^B)^2]$ then one obtains the following expression for the Gaussian relaxation time constant:

$$\tau_{\mathrm{F}}^{\mathrm{B}} = \sqrt{\frac{-2C_{\mathrm{FF}}^{\mathrm{B}}(t=0)}{\ddot{C}_{\mathrm{FF}}^{\mathrm{B}}(t=0)}} \tag{12.11}$$

where (and hereafter) the double dots signify the double derivative of the function in question with respect to time. This quantity is given by the following expression:

$$\ddot{C}_{\mathrm{FF}}^{\mathrm{B}}(t) = (4\pi\rho k_{\mathrm{B}}T) \cdot \int_0^{\infty} dr\, r^2 \ddot{G}_{12}(r,t) \nabla_r^2 v_{12}(r) \tag{12.12}$$

We now derive an expression for $\ddot{G}_{12}(r, t)$ that is valid at short times. From the definition of $G_{12}(r, t)$, the expression for $G_{12}(r, t + \Delta t)$ can be written as [289]

$$G_{12}(r, t+\Delta t) = \langle \delta[r - r(t+\Delta t)] \rangle = \langle \delta[r - r(t) - \Delta r(t)] \rangle \tag{12.13}$$

Expanding the above δ-function with respect to powers of Δr, we obtain

$$G_{12}(r, t+\Delta t) = \langle \delta[r - r(t)] \rangle - \Delta r \left\langle \frac{\partial}{\partial r} \delta[r - r(t)] \right\rangle + \frac{(\Delta r)^2}{2} \left\langle \frac{\partial^2}{\partial r^2} \delta[r - r(t)] \right\rangle - \dots \tag{12.14}$$

We now define the increment, $\Delta G_{12}(r, t)$, over $G_{12}(r, t)$ as follows

$$\Delta G_{12}(r, t) = G_{12}(r, t+\Delta t) - G_{12}(r, t) \tag{12.15}$$

which, after using Eq. (12.14) and the definition of $G_{12}(r, t)$, produces the following expression

$$\Delta G_{12}(r, t) = -\Delta r \left\langle \frac{\partial}{\partial r} \delta[r - r(t)] \right\rangle + \frac{(\Delta r)^2}{2} \left\langle \frac{\partial^2}{\partial r^2} \delta[r - r(t)] \right\rangle - \dots \tag{12.16}$$

Now, since $\Delta r \to 0$ as $\Delta t \to 0$, we neglect all the terms in Eq. (12.16) containing $(\Delta r)^2$ and higher powers of it and obtain the following expression for $\Delta G_{12}(r, t)$:

$$\Delta G_{12}(r, t) = -\Delta r(t) \nabla_r G_{12}(r, t) \tag{12.17}$$

Now in the same way if we define the increment over $\Delta G_{12}(r, t)$ as

$$\Delta\Delta G_{12}(r, t) = \Delta G_{12}(r, t+\Delta t) - \Delta G_{12}(r, t) \tag{12.18}$$

where $\Delta G_{12}(r, t+\Delta t)$ is given by

$$\Delta G_{12}(r, t+\Delta t) = -[\Delta r(t+\Delta t)]\left[\frac{\partial}{\partial r}\langle \delta(r - r(t) - \Delta r(t)\rangle\right] \tag{12.19}$$

Now expanding Eq. (12.19) in the same way as was done for Eq. (12.14), neglecting the terms contaning $(\Delta r)^2$ and higher powers of it, and taking into consideration the fact that the average velocity is zero, we arrive at the following exact expression:

$$\ddot{G}_{12}(r, t) = -\ddot{r}(t)\nabla_r G_{12}(r, t) \tag{12.20}$$

We now define an effective force as $F(r) = \mu\ddot{r}(t)$, where μ is the reduced mass of the solute–solvent composite system, and use the equality $G_{12} = (r, t = 0) = g_{12}(r)$ to rewrite Eq. (12.20) in the following form:

$$\ddot{G}_{12}(r, t = 0) = -\frac{F(r)\nabla_r g_{12}(r)}{\mu} \tag{12.21}$$

In the next step, we need to calculate the force acting on the solute particle. There are two alternative procedures to obtain this force. First, one may approximate this by taking the derivative of the Kirkwood's potential of mean force $W(r)$ related to $g(r)$ by $W(r) = -k_B T \ln g(r)$ [174]. This is the form used by Skinner and co-workers [108]. This is inaccurate, however, in the short time during which force is determined by the direct pair-wise binary interactions. Consequently, the expression for $F(r)$ is assumed to be given by

$$F(r) = -\nabla_r v_{12}(r) \tag{12.22}$$

Note that at long times one should use the potential of mean force to obtain $F(r)$.

To calculate the Gaussian time constant τ_F^B from Eq. (12.11), we need the expression for $\ddot{C}_{FF}^B(t = 0)$. Following the same procedure as outlined above, the analytic expression was derived as

$$\ddot{C}_{FF}^B(t = 0) = \frac{4\pi\rho k_B T}{\mu}\int_0^\infty dr\, r^2 \nabla_r g_{12}(r)\nabla_r v_{12}(r)\nabla_r^2 v_{12}(r) \tag{12.23}$$

We find that the time constants predicted by these two methods [Eqs. (12.8) and (12.11)] differ by only about 15%. Although the first one is expected to be more nearly accurate, the second approach is far simpler and can be used to obtain an estimate of the force–force time correlation function.

B. Vibrational Energy Relaxation: Role of Biphasic Frictional Response

We present the calculated rate of the VER in Table XI at $\rho^\star = 1.05$ and $T^\star = 2.5$. The atomic mass and size of the diatomic solute are the same as those of an argon atom. The frequency-dependent bond friction is calculated from the mode-coupling theory. Figure 44 shows the comparison of

TABLE XI
Vibrational Energy Relaxation Rates[a]

	$1/T_1$, ps^{-1}		
$\bar{\nu}$, cm^{-1}	Theory	Simulation	INM
39.7	11.01	11.26	12.69
68.7	8.1	9.64	12.5
145.0	2.9	3.16	3.26
221.6	0.5	0.99	0.05

[a] Calculated by using the Eq. (6.1). The system is a homonuclear diatomic solute dissolved in argon at $\rho^\star = 1.05$ and $T^\star = 2.5$, with $\sigma_{\mathrm{Ar}} = 3.41$ Å and $m_{\mathrm{Ar}} = 39.5$ amu. The atomic size and the atomic mass of the solute are the same as those of an argon atom. For comparison, the INM results of Stratt and co-workers [113] and the simulation results of Berne and co-workers [288] are also shown.

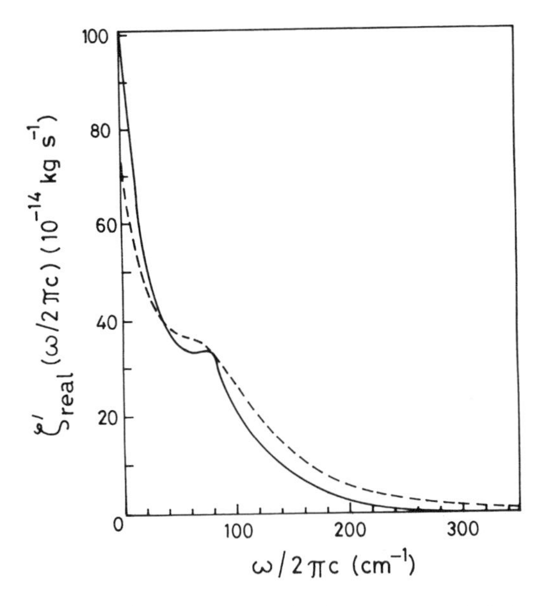

Figure 44. Half of the real part of the calculated total friction plotted against frequency. The calculated friction (*solid line*) is compared to the simulated friction (*dashed line*) [288a]. The solvent considered is argon at $T^\star = 2.5$ and $\rho^\star = 1.05$. Note that $\zeta'_{\mathrm{real}} = \frac{\zeta_{\mathrm{real}}}{2}$.

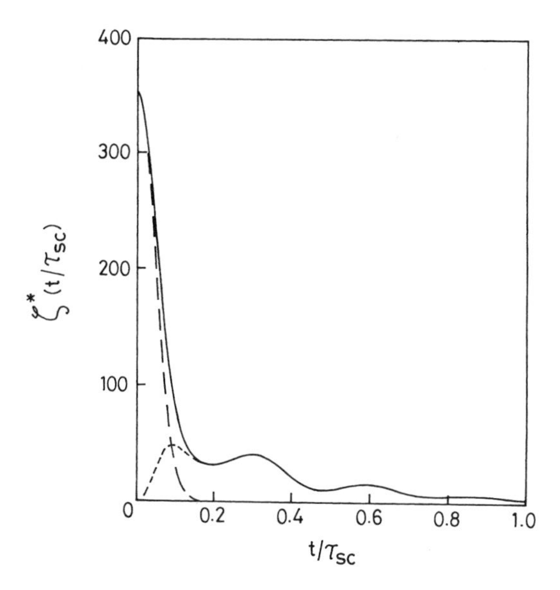

Figure 45. Comparison of the binary and the collective components of the time-dependent friction $\zeta^*(t/\tau_{sc})$, which is plotted as a function of time. The *solid line* denotes the decay of the total friction with time, the *long dashed line* represents the decay of the binary part of the total friction, and the *small dashed line* shows the decay of the collective part of the total friction. Note that the friction plotted here is scaled as follows: $\zeta^*(t/\tau_{sc}) = [\zeta(t/\tau_{sc})] \cdot (\tau_{sc}^2/m)$. The solvent considered is argon at $T^* = 2.5$ and $\rho^* = 1.05$. Note that the time is scaled by the quantity $\tau_{sc} = \sqrt{\frac{m\sigma^2}{k_B T}}$, which is equal to 1.3636×10^{-12} s.

the calculated friction and that obtained from the simulation studies of Straub and co-workers [288a]. The plot shows that there are some oscillations present in the friction calculated from the mode-coupling theory, which actually worsens the otherwise good agreement. The results obtained from the INM studies by Stratt and co-workers [113] and those from the simulations of Berne and co-workers [288a] are also presented in Table XI. The comparison shows that at high frequency (studied here), the INM approach underestimates the rate of the vibrational energy relaxation.

Figure 45 shows the time dependence of this friction for a Lennard–Jones system at the reduced density $\rho^* = 1.05$ and reduced temperature $T^* = 2.5$. This binary part decays on an extremely fast time scale, and the decay is mostly over by about 200 fs. The remaining contribution from the frequency-dependent friction comes from the slower density and transverse current relaxations of the liquid. Note that although the last two components can make a significant contribution to the total zero frequency friction (which is related to the diffusion coefficient), they play no role in

TABLE XII
Solvent Density Dependence of VER Rates Calculated at Four Frequencies[a]

	$1/T_1$, ps^{-1}		
$\bar{\nu}(cm^{-1})$	$\rho^\star = 0.85$	$\rho^\star = 0.95$	$\rho^\star = 1.05$
39.7	7.53	9.45	11.44
68.7	6.19	8.16	9.8
145.0	1.94	2.4	2.6
221.6	0.283	0.365	0.4

[a] The system studied is a homonuclear diatomic solute dissolved in a Lennard–Jones atomic solvent at $T^\star = 2.5$. The diameter and the mass of an atom of the atomic liquid are taken to be the same as those of an argon atom ($\sigma_{Ar} = 3.41$ Å and $m_{Ar} = 39.5$ amu.) The size and the mass of an atom of the homonuclear diatomic solute are also the same as those of an argon atom.

the short-time dynamics. Since VER of a molecule in a liquid couples only to the high-frequency response of the liquid, it is only the binary part of the friction that is relevant for vibrational energy relaxation. The different time scales involved in the time-dependent friction are also shown in Figure 45.

Thus, to obtain the solvent dependence of the VER, we must calculate the binary part as well as the collective density part. It is clear from Eqs. (12.9)–(12.12) that the binary part is sensitive not only to the details of the interaction potential between the vibrating solute and the surrounding solvent molecules but also to the solute–solvent radial distribution function. Because the expression for the binary friction is reliable, we find that the for Lennard–Jones system, the VER depends also on the energy parameter ε.

Table XII shows the density dependence of the vibrational energy relaxation at a constant temperature ($T^\star = 2.5$). We calculated the rate of the vibrational energy relaxation by using the Eq. (12.1) in which the bond friction is calculated from the MCT. The system is a homonuclear diatomic solute dissolved in Lennard–Jones argon in which the mass and size of an atom of the solute are those of an argon atom. The results are tabulated for four frequencies at each density. A comparative study among the density-dependent VER rates clearly reveals that the rate increases almost linearly with density at all frequencies. The increase in rate with density can be explained from the binary interaction picture. The frequency of the effective binary collision increases as the density increases. This, in turn, reduces the time scale of the decay of the time-dependent binary friction, leading to a more rapid energy transfer from one vibrational state to the other, which gives rise to an overall increase in the VER rate. We would like to mention here that within the limited range of density and temperature investigated, we find an approximate $\rho T^{\frac{1}{2}}$ dependence of the VER rate, which appears to be in agreement with the earlier work of Hills [290].

C. Vibrational Relaxation at High Frequency: Quantum Effects

The calculation of the VER rate for the high-frequency modes is considerably more difficult than for the low-frequency modes [291–293]. Unfortunately, there is yet no reliable method to treat the VER of a high-frequency mode, for several reasons. First, for high-frequency modes, the system cannot be treated classically, and the quantum effects become important. The coupling of the vibrating system with the bath also becomes nonlinear when the probing frequency is large. At high frequency, multiphonon processes affect the VER rate significantly [126]. Second, the solvent density of states at high frequency is often small, leading to small rates from the classical Landau–Teller expression. Thus the relaxation becomes too slow to be calculated directly from the nonequilibrium molecular dynamic simulations. Third, one needs to Fourier transform the FFTCF at a frequency much higher than the characteristic frequencies of the correlation function. This means measuring a weak signal in presence of a large signal : noise ratio. These difficulties have made study of VER involving large frequencies nontrivial. Recently, Nitzan and co-workers [291] and Skinner and co-workers [293] made useful attempts in this direction, by using a semiclassical approach. In this approach, the vibrating mode is treated quantum mechanically, whereas the solvent molecules (bath) are treated classically. In this semiclassical scheme, the quantum effect is often introduced through a quantum correction factor [291–293].

The Hamiltonian for the system (vibrating mode plus the solvent) can be written as follows [119]:

$$H = H_0 + H_{\mathrm{B}} + V \tag{12.24}$$

where H_0 is the Hamiltonian for the vibrational degrees of freedom, assumed to be harmonic in the present discussion, H_{B} is the bath Hamiltonian, which includes the rotational and the translational degrees of freedom; and V couples the system to the bath. By using the transition rate given by Fermi's golden rule, one can show that the transition rate, τ_{ij}^{-1} from the vibrational level i to j is given by [119]

$$\frac{1}{\tau_{ij}} = \frac{2\hbar^{-2}}{1+\exp[-\beta\hbar\omega_{ij}]}\int_{-\infty}^{\infty} dt\ \exp[i\omega_{ij}t]\cdot\left\langle\frac{1}{2}[V_{ij}(t), V_{ji}(0)]_+\right\rangle \tag{12.25}$$

where $[A, B]_+$ is the symmetrized anticommutator defined by

$$[A, B]_+ = AB + BA \tag{12.26}$$

$V_{ij}(t)$ is an operator in the bath degrees of freedom defined by [119]

$$V_{ij}(t) = \exp\left[-\frac{iH_B t}{\hbar}\right] V_{ij} \tag{12.27}$$

Evaluation of the quantum mechanical correlation function in the anticommutator is prohibitively difficult. Attempts have been made to replace this correlation by a classical one, because the bath degrees of freedom (rotations and translations) may be treated classicaly. A simple method would be to replace the anticommutator directly by a classical correlation function, leading to

$$\frac{1}{\tau_{ij}} = \frac{2\hbar^{-2}}{1+\exp[-\beta\hbar\omega_{ij}]} \int_{-\infty}^{\infty} dt \, \exp[i\omega_{ij}t] \times \langle V_{ij}^{\text{class}}(0) V_{ij}^{\text{class}}(t) \rangle \tag{12.28}$$

In the subsequent step, the coupling potential V is expanded in the normal modes $\{Q_\alpha\}$

$$V = V_0 + \sum_\alpha F_1^\alpha Q^\alpha + \frac{1}{2}\sum_\alpha \sum_\beta F_2^{\alpha\beta} Q^\alpha Q^\beta + \ldots \tag{12.29}$$

where

$$F_1^\alpha \equiv \left. \frac{\partial V}{\partial Q^\alpha} \right]_{Q=0}$$

and

$$F_2^{\alpha\beta} \equiv \left. \frac{\partial^2 V}{\partial Q^\alpha \partial Q^\beta} \right]_{Q=0}$$

If one keeps only the first term (i.e., V is linear in Q), then one recovers the Landau–Teller expression in the classical limit. The quantum effects in Eq. (12.28) are somewhat trivial, except when inharmonicity is significant.

Bader and Berne [292] discussed several approximate schemes to include quantum effects directly at the level of the anticommutator of the force–force correlation function. They also considered several forms of nonlinear coupling between the vibrational coordinate and the solvent to treat the multiphonon process involved in the VER of high-frequency modes.

Recently, Egorov and Skinner [293] addressed the same question for vibrational energy relaxation in oxygen by using a classical molecular dynamics simulation of liquid oxygen to calculate the classical force time correlation function. Several approximate schemes for including the quantum effects were considered. A large enhancement of rate owing to the quantum effects was observed [293].

D. Conclusions

We found that the vibrational energy relaxation is entirely dominated by the binary part of the total frictional response of the solvent and is decoupled from the macroscopic friction. We saw in Section XI that the same binary component determines the initial part of the nonpolar solvent response. As discussed earlier, this binary response originates from the solute–solvent cage dynamics; therefore, the nonpolar solvation dynamics and the vibrational energy relaxation are intimately connected.

The agreement between theory and simulation (Table XI) indicates that the binary part seems to explain the rate of the VER. Here we have assumed that the bond between the atoms of the homonuclear diatomic is rigid and that there is no coupling between the translational and rotational motions. In real systems, however, the bond may not be rigid and the coupling between the rotational and the translational motions may become rather important. The study of VER becomes even more challanging and interesting when the energy relaxation occurs at high frequency. For high-frequency modes, the classical Landau–Teller description provides an inaccurate description, since the quantum effects become significant in this case.

XIII. VIBRATIONAL PHASE RELAXATION IN LIQUIDS: NONCLASSICAL BEHAVIOR OWING TO BIMODAL FRICTION

As mentioned in Section I, both vibrational phase and energy relaxations are the vital probes for studying the interaction of a chemical bond with the surrounding solvent molecules. Because many chemical reactions take place in the liquid phase and because the solvents act as both source and sink, this information is of great importance. In fact, the study of vibrational relaxations can even provide information about the inharmonic coupling between different vibrational degrees of freedom (such as bending and stretching). Generally, this information is not easily available from other sources.

VPR in small molecules is usually much faster than vibrational energy relaxation. Although it has been known for a long time that VER is sensitive only to the high-frequency (or short-time) response of the liquid, it was believed that such short-time dynamics were not relevant for vibrational dephasing. Recent studies with ultrafast laser spectroscopy seem to indicate a different picture.

In this section, we present a brief review of VPR, with the emphasis mainly on the role of the initial Gaussian component of the biphasic solvent frictional response on vibrational dephasing.

A. Background Information

Let us assume that a given vibrational mode (with harmonic frequency ω_0) of all the molecules in a liquid is prepared at a given phase initially by the application of an ultrafast laser pulse. This phase coherence between different molecules is destroyed by two independent mechanisms. The first one involves nearly elastic collisions with the surrounding solvent molecules. This interaction leads to small frequency shifts $\Delta\omega(t)$ from the average frequency $\overline{\omega}$ of the solvent molecules in liquid. Thus, the instantaneous frequency of a vibration of a given mode of a particular molecule is given by [294]

$$\omega(t) = \overline{\omega} + \Delta\omega(t) \tag{13.1}$$

where $\Delta\omega(t)$ represents the stochastic modulation of $\omega(t)$ as a result of interactions with the environment. Note that $\overline{\omega}$ contains the shift from the gas phase frequency ω_0. When $\Delta\omega(t)$ is characterized by statistical properties common to all the molecules, the mechanism is called the homogeneous mechanism of dephasing [294].

In dense liquids there are motions that are slow compared to the time of vibrational dephasing. For example, the exchange of different species in a binary mixture is a slow process. Therefore, different statistical distributions of a particular species around the vibrating solute may lead to different frequencies in different molecules; i.e., $\overline{\omega}$ itself would be different for different molecules. This is called the inhomogeneous mechanism of vibrational dephasing, and it leads to the loss of vibrational coherence among different molecules. The relative importance of homogeneous and inhomgeneous mechanisms for vibrational dephasing has been an active area of research for several decades.

As mentioned in Section I, the traditional method for studying the VPR (vibrational dephsaing) is IR and isotropic Raman line-shape analyses [120–125]. If the time constant of the vibrational dephasing is denoted by τ_v, then the full width at half maxima (FWHM) of the isotropic Raman line shape is equal to $\frac{2}{\tau_v}$. For slow dephasing, such as in N_2, this is a valid scheme. The dephasing times here are typically >100 ps [120,121].

There are several other systems in which the vibrational dephasing is much faster, such as the C–I stretching in CH_3I. Here the dephasing time could be on the order of 2–4 ps [23]. The dephasing of C–I stretching also shows subquadratic quantum number dependence of overtone dephasing. This is interesting because the classical Kubo–Oxtoby theory [120–122] predicts quadratic quantum number dependence.

Recently, a nonlinear optical spectroscopic technique was used to study vibrational dephasing directly in the time domain [23]. It was observed that even the high-frequency modes, such as C–H stretching in $CHCl_3$, exhibit subquadratic quantum number dependence. The origin of such behavior is yet to be understood.

As discussed below, when the time correlation function of the fluctuating frequency shifts ($\langle \Delta\omega(0)\Delta\omega(t) \rangle$) decays in the same time scale as that of the vibrational dephasing, then the traditional approach to the problem breaks down. In fact, in this case the initial decay of the FFTCF plays an important role and can provide a microscopic explanation of the observed sub-quadratic quantum number dependence of the overtone dephasing of C–I stretching in CH_3I.

B. Kubo–Oxtoby Theory

Most theories of vibrational dephasing start with Kubo's stochastic theory of line shape [294]. In his pioneering work, Oxtoby applied Kubo's theory of vibrational dephasing and demonstrated that inharmonicity could play an important role in enhancing dephasing rates by many orders of magnitude [120–122]. The final expression of Oxtoby involves a force–force time correlation function that acts on the normal coordinate. This force is coming from the surrounding solvent molecules. Oxtoby related the dephasing rate to the solvent viscosity by a hydrodynamic argument [120–122].

The broadened isotropic Raman line shape $I(\omega)$ is the Fourier transform of the normal coordinate time correlation function by [23a,120]

$$I(\omega) = \int_0^\infty \exp(i\omega t)\langle Q(t)Q(0) \rangle \tag{13.2}$$

where ω is the Laplace frequency conjugate to time t. The experimental observables are either the line-shape function $I(\omega)$ or the normal coordinate time correlation function $\langle Q(0)Q(t) \rangle$, as in the time-domain experiments of Tominaga and Yoshihara [23]. The normal coordinate time correlation is related to frequency modulation time correlation function by [120–122]

$$\langle Q(t)Q(0) \rangle = Re\ \exp(i\omega_0 t)\left\langle \exp\left[i\int_0^t dt'\ \Delta\omega_{mn}(t') \right] \right\rangle \tag{13.3}$$

where ω_0 is the vibrational frequency and $\hbar\Delta\omega_{mn}(t) = V_{nn}(t) - V_{mm}(t)$ is the fluctuation in energy between vibrational levels of n and m, where n and m represent vibrational quantum numbers. V_{nn} is the Hamiltonian matrix element of the coupling of the vibrational mode to the solvent bath; and $\Delta\omega(t)_{mn}$ is, therefore, the instantaneous shift in the vibrational frequency owing to interactions with the solvent molecules. One usually performs a cumulant expansion [294] of Eq. (13.3), which leads to an expression of the normal coordinate time correlation function in terms of the frequency modulation time correlation function.

Kubo–Oxtoby analysis starts with a general Hamiltonian of the following form [120–122]:

$$H = H_{\text{vib}} + H_{\text{B}} + V \tag{13.4}$$

where H_{vib} is the Hamiltonian for the vibrational degrees of freedom of the isolated molecule. H_{B} is the Hamiltonian for the rotational and translational degrees of freedom. These two degrees of freedom act jointly as a bath, and Oxtoby treated them classically. V represents the inharmonic oscillator–medium interaction.

Oxtoby [120–122] used the following Hamiltonian for the vibrating mode:

$$H_{\text{vib}} = K_{11}Q^2 + K_{111}Q^3 \tag{13.5}$$

where K_{111} is the coefficient that gives rise to the inharmonicity in the vibration.

If V is the inharmonic oscillator–medium interaction, then expanding V in the vibrational coordinate Q using Taylor's series,

$$\hbar\Delta\omega_n(t) = (Q_{nn} - Q_{00})\left(\frac{\partial V}{\partial Q}\right)_{Q=0}(t) + \frac{1}{2}(Q_{nn}^2 - Q_{00}^2)\left(\frac{\partial^2 V}{\partial^2 Q}\right)_{Q=0}(t) + \ldots \tag{13.6}$$

An important ingredient of Oxtoby's work was the decomposition of the force on the normal coordinate $-\left(\frac{\partial V}{\partial Q}\right)$ in terms of the force on the atoms involved.

Oxtoby assumed that the forces acting on the different atoms of the diatomic were uncorrelated and that the area of contact of each atom with the solvent was a half-sphere. He then derived the following expression for the frequency time correlation function [23a,120–122]

$$\langle\Delta\omega_{mn}(t)\Delta\omega(0)\rangle = \frac{n^2}{2}\sum_i\left[\frac{3(-K_{111})l_{ik}}{\omega_0^3 m_i^{1/2}} + \frac{l_{ik}^2}{2\omega_0 L m_i}\right]^2 \langle\mathbf{F_i(t)F_i(0)}\rangle \tag{13.7}$$

where L is the characteristic potential range; $\langle\mathbf{F_i(t)F_i(0)}\rangle$ represents the force–force correlation function on the atom i moving along the direction of vibration; m_i is the mass of the ith atom; and l_{ik} is related to a characteristic vector $\mathbf{l_{ik}}$ along the normal mode Q_k, as $\mathbf{l_{ik}} = l_{ik}u_{ik}$. For a diatomic molecule, $\mathbf{l_{ik}} = \sqrt{(m_i/\mu)\gamma_i}$, where $\gamma_i = m_i/(m_i + m_j)$ and μ represents the reduced mass of the system.

The inharmonicity parameter K_{111} was obtained as follows. First, it was assumed the vibrational bond energy was given by the Morse potential in the following form

$$V(r) = D_e\{1 - \exp[-B(r - r_e)]\}^2 \tag{13.8}$$

where B and D_e are the Morse potential fit parameters and r_e is the equilibrium bond length. Then a Maclaurin series expansion of the potential about the equilibrium position of the vibration and a term by term comparison with Eq. (13.5) produced the expression for K_{111}.

Eq. (13.7) is the expression used in many studies of vibrational dephasing and in computer simulations. Note that does not include the vibrational–rotational contribution to dephasing and the resonant energy transfer between different molecules. These are somewhat difficult to model theoretically but can always be included in a simulation work, as demonstrated by Oxtoby and co-workers [120].

The presence of n^2 in Eq. (13.7) is why the vibrational dephasing rate is usually assumed to exhibit the quadratic quantum number dependence for overtones and hot bands. Although $\langle \Delta\omega_{mn}(t)\Delta\omega(0)\rangle$ may have an n^2 dependence, the average dephasing rate $\langle \tau_v \rangle^{-1}$ can show subquadratic dependence when $\langle Q(t)Q(0)\rangle$ follows a Gaussian decay at short times, with the form

$$\langle Q(t)Q(0)\rangle = \exp[-n^2 t^2/\tau^2] \tag{13.9}$$

The decay of the vibrational correlation function in dense liquids, however, is expected to be more complex, since the frictional response of dense liquid is strongly biphasic. Therefore, a switchover from quadratic to subquadratic quantum number dependence will be largely controlled by the respective amplitude of each of the components (Gaussian and exponential) of the bimodal frictional response.

Recently Gayathri and co-workers [23a] performed a detailed study of overtone dephasing of C–H stretching in CH_3I, in which they used the mode-coupling theory to calculate the friction on the normal coordinate. The picture that emerged from their theoretical study is as follows.

1. The subquadratic quantum number dependence owing to the Gaussian decay of the force–force time correlation function can occur only when the time scale of decay of the frequency time correlation function and the normal coordinate time correlation function are comparable and these two functions overlap.

2. This overlap is possible only when the harmonic frequency is not too large, the inharmonicity is significant, and the mean-square fluctuation $\langle \Delta\omega^2 \rangle$ is large.

Next, we briefly discuss the work of Gayathri and co-workers [23a]. In this calculation, the FFTCF was obtained by using the mode-coupling theory [26,26a,26b,249–252].

C. Mode-Coupling Theory Calculation of the Force–Force Time Correlation Function

As discussed in Section XII, the MCT provides an accurate description of both the short- and long-time dynamics of the liquid. In the MCT, the separation of time between the binary collision and the repeated recollisions is used to decompose the time-dependent friction into short-time and long-time parts. The resulting expressions were given by Eqs. (12.3) and (12.4). Another advantage of the mode-coupling theory is that it provides an accurate estimate of the friction at time $t = 0$. This is important in the present problem.

The results of the MCTanalysis were as follows. It was found that the pronounced Gaussian decay of the FFTCF made the decay of the normal coordinate time correlation function for the overtone dephasing (involving quantum nunber 0 and n) also partly Gaussian. This Gaussian nature of the normal coordinate time correlation function becomes pronounced as the quantum number n increases. This leads to a pronounced subquadratic dependence of the rate of overtone dephasing on quantum number n.

The time-dependent friction profiles [$\langle F(t)F(0)\rangle$ vs. t plots] that were obtained by Gayathri and co-workers [23a] using the MCT for CH_3 and I are shown in Figures 46 and 47, respectively. In both cases, the friction on the atom shows a strong bimodal response in the $\langle F(t)F(0)\rangle$ profile; Gaussian behavior in the initial time scale followed by a slowly relaxing component. There is even a rise in friction in the intermediate time scale. This arises from the coupling of the solute motion to the collective density relaxation of the solvent [249-252]. As mentioned, the Gaussian component arises from the binary collisions, and the slower part arises from correlated recollisions. However, the frictions on CH_3 and I were found to be quite different; it was much higher for I than for CH_3. This is expected because, although the LJ diameters of these spheres are nearly equal, their individual masses are considerably different ($CH_3 = 15$ g/mol and $I = 126$ g/mol). Thus the positional coordinate Q corresponding to the equilibrium point is much closer to the iodine atom. As a result, the iodine atom is a lot more static than CH_3. The large value of the friction for I also arises from its large ε value compared to the small value for CH_3, because ε is also a measure of the force acting on the molecules. Therefore, CH_3 is more involved in collisions and is freer to move, since it is smaller and lighter; so the friction on it is substantially reduced. Furthermore, the contribution of the heavy atom gets reduced because of the presence of the mass term in the denomi-

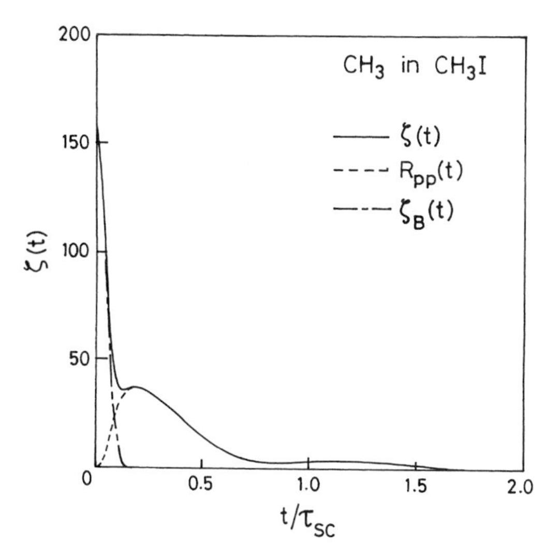

Figure 46. The calculated frictions plotted as a function of time for the CH_3 system at the reduced temperature $T^*(= k_B T/\varepsilon) = 1.158$ and the density of CH_3I, $\rho^* = 0.91$. The time-dependent friction is scaled by $m_i \tau_{sc}^{-2}$, where $\tau_{sc} = [m_i \sigma_j^2 / k_B T]^{\frac{1}{2}} \simeq 1.1$ ps. The plot shows a strong Gaussian component in the initial time scale for the binary part $\zeta_D(t)$ and slower damped oscilatory behavior for the $R_{\rho\rho}(t)$ part.

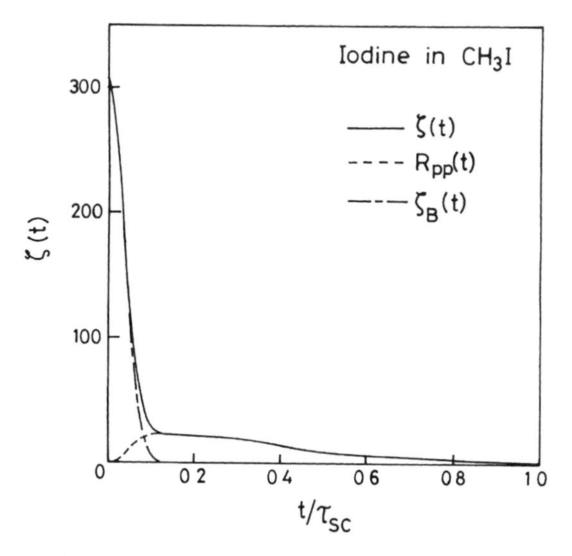

Figure 47. A similar plot as shown in Figure 46 for the I system at the reduced temperature $T^*(= k_B T/\varepsilon) = 0.363$ and the density of CH_3I, $\rho^* = 0.91$. The time-dependent friction is scaled by $m_i \tau_{sc}^{-2}$, where $\tau_{sc} = [m_i \sigma_j^2 / k_B T]^{\frac{1}{2}} \simeq 0.1$ ps. The friction is much higher for I than for CH_3.

nator of prefactor term corresponding to each atom in Eq. (13.7). Thus the effective friction on the vibrational coordinate becomes less than the earlier estimates [120–122] of the same. These authors [23a] also studied the dephasing of C–H stretching in $CHCl_3$ and found the same trend as observed for CH_3I.

D. Subquadratic Quantum Number Dependence of Overtone Dephasing

The study of Gayathri and co-workers [23a] revealed that the overtone dephasing rates were substantially subquadratic (close to $3n$ in the case of CH_3I and $1.5n$ in the case of $CHCl_3$) toward the higher quantum levels. The results were in good qualitative agreement with the experimental observations reported for the C–I stretching of CH_3I in hexane [295] and C–D stretching of CD_3I [23] (Table XIII). In particular, the dephasing times obtained for the 1 level of the C–I stretching mode of CH_3I and the C–H mode in $CHCl_3$ were about 2.6 and 1.4 ps, respectively and were close to the experimentally reported value of 2.3 ps for the C–I stretching in CH_3I [23a] and in the reported range of 1.1–1.38 ps for the C–H stretching in $CHCl_3$ [23a]. In view of the approximations involved in the modeling, this agreement might be fortuitous. However, one can at least believe that the theoretical approach undertakes by Gayathri and co-workers [23a] can reproduce the experimental results semiquantitatively.

The subquadratic n dependence clearly arises from the nonexponential component of $\langle Q(t)Q(0)\rangle$ in the initial time scale which increases with increase in the quantum number n, which strongly reflects the presence of

TABLE XIII
Theoretically Obtained Vibrational Dephasing Times for neat CH_3I and neat $CHCl_3$ as a Function of the Quantum Number n [a]

	CH_3–I in CH_3I, ps		C–H in $CHCl_3$, ps	
n	$\tau_{v,n}$ [b]	$\tau_{v,1}/\tau_{v,n}$	$\tau_{v,n}$ [b]	$\tau_{v,1}/\tau_{v,n}$
1	2.6265	1.0000	1.4050	1.0000
2	0.9224	2.8474	1.0245	1.3713
3	0.4983	5.2710	0.6746	2.0828
4	0.3312	7.9304	0.4477	3.1384
5	0.2439	10.7678	0.3174	4.4271
6	0.1908	13.7667	0.2403	5.8458
7	0.1553	16.9101	0.1910	7.3559
8	0.1302	20.1754	0.1570	8.9478
9	0.1116	23.5378	0.1323	10.6178
10	0.0974	26.9745	0.1136	12.3631

[a] The results show a strong linear dependence on n in the higher quantum number range.
[b] The dephasing time for the nth level.

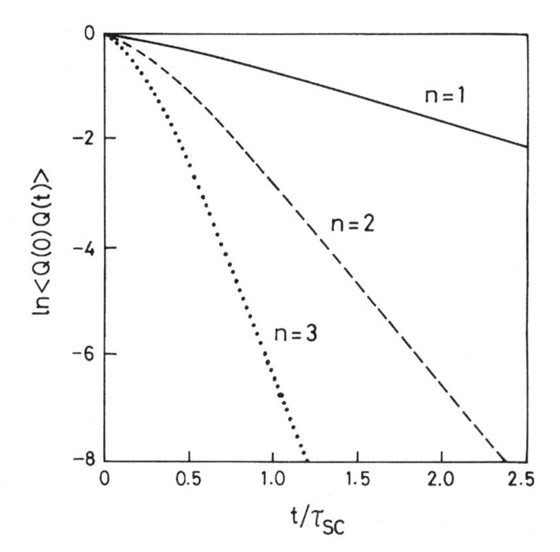

Figure 48. Theoretically obtained plots of ln $\langle Q(t)Q(0)\rangle$ versus t (where t is scaled by 1.1 ps) for the first three quantum levels. The results show an increasing Gaussian behavior in the short-time scale with increasing quantum number n.

the Gaussian components of binary friction (Fig. 48). This is responsible for the nearly linear n dependence in the higher levels because of dominant Gaussian behavior.

E. Vibrational Phase Relaxation Near the Gas–Liquid Critical Point

There is considerable interest in understanding the dephasing process near a phase transition, particularly near the gas–liquid critical point. Early experimental studies of Clouter and Keifte [296] on N_2 and O_2 showed interesting variations. Although the change in the dephasing rate was small when the thermodynamic conditions were changed from near the melting point to the boiling point, a large increase was observed as the critical point was approached.

XIV. LIMITING IONIC CONDUCTIVITY IN ELECTROLYTE SOLUTIONS: A MOLECULAR THEORY

What determines the ionic conductivity of an electrolyte solution has remained a problem of great interest to chemists for more than a century [127–132] Such long-standing interest stems not only from its relevance in many chemical and biological applications but also from the many fascinating, often anomalous, behavior that ionic conductivity exhibits in a large number of solvents. Most often discussed of these properties are the concentration and the nonmonotonic ion size dependencies. Even after a century-old debate and discussion, neither of these problems has been satisfactorily resolved. The mobility of an ion in a polar solvent is determined by its complex interactions with the surrounding polar molecules; these interactions are long ranged and anisotropic. In addition, the dynamics of polar liquids have been poorly understood until recently.

There were several significant developments in the recent past in understanding the dynamics of dense liquids that make the study of this fundamental problem quite interesting. Perhaps the most important development is the discovery that the polar solvation dynamics in most common solvents are strongly biphasic with an initial ultrafast component that is in the femtosecond regime and that often contributes 60–80 % to the total energy relaxation [27,27a,29,65,66,243]. The discovery of this ultrafast component raises several interesting questions. For example, what is the role of this component in determining the mobility of the ions? Both the solvation dynamics of an ion and the dielectric friction on it are expected to be intimately related. The second notable development is in the microscopic understanding of the relation between diffusion and viscosity in dense liquids [26,26a,26b,174,175,249–252]. In the study of ionic conductivity, one usually assumes that diffusion of ions is related to the solvent viscosity by Stokes's law, which is unsatisfactory for small ions, such as Li^+ and Na^+. Recent theoretical developments [26,26a,26b,174,175,249–252] can now provide microscopic description of diffusion of small solutes in dense liquids. The starting point of most discussions on ionic conductivity is Kohlrausch's law which is expressed as [127–130]

$$\Lambda_m = \Lambda_0 - \kappa\sqrt{c} \tag{14.1}$$

where Λ_m represents the equivalent molar conductivity and Λ_0 its limiting value at infinite dilution. The latter can be determined by applying Walden's rule, which states that for a particular ion, the product of Λ_0 with solvent viscosity η_0 should be constant [127–130]:

$$\Lambda_0\eta_0 = \text{constant} \tag{14.2}$$

Even though approximate, Eqs. (14.1) and (14.2) are the two most important statements on ionic conductivity of an electrolyte solution. The first one has been explained in terms of the Debye–Huckel–Onsager theory [297], which also provides an expression for the prefactor κ, which depends on, among other things, the limiting ionic conductivity Λ_0. Eq. (14.1) has been confirmed for low concentration. Eq. (14.2) can be rationalized in terms of the well-known Stokes's law [127–130], which predicts that the friction on the ion is proportional directly to the viscosity η_0 and inversely to the crystallographic radius of the ion r_{ion}. The use of Einstein's relation between the friction and the diffusion coefficient (which is essentially Λ_0) then produces Eq. (14.2). Experimental results [127–131], however, indicate that the ionic mobilities in polar solvents do not always decrease monotonically with increasing radius. Instead, there is often a maximum as $\Lambda_0\eta_0$ is plotted against r_{ion}^{-1}, as shown in Figure 2. In fact, the breakdown of Walden's rule (and of Stokes's law) has been observed for all the solvents studied and can be regarded as universal.

What makes the experimental results deviate so strongly from Walden's rule? Two completely different explanations have been put forward. The first and the oldest one is the solvent-berg model [127–131]. The phenomenological solvent-berg model assumes the formation of a rigid solvent cage around a small ion, which leads to an increase of the effective radius of the ion. This, in turn, reduces the conductivity of the ion. In addition, the maximum in Λ_0 near Cs^+ is explained in terms of the orientational structure breaking of the solvent by the ion [131]. However, this approach completely fails to provide a coherent quantitative description of the limiting ionic conductivity.

The second, more successful, model is based on a continuum description of the solvent [298–303]. Because of the long-range nature of the ion–dipole interaction potential, it was originally believed that this interaction could be replaced by the interaction of the ion with a continuum solvent, and the molecularity of interaction might not be important. This model was originally introduced by Born, who modified the usual Stokes–Einstein hydrodynamic model of diffusion by coupling the ionic field of the solute with the bulk polarization mode of the solvent [298]. According to his picture, the ionic motion disturbs the equilibrium polarization of the solvent and the relaxation of the ensuing nonequilibrium polarization dissipates energy, thereby enhancing the friction on the ion. He coined the term dielectric friction to describe this extra dissipative mechanism and expressed the total friction (ζ_{total}) experienced by the ion moving through the viscous continuum as follows:

$$\zeta_{\text{total}} = \zeta_0 + \zeta_{\text{DF}} \tag{14.3}$$

where ζ_{DF} is the dielectric friction. ζ_{bare} is the friction arising from Stokes's law owing to the zero frequency shear viscosity η_0 of the solvent. This model was further developed by Boyd [299] and Zwanzig [300]. The final expression (by Zwanzig) for dielectric friction leads to an overestimation of friction for small ions. In an attempt to rectify this lacuna, Hubbard and Onsager (H–O) studied the ionic mobility problem in great detail [301–302] within the framework of the continuum picture. They proposed a theory that can be regarded (in the language of Wolynes) [304] as the "ultimate achievement in a purely continuum theory of ionic mobility."

It is clear from Figure 2 that, although the simple treatment of Zwanzig [300] can explain the observed nonmonotonic dependence of Λ_0 on r_{ion}^{-1}, it fails to reproduce the experimentally observed ionic mobilities, because it overestimates the dielectric friction. The Hubbard–Onsager theory [301] is satisfactory up to intermediate-sized ions but fails to describe the sharp decrease for small ions.

Clearly, the above continuum model-based theories fail to describe the ion transport in polar solvents. There are many reasons for this failure, which has been extensively discussed in the literature [303–307]. The most important is the representation of the real solvent by a viscous dielectric continuum. No molecularity of the solvent was considered. In addition, the description of solvent dynamics was vastly inadequate.

The microscopic theory presented here is based on a simple physical picture. Consider a tagged, singly charged ion in a dipolar liquid. For spherical solute ions, the interaction between the ion and the dipolar liquid molecules can be separated into two parts [304–307]. The first part originates from a short-range, spherically symmetric potential that is primarily repulsive. This gives rise to a friction that can be described (with certain limitations) by Stokes's law. This is nonpolar in nature and has been referred to as the bare friction ζ_0. The second part originates from the long range ion–dipole interaction and is referred to as the dielectric friction ζ_{DF}. The latter is dominated by the long wavelength solvent polarization fluctuations. Here it is particularly important to note that these long wavelength polarization fluctuations are the ones primarily responsible [159–164] for the ultrafast Gaussian solvation dynamics observed in experiments. As the size of the ion decreases, ζ_0 decreases but ζ_{DF} increases rapidly. The diffusion coefficient of the ion is given by $D_T^{\mathrm{ion}} = k_B T/\zeta$, where $\zeta = \zeta_0 + \zeta_{\mathrm{DF}}$. It is, therefore, the dielectric friction part that is responsible for the observed anomalous ion mobilities in the dipolar solvents.

A yet ill-understood problem of ion–solvent dynamics is the correlation between the two rather different pictures: namely, the dielectric friction and the solvent-berg models. For small ions in slow liquids, the solvent-berg is expected to provide a realistic picture [304–306]. In this section, we show that a self-consistent treatment can indeed be developed to describe both limits. As noted, the theory reveals a dynamical cooperativity between the ions and solvent's motion, mediated through a nonlinear coupling.

Although only the zero frequency dielectric friction, $\zeta_{\mathrm{DF}}(\omega \to 0)$ is required for finding the limiting ionic conductivity in solution, the frequency or time-dependent dielectric friction is often required in theoretical studies of other problems. For example, in the study of intramolecular proton

(H^+) transfer reaction [308] and in vibrational relaxation [119–122] in dipolar liquids, the frequency dependence of dielectric friction plays a crucial role. In this section, an explicit calculation of $\zeta_{DF}(t)$ is presented.

The present theoretical study gives several interesting new results and provides insight into the problem of ion–solvent dynamics. We find that the ultrafast solvents modes are indeed important in determining the ionic mobility in dipolar solvents. An important aspect of the present theory is the recovery of a dynamic version of the classical solvent-berg model from microscopic considerations. The theory naturally gives rise to a smaller translational diffusion of the solvent molecules in first solvation shell via a nonlinear coupling between the ionic field and the solvent translational modes. It is found that $\zeta_{DF}(t)$ exhibits bimodal dynamics. The relation between the solvation dynamics of an ion and its conductivity is also clarified below.

The organization of the rest of this section is as follows. Next we discuss the theoretical formulation, followed by the calculational details. The relation between the solvation dynamics of an ion and its mobility is clarified in Section XIV.C. The quenching of the solvent translational modes owing to ionic field is presented as well.

A. Theoretical Formulation

The calculation of the diffusion coefficient or the friction on a molecule in a dense liquid is a difficult problem, even for a model liquid consisting of spheres that interact by a simple Lennard–Jones potential [174,175,249–252]. In the present case, in which the ions interact with dipolar molecules via short- and long-range complex polar interactions, the calculation of the friction is indeed highly nontrivial. Thus progress can be made only by making simplifying assumptions, without, however, sacrificing the essential aspects of space and time dependence of the relevant two-point space–time correlation functions. Fortunately, we shall deal only with rigid positive ions here, which makes the problem somewhat tractable. Below, we describe the general microscopic formulation of the problem, both of the short-range local friction ζ_0 and of the dielectric friction ζ_{DF} from the ion–dipole interaction. The total friction ζ is the sum of ζ_0 and ζ_{DF}. The translational diffusion coefficient of the ion D_T^{ion} is obtained from the total friction by using the well-known Einstein relation $D_T^{ion} = k_B T/\zeta$. The limiting ionic conductivity is obtained from the diffusion coefficient by use of the Nernst–Einstein relation [127–130]. As will be made clear in the subsequent discussions, the present formulation is close to the mode-coupling theoretic formulation, which is currently popular in the theory of glassy liquids [26,26a,26b].

Let us consider a dilute solution of strong uniunivalent electrolyte in a dipolar solvent. We shall approximate the solvent molecules by dipolar spheres with a point dipole at the origin. The interaction potential between the ions and the solvent molecules is assumed to consist of a Lennard–Jones potential and an ion–dipole interaction term, given by the following expression:

$$V_{\text{ion-dipole}}(\mathbf{r}, \mathbf{\Omega}) = U_{\text{LJ}}(\mathbf{r}) + U_{\text{id}}(\mathbf{r}, \mathbf{\Omega}) \tag{14.4}$$

where $\mathbf{r}$ and $\mathbf{\Omega}$ are the vector distance between the ion and the dipolar molecule and the orientation of the dipole in the space frame, respectively. $U_{\text{LJ}}(r)$ is the space r dependent Lennard–Jones interaction potential, which is given as follows [174]:

$$U_{\text{LJ}}(r) = 4\varepsilon\left[\left(\frac{\sigma_{12}}{r}\right)^{12} - \left(\frac{\sigma_{12}}{r}\right)^{6}\right] \tag{14.5}$$

where σ_{12} is the distance between the closest approach of the two species in question and ε is the well depth. The ion–dipole interaction term $U_{\text{id}}(\mathbf{r}, \mathbf{\Omega})$ is expressed as follows [206]:

$$\begin{aligned} U_{\text{id}}(\mathbf{r}, \mathbf{\Omega}) &= \infty, \; r < r_{\text{id}} \\ U_{\text{id}}(\mathbf{r}, \mathbf{\Omega}) &= -\frac{z_i Q \mu(\Omega).\hat{r}}{r^2}, \; r > r_{\text{id}} \end{aligned} \tag{14.6}$$

where r_{id} is the distance of the closest approach between the ion and a solvent molecule, Q is the protonic (or electronic) charge, z_i is the valency on the ion, and $\mu(\Omega)$ is the orientation-dependent dipole moment. The interaction among the solvent molecules is also assumed to consist of two terms—an LJ and a dipole–dipole interaction term:

$$V_{\text{dipole-dipole}}(\mathbf{r}, \mathbf{\Omega}) = U_{\text{LJ}}(\mathbf{r}) + U_{\text{dd}}(\mathbf{r}, \mathbf{\Omega}) \tag{14.7}$$

where the dipole–dipole interaction term U_{dd} can be given as follows [201]:

$$\begin{aligned} U_{\text{dd}}(\mathbf{r}, \mathbf{\Omega}) &= \infty, \; r < r_{\text{dd}} \\ U_{\text{dd}}(\mathbf{r}, \mathbf{\Omega}) &= -\frac{\mu^2 D_{12}}{r^3}, \; r > r_{\text{dd}} \end{aligned} \tag{14.8}$$

where $D_{12} \equiv \hat{\mu}(\Omega_1).(3\hat{r}\hat{r} - \mathbf{I}).\hat{\mu}(\Omega_2)$, with $\mathbf{I}$ being a 3×3 unit tensor, and $\hat{\mu}(\Omega)$ is a unit vector in the direction of the variable Ω.

On rather general ground, the friction on a tagged particle can be expressed as a time integral over the force–force time autocorrelation function $C_{\text{FF}}(t)$; the force is the random force acting on the tagged particle at any time. The FFTCF naturally exhibits complex dynamics and is highly non-trivial to calculate from first principles. What makes things somewhat simpler is the separation of time scales that is naturally present in a dense liquid. The initial short time decay of $C_{\mathrm{FF}}(t)$ is dominated by short-range collisional contribution. This is essentially two particles in nature and can be accurately calculated. The long-time part is rather complex, especially for the present problem in which we need to consider not only the effects from the Lennard–Jones-type interactions but also from the long-range ion–dipole interactions. The main point to note here is that this decomposition of the FFTCF is based only on the separation of time scales and is rather general. This decomposition of the total $C_{\mathrm{FF}}(t)$ is shown in Figure 45.

It was pointed out by Wolynes [304–306] that the neglect of the cross-terms $\langle F^H(0).F^S(t)\rangle$ and $\langle F^S(0).F^H(t)\rangle$ in calculating the total $C_{\mathrm{FF}}(t)$ means the neglect of certain hydrodynamic interactions, such as the effect of the flow field around the moving ion on the dynamic response of the solvent. The computer simulation studies of Berkowitz and Wan [309] indicated that these cross-correlations can be important. If it is assumed that the soft force consists only of ion–dipole interactions, then in the present theory the cross-correlations should enter through terms such as $\langle a_{00}(-\mathbf{k}).a_{10}(\mathbf{k})\rangle$. It has been shown elsewhere [307] that these terms identically become zero within a linearized equilibrium theory (such as MSA, LHNC) of dipolar liquids; however, these terms can be nonzero in a nonlinear theory. On the other hand, if the soft force contains a radially symmetric attractive term, then these cross-correlations can be rather important.

Because we are interested in the friction acting on a tagged ion in the limit of infinitely dilute solution of strong electrolytes only, we need to incorporate the distorted structure of the solvent around the ion. This distortion again has two aspects. The first is the spatial (angle-independent) distortion owing to the combined effect of the size and charge of the ion. In a dense liquid in which the spatial structure is largely determined by the harsh repulsive part of the intermolecular potential, this local spatial distortion is determined by the relative sizes of the ion and the solvent molecules. The second aspect is the distortion of the orientational correlation between the dipolar molecules owing to the ion and the existence of the nontrivial static and dynamic correlations between the ion and the dipolar solvent molecules. The latter plays a crucial role in determining dielectric friction. For small ions such as Li^+, dielectric friction dominates the total friction. Thus the friction on an ion is determined by a host of complicated factors. The viscosity, on the other hand, is a purely solvent property. We next describe the calculation procedure of the local and dielectric frictions.

1. *Calculation of the Local Friction*

The local friction acting on a moving ion can be calculated in the following way. First, we assume that the charge on the solute ion is switched off completely. Then, following the standard prescription of renormalized kinetic theory [26a], the total friction acting on this uncharged solute can be expressed as a sum of three contributions coming from well-separated time scales. The first contribution comes from the short-range repulsive interaction, and this is essentially collisional in nature. This is termed as Γ_B, since it originates from the binary collision. The other part of the total friction originates from the coupling of the solute's motion to the density fluctuation of the solvent. The friction that comes from the solvent density fluctuation is denoted as $\Gamma_{\rho\rho}$. The decoupling of Γ_B from $\Gamma_{\rho\rho}$ is clearly an approximation based on the different time scales associated with these dynamic processes in dense liquids. Γ_B is determined by the static, local correlations present in a dense liquid, and there exists a well-defined expression for this, which is somewhat complex. Because in this section we are interested only in the zero frequency friction and because Γ_B is usually rather small compared to other contributions, we shall approximate Γ_B by the Enskog friction, which is given by the following well-known expression [174,175]:

$$\Gamma_E = \frac{8}{3m}\sqrt{2\pi s k_B T}[\rho\sigma_{12}^2 g_{12}(\sigma_{12})] \tag{14.9}$$

where s is the reduced mass of the solute–solvent composite system; $\sigma_{12} = \frac{\sigma_1+\sigma_2}{2}$, with σ_1 and σ_2 being the diameters of the solute and the solvent molecules, respectively; and $g_{12}(\sigma_{12})$ is the value for the radial distribution function at the contact. This is calculated using the pressure equation [285] and assuming the solute and the solvent molecules are hard spheres.

The second contribution to the total local friction, $R_{\rho\rho}$, comes from the solvent density fluctuation. This can be calculated by using the mode-coupling expression [174,175,249–252] given by Eq. (12.4). The other quantities necessary for the calculation of $R_{\rho\rho}(t)$ were discussed in Sections XII and XIII.

In our calculations, we assume that the total local friction is given by $\zeta_0 = \Gamma_B + R_{\rho\rho}$. Note that in all the earlier studies it was assumed that $\zeta_0 = 4\pi\eta r_{ion}$. This is clearly an approximation, since the validity of hydrodynamics becomes questionable for describing the dynamics of particles smaller than the size of the solvent molecule [304–306]. In a separate study, we carried out detailed calculations to check the relation between the friction and the viscosity, in which the viscosity is also calculated by using MCT expressions [26,26a,174,175,249–252]. Although we found that the Stokesian hydrodynamics breaks down for solutes of sizes comparable to the size of the solvent molecules, the hydrodynamic relation between friction and viscosity (with slip boundary condition) continues to hold with surprising accuracy. Figure 49 shows the calculated local friction plotted against the calculated viscosity. It can be seen that not only the curve is linear but the slope is also close to 4π. This is because both the friction and the viscosity are deter-

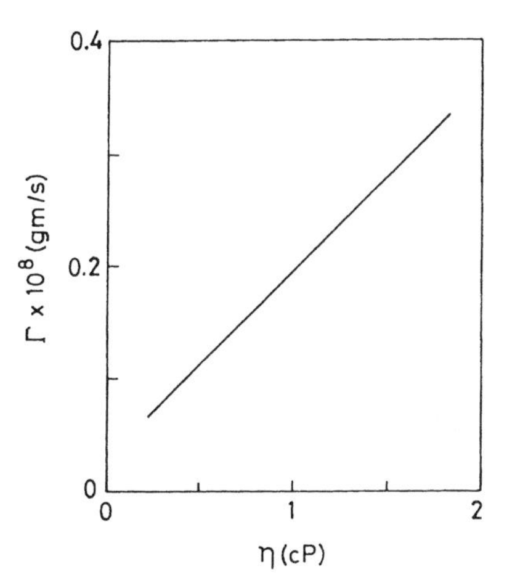

Figure 49. Calculated values of the local friction Γ are plotted against the calculated viscosity η. The calculations of both Γ and η are carried out by using the MCT [249]. Calculations are performed for the same solute–solvent size and mass ratios, at 120 K. The diameter and the mass of a solvent molecule were taken as 4 Å and 100 amu, respectively.

mined in a dense liquid by the local short-range intermolecular correlations. To avoid confusion, let us clarify: by the breakdown of Stokesian hydrodynamics, we mean that the contribution to the friction comes not from the current–current correlation function but rather from Γ_B and $R_{\rho\rho}$. As shown in Refs. [249–252], the contribution of the current term is negligibly small for all solute sizes comparable to the solvent molecules. Thus we shall approximate the local friction by the hydrodynamic relation $\zeta_0 = 4\pi\eta r_{\text{ion}}$. Not only is this numerically accurate but it is robust, because for small ions the total friction is overwhelmingly dominated by the dielectric friction [307,310–312].

2. *Calculation of the Dielectric Friction*

In this section, we derive a microscopic expression of the dielectric friction ζ_{DF}. The theoretical formulation presented here is based on the microscopic treatment of dielectric friction proposed recently by Bagchi [307]. The advantage of this treatment is that the resulting theory is nearly analytical and transparent. By construction, the dielectric friction arises from the force due to the electrical part of ion–dipole interaction only. Therefore, we start with the well-known Kirkwood formula for dielectric friction [205]:

$$\zeta_{\text{DF}} = \frac{1}{3k_B T}\int_0^\infty dt\langle \mathbf{F}_{\mathbf{id}}.(0)\mathbf{F}_{\mathbf{id}}(t)\rangle \tag{14.10}$$

where $\mathbf{F}_{\mathrm{id}}(t)$ is the force acting on the ion as a result of the ion–dipole interaction only and the angle brackets, stand for the ensemble averaging. The total friction is obtained by adding this friction to the bare friction. As already discussed, such a separation of friction is reasonable, as the interaction potential can be separated into a short-range orientation independent part (assumed here to be given by the Lennard–Jones 6–12 potential) and a long-range soft part (coming from electrical part of ion–dipole interaction). For small ions, ζ_0 is much smaller than ζ_{DF}; therefore, the error made in such a separation is likely to be not significant [306].

The force that is responsible for dielectric friction is long range in nature because it originates from the ion–dipole interaction. Therefore, this interaction can be described by a time-dependent mean field (TDMF) theory, and the force can be obtained by using a Ginzburg–Landau-type free energy functional [143]. Because the liquids we consider are all at high density, this free energy can be given by the well-known density functional theory [15,15a]. This approach has been enormously successful in recent years for calculating the transport properties of strongly correlated random systems, such as dense liquids [26]. The density functional theory (DFT) gives the following expression for the free energy of the solute–solvent system [15,15a]

$$
\begin{aligned}
\beta\mathcal{F}[n_{\mathrm{ion}}(\mathbf{r}), \rho(\mathbf{r},\mathbf{\Omega})] = & \int d\mathbf{r}\, d\mathbf{\Omega}\, \rho(\mathbf{r},\mathbf{\Omega})\left[\ln\frac{\rho(\mathbf{r},\mathbf{\Omega})}{\rho^0/4\pi} - 1\right] \\
& + \sum_{\alpha,\beta}\int d\mathbf{r}\, n_\alpha(\mathbf{r})\, [\ln n_\alpha(\mathbf{r}) - 1] \\
& - \frac{1}{2}\sum_{\alpha,\beta}\int d\mathbf{r}\, d\mathbf{r}'\, c_{\mathrm{ii}}(\mathbf{r},\mathbf{r}')\, \delta n_\alpha(\mathbf{r})\, \delta n_\beta(\mathbf{r}') \\
& - \frac{1}{2}\sum_{\alpha}\int d\mathbf{r}\, d\mathbf{r}'\, c_{\mathrm{id}}(\mathbf{r},\mathbf{r}'_{\Omega})\, \delta n_\alpha(\mathbf{r})\, \delta\rho(\mathbf{r}',\mathbf{\Omega}) \\
& - \frac{1}{2}\int d\mathbf{r}\, d\mathbf{\Omega}\, d\mathbf{r}'\, d\mathbf{\Omega}'\, c_{\mathrm{dd}}(\mathbf{r},\mathbf{\Omega};\mathbf{r}',\mathbf{\Omega}')\, \delta\rho(\mathbf{r},\mathbf{\Omega})\, \delta\rho(\mathbf{r}',\mathbf{\Omega}') \\
& + \text{higher-order terms}
\end{aligned}
\tag{14.11}
$$

where $\delta\rho(\mathbf{r},\mathbf{\Omega}) = \rho(\mathbf{r},\mathbf{\Omega}) - \frac{\rho_o}{4\pi}$ represents the fluctuation in the solvent number density; $\delta n_\alpha(\mathbf{r}) = n_\alpha(\mathbf{r}) - n_\alpha^o$ denotes the fluctuation in the number density of ionic species α; ρ_o and n_α^o represent the average number density of the solvent and the ion, respectively; $c_{\mathrm{dd}}(\mathbf{r},\mathbf{\Omega};\mathbf{r}',\mathbf{\Omega}')$ is the two-particle direct correlation function between two solvent molecules at positions $\mathbf{r}$ and $\mathbf{r}'$ with orientations $\mathbf{\Omega}$ and $\mathbf{\Omega}'$, respectively; and c_{ii} and c_{id} are the ion–ion and the ion–dipole direct correlation functions, respectively. These direct correlation functions contain detailed microscopic information about the spatial and the orientational correlations present in the molecular liquid. In the present treatment, we ignore the effects of the higher-order density fluctuations.

Eq. (14.11) is then minimized (as performed in Section III) to obtain the following expression for the equilibrium density of the ionic species α:

$$n_\alpha^{eq}(\mathbf{r}) = n_\alpha^o \exp[-V_{\text{eff}}(\mathbf{r})/k_B T] \tag{14.12}$$

where the effective potential, $V_{\text{eff}}(\mathbf{r})$ on the ion at a position $\mathbf{r}$ derives contributions from the interaction of this with other ions and from its interaction with the surrounding solvent molecules. In the limit of infinite dilution, $V_{\text{eff}}(\mathbf{r})$ is given by

$$V_{\text{eff}}(\mathbf{r}) = -k_B \int d\mathbf{r}'\, d\mathbf{\Omega}'\, c_{id}(\mathbf{r}, \mathbf{r}', \mathbf{\Omega}')\, \delta\rho(\mathbf{r}', \mathbf{\Omega}') \tag{14.13}$$

Eq. (14.13) then leads to the following expression for the force acting on the ion:

$$\mathbf{F}(\mathbf{r}) = k_B \int d\mathbf{r}'\, d\mathbf{\Omega}'\, c_{id}(\mathbf{r}, \mathbf{r}', \mathbf{\Omega}')\, \delta\rho(\mathbf{r}', \mathbf{\Omega}') \tag{14.14}$$

Because the time dependence of the force acting on the ion arises from the time-dependent density fluctuation of the solvent, Eq. (14.14) can be generalized to obtain the following expression for the time-dependent force on the ion:

$$\mathbf{F}(\mathbf{r}, t) = k_B \int d\mathbf{r}'\, d\mathbf{\Omega}'\, c_{id}(\mathbf{r}, \mathbf{r}', \mathbf{\Omega}')\, \delta\rho(\mathbf{r}', \mathbf{\Omega}') \tag{14.15}$$

Eq. (14.15) describes the time-dependent force acting on a fixed ion; however, the translational motion of the ion can open up an extra decay channel for the force relaxation. This was observed for solvation dynamics, described in Section IV. When the self-motion of the ion is included, the DFT provides the following expression for the time-dependent force density (arising from the long-range, ion–dipole interaction) on the ion [310–313]:

$$\mathbf{F}(\mathbf{r}, t) = k_B T n_{\text{ion}}(\mathbf{r}, t) \nabla \int d\mathbf{r}'\, d\mathbf{\Omega}'\, c_{id}(\mathbf{r}, \mathbf{r}', \mathbf{\Omega}')\, \delta\rho(\mathbf{r}', \mathbf{\Omega}', t) \tag{14.16}$$

where $n_{\text{ion}}(\mathbf{r}, t)$ is the number density of the ion. Next, the density and the direct correlation function are expanded in the spherical harmonics. We then use the standard Gaussian decoupling approximation to obtain the following microscopic expression for the frequency-dependent dielectric friction

[303,310–312]:

$$\zeta_{\mathrm{DF}}(\mathbf{k}, z) = \frac{2k_{\mathrm{B}}T\rho_0}{3(2\pi)^2}\int_0^{\infty} dt\, e^{-zt}\int_0^{\infty} dq\, q^4 S_{\mathrm{ion}}(\mathbf{k}-q, t)|c_{\mathrm{id}}^{10}(q)|^2 S_{\mathrm{solv}}^{10}(q, t) \tag{14.17}$$

where $c_{\mathrm{id}}^{10}(q)$ and $S_{\mathrm{solv}}^{10}(q, t)$ are the longitudinal (i.e., 10) components of the ion–dipole DCF and the orientational dynamic structure factor of the pure solvent, respectively. In defining these correlation functions, the wavenumber is taken parallel to the z-axis, ρ_0 is the average number density of the solvent, $S_{\mathrm{ion}}(q, t)$ denotes the self-dynamic structure factor of the ion. The $\mathbf{k} = 0$ and $z = 0$ limit of Eq. (14.17) provide the macroscopic friction. Note that notation for dynamic structure factors here have been changed to $S_{\mathrm{solv}}^{10}(q, t)$, to keep the notation tractable and to separate the two calculations of local and dielectric frictions.

Eq. (14.17) is the expression used to calculate the magnitude of dielectric friction. This is a nonlinear, microscopic expression for dielectric friction. Note that this equation is nonlinear, as it involves $\zeta_{\mathrm{DF}}(z)$ on both sides. Thus it must be solved self-consistently. This kind of approach is well known in the existing literature of the MCT [26,26a,26b,174,175]. To obtain ζ_{DF} $[\equiv \zeta_{\mathrm{DF}}(z = 0)]$ from Eq. (14.17), we need to specify both $S_{\mathrm{solv}}^{10}(q, t)$ and $S_{\mathrm{ion}}(q, t)$; the latter is given by Eq. (4.16).

The orientational dynamic structure factor of the solvent is given by Eq. (4.17). As discussed earlier the most important quantities in the dynamic solvent structure factor are the wavenumber- and frequency-dependent rate of the orientational solvent polarization relaxation $\Sigma_{10}(q, z)$. The calculation of the latter involves the calculations of the rotational memory kernel, the translational memory kernel, and the static orientational structure factor.

B. Calculation Procedure

In this section, we describe the calculations of the wavenumber- and frequency-dependent generalized rate of the orientational solvent polarization relaxation $\Sigma_{10}(q, z)$ and the necessary orientational static pair correlation functions. This discussion is brief because the details were described earlier.

1. Calculation of the Wavenumber- and Frequency-Dependent Generalized Rate of Solvent Polarization Relaxation

As discussed, the calculation of the generalized rate of polarization relaxation $\Sigma_{10}(q, z)$ is a nontrivial excercise. It contains two friction kernels: the rotational kernel $\Gamma_{\mathrm{R}}(q, z)$ and the translational kernel $\Gamma_{\mathrm{T}}(q, z)$. $\Gamma_{\mathrm{R}}(q, z)$ is calculated from $\varepsilon(z)$ by using Eq. (2.12). The calculation of $\Gamma_{\mathrm{T}}(q, z)$ is new and was discussed in greater detail [311,312].

a. Solvent Translational Friction. We next describe the calculation of the solvent translational frictional kernel $\Gamma_T(q, z)$. Note first that the solvent translation is important in determining the dielectric friction, because it accelerates the relaxation of $S^{10}_{solv}(q, z)$ at intermediate to large wavenumbers $(q\sigma \geq 2\pi)$; this should be kept in mind in the subsequent discussion. The translational kernel of the bulk solvent molecules $\Gamma^0_T(q, z)$ can be easily obtained from the translational dynamic structure factor of the solvent [166] $S(q, z)$. The latter is assumed to be given by the following well-known expression [166]:

$$S(q, z) = \frac{S(q)}{z + D^{solv}_T q^2/S(q)}, \tag{14.18}$$

where D^{solv}_T is the translational diffusion coefficient of a solvent molecule. For a solvent molecule in the bulk, this can be calculated from the bulk translational friction using the Einstein relation $D^{solv}_T(\text{bulk}) = k_B T/\Gamma^0_T(k = 0, z = 0)$. The translational bulk friction on a solvent molecule, $\Gamma^0_T(q = 0, z = 0)$ may then obtained from Stokes's relation, which connects the solvent viscosity η_0 with the friction as follows: $\Gamma^0_T(q = 0, z = 0) = 2\pi\eta_0\sigma$. Note that Eq. (14.18) is reliable at intermediate wavenumbers. As mentioned earlier, this is not a restriction in the present case, because translation is important at these wavenumbers only.

The translational kernel that enters into the present description is rather different from that of the bulk. As is widely believed and discussed at length by Wolynes, a solvent cage will form around a small ion in slow solvents, because the solvent molecules in the first shell are made less mobile by the strong electric field of the ion. The equilibrium aspect of this correlation is taken into account, at least partly, through the c_{id} term—the treatment of the dynamical inhomogeneity induced by the ion is more difficult. Note that the Γ_R is also affected; but this effect is more important for Γ_T, because the latter is relevant for the nearest-neighbor molecules only. Thus, for small ions, the solvent translational frictional kernel cannot be approximated by its bare value $\Gamma^0_T(q = 0, z = 0)$.

The increase of the translational friction on the neighboring solvent molecules can be quantified in the following fashion. The force on a tagged solvent molecule is written as

$$F^{sol} = F^{solv} + F^{ion} \tag{14.19}$$

where F^{solv} is the usual force on a solvent molecule owing to its interactions with the other solvent molecules and F^{ion} is the force owing to the presence of the ion at a short distance. We again use the Kirkwood formula to obtain the total translational friction on a solvent molecule:

$$\Gamma_{\mathrm{T}} = \Gamma_{\mathrm{T}}^{\mathrm{solv}} + \Gamma_{\mathrm{T}}^{\mathrm{ion}} \tag{14.20}$$

where $\Gamma_{\mathrm{T}}^{\mathrm{solv}}$ is the friction on the tagged solvent molecule due to the surrounding solvent molecules and $\Gamma_{\mathrm{T}}^{\mathrm{ion}}$ is that due to the presence of the ion. Here we have assumed that F_{solv} and F_{ion} are uncorrelated, an approximation not expected to be reliable. Furthermore, we shall approximate $\Gamma_{\mathrm{T}}^{\mathrm{solv}}$ by Γ_{T}^{0}, defined in Eq. (14.18). In the following discussion we describe the calculation of $\Gamma_{\mathrm{T}}^{\mathrm{ion}}$.

We again use the time-dependent density functional theory to obtain the force F^{ion} on a tagged solvent molecule at a position $\mathbf{r}$ with orientation $\boldsymbol{\Omega}$, which is given by

$$\mathbf{F}^{\mathrm{ion}}(\mathbf{r}, \boldsymbol{\Omega}, t) = k_{\mathrm{B}}T\, \delta\rho_{\mathrm{solv}}(\mathbf{r}, \boldsymbol{\Omega}, t)\nabla \int d\mathbf{r}'\, c_{\mathrm{di}}(\mathbf{r}, \mathbf{r}', \boldsymbol{\Omega}) n_{\mathrm{ion}}(\mathbf{r}', t) \tag{14.21}$$

where $c_{\mathrm{di}}(\mathbf{r}, \mathbf{r}', \boldsymbol{\Omega})$ is the dipole–ion DCF between the solvent molecule at $\mathbf{r}$ and the ion at $\mathbf{r}'$. We assume $c_{\mathrm{di}} \equiv c_{\mathrm{id}}$. Note here that the integration is taking care of the different position vectors $\mathbf{r}'$ of the ion moving from place to place. Straightforward, but lengthy, algebra leads to the following expression for $\Gamma_{\mathrm{T}}^{\mathrm{ion}}$

$$\Gamma_{\mathrm{T}}^{\mathrm{ion}} = \frac{k_{\mathrm{B}}T}{12\pi V} \int_0^{\infty} dt\, e^{-zt} \int_0^{\infty} dq\, q^4 |c_{\mathrm{id}}^{10}(q)|^2 S_{\mathrm{ion}}(q, t) S_{\mathrm{solv}}^{10}(q, t) \tag{14.22}$$

Several comments about Eq. (14.22) are in order.

1. The magnitude of $\Gamma_{\mathrm{T}}^{\mathrm{ion}}$ depends on the generalized rate $\Sigma_{10}(q, z)$. Thus it is actually a nonlinear equation and must to be solved self-consistently; the latter is again partly determined by the full Γ_{T}.
2. The magnitude of $\Gamma_{\mathrm{T}}^{\mathrm{ion}}$ increases rapidly as $\Sigma_{10}(q, z)$ decreases. Thus the friction on the neighboring solvent molecules due to the ionic field will be large if the orientational and translational relaxations of the solvent are themselves slow. This is an important result, as it supports the solvent-berg picture for slow liquids, such as alcohols and amides. For fast liquids, like water and acetonitrile, $\Gamma_{\mathrm{T}}^{\mathrm{ion}}$ is not significant; and the solvent-berg model appears not to be valid. But for slow liquids, such as alcohols, amides, and water, at low temperature, $\Gamma_{\mathrm{T}}^{\mathrm{ion}}$ is significant for small ions, such as Na^+ and Li^+.
3. In the derivation of Eq. (14.22), it was assumed that the ionic field quenches the rotational motion of the neighboring solvent molecules. This is again expected to be reliable only for the nearest-neighbor molecules

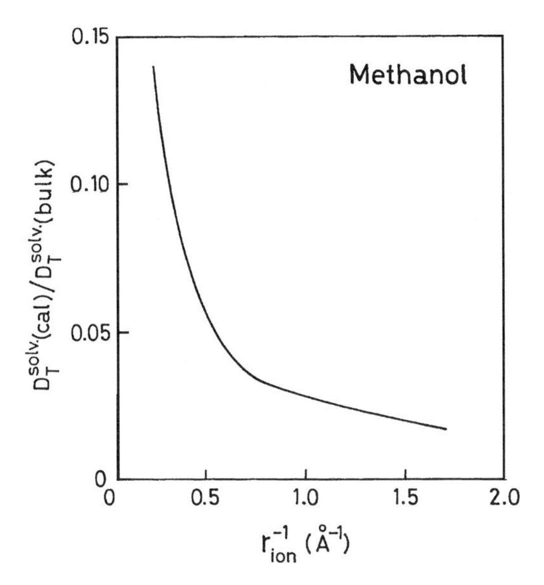

Figure 50. Ionic field induced quenching of the translational motions of the solvent molecules that are nearest neighbors to the diffusing ion. The ratio of the calculated solvent diffusion coefficient D_T^{solv}(cal) and the bulk solvent diffusion coefficient D_T^{solv}(bulk) of methanol is plotted as a function of the inverse of the crystallographic ionic radius r_{ion}^{-1}. The calculational procedure is described in detail in Section XIV.B.

and for small ions. A full calculation that shows the distance dependence of this important effect is not yet available.

4. Note that Eq. (14.20) differs from the analogous Eq. (12.4) for the isotropic density contribution to the local friction in one important aspect: the absence of the inertial term in the former. The reason for this is that the contribution of the dielectric friction comes essentially from coupling to the orientational density relaxation.

In our calculation of the dielectric friction on the solute ion, we used the Γ_T given by Eq. (14.20). The calculated Γ_T is found to be larger than the bare friction Γ_T^0. The result of this calculation is shown as a function of the ion size in Figure 50, which plots the ratio of the calculated : bulk (unperturbed) diffusion coefficient D_T^{solv} : D_{bulk} in methanol. Note the strong dependence on the ion size and the large reduction in the magnitude of the diffusion coefficient because of the ionic field. This is the general result obtained for all the solvents; the reduction becomes more pronounced with the polarity of the solvent. As mentioned, this calculation is reasonable only for the nearest-neighbor molecules, which are the most important as far the translational contribution to the generalized rate $\Sigma_{10}(q, z)$ is concerned.

We now use the modified solvent translational friction Γ_T in the following expression $D_T^{solv} = k_B T / \Gamma_T$ to calculate the modified solvent translational diffusion coefficient D_T^{solv}, which was used in Eq. (14.18) to obtain $S(q, z)$. We then connect the translational kernel of the pure solvent to the dynamic solvent structure factor as follows [166]:

$$\frac{k_B T}{m\sigma^2[z + \Gamma_T(q, z)]} = \frac{S(q)[S(q) - zS(q, z)]}{q^2 S(q, z)} \tag{14.23}$$

2. *Calculation of the Static, Orientational Correlation Functions*

The important static correlations required in the calculations are the ion–dipole and the dipole–dipole two-particle direct correlation functions. Fortunately, several methods are available for calculating these functions. The important point to remember in these calculations is that the correlation functions have well-known properties both in small and in long distances of separations. In the usual terminology of equilibrium theory of polar liquids, it can be stated that these correlation functions are known at small and large wavenumbers. This is an important advantage, because these limits make important contributions to the dielectric friction on an ion. For acetonitrile, water, and methanol, the longitudinal component of the wavenumber-dependent dielectric function $\varepsilon_{10}(q)$ is calculated from the XRISM calculation of Raineri and co-workers [157]. Subsequently, we obtained $f_{110}(q)$ by using Eq. (2.11).

We next describe the calculation of the ion–dipole DCF $c_{id}(k)$, which we obtain by Fourier transforming the expression of microscopic polarization $P_{mic}(r)$ given by Chan and co-workers [206]. The calculational details in this regard were discussed in Section IV.

Once the static and dynamic parameters are calculated, we put the quantities in Eq. (14.17) and calculate ζ_{DF} self-consistently. We then use the Einstein relation to obtain the translational diffusion coefficient of the ion. The equivalent (limiting) conductance at infinite dilution Λ_0 is obtained from the calculated D_T^{ion} by using the following well-known Nernst–Einstein relation [127–130]

$$\Lambda_0 = \frac{z^2 F^2 D_T^{ion}}{RT} \tag{14.24}$$

where z is the valency on the ion, F is the amount of electricity carried by 1 g equiv of the conducting ion, R is the universal gas constant, and T is the absolute temperature. We have not used any adjustable parameter at any stage of the calculation.

C. Numerical Calculations

Here we discuss briefly the interconnections between the solvation dynamics of an ion and its mobility in polar solvents. We also show here the quenching of the translational motion of the nearest-neighbor solvent molecules owing to the ionic field. The origin of the size dependence of the microscopic friction is also discussed.

1. *Relation between the Ionic Conductivity and Solvation Dynamics*

In Section IV, we noted that the general microscopic expression for studying the time-dependent solvation process of a newly created ion is given by

$$S(t) = \frac{\int_0^\infty dq\, q^2 |c_{id}^{10}(q)|^2 [1 - 1/\varepsilon_L(q)] \mathcal{L}^{-1}[z + \Sigma_{10}(q, z)]^{-1}}{\int_0^\infty dq\, q^2 |c_{id}^{10}(q)|^2 [1 - 1/\varepsilon_L(q)]} \tag{14.25}$$

Note the similarity of the numerator of Eq. (14.25) and that of Eq. (14.17) for dielectric friction.

The $S(t)$ in Eq. (14.25) shows that the initial part of ionic solvation dynamics is dominated primarily by the small wavenumber (i.e., $q \to 0$) processes (Fig. 41). The large wavenumber fluctuations are significant only at the longer time, which was detailed in Section IX. On the other hand, the force–force correlation function (the time integral of this quantity gives the dielectric friction) probes the large wavenumber processes more strongly, because it has a quartic wavenumber dependence [Eq. (14.17)]. As a result, the molecular length scale processes are more important in determining the ionic mobility than they are in solvation dynamics. This, in turn, implies that the dynamics and the structure of the solvent around the ion are more effective in controlling the motion of the ion than in determining its solvation dynamics. We will elaborate this point later when the numerical results for Λ_0 in alcohols are be presented We next turn our attention to the study of the time dependence of the dielectric friction.

2. *Size Dependence of Dielectric Friction*

As discussed in Section I, the size dependence of the time-dependent dielectric friction $\zeta_{Df}(t)$ can be important for various chemical and vibrational relaxations. Figure 51 shows the time-dependent dielectric friction for two ions of different sizes in methanol. Note the bimodal nature of $\zeta_{DF}(t)$. We calculate the time dependent dielectric friction as follows:

$$\zeta_{DF}(t) = \mathcal{L}^{-1}[\zeta_{DF}(z)] \tag{14.26}$$

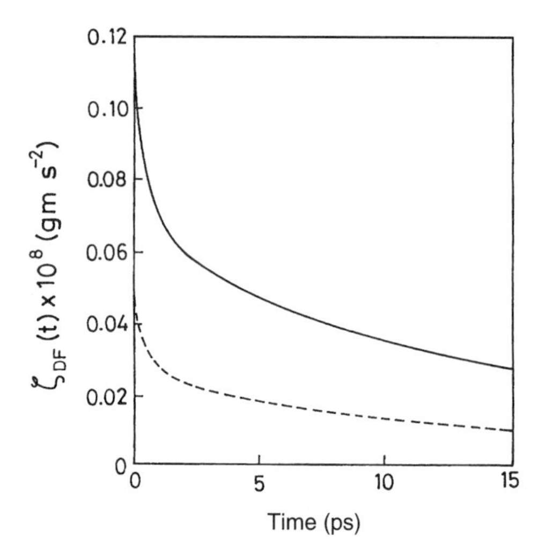

Figure 51. The solute size dependence of the time-dependent dielectric friction $\zeta_{DF}(t)$ in methanol, which is plotted as a function of time t for two ions. The *solid line* represents the theoretical results for Li^+ ($r_{ion} = 0.62$ Å), and the *dashed line* represents the results for Cs^+ ($r_{ion} = 1.76$ Å). Note the bimodal nature of $\zeta_{DF}(t)$. The necessary parameters for the calculation are given in Tables VI and VII.

where $\zeta_{DF}(z)$ is given by Eq. (14.17). The Laplace inversion $\mathcal{L}^{-1}$ is carried out by using the Stehfest algorithm. It is clear that the time-dependent friction, has a strong size dependence, which is in contrasted to solvation dynamics, for which the size dependence of the probe is essentially absent.

D. Conclusions

Let us first summarize the main results of this section. A self-consistent microscopic theory is presented for the limiting ionic conductance of strong 1:1 electrolyte solutions in dipolar liquids. The relation between the polar solvation dynamics of an ion and its mobility is clarified. The theory also explains how a dynamic version of the classical solvent-berg model can be recovered for small ions in the limit of slow liquids. The present theory also explains why the size dependence of ionic mobility is so strong whereas that of solvation dynamics is essentially absent.

XV. LIMITING IONIC CONDUCTIVITY IN AQUEOUS SOLUTIONS: TEMPERATURE DEPENDENCE AND SOLVENT ISOTOPE EFFECTS

As discussed in Section XIV, the limiting ionic conductance Λ_0 of small rigid symmetrical ions in common dipolar solvents is an important entity of the liquid phase chemistry [127–131]. Despite its importance, our understanding of the factors that determine Λ_0 is still poor. The complexity of the problem drew the attention of great scientists like Born, Debye, and Onsager; but even then many of the basic aspects of the problem are not yet well understood. The reason for the lack of progress is attributed to the complex nature of the ion–solvent and solvent–solvent interactions and the complex dynamics of the solvents.

In this section, we apply the molecular theory developed in the previous section to calculate the limiting ionic conductivity of monopositive ions in aqueous solution. Here we show that the theory described in Section XIV can predict the limiting ionic conductivity and its temperature dependence rather successfully. We also investigate the effects of solvent isotopic substitution on limiting ionic conductivity Λ_0 in aqueous solution. The value of the limiting ionic conductance is determined by the interactions among the ion and solvent molecules and the relative dynamics of the solute–solvent system. Λ_0 itself shows several interesting, even anomalous, behaviors, which are nontrivial to explain:

1. Λ_0 shows a maximum when plotted against the inverse of the crystallographic ionic radius r_{ion}^{-1}. This particular feature is shown in Figure 52, which also depicts the complete breakdown of Stokes's law for small ions, like Li^+ and Na^+.
2. Λ_0 shows a strong temperature dependence. For monovalent simple ions (e.g., tetra-alkyl ammonium ions and alkali metal ions), the temperature coefficient of Λ_0 is almost 2% per degree [128].
3. Λ_0 exhibits a significant solvent isotope effect. Experimental results reveal that Λ_0 of a particular ion in D_2O is 20% less than that in H_2O. This reduction of mobility is universal for all monopositive ions, irrespective of their size [131].

None of the above results can be explained in terms of the Stokes–Einstein relation, which relates the diffusion coefficient (or the conductivity) of the ion to the viscosity η_0 of the medium. Traditionally, there have been two general approaches for rationalizing the breakdown of Stokes's law. The phenomenological solvent-berg model [131] assumes the formation of a rigid solvent cage around a small ion, which leads to an increase of the effective radius of the ion. This, in turn, leads to a sharp decrease of the conductivity (Fig. 52). In addition, the maximum in Λ_0 near Cs^+ is explained in terms of the orientational structure breaking of the solvent by the ion [131]. This approach, however, completely fails to provide a coherent quantitative description of Λ_0. The second approach was initiated by Born [298], who suggested that, because of the increased dissipation of momentum owing

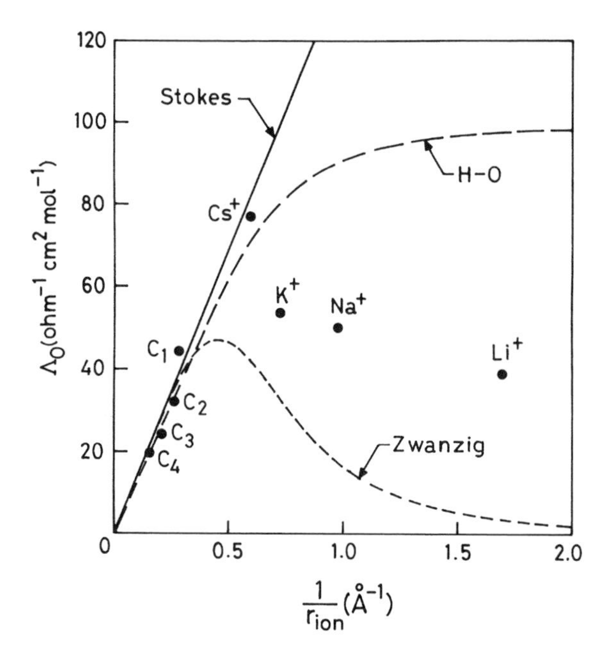

Figure 52. Comparison of the experimental results of Λ_0 and those from continuum theories. The experimental values of the limiting ionic mobility of rigid, monopositive ions in water at 298 K are plotted as a function of the inverse of the crystallographic ionic radius r_{ion}^{-1}. The experimental results are denoted by the *solid circles*. The *solid line* represents the predictions of Stokes's law (with a slip boundary condition), the *large dashed line* represents the Hubbard–Onsager theory, and the *small dashed line* is the theory of Zwanzig (with a slip boundary condition). Note that Stokes's law is valid for tetra-alkyl ammonium ions.

to the long range ion–solvent interactions, the ion experiences an additional friction over and above the prediction of Stokes's law. The friction acting on the moving ion can, therefore, be written as a sum of two contributions

$$\zeta = \zeta_{\text{bare}} + \zeta_{\text{DF}} \tag{15.1}$$

where ζ_{bare} is the friction caused by the short-range nonpolar interactions and ζ_{DF} is the dielectric friction originating from the long-range polar interactions. Conventionally, ζ_{bare} is approximated by the Stokes relation with a proper boundary condition. The main emphasis of this approach is the calculation of the dielectric friction. This is, of course, nontrivial. Initially, ζ_{DF} was obtained by continuum models, but more recently a microscopic approach has been initiated [304–307]. In the following we first briefly describe the main results of the continuum models.

As already discussed, the first consistent electrohydrodynamic calculation of the dielectric friction was presented by Zwanzig [300]. It leads to a simple expression for ζ_{DF} in terms of the static dielectric constant ε_0 and the Debye relaxation time τ_D. The resulting expression can explain the nonmonotonic size dependence of Λ_0 but overestimates the dielectric friction by a factor of 3–5 for small ions, like Na^+ and Li^+ (Fig. 52). In a different continuum approach [301-302], Hubbard and Onsager derived an expression for the total friction acting on the ion by generalizing the Navier–Stokes equations for hydrodynamic flow to include the polarization relaxation of the solvent in the vicinity of the moving ion. The resulting hydrodynamic equations were then solved with the constraint of invariance of the energy dissipation with respect to rigid body kinematic transformation (rotation and translation). The Hubbard–Onsager continuum electrohydrodynamic approach constitutes a beautiful treatment of macrodynamics, and it predicts the mobility of large ions correctly. However, it severely underestimates the value of ζ_{DF} and thus fails to provide a quantitative description of the ion transport mechanism (Fig. 52).

In a complete breakaway from the continuum models, Wolynes proposed a theory to obtain the dielectric friction ζ_{DF} from the force–force time correlation function; the force on the ion was obtained from microscopic quantities, such as the radial distribution function [304–306]. The theory was rather successful in describing many aspects of the ionic mobilities in water and acetonitrile [306]. Several limitations of this approach were removed in a subsequent theory, which pays proper attention to the various static and dynamic aspects of the ion–solvent composite system [303, 310-312]. The notable feature of the extended theory is the self-consistent treatment of the self-motion of the ion and the biphasic polar solvent response. The results were in satisfactory agreement with all the known results, not only for water and acetonitrile [310] but also for monohydroxy alcohols [311]. Most notably, the nonmonotonic dependence of Λ_0 on r_{ion}^{-1} was correctly reproduced for all these solvents. No theoretical studies on temperature dependence or the solvent isotope effect on limiting ionic conductivity has been carried out. As mentioned, the recently discovered ultrafast component in solvation dynamics is expected to play an important role in determining the solvent isotope effect and the temperature and pressure dependencies of Λ_0.

Nevertheless, the strong temperature dependence of Λ_0 in water is certainly paradoxical. On increasing the temperature from 283 to 318 K, the density of water decreases by only about 1%, the static dielectric constant by about 15%, and the Debye relaxation time and the solvent viscosity by about 50% each [314]. On the other hand, Λ_0 for Li^+ increases by 120%, from 26.37 at 283 K to 58.02 at 318 K. The same trend of increase is observed not only for Cs^+, Na^+, and Li^+ but also for the relatively large tetra-alkyl ammonium ions [131]. As this large change cannot be easily accounted for within the existing continuum model theories, explanations were offered by invoking the partial breakdown of the hydrogen-bonded network at high temperatures and the formation of solvent-berg at low temperatures [131]. Unfortunately, such pictures are difficult (if not impossible) to quantify, and there is no experimental evidence of significant structural change between, for example, 283 and 298 K, where Λ_0 changes by about 50%. Thus the explanation of the temperature coefficient of Λ_0 has remained largely unsolved.

In contrast to the temperature dependence, the solvent isotope effect on limiting ionic conductivity is less anomalous [131]. For Na^+, the change in Λ_0 is about 20% when the solvent is changed from H_2O to D_2O. This is comparable to the viscosity change of the liquid and, in principle, could be explained by hydrodynamics—via the continuum models—except that they all give a completely wrong magnitude of Λ_0.

We used the microscopic theory developed in Section XIV to calculate the temperature-dependent limiting ionic conductivity. This self-consistent theory provides a good description for both the temperature dependence of Λ_0 and the solvent isotope effects on limiting ionic conductivity. The strong temperature dependence arises from a collection of several small microscopic effects, all acting in the same direction in a concerted fashion. These effects include a change in the ion–dipole direct pair correlation function and in the dynamics of the solvent. The theory also provides a fairly satisfactory description of the solvent isotope effect.

We use Eq. (14.17) to calculate the dielectric friction ζ_{DF}. This is then added to the bare friction ζ_0, which is obtained by using the Stokes' relation with a slip boundary condition. This gives the translational diffusion coefficient of the ion as follows: $D_T^{ion} = \frac{k_B T}{[\zeta_0 + \zeta_{DF}]}$. Subsequently, Eq. (14.17) is used to obtain the limiting ionic conductivity. The viscosity of the solvent is taken from experiments.

The organization of the rest of this section is as follows. In the next part we briefly describe the calculational procedure. We present numerical results of the temperature dependence of the limiting ionic conductivity in Section XV.B. Next, we discuss the results on the solvent isotope effect.

A. Calculation Procedure

As noted, the calculation of ζ_{DF} from Eq. (14.17) involves the calculations of the wavenumber-dependent ion–dipole direct correlation function $c_{id}(q)$, the solvent–solvent static orientational correlations $\varepsilon_{10}(q)$, the solute dynamic structure factor $S_{ion}(q, t)$, and the generalized rate of solvent polarization relaxation $\Sigma_{10}(q, z)$. The solute dynamic structure factor is calculated by using Eq. (4.16). The calculation of the other quantities is described below.

TABLE XIV
Solvent Static Parameters at Three Temperatures

Solvent	Temperature, K	Diameter, Å	μ, D	ρ, g/mL	η_0, cp
H_2O	283	2.8	1.850	0.9997	1.3070
	298	2.8	1.850	0.9970	0.8904
	318	2.8	1.850	0.9902	0.5960
D_2O	298	2.8	1.855	1.1045	1.0970

1. *Calculation of the Ion–Dipole Direct Correlation Function*

As described in Section IV, we obtain the ion–dipole direct correlation function $c_{id}(q)$ from the work of Chan and co-workers [206]. We have, of course, used it in the limit of zero ionic concentration. The wavenumber-dependent dielectric function is obtained by using the XRISM calculation of Raineri and co-workers [157]. The solvent static parameters that are necessary for calculating c_{id} are given in Table XIV.

2. *Calculation of the Wavenumber- and Frequency-Dependent Generalized Rate of Solvent Polarization Relaxation*

As discussed, $\Sigma_{10}(q, z)$ is the dynamic response function of the solvent, which is a measure of the rate of orientational solvent polarization density relaxation, given by Eq. (2.9). The calculation of the dynamic response function $\Sigma(q, z)$ is a nontrivial excercise. It contains the rotational and the translational friction kernels.

a. Rotational Friction. We calculate the rotational friction $\Gamma_R(q, z)$ by directly using the experimental results on dielectric relaxation and far-IR line-shape measurements in water. The relation that connects $\Gamma_R(q, z)$ to the dielectric relaxation through the frequency-dependent dielectric function $\varepsilon(z)$ is given by Eq. (2.12). In the present calculations, $\Gamma_R(q, z)$ for water was obtained using the above relation in the following way. The frequency-dependent dielectric function $\varepsilon(z)$ in the low frequency regime is described by two consecutive, well-separated Debye dispersions. The dielectric dispersions involved in Debye relaxations are given in Tables III and XV. The temperature-dependent Debye relaxation times are obtained by scaling the room temperature relaxation times with the temperature-dependent viscosity of water, for which where the intermediate dielectric constants are taken as those at room temperature. The full expression of $\varepsilon(z)$ for water is given by Eq. (5.3).

TABLE XV
The Temperature-Dependent Dielectric Relaxation Parameters of Water[a]

Temperature, K	ε_1	τ_1, ps	ε_2	τ_2, ps	ε_3
283	83.83	12.145	6.18	1.498	4.49
298	78.3	8.32	6.18	1.02	4.49
318	71.51	5.538	6.18	0.683	4.49

[a] $\varepsilon_1 = \varepsilon_0$ is the static dielectric constant of the solvent; $\varepsilon_3 = \varepsilon_\infty$ is the infinite frequency dielectric constant obtained by fitting the low-frequency relaxation to a sum of two Debye dispersions. The high-frequency dielectric dispersions of water are given in Table III. In our calculations of Λ_0 in water at 283 and 298 K, we ignored the temperature dependence of the high-frequency librational and intermolecular vibrational band. The dielectric relaxation data of heavy water are given in Tables II and III.

b. Solvent Translational Friction. The solvent translational motion can enhance the rate of solvation and mobility of an ion by accelerating the rate of the solvent polarization relaxation. In the case of ionic mobility, the most important and effective translational modes are those of nearest-neighbor solvent molecules. Naturally, the solvent translational motion near the ion will be rather different from those in the bulks, since the strong ionic–dipole interaction quenches the free translational motion of the nearest neighbors. This is an example of the back reaction of the solute on the solvent, which leads to an interesting dynamic cooperativity, which is intrinsically nonlinear in nature. We have calculated the translational diffusion coefficient of the nearest-neighbor solvent molecules through a nonlinear equation, which couples the solvent translational mode with that of the ion (see Section XIV.B).

B. Numerical Results

1. Temperature Dependence of the Limiting Ionic Conductivity in Water

In this section we present the numerical results on the temperature-dependent limiting ionic conductivity, Λ_0. The calculated limiting ionic conductivity at 283 K is shown in Figure 53, where Λ_0 is plotted as a function of the inverse of the crystallographic ionic radius r_{ion}^{-1}. The available experimental results [131] are also shown. The comparison clearly indicates a fair agreement between the theoretical predictions and the experimental results. In particular, the nonmonotonic size dependence is correctly reproduced by the present molecular theory. This is indeed satisfactory if one considers the complex nature of the solvent and the approximations involved. There are, however, still some minor discrepancies. The theory predicts a peak value for Λ_0 that is smaller than the experimental value by about 10%. Moreover, the

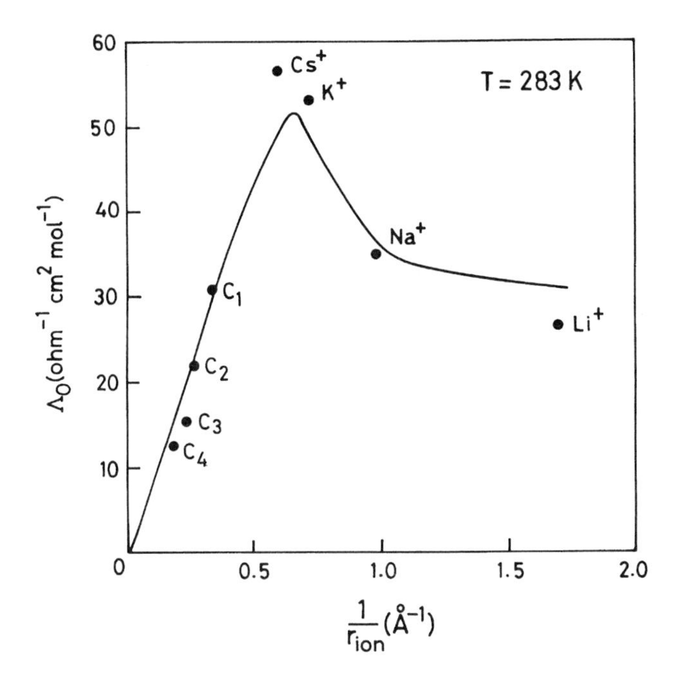

Figure 53. The values of the limiting ionic conductivity Λ_0 of rigid, monopositive ions plotted as a function of the inverse ionic radius r_{ion}^{-1} in water at 283 K. The *solid line* represents the predictions of the present microscopic theory. The *solid circles* denote the experimental results.

theory predicts a shift in the peak position where the theoretical peak in Λ_0 corresponds to K^+ while that in the experiment is for Cs^+. For Li^+, the calculated Λ_0 is about 15% greater than that of the experimental results.

Figure 54 presents the calculated limiting ionic conductivity for water at 298 K. The relevant experimental results [131] are also shown. The agreement here is excellent. Figure 55 compares the theoretical predictions on limiting ionic conductivity with those from the experiments by plotting Λ_0 against r_{ion}^{-1} at 318 K. The relevant experimental results [131] are also shown. Note that the theory predicts the peak value of Λ_0 quite successfully but again fails to describe the experimental results for Na^+ and Li^+ quantitatively.

The fair agreement between the theoretical predictions and the experimental results indicates that the present theory can capture essentially all the static and dynamic aspects of the solute–solvent system correctly at different temperatures. We would like to emphasize again that this agreement was achieved without the use of any adjustable parameter.

The reasons for the remaining discrepancies are not clear. It is likely that the MSA model used to obtain the static pair correlations does not describe

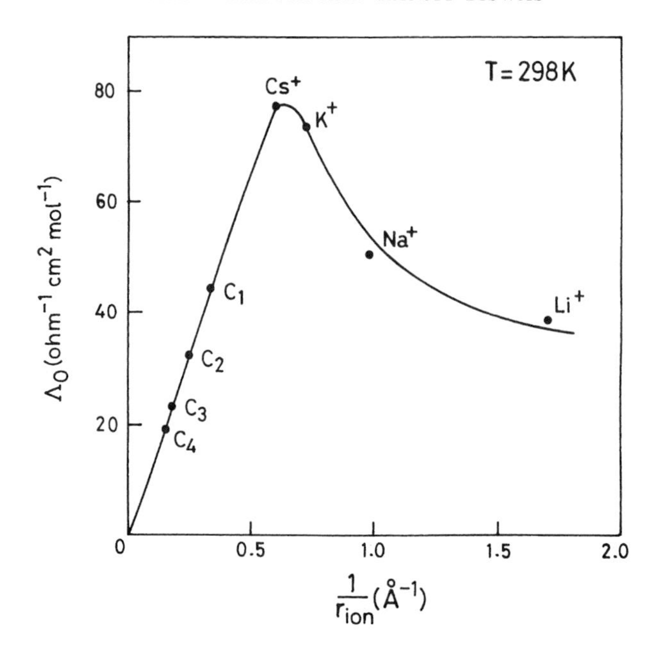

Figure 54. The values of the limiting ionic conductivity Λ_0 plotted as a function of the inverse crystallographic ionic radius r_{ion}^{-1} in water at 298 K. The symbols are described in Figure 53.

the real solvent accurately. Another reason may be the use of the dielectric relaxation data which at 283 and at 318 K were obtained from those at 298 K by scaling the relaxation times linearly with the viscosity. Any slight incompatibility in these data will be magnified in the calculation of Λ_0. This is more so for the relatively small ions, because they couple with the dynamic response of the solvent more strongly than do larger ions.

2. *Origin of the Observed Temperature Dependence of the Limiting Ionic Conductivity*

Let us now comment on the physical origin of the strong temperature dependence of Λ_0. The present theory takes into account the temperature effect through various molecular parameters, each contributing a small effect with the change in the temperature. These effects act in a concerted fashion to a single direction: either to decrease or to increase the limiting ionic conductivity, depending on the direction in which the temperature is changed. For example, the increase in the temperature from 283 to 318 K gives a ~10% reduction in the polarity ($3Y$) parameter. This translates into the similar reduction in the orientational static correlations and ion–dipole direct correlation functions. The viscosity of the solvent also reduced by ~50%.

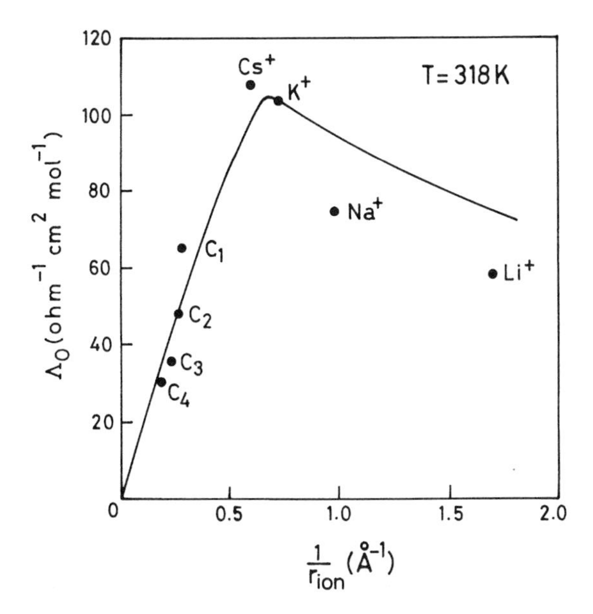

Figure 55. The values of the limiting ionic conductivity Λ_0 of rigid, monopositive ions plotted as a function of the inverse crystallographic ionic radius in water at 318 K. The *solid line* represents the predictions of the present theory, and the *solid circles* denote the experimental results.

This reduction in the solvent viscosity changes Λ_0 in two ways — changing the bare friction and the dielectric friction. The reduced viscosity also decreases the Debye relaxation times, which were used to evaluate the dynamic solvent structure factor, $S^{10}_{\text{solv}}(q, t)$ through the calculation of $\Sigma_{10}(q, z)$. To understand the dynamics of solvent response, Figure 56 shows the calculated dynamic solvent structure factors (normalized) $S^{10}_{\text{solv}}(q, t)$, for two temperatures. Note that $S^{10}_{\text{solv}}(q, t)$ is obtained numerically by Laplace inverting, and we show the results for intermediate wavenumbers only. It is clear from Figure 56 that the response function at high temperature decays more rapidly than at the low temperature. This, in turn, produces less dielectric friction at the high temperature.

It is interesting to see how so many changes act in the same direction. Thus we are now in a position to understand the anomalous temperature coefficient of the limiting ionic conductance in aqueous electrolyte solutions.

3. *Solvent Isotope Effect: Limiting Ionic Conductivity in Heavy Water*

Below we present the theoretical results on ionic conductivity in heavy water at 298 K. The necessary static parameters and dielectric relaxation data

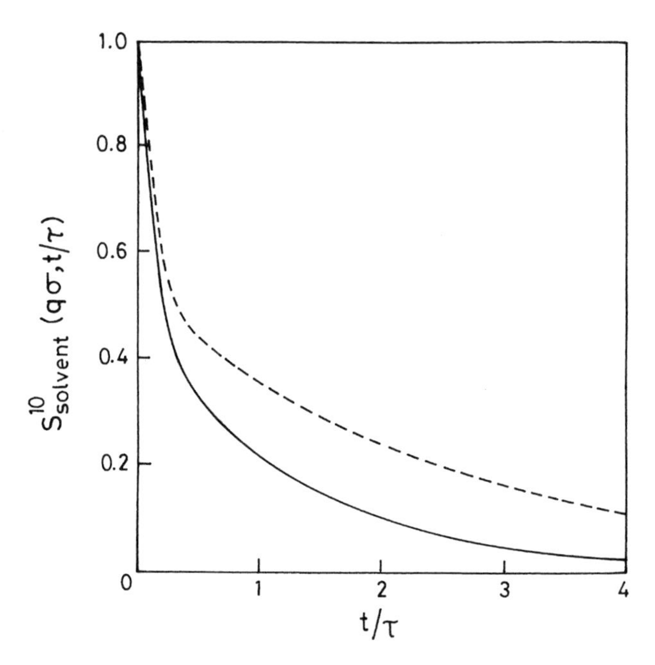

Figure 56. The rate of the decay of the orientational dynamic solvent structure factor $S^{10}_{\mathrm{solv}}(q, t)$ as a function of time t for water at two different temperatures. The *solid line* and *dashed lines* represent the decay of $S^{10}_{\mathrm{solv}}(q\sigma, t/\tau)$ at 318 and 283 K, respectively. Note that the numerical results were obtained with $q\sigma = 6.3$. The time is scaled by the quantity $\tau = 1 \times 10^{-12}$ s.

needed for the calculation of Λ_0 in D_2O are given in Tables II and III. The other static parameters are described in Section V.

The calculated results for the limiting ionic conductivity Λ_0 in heavy water at 298 K are shown in Figure 57. The available experimental results [131] for D_2O are also shown. The theoretical results are again in good agreement with those from the experimental ones. Note that the theory can successfully predict the peak value of Λ_0 in D_2O. The experimental observation that the solvent isotope effect reduces the mobility by about 20% is correctly reproduced here.

In view of the present isotope effect, it is interesting to recall the significant isotope effect observed in electron mobility [89]. The latter is, of course, a more difficult problem, because a fully quantum mechanical treatment is required to understand the nature of the solvent isotope effect on electron mobility [89].

C. Conclusions

In this section, we presented a microscopic calculation that, for the first time, explains the anomalous temperature dependence of the limiting ionic con-

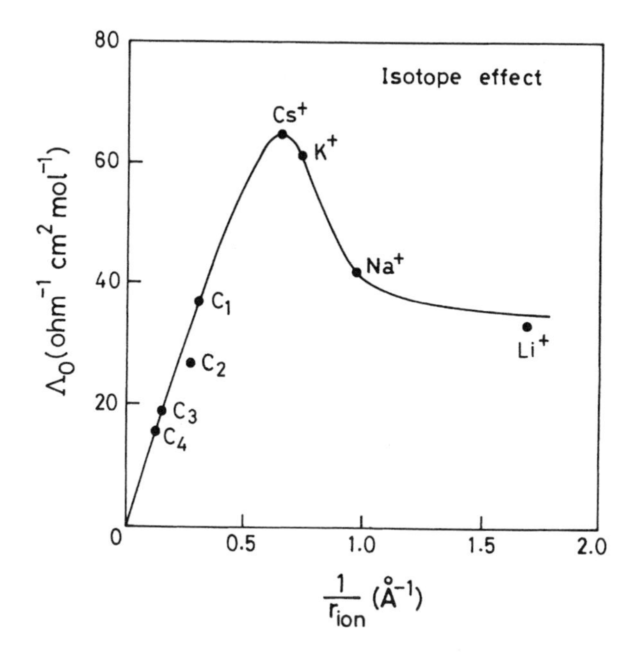

Figure 57. The effect of isotopic substitution in solvent on limiting ionic conductivity in electrolyte solution. The values of the limiting ionic conductivity Λ_0 are plotted as a function of the inverse of the crystallographic ionic radius in heavy water at 298 K. The prediction of the present molecular theory is represented by the *solid line* and predictions from experiments are shown by the *solid circles.*

ductivity in aqueous solutions. The strong temperature dependence is shown to arise from a collection of small effects all acting in the same direction. Thus one need not invoke any unquantifiable physical concepts, such as the formation or breaking of solvent-berg, to explain the experimental results. The theory can also explain the significant solvent isotope effect, which has been known for a long time but has not hitherto been explained quantitatively. The nonmonotonic size dependence of limiting ionic conductivity at various temperature has also been correctly described in terms of the dielectric friction.

The theory presented here is based on a simple idea, although its implementation requires the consideration of a rather large number of factors. One needs not only a detailed knowledge of the ion–dipole and dipole–dipole pair correlation functions of the solution but also a detailed description of the dielectric dispersion of the pure solvent. Even after these are obtained, one must calculate the rotational and the translational generalized frictions, which are both frequency and wavenumber dependent. However, all the functions involved are well behaved and the self-consistent calculation of ζ_{DF} is straightforward. In fact, given the complexity of the problem, it is difficult to imagine that a simpler theory than the present one can be successful.

XVI. IONIC MOBILITY IN MONOHYDROXY ALCOHOLS

In this section we present the numerical results of the limiting ionic conductivity of monovalent ions in methanol, ethanol, and propanol. We use the microscopic theory developed in Section XIV to predict the limiting ionic conductivities in these solvents. The motivation of the present study comes from the following facts. First, alcohols constitute an important class of solvents for which ionic mobilities are well known. Second, the most reliable dielectric relaxation data for these alcohols are now available. Third, recent experimental [65–66,243] and theoretical [208] studies of polar solvation dynamics in these alcohols reveal the importance of the high-frequency modes and that the observed dynamics are much faster than what was expected. As noted, the earlier theoretical studies of ionic mobility in alcohols used only the slow solvent relaxation and considered only a single Debye relaxation. Recent experimental studies [65,66,243] have shown that the initial polar response of the solvent is largely underdamped. And finally, the present theory is successful in predicting the ionic mobilities in ultrafast, underdamped solvents, such as water [310,312] and acetonitrile [310], as has been discussed.

In the next section we briefly describe the calculation details of some relevant quantities. Numerical results are given in Section XVI.B.

A. Calculation Procedure

In this section, the calculations of the wavenumber- and frequency-dependent generalized rate of orientational solvent polarization relaxation $\Sigma_{10}(q, z)$ and the necessary static, orientational correlation functions are described.

1. Calculation of the Wavenumber- and Frequency-Dependent Generalized Rate of Solvent Polarization Relaxation

As discussed, the generalized rate of polarization relaxation $\Sigma_{10}(q, z)$ contains two friction kernels: the rotational kernel and the translational kernel. Eq. (2.9) gives the full expression for $\Sigma_{10}(q, z)$. The memory kernels are calculated as follows.

a. Rotational Kernel. The rotational kernel $\Gamma_R(k, z)$ is calculated directly from the frequency-dependent dielectric function $\varepsilon(z)$ by using Eq. (2.12). The experimental $\varepsilon(z)$ values for monohydroxy alcohols are described by

Eq. (6.4). The necessary static parameters and the experimental dielectric relaxation data are tabulated in Tables VI and VII, respectively.

b. Solvent Translational Friction. The calculation of $\Gamma_T(q, z)$ for slow solvents is rather involved. As noted, the partial quenching of the motions of the nearest-neighbor solvent molecules in these solvents can give rise to a dynamic cooperativity through a nonlinear coupling of the motions of the ion and solvent. This can have a profound effect on the ionic mobility in otherwise slow solvents. The nonlinear coupling of the motions of the ion and the solvent molecules via ionic field gives rise to the dynamic version of the solvent-berg model. Therefore, we calculated $\Gamma_T(q, z)$ as described in Section XIV.B.

2. Calculation of the Static, Orientational Correlation Functions

The static correlations in these liquids at the intermediate (i.e., $q\sigma \simeq 2\pi$) to large (i.e., $q\sigma > 2\pi$) wavenumbers play important roles in slowing down the rate of orientational polarization relaxation at long time. For methanol, the correlations were taken from the XRISM calculation of Raineri and co-workers [157]. For ultrafast solvents like methanol, however, the solvent polarization relaxation is primarily governed by the long wavelength modes; for ethanol and propanol, the solvent structure at small wavelengths (i.e., $q\sigma \geq 2\pi$) may also be important. For ethanol and propanol, we calculated the static, orientational correlation functions from the MSA model [188]. We have, of course, corrected them at both the $q \to 0$ and the $q \to \infty$ limits. The calculation procedure in this regard was described in Section VI.

The ion–dipole DCF, $c_{id}(q)$ is taken directly from the solution of Chan and co-workers [206], who used the mean spherical model of an electrolyte solution. We have, of course, used it in the zero concentration limit. The calculation details were given in Section V.

Once these static and dynamic parameters are calculated, we put the quantities in Eq. (14.17) and calculate ζ_{DF} self-consistently. The bare friction ζ_0 is calculated from Stokes's relation using a slip boundary condition. We then use the Einstein relation to obtain the translational diffusion coefficient of the ion D_T^{ion}. Subsequently, Eq. (14.17) is employed to obtain the equivalent (limiting) ionic conductance at infinite dilution Λ_0. We used $T = 298$ K in all the calculations reported in this section.

B. Numerical Results

We next present the numerical results on ionic mobility. As noted, the experimental results on ionic mobilities are most conveniently represented by plotting the Walden product $\Lambda_0\eta_0$ against the inverse of the crystallographic radius r_{ion}^{-1}. Λ_0 is calculated from the well-known Nernst–Einstein relation

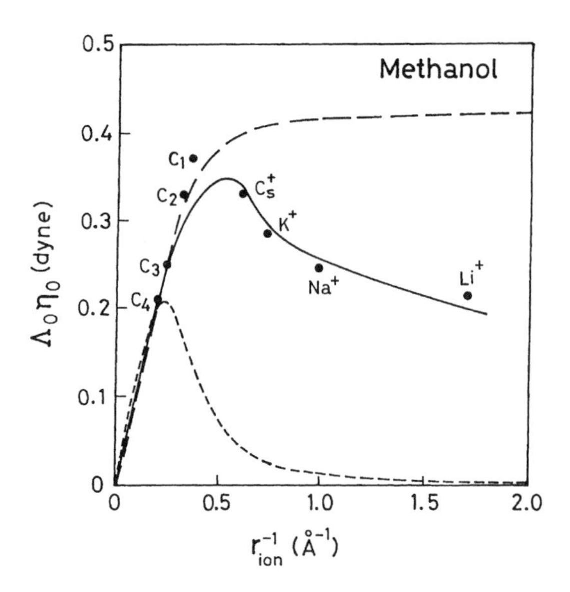

Figure 58. The values of the Walden product $\Lambda_0\eta_0$ of rigid, monopositive ions plotted as a function of the inverse ionic radius r_{ion}^{-1} in methanol at 298 K. The *solid line* represents the predictions of the present microscopic theory; the *solid circles* denote the experimental results; and the *large dashed line* and the *short dashed line* represent the Hubbard–Onsager and the Zwanzig theories, respectively.

from the diffusion coefficient $D_{\text{T}}^{\text{ion}}$ using Eq. (14.17). The viscosity of the solvent η_0 is taken from experiments.

1. *Methanol*

Figure 58 presents the results for methanol, for which the calculated Walden product is plotted against r_{ion}^{-1}. The available experimental results [131] are also shown. The theoretical predictions are in surprisingly good agreement with the experimental results, given the complexity of the problem and the approximations involved. The main disagreement between the theory and the experiment is in the height of the maximum: the calculated value of $\Lambda_0\eta_0$ is about 10% smaller than the experimental value. Although the reason for this discrepancy is not clear to us, it does not appear to be connected with the translational modes or the solvent-berg picture.

The Walden products calculated from the continuum theory of Zwanzig [300] and the Hubbard–Onsanger theory [301] are also shown in Figure 58. The results obtained from Hubbard–Onsanger theory overestimate the mobility for small ions, showing a saturation for small ions at large values.

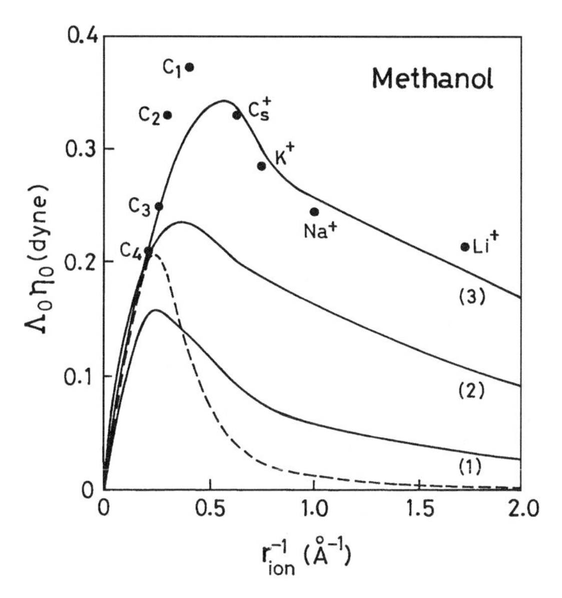

Figure 59. The effects of the sequential addition of the ultrafast component of the solvent orientational motion on the limiting ionic mobility in methanol at 298 K. The values of the Walden product $\Lambda_0\eta_0$ are plotted as a function of the inverse ionic radius. The curves labeled *1*, *2*, and *3* are the predictions of the present molecular theory: the first one, the first two, and all three Debye relaxations of the experimentally obtained dielectric relaxation data by Kindt and Schmuttenmaer. The experimental results of the Walden product for different ions are denoted by the *solid circles*; the *dashed line* represents the predictions of the theory of Zwanzig.

As evident from the figure, the continuum model-based theory of Zwanzig overestimates the dielectric friction. The reason for this failure may actually be from neglecting the fast, inertial modes that are present in methanol. In fact, if the fast modes are neglected, then the present theory predicts ionic mobilities that are comparable to those given by Zwanzig's expression [300]. This is shown in Figure 59, which shows the effects of the sequential addition of the ultrafast components of solvent modes on the limiting ionic conductivity in methanol. Note again that if we neglect all the ultrafast components in dielectric relaxation completely and consider only the first Debye relaxation of methanol, we obtain limiting conductivities that are comparable to those predicted by Zwanzig. This comparison shows that the ultrafast components of solvent modes are indeed important for a proper explanation of the molecular mechanism of ion transport in these liquids, which implies that it might be interesting to generalize Zwanzig's treatment to non-Debye (or multi-Debye) solvents. It is also interesting to note that the viscosity plays no role in the dependence of Λ_0 on the ultrafast modes. The same behavior was found for water.

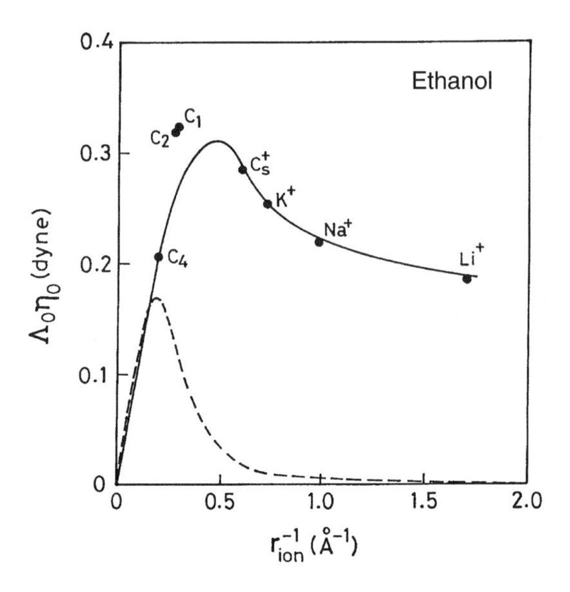

Figure 60. The values of the Walden product $\Lambda_0\eta_0$ of rigid, monopositive ions plotted as a function of the inverse ionic radius in ethanol at 298 K. The *solid line* represents the predictions of the present theory, the *solid circles* denote the experimental results, and the *dashed line* represents the results obtained from Zwanzig's theory.

2. Ethanol

We next present the results on ionic mobilities in ethanol. Figure 60 shows the plot of the calculated $\Lambda_0\eta_0$ values against r_{ion}^{-1} for ethanol. For comparison, the experimental results [131] and those from Zwanzig's theory are also shown. The agreement of the predictions from the present molecular theory and the experimental results is quite satisfactory, except the value at the peak, which is again smaller by 10%. The breakdown of Zwanzig's theory is also obvious here. The reason is the same as discussed for methanol.

3. Propanol

Figure 61 shows the calculated Walden product $\Lambda_0\eta_0$ plotted against the inverse of the crystallographic radius r_{ion}^{-1} for propanol. The results calculated from Zwanzig's theory are also shown. Unfortunately, we cannot compare our present theoretical predictions on ionic mobility in propanol directly to the experimental results, since experimental results for univalent single ions in propanol are not available.

The total equivalent conductances Λ_0^{total} of several uniunivalent electrolytes at zero ionic concentration in propanol are known experimentally [315]. These

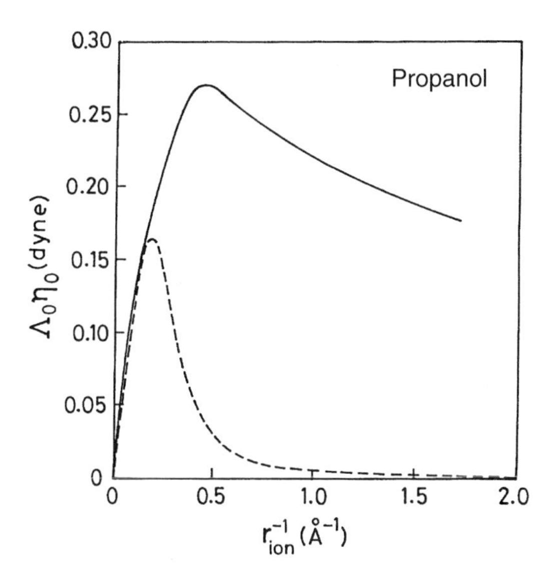

Figure 61. The variation of the Walden product $\Lambda_0\eta_0$ with the ionic radius in propanol at 298 K. The predictions of the present molecular theory are represented by the *solid line* and those from Zwanzig's theory are shown by the *dashed line.*

can also be calculated from our present theory if we use the Kohlrausch law of independent ion migration [127–130]. According to this law, the total equivalent conductance of an electrolyte solution at infinite dilution can be obtained by adding up the individiual conductances of the constituent ions:

$$\Lambda_0^{\text{total}} = \Lambda_0^{\text{cation}} + \Lambda_0^{\text{anion}} \tag{16.1}$$

Because our theory is sensitive only to the amount of charge on the ion and not to the nature of it (positive or negative), we can easily calculate the individual ionic conductances using Eq. (16.1). The only parameter we need for characterizing a particular ion is its crystallographic radius, which is taken from the literature [316]. The total limiting conductances thus obtained for several uniunivalent electrolytes are shown in Table XVI, which also presents the experimental results [315]. The agreement is again satisfactory.

We have also calculated the limiting ionic conductivities of F^-, Cl^-, Br^-, and I^- ions in methanol and ethanol. The theoretical predictions are given in Tables XVII and XVIII. For comparison, the tables also show the experimentally obtained limiting ionic conductivities of these halide ions [131] in methanol and ethanol. The agreement is satisfactory and may

TABLE XVI
The Total Limiting Ionic Conductivity for Strong Electrolytes in Propanol at 298 K[a]

Electrolyte	Λ_0^{total}	
	Theory	Experiment
NaI	22.2	23.92
KI	23.1	25.75
$(C_2H_5)_4NI$	23.5	28.5
$(C_3H_1)_4NI$	22.3	25.9
$(C_3H_1)_4NBr$	22.2	24.5

TABLE XVII
The Walden Product for Halide Ions in Methanol at 298 K

Ion	$\Lambda_0\eta_0$	
	Theory	Experiment
F^-	0.29	—
Cl^-	0.30	0.284
Br^-	0.31	0.306
I^-	0.32	0.341

TABLE XVIII
The Walden Product for Halide Ions in Ethanol at 298 K

Ion	$\Lambda_0\eta_0$	
	Theory	Experiment
F^-	0.24	—
Cl^-	0.25	0.235
Br^-	0.26	0.26
I^-	0.27	0.281

be rather fortuitous or may imply that the ionic mobilities in alcohols are less sensitive to the sign of the ionic charge than they are for water.

C. Discussion

Let us first summarize the main results of the present work. We have presented fully microscopic and self-consistent theoretical calculations of ionic mobilities in monohydroxy alcohols. The calculated ionic mobilities of univalent alkali metal ions and quaternary alkyl ammonium ions at room temperature are in good agreement with the long-known experimental results. The nonmonotonic size dependence of limiting ionic conductivities in these alcohols is explained satisfactorily by the present theory in terms of the enhanced dielectric friction for small ions. This is encouraging if we keep in mind that no theoretical or computer simulation studies had been carried out for these systems before the present work was undertaken.

Another interesting outcome of the present study is the recovery of a solvent-berg model for smaller ions in these solvents and for water at low temperatures. It can be intuitively understood that the mobility of an ion will be reduced by the solvent-berg formed around it if the solvent possesses slow relaxation times; however, the solvent-berg model that we recovered for slow liquids is the dynamic solvent-berg model in which the mobilities are affected by a nonlinear coupling of the solute and solvent motions through the translational kernel of the solvent. The prediction that the total translational friction on the neighboring solvent molecules is much larger than that on the bulk molecules can be tested by computer simulations. The dynamic coupling between the solute and the solvent motion that was found to be important for ionic motion in otherwise slow solvents may also be useful for predicting dielectric friction on a rotating dipole.

The reason for the success of the present theoretical scheme can be traced back to the following factors. First, it is essential to perform a self-consistent calculation of the ionic conductivity — this makes the present scheme robust. Such a self-consistent calculation is only possible when the self-motion of the ion is incorporated consistently. The lack of inclusion of the self-motion was a major drawback of all the earlier studies. The second important ingredient is an accurate treatment of the ion–dipole direct correlation function, which is actually nontrivial. We obtained it from the work of Chan and co-workers [206] but it requires some amount of numerical work. This was also not done in any of the earlier studies. The third important ingredient is a proper translation of the dynamic response. Not only the orientational motion but the translational motions of the solvent molecules also are important in determining the magnitude of the dielectric friction. Finally, one must include the nonlinear effect of the ionic field of the ion itself on the surrounding solvent molecules. For slow and strongly polar liquids, like alcohols and formamide, the ionic field nearly immobilizes the surrounding solvent molecules. Thus one essentially recovers the solvent-berg model. Therefore, the theoretical formalism becomes more involved; however, given the complexity of the problem it is unlikely that a simpler theory than this can be successful in explaining the experimental results.

It is interesting to note that the simple theory presented here can reproduce the experimentally observed ionic mobilities in solvents with such disparate molecular shapes, sizes, and charge distributions as water, acetonitrile

[310], and alcohols. The present theory is, of course, largely insensitive to various subtle aspects of different solute–solvent and solvent–solvent interactions. This is in contrast with the recent computer simulation studies of Lee and Rasaiah [317,318]. An interesting finding of Lee and Rasaiah's work is the marked dependence of their results on the solute–solvent and solvent–solvent intermolecular potentials. The potential energy functions used in their simulation studies are considerably more complicated than the hard sphere plus ion–dipole and dipole–dipole potentials used to calculate the static correlations like $c(110; q)$ and $c_{\text{id}}^{10}(q)$ under the mean spherical approximation (corrected both at the $q = 0$ and the $q \to \infty$ limits). Therefore, the good predictive ability of the present theory can be attributed to the use of experimental data to describe the frequency-dependent dielectric function of the solvent together with the self-consistent scheme discussed above. This suggests that, for cations, the mobility is determined primarily by the dynamics of the solvent, for which the details of the specific ion–solvent interactions are not crucial.

This scenario becomes more intriguing when one considers the case of the anionic conductivities. At room temperature, the ionic mobilities of halide ions in aqueous solution demonstrate a different curve than that for cations when plotted as a function of the crystallographic radius of the ion. This experimental trend was also reproduced semiquantitatively by the recent simulation studies of Lee and Rasaiah [317,318]. All these studies seem to indicate that there is a lack of symmetry in the static and dynamic solvent–cation and solvent–anion correlations. The present theory, however, cannot explain the distinct maximum observed for the halide ions in aqueous solution, because it uses the MSA model to obtain the ion–solvent direct correlation function, which is insensitive to the sign of the ionic charge. The present theory, in principle, can be extended to explain the differences by using the proper charge representation of the solvent molecule, which means one must abandon the point dipole picture of the solvent molecules. Although this is certainly worthwhile, the procedure would involve extensive numerical work.

As mentioned, the theoretical study could not be attempted in many cases because of the lack of experimental data not only on Λ_0 but also on the dielectric relaxation. Thus we suggest a renewed effort to obtain these data for common liquids and for mixtures that may show more complex dynamic behavior.

An important unsolved problem in this field is the reduction of the molecular theory to the Hubbard–Onsager expression. As discussed, to achieve this we need to derive a microscopic expression for the contribution of the orientational current term to the dielectric friction. This remains a nontrivial problem. We shall come back to this point in the next section.

XVII. MICROSCOPIC DERIVATION OF THE HUBBARD–ONSAGER EXPRESSION OF LIMITING IONIC CONDUCTIVITY

As discussed, the relative success of the microscopic theory leads one to ask the following question: Is it possible to reduce the microsocpic expression of the dielectric friction to that of Hubbard–Onsager [301]? The answer seems to be negative, which is in contrast with the case of the rotational dielectric friction for which the molecular expression neatly goes over to the continuum model expression under proper conditions [319]. The failure of the molecular expression to reproduce the correct continuum model expression remains as an unsolved problem in this field.

Here we note that the microscopic theory developed earlier can never be reduced to the continuum model, simply because the two theories calculate two entirely different contributions to the friction. Whereas the continuum model calculates the contribution arising from the polarization current term, molecular theories calculate the contribution from the density term. Our final expression for the total friction (derived below) is given by [320]

$$\frac{1}{\zeta} = \frac{1}{\zeta_{\text{bin}} + \zeta_{\rho\rho} + \zeta_{\text{mic,DF}}} + \frac{1}{\zeta_{\text{hyd}} + \zeta_{\text{hyd,DF}}} \tag{17.1}$$

where ζ_{hyd} and $\zeta_{\text{hyd,DF}}$ are the hydrodynamic friction (without the polar contribution) and the hydrodynamic (polarization current) dielectric friction. A derivation of this term is provided later. Note here that the relative contributions of these terms are determined by several factors, the most important being the ion : solvent size ratio. For large ions, the second term in Eq. (17.1) dominates, and one expects to recover the Hubbard–Onsager expression. For small ions (Li^+ and Na^+), on the other hand, the first term dominates and the microscopic expression provides the correct description.

There is a serious corollary of the above result though. In the microscopic theory, the contribution of $\zeta_{\text{bin}} + \zeta_{\rho\rho}$ is equated with $4\pi\eta_0 r_{\text{ion}}$, where η_0 is the solvent viscosity and r_{ion} is the radius of the ion. This is clearly inconsistent, even though for some intermediate-sized ions (e.g., Cs^+ in water), the two terms are numerically close. This inconsistency becomes glaring for small ions (e.g., Li^+). We shall come back to this point.

Eq. (17.1) provides the marriage between the two (which were until now) disjoint approaches to limiting ionic conductivity. It clearly shows that the friction on a large ion is dominated by the hydrodynamic friction, because the microscopic forces become much larger than the former. Exactly the opposite is true for small ions.

A. Strategy for Deriving Continuum Results from Molecular Theories

The success of a molecular theory is often judged by its ability to reproduce the correct phenomenologic theory. In the present context this implies that the microscopic approach adopted here should be able to reproduce the continuum model results. We shall give two examples to stress this point: (1) the dielectric friction on a rotating dipole for which the microscopic expression can be successfully reduced [319] to a continuum model expression [321]; molecular hydrodynamics; and (3) the MCT expression for translational friction, which correctly goes over to the Stokes relation (with the slip boundary condition) when the size of the tagged particle is much larger than that of the solvent molecules [322,249]. In all these cases, the following strategy was used.

1. The response of the liquid is Markovian with only one relaxation time. For rotational dielectric friction, this is the Debye relaxation time of the dielectric function, whereas for translational friction, it is the viscous relaxation time of the transverse current autocorrelation function.

2. The pair correlation function between the solute and the solvent molecules is replaced by the asymptotic limiting result. For example, for rotational dielectric friction, it is the dipole–dipole direct correlation function $c_{\mathrm{d-d}}(k)$ that is replaced by its asymptotic form. In the Gaskell–Miller [322] reduction of the mode-coupling expression to the slip hydrodynamic result, it is the form factor that takes the form of a step function. Subsequently, the same derivation has been carried out in a more microscopic manner in which the relevant vertex functions are replaced by simple forms [249].

3. One may neglect the self-motion of the solute. This strategy, however, fails completely when one tries to derive the continuum model expressions from the microscpic theory of Wolynes [304–306]. As noted by Wolynes himself, the main problem is that the relevant correlation functions vary too strongly for any continuum limit to exist.

B. The Microscopic Derivation

There are several ways to derive Eq. (17.1). The simplest one is to appeal to fluctuating hydrodynamics [323], which provide the following equations for the time dependence of the tagged particle number density $n_s(\mathbf{r}, t)$ and the current term $\mathbf{J}(\mathbf{r}, t)$:

$$\frac{\partial n_s}{\partial t} = -\nabla.\mathbf{J}(\mathbf{r}, t) \tag{17.2}$$

$$\mathbf{J}(\mathbf{r}, t) = -D_{\text{mic}} \nabla n_s(\mathbf{r}, t) + \mathbf{v}(\mathbf{r}, t) \tag{17.3}$$

where D_{mic} is the microscopic diffusion term determined by the short-range interactions between the ion and the solvent molecules. The term J_{R} is the source of thermal fluctuations and will be neglected here.

Eqs. (17.2) and (17.3) each have a nice physical meaning. The position of the tagged particle can change either by interactions with its immediate neighbors, which will cause a random walk of the particle, or by the coupling of the particle to the natural currents of the liquid. These random walks are determined by processes that occur at different length scales. Although the former is determined primarily by structural relaxation, and hence by the molecular hydrodynamics, the latter is determined by the usual Navier–Stokes hydrodynamics, and hence by the viscosity of the medium. Eq. (17.3) leads to the following decomposition of the self-diffusion coefficient:

$$D = D_{\text{mic}} + D_{\text{hyd}} \tag{17.4}$$

Note that essentially the same expression was derived by others [324,325].

It still remains to determine D_{mic} and D_{hyd}, which is the real nontrivial problem. To determine them, we turn to the MCT, which by itself also leads to the form given by Eq. (17.4). The starting point of our derivation is the following expression for the time-dependent friction [26a]:

$$\zeta(t) = \frac{\beta}{mV} \int d1\; d2\; d1'\; d2'\; [\hat{q} \cdot \nabla_{r_1} v(\mathbf{r}_1 - \mathbf{r}_2, \Omega_2)] G^s(12; 1'2'|t) [\hat{q} \cdot \nabla'_{r_1} v(\mathbf{r}'_1 - \mathbf{r}'_2, \Omega'_2)] \tag{17.5}$$

where $v(\mathbf{r_1} - \mathbf{r_2}, \Omega_2)$ is the interaction potential between the ion and a dipolar solvent molecule; $d1 = d\mathbf{r_1}\, d\mathbf{p_1}$ and $d2 = d\mathbf{r}_2\, d\mathbf{p}_2 d\mathbf{\Omega}_2$, where $\mathbf{r}, \mathbf{p}, \Omega$ denote the position, momentum, and orientation of either a tagged solute (subscript 1) or of a dipolar solvent molecule (subscript 2); and $G^s(12; 1'2'|t)$ is the resolvent operator, which describes the correlated time evolution of both the tagged particle and the surrounding solvent molecules. That is, as the ion moves from $(\mathbf{r}_1, \mathbf{p}_1)$ at time t to $(\mathbf{r}'_1, \mathbf{p}'_1)$ at t', the solvent molecules at $(\mathbf{r}_2, \mathbf{p}_2, \mathbf{\Omega}_2)$ move to $(\mathbf{r}'_2, \mathbf{p}'_2, \mathbf{\Omega}'_2)$. Thus, Green's function describes the time evolution of the two-particle (coupled solute–solvent) system.

The subsequent steps follow the treatment of Sjogren and Sjolander [26a] closely. One first separates the binary interaction term. For this derivation we include only the isotropic part of the short-range interaction in the binary term. The frictional contribution of this binary term is also nontrivial for continuous potentials, but it can be dealt with. The rest of Green's function is then expanded into the hydrodynamic modes, which are the conserved variables. For solutes, they are the density and the momenta, because the energy fluctuations are neglected. For solvent, the number density includes the orientational terms as well.

As stated, we are neglecting the angular momentum relaxation of the solvent molecules; therefore, the analysis required here is essentially the same as that presented in Ref. [26b]. The final expression of the friction can be written as

$$\zeta(z) = \zeta_{\text{mic}}(z) - \left(\frac{\zeta_{\text{mic}}(z)}{\zeta_{\text{curr}}(z)}\right)\zeta(z) \tag{17.6}$$

$\zeta_{\text{mic}}(z)$ is the microscopic friction, given by

$$\zeta_{\text{mic}}(z) = \zeta_{\text{bin}} + \zeta_{\rho\rho}(z) + \zeta_{\text{PP}}(z) \tag{17.7}$$

where $\zeta_{\rho\rho}(z)$ is the contribution from the isotropic density fluctuation that would be present even in a nondipolar liquid and $\zeta_{\text{PP}}(z)$ is the contribution from the polarization fluctuation. We can now identify ζ_{PP} with the dielectric friction of the molecular theories, i.e., with $\zeta_{\text{mic,DF}}$ of Eq. (17.1). $\zeta_{\text{curr}}(z)$ is the natural hydrodynamic term containing contributions from the current flows. Eq. (17.11) can be solved to obtain

$$\frac{1}{\zeta(z)} = \frac{1}{\zeta_{\text{mic}}(z)} + \frac{1}{\zeta_{\text{curr}}(z)} \tag{17.8}$$

This is essentially the same as Eq. (17.1), so our derivation is over. But we still need to show that all the terms are available and that the current term goes over to the Hubbard–Onsager–Zwanzig form, which is detailed below.

The expression for the two density terms can be obtained directly from the density functional theory, which leads to the following expression of the force density on a tagged ion [304–306]:

$$\mathbf{F}(\mathbf{r}, t) = k_B T n_{\text{ion}}(\mathbf{r}, t)\nabla \int d\mathbf{r}'\, d\boldsymbol{\Omega}'\, c_{\text{id}}(\mathbf{r}, \mathbf{r}', \boldsymbol{\Omega}')\, \delta\rho(\mathbf{r}', \boldsymbol{\Omega}', t) \tag{17.9}$$

The frictional contribution can be obtained by using Kirkwood's formula [205]. This leads to the following expressions for $R_{\rho\rho}$ and R_{PP}

[26a,249,303,310–312]:

$$R_{\rho\rho}(t) = \frac{\rho k_{\mathrm{B}} T}{[m(2\pi)^3]} \int d\mathbf{q}' \, (\hat{q}.\hat{q}')^2 (q')^2 F^s(q', t)[c_{\mathrm{id}}^{\mathrm{iso}}(q')]^2 F(q', t) \tag{17.10}$$

$$R_{\mathrm{PP}}(t) = \frac{2k_{\mathrm{B}} T\rho}{3(2\pi)^2} \int d\mathbf{q} \, q^2 S_{\mathrm{ion}}(q, t)[c_{\mathrm{id}}^{10}(q)]^2 S_{\mathrm{solv}}^{10}(q, t) \tag{17.11}$$

where $c_{\mathrm{id}}^{\mathrm{iso}}(q)$ and $c_{\mathrm{id}}^{10}(q)$ are the isotropic and the longitudinal components of the ion-dipole direct correlation functions, respectively [188] and $S_{\mathrm{solv}}^{10}(q, t)$ is the orientational dynamic structure factor of the pure solvent. In defining these correlation functions, the wavenumber is taken parallel to the z-axis, ρ_0 is the average number density of the solvent and $F_s(q, t)$ denotes the self-dynamic structure factor of the ion. In Eq. (17.10), $\hat{q}$ is a unit vector in the laboratory fixed frame.

We next calculate the current term. This can again be obtained from the elegant treatment of this term by Gaskell and Miller [322]. The current contribution to the friction can be obtained by projecting the velocity field of the ion on the binary term $\rho_{q0} j_q$, where ρ_{q0} is the ion density and j_k is the current of the liquid. The subsequent steps are well documented, so we present only the final expressions.

$$R_{tt}(z) = \frac{1}{3\rho(2\pi)^3} \int d\mathbf{q} \, F^s(q, t) \times \frac{2\hat{f}(q)}{q^2} C_{tt}(q, t) \tag{17.12}$$

where f is a form factor that takes into account the finite size of the solute.

The following steps are now employed in the evaluation of the current term in the hydrodynamic limit. First, the current–current time correlation function is given by the following simple expression

$$C_{tt}(k, t) = C_{tt}(k, t = 0) \exp(-\eta k^2 t) \tag{17.13}$$

where η is the viscosity of the medium.

Second, the viscosity of the medium is a sum of two terms. The first term is the usual shear viscosity of the pure liquid, and the second term is the enhancement of the viscosity owing to the presence of the ion. This is a visco-electric effect. The enhancement of the fluid viscosity is given by the following well-known expression:

$$\eta = \eta_0 \left[1 + \frac{\tau_D}{16\pi\eta_0}](\varepsilon_0 - \varepsilon_\infty)^2 E^2 \right] \tag{17.14}$$

where E is the screened electric field of the ion. The enhanced electric field in the vicinity of the ion is the picture originally used by Fuoss [304]. Thus one must deal with a position-dependent viscosity, given by

$$\eta(r) = \eta_0\left[1 + \frac{\tau_D(\varepsilon_0 - \varepsilon_\infty)q^2}{16\pi\eta_0\varepsilon_0^2 r^4}\right] \tag{17.15}$$

Third, the position-dependent viscosity $\eta(r)$ is replaced by $\eta(r = R_{\text{ion}})$, i.e., by the value at the surface. This approximation can be improved but calls for numerical work, which is not necessary here.

Fourth, as mentioned, the self-motion term is neglected. That is, set $F_s(k, t)$ equal to unity.

We next follow the treatment of Gaskell and Miller to obtain the hydrodynamic approximation of the current term. The following expressions are obtained for the two hydrodynamic boundary conditions:

$$\zeta_{\text{curr}} = 4\pi\eta R_{\text{ion}} + \frac{1}{4}\frac{q^2}{R_{\text{ion}}^3}(1 - \varepsilon_\infty/\varepsilon_0)\frac{\tau_D}{\varepsilon_0} \tag{17.16}$$

for the slip boundary condition and

$$\zeta_{\text{curr}} = 6\pi\eta R_{\text{ion}} + \frac{3}{8}\frac{q^2}{R_{\text{ion}}^3}(1 - \varepsilon_\infty/\varepsilon_0)\frac{\tau_D}{\varepsilon_0} \tag{17.17}$$

for the stick boundary condition.

One can now easily identify the dielectric friction owing to the current term. The expressions are identical to the forms obtained by Hubbard and Onsager except for a prefactor that is somewhat larger.

It is possible to carry out a full evaluation of the current term, but at the expense of extensive numerical work; because even after making the continuum model–like approximations, we need to solve for the current time correlation function in terms of position-dependent viscosity. When this is done, the value of the friction is expected to decrease and become closer to that of Hubbard and Onsager.

C. Conclusions

In this section we presented a microscopic derivation of the dielectric friction on a moving ion in a dipolar liquid. The final expression [Eq. (17.1)] provides a marriage of the two different theories of the dielectric friction. Eq. (17.1) is more accurate than either of the two approaches but calls for the use of proper evaluation of the respective contributions.

What are the consequences of Eq. (17.1)? First, it provides the much needed description of the crossover from the structural relaxation–dominated regime for small ions to the hydrodynamic-dominated regime for large ions. In this crossover, each of the five terms of Eq. (17.1) plays important role. This was not appreciated before. Second, the scenario is really interesting for ions that are slightly larger than that of the solvent. For example, the friction on the lower members of quarternary alkyl ammonium ions in water and acetonitrile may involve nearly equal contributions from the ζ_{mic} and ζ_{hyd} terms. This raises the interesting idea that the relative importance of the microscopic and hydrodynamic frictions is determined by the magnitude of the two dielectric frictions. This certainly deserves to be tested numerically. One should also consider the implication of Eq. (17.1) for the time (or frequency) dependence of the dielectric friction.

XVIII. DYNAMICS OF SOLVATION IN ELECTROLYTE SOLUTIONS

The dynamic properties of electrolyte solutions at finite ion concentration have been a subject of long-standing interest. The time-dependent interaction of a charge distribution with the surrounding ions and solvent molecules has played a central role in the investigations of chemical reaction dynamics in electrolyte solutions. Solvation in electrolyte solution is one such chemical process that involves motion of both ions and solvent molecules, and recently solvation dynamics studies were employed to explore the dynamics of electrolyte solutions at the molecular level.

Huppert and co-workers [326] and Chapman and Maroncelli [327] carried out experimental studies of solvation dynamics in electrolyte solutions of varying ion concentrations ranging from 10^{-3} to 3 M. It was found that the relaxation of ion–solution interaction energy can be separated into its solvent and ionic components. The solvent response was much faster than the ionic relaxation. Neria and Nitzan [328] carried out a detailed computer simulation study of solvation dynamics in ionic solutions. They found that there is a fast Gaussian decay characterized by the Gaussian time constant τ_G that is followed by a slow exponential-like decay. It was also found that the Gaussian time constant is practically independent of ion concentration, whereas the long-time exponential time constant depends rather strongly on ion concentration and decreases when ion concentration is increased. On the theoretical side, van der Zwan and Hynes [329] studied the role of ion atmosphere relaxation on dipole solvation dynamics by employing a primitive model of the solution and Debye–Falkenhagen theory [330] of ion atmosphere relaxation. Because they used a primitive model, the effects of solvent dynamics can be considered in their theory.

Recently, Chandra and co-workers [331–333] presented a molecular theory of ion solvation dynamics in electrolyte solutions that properly includes the molecularity of both solvent and ions. In their study, a fast initial decay was found in the solvation dynamics, which was then followed by a slow exponential decay, in agreement with the observations of experiments and computer simulations [331–333]. According to the theory of Chandra and co-workers [331–333], at short times the relaxation of $S(t)$ is given by

$$S(t) \approx e^{-t^2/2\tau_G^2} \tag{18.1}$$

whereas at long times, $S(t)$ is given by

$$S(t) \approx e^{-t/\tau_M} \tag{18.2}$$

where

$$\tau_G^2 = \frac{\varepsilon_0 - 1}{3y\varepsilon_0} \frac{I}{2k_B T} \tag{18.3}$$

and

$$\tau_M = \frac{4\pi\varepsilon_0}{\sigma} \tag{18.4}$$

where σ is the conductivity of the solution. Note that τ_G is primarily a property of the solvent. Although the static dielectric constant ε_0 depends on ion concentration to some extent, the ratio $(\varepsilon_0 - 1) : \varepsilon_0$ is always close to unity for highly or even moderately polar solvents; hence the initial Gaussian decay depends only weakly on ion concentration.

The long-time decay of the solvation function, on the other hand, is determined by the Maxwell relaxation time τ_M which is inversely proportional to the conductivity σ. In general, the conductivity increases with ion concentration, and the theory of Chandra and co-workers [331–333] could correctly explain the decrease of long-time slow relaxation rate with ion concentration. These predictions were made by considering only the long wavelength responses of the solvent and ions of the solution. Subsequently, Mahajan and Chandra [334] presented a more generalized theory of ion solvation dynamics in electrolyte solutions by considering finite wave vector responses of the solution. It was found that although the short-time decay is associated with the ultrafast solvent motion, the long-time decay involves both ion and solvent motions, implying that no true separability of dynamic responses exists in the solution. This happens because an ion motion necessarily involves solvent rearrangement inside the solution. Also, the predicted ion concentration dependence of solvation dynamics is consistent with the experimental results of solvation dynamics in solutions of varying ionic strength.

XIX. DIELECTRIC RELAXATION AND SOLVATION DYNAMICS IN ORGANIZED ASSEMBLIES

Many important biological processes occur on surfaces and within of cavities, proteins, lipid membranes, zeolites, and micelles that involve solvation and movement of charged species. Water is involved in almost all the systems of interest. Current interests in solvation dynamics and ionic mobility have been driven at least partly by the desire to formulate simple models for these complex systems. An example in point is the recent work of Fleming and co-workers [335], who used nonlinear spectroscopic methods, first applied to study simple liquids, to investigate the dynamics of photosynthetic electron transfer.

It is no surprise, therefore, that much attention has been focused in recent years on the study of dynamics in complex system. A universal nature of these complex systems is that they impose restrictions on the natural dynamics of the solvent involved, which is often water. Therefore, the dynamics of water in restricted environments deserve to be studied in great detail. Naturally, one again uses the simple experimental systems to mimic the role of the restricted environment. Common examples of such restricted environments that have been studied rather exhaustively are the α, β, and γ conformers of cyclodextrin. These compounds form inclusion complexes with smaller molecules and can exist in aqueous solution. Note that these environments are microscopically hetergeneous regions in which the local properties are significantly different from those of the bulk. Solvation studies of various conformers of cyclodextrin are, therefore, important for a number of reasons.

A. Solvation Dynamics in the Cyclodextrin Cavity

1. *Experimental Observations*

Recently, the solvation dynamics of a coumarin dye molecule (C-480) within the γ-cyclodextrin ring was studied both experimentally and by computer simulation [335]. The experimental results show that the initial ultrafast Gaussian component of the solvation time correlation function in the C-480/γ-cyclodextrin inclusion complex has almost the same time constant ($\tau_G \simeq 55$ fs) and amplitude as those in pure water [29]. This result is somewhat surprising, because the cavity certainly affects the dynamics of the first solvation shell strongly. The long-time decay of $S(t)$ is, however, much slower in the inclusion complex than that in pure water. The slow, exponential-like decay of $S(t)$ has three characteristic time constants: $\tau_{e,1} = 13$ ps, $\tau_{e,2} = 109$ ps, and $\tau_{e,3} = 1200$ ps [335]. These decay time constants are much slower than those for the long-time decay of $S(t)$ in pure water [29].

The experimental results thus indicate that the environment can have rather unexpected effects on the solvation dynamics. A theoretical framework is necessary to understand such effects. Within the cavity, the solvent structure is expected to be significantly distorted, and even a phenomenologic description of such a system is difficult. One can, however, explore the dynamics of such a system by using existing knowledge and initiating theoretical studies. This section is devoted to achieving this goal. First, we have studied the solvation dynamics of a nascent ion using the continuum theory based on the multishell model developed by Castner and co-workers [43]. It has been assumed here that the ion is surrounded by two concentric solvent shells, which are representatives of the γ-cyclodextrin cavity and bulk water, each having its own distinct dielectric behavior. Second, we study the solvation dynamics of the ion in cyclodextrin by using the MHT [15,15a]. This is done by freezing of the translational modes of the solvent molecules to mimic the experimental situation, in which the hydrogen bonding with the γ-cyclodextrin ring and the hindrance imposed by the cavity walls restrict the translational motion of the solvent molecules incorporated with the C-480 molecule within the γ-cyclodextrin cavity.

2. *Theoretical Approach*

Recently, a detailed study of the solvation dynamics of a charged coumarin dye molecule in γ-cyclodextrin and water was carried out by using two theoretical approaches [63]. The first approach is based on the multishell continuum model (MSCM). In this model, one assumes that the medium around the solute probe consists of several distinct regions of different characteristics. For the cyclodextrin cavity, the three distinct regions are the cyclodextrin itself, the water trapped inside it, and the bulk water. The model predicts the time scales of the dynamics rather well, provided an accurate description of the frequency-dependent dielectric function is supplied. The reason for this rather surprising agreement is twofold: First, there is a cancellation of errors; and second, the two-zone model mimics the heterogeneous microenvironment surrounding the ion rather well. The second approach is based on the MHT, in which the solvation dynamics are studied by restricting the translational motion of the solvent molecules enclosed within the cavity. The results from the molecular theory are also in good agreement with the experimental results. Our study indicates that in the present case the restricted environment affects only the long-time decay of the solvation time correlation function. The short-time dynamics are still governed by the librational (and/or vibrational) modes present in bulk water.

B. Solvation Dynamics in Miceller Systems

Bhattacharyya and co-workers [336,337] reported experimental stduies of the time-dependent fluorescence Stokes shift of polar dyes in both ordinary and reverse micelles. Their results showed that the solvation dynamics acquire a slow component in these systems, which could be attributed to bound water. Recently, a theoretical study [338] of the solvation dynamics of excited coumarin-480 dye in a reverse micellar system (aerosol OT in heptane) was carried out to understand the experimental results of Bhattacharyya and co-workers. Reverse micelle presents a unique example of a small finite system in which the solvation dynamics can be markedly slower than those in a macroscopic system and offers an opportunity to study the nature of the bound water molecules present in the water pool of the micelle. A MSCM calculation was presented that provided an almost quantitative agreement with the recently published experimental results. It was observed that for both the low and the high water content states of the micelle, almost 70% of the solvation is complete within $\sim$10 ns. At the low water content, the solvation energy time correlation function $S(t)$ decays exponentially on a nanosecond scale and is explained in terms of the slow relaxation of the water molecules bound to the surfactant head groups. For the high water content state, the decay of $S(t)$ is biexponential and is somewhat faster than the decay of $S(t)$ for the low water content state. The faster rate in the former case is explained in terms of the free water molecules present near the center of the pool, which relax at a faster rate than the water molecules that are bound to the head groups of the surfactants at the peripheral region of the pool.

C. Dielectric Relaxation and Solvation Dynamics in Biological Water

Many biological systems, such as proteins and enzymes, are inactive without water. For a complete knowledge of the function of such systems, an understanding of the structure and dynamics of the aqueous environment surrounding the concerned biomolecule is essential. The properties of water molecules in the vicinity of a biomolecule differ appreciably from the bulk water [339]. The water molecules enclosed within the solvation shell present in the immediate vicinity of the biomolecule are termed biological water. The dynamics and structure of the biological water near proteins and DNA and in reverse micelles have been the subject of intense research [339,340]. Dielectric spectroscopy and NMR are the two most extensively used techniques for understanding the interaction of water with proteins. Both these methods essentially probe molecular orientational relaxation. It is now known that the hydration shell surrounding a protein molecule comprises different types of water [339,340]. Few water molecules remain rigidly bound to the protein for a long time. In the immediate vicinity of the surface of the protein, there are water molecules that experience a much faster rotational and translational diffusion rate than do water molecules directly bound to the biomolecule. Thus biological water is believed to consist of two kinds of water molecules — bound and free — depending on their momentary states of existence. There is, of course, a dynamic exchange between the two species.

Based on the experimental studies on the dynamic behavior of water near biomolecules, it has been established that the frequency-dependent dielectric constant of the combined biomolecule–water system can be written as a sum of four dispersion terms [341]:

$$\varepsilon(\omega) = \varepsilon_\infty + \sum_{j=1}^{4} \frac{\Delta_j}{1 + i\omega\tau_j} \tag{19.1}$$

where ε_∞ denotes the infinite frequency dielectric constant of bulk water, Δ_j the relative weight of a given relaxation type, and τ_j the respective time constant. Δ_1 and τ_1 are the relative weight and time constants associated with the orientational motion of the biomolecule. For a typical protein solution like the myoglobin–water system, τ_1 is about 74 ns [342]. (Δ_2, τ_2) and (Δ_3, τ_3) correspond to the relaxation of biological water associated with the protein. Although these relaxation phenomena are quite different, they have approximately equal weights. The relaxation times are about $\tau_2 = 10$ ns and $\tau_3 = 40$ ps, respectively. This behavior is nearly universal, and is typically referred to as the bimodality of the reorientational response of biological water. Finally, Δ_4 corresponds to the relative weight of the rotational relaxation of bulk water, and τ_4 is the corresponding relaxation time, equal to 8.3 ps. This bimodal behavior is typical and has been observed with DNA, water enclosed within the cavities of cyclodextrin, and in the aqueous medium of reverse micelles.

Many workers have reviewed the results of dielectric measurements on protein–water systems [339–342]. The earliest measurement was made by Oncley [343], who concluded that the carboxyhemoglobin molecule is associated with a rotational relaxation time constant of 84 ns. The most detailed characterization of a protein–water solution was provided by Grant and co-workers [341], who showed that protein-bound water exhibits a range of relaxation time constants. They concluded that these time constants are associated with distinct processes, such as the rotational motions of the protein, biological water and bulk water. Pethig [340] concluded that the primary hydration layer is strongly bound and rotationally hindered and that the microwave dielectric behavior is predominantly influenced by the thermally activated water in the secondary layer of biological water.

Recently, a microscopic explanation of the bimodal relaxation of biological water was presented in which the time-dependent relaxation of biological water is described in terms of a dynamic equilibrium between the free and bound water molecules [338]. It was assumed that only the free water molecules undergo orientational motion; the bound water contribution enters only through the rotation of the biomolecule, which is also considered. The dielectric relaxation is then determined by the equilibrium constant between the two species and the rate of conversion from bound to free state and vice versa. The dielectric relaxation in such complex biomolecular systems, however, depends on several parameters, such as the rotational time constant of the protein molecule, the dimension of the hydration shell, the strength of the hydrogen bond, and the static dielectric constant of the water bound to the biomolecule. The theory included all these aspects in a consistent way. The results were in good agreement with many known results.

This theoretical study was recently extended to investigate dielectric relaxation and solvation dynamics of aqueous protein solutions [342]. The theory suggested that the enhanced dielectric constant for myoglobin solution over that of the bulk water might arise from the order around the hydrated shell. The same phenomenon is expected to be responsible for the anomalous dynamics of protein solutions.

XX. FUTURE PROBLEMS

In this chapter, we discussed several different but related problems of chemical dynamics in solution, including the polar and nonpolar solvation dynamics and ion diffusion and vibrational relaxations (both energy and phase) in dense liquids. Wherever possible, we compared theoretical results with those from experiments and simulations. Although good agreement was obtained in many cases, many problems remain unresolved. In this last section, we take the opportunity to discuss several interesting unsolved problems that may be studied in future.

A. Dielectric Relaxation and Solvation Dynamics in Organized Assemblies

As discussed, orientational and translational motions of water molecules in the restricted environments are dramatically different from those in the bulk. Dielectric relaxation and solvation dynamics may provide two useful techniques for studying water motions in restricted assemblies. Specific interesting problems include zeolites, micelles, protein solutions, and DNA. Some experimental results already exist for these systems. Dielectric relaxation of aqueous protein solutions show bizzare concentration dependence, the origin of which has been addressed only recently [342]; much remains to be explored and understood. There is the curious universal bimodal response of water in micelles and protein solutions by which the time constants of relaxation of the two components differ by almost two orders of magnitude. Although an interpretation has been offered in terms of the dynamic equilibrium between the free and bound water near the surface, much work is left to be done. Actually, these complex problems require much more experimental work than that which is available at present.

B. Effects of Ultrafast Nonpolar Solvent Response on Electron Transfer Reactions

In Section VI, we noted that different experiments on solvation dynamics in monohydroxy alcohols (methanol, butanol) [65,66,243] have reported different results. Although Joo and co-workers [243] observed a nearly universal ultrafast component for all the alcohols, Horng and co-workers [66] and Bingemann and co-workers [65] found the same only for methanol. The origin of such an ultrafast component in all the alcohols has not yet been unambiguously established. Recently, a tentative explanation was offered in terms of a nondipolar, nearest-neighbor solute–solvent interaction [208]; however, if the nonpolar, solute–solvent binary dynamics are indeed responsible for the ultrafast component, then one would like to ask the following question. What are the effects of the ultrafast, nonpolar, solvent dynamic modes on the electron transfer reactions occurring in these solvents? Naturally, the Marcus model [254,255] must be generalized appropriately to investigate the effects of such ultrafast, nonpolar, solvent dynamic modes on electron transfer reactions. Some attempts toward this have been made [17], although a detailed study is warranted.

C. Dielectric Relaxation and Solvation Dynamics in Mixtures

Dielectric relaxation and solvation dynamics are of great interest, because many useful chemical solvents are often mixtures of two polar liquids (e.g., water and DMSO). Little is known about the ultrafast dynamics of these mixtures. In cases of mixtures with water, the other solvent may destroy the extensive hydrogen-bonded network, e.g., this might happen in water–acetonitrile mixtures. Thus the 193 cm^{-1} band may rapidly become weak when the mole fraction of the second constituent exceeds a certain critical value. In any case, the dielectric relaxation of binary mixtures are known to exhibit strong nonideal behavior. Then there is the well-known case of preferential solvation in binary mixtures, which needs to be understood.

Chandra and Bagchi addressed both the problem of the dielectric relaxation and solvation dynamics in binary mixtures [72]. The theoretical studies were, however, restricted to idealized situations. Even then, the theory could explain both the non-ideality of dielectric relaxation and solvation dynamics from a microscopic picture. The work of Chandra and Bagchi [72] can be extended to address real mixtures, such as water and acetonitrile.

D. Concentration Dependence of Ionic Mobility

The concentration dependence of ionic mobility in strong electrolyte solutions exhibits a rich and complex behavior. The solvent-mediated ion–ion and the ion–solvent Coulombic interactions in concentrated solutions profoundly affect the electrical conductivity of the medium. The square root of concentration dependence of ionic mobility $\sqrt{c}$ predicted by the Debye–Huckel–Onsager (DHO) theory [127–130] has only limited validity (valid for $c \leq 10^{-3}$ mol/L). For moderately concentrated solutions ($c \leq 1$ mol/L), an empirical fit, known as Shedlovsky equation [8], was proposed to analyze the conductivity. In addition to the usual $\sqrt{c}$ term in the DHO theory, this equation contains logarithmic and quadratic concentration c^2 dependence. The origin of these higher-order terms has not yet been fully understood. A microscopic theory that includes the dynamics of the solution and retains the molecularity of the system would be of great help in understanding the molecular origin of the diverse concentration dependence of ionic conductivity in electrolyte solution.

Recent theoretical investigations [344] have shown that the relative contributions from the rotational and translational modes can have important consequences in the relaxation of the ion atmosphere in an electrolyte solution. This is an aspect that has not been addressed at all in the Debye–Huckel–Onsager theory of ionic conductivity. Of course, this effect is important only at moderate to large concentrations. The resulting molecular theory provides an expression for the force–force correlation function (or the dielectric friction), which has a structure similar to that of Eq. (14.17) and contains a large number of terms [344]. However, the $\sqrt{c}$ dependence of Λ should be obtained in the correct asymptotic limit. Work in this direction is in progress.

E. Limiting Ionic Conductivity of Halide Anions

It is known that the limiting ionic conductivity of halide ions in water, when plotted as a function of inverse of the crystallographic radius, constitutes a curve that is different from that of the alkali metal ions [131]. This was explained by Lee and Rasaiah [317,318], who showed that the difference could arise from a lack of symmetry between the cation–water and anion–water interactions. The effects of such short-range interactions are yet to be included in a molecular description. This is certainly a problem that deserves further attention.

F. Limiting Ionic Conductivity in Water–Alcohol Mixtures

Limiting ionic conductivity in water–alcohol (e.g., a water and *t*-butyl alcohol) mixture [345] exhibits exotic composition dependence. For example, an inverted parabola is obtained when the limiting ionic conductivity of K^+ (or any monopositive alkali metal) ion is plotted against the mole fraction of alcohol. Here the preferential solvation may play a role in the parabolic dependence of ionic mobility that cannot be understood in terms of the existing continuum models. Therefore, a microscopic theory is needed to investigate this exotic behavior of the limiting ionic conductivity in water–alcohol mixture.

G. Viscosity of Aqueous Solutions of Strong Electrolytes

It is known that some salts increase the viscosity of water, whereas others do the reverse [346,347] when added in low concentrations ($c < 0.2$ normal). For example, barium chloride increases the viscosity of water [346], whereas cesium nitrate and lithium nitrate diminish the viscosity of the medium [348]. The negative curvature of the viscosity concentration curve for aqueous solutions is a rather general observation for many salts. It is also observed that the negative curvature becomes more pronounced at low temperature and in more dilute solutions [346]. Although these observations have been known for a long time, no molecular theory exists that connects the salt concentration with the viscosity of the medium and offers a microscopic explanation of this interesting behavior.

H. Solubility and Solvation Dynamics in Supercritical Water

SCW exhibits anomalous solvent properties; the most important is its ability to solubilize organic solutes of various sizes such as benzene, toluene, and oils [68–70]. The critical point of water is located at $P_c = 22.1$ MPa and $T_c = 647$ K, with $d_c = 0.32$ g/cm^3. Most of the experimental studies on solubility in supercritical water are performed in the liquid-like density region (0.6–0.9 g/cm^3), with the applied pressure typically in the 40–60 MPa range. The dielectric constant of water at T_c is around 6. This makes water behave like an organic solvent; therefore, organic substances become highly soluble in water at these supercritical conditions. This is certainly an anomalous solvent behavior of SCW compared to that of normal water at ambient conditions. Although the unusual solvent properties of SCW have been known for some time, no microscopic explanation is available for its anomalous solubility. Recently, a theoretical study was carried out to understand the exotic solubility of SCW [349].

The problem of solvation dynamics and ionic conductivity of supercritical water are two related important problems that need to be considered urgently.

Balbuena and co-workers [350] carried out molecular dynamics simulation studies to understand the ion transport mechanism and water reorientation dynamics in supercritical water. Their results are interesting, and a proper theoretical work is warranted in this direction.

I. Dynamic Response Functions from Nonlinear Optical Spectroscopy

Although a large number of nonlinear optical techniques (photon echo; four wave mixing; transient hole burning; Raman echo, third-, fifth-, and seventh-order responses) are being used to probe the solute–solvent dynamics directly in the time domain, a theoretical understanding of the relevant dynamic response function is still missing [351-352]. We will use the example of transient hole burning. Here the dynamics of the medium is probed via the vibrational relaxation of a bond, which is coupled to the medium dynamics. Again, it is only recently that a full model of this problem has been attempted [352]. The same scenario exists for almost all the other techniques—we do not even have a full calculation for the third-order off-resonant response function for any realistic liquid. One expects the nonlinear optical techniques to be really useful in the study of the mixtures in which different time scales originate from different sources.

These represent only a few of the many unsolved problems in the rich field of chemical dynamics in the solution phase. With the availability of sophisticated nonlinear optical techniques, a much better understanding of the details of the solvent dynamic response is beginning to emerge. This, in turn, will help our understanding of the dynamic solvent effects on various molecular relaxation processes in solution. We can look forward to an exciting future of these and related problems of classical physical chemistry, which is currently undergoing a rejuvenation.

APPENDIX A. Derivation of $\mathbf{F}_{10}(t)$ and $\mathbf{F}_{11}(t)$

Here we derive the expressions for $F_{10}(t)$ and $F_{11}(t)$ needed for the calculation of $S_E(t)$ using Eq. (4.20). The spherical harmonic expansion and subsequent integration over $\mathbf{\Omega}$ takes Eq. (4.19) to the following expression:

$$E_{\mathrm{sol}}(\mathbf{k}, \mathbf{\Omega}, t) = -k_B T \sum_{\ell_1, m_1, \ell_2, \ell_2', m_2} (-1)^{m_2} Y_{\ell_1 m_1}(\mathbf{\Omega}) Y_{\ell_2 m_2}(\mathbf{\Omega}) \times \int d\mathbf{q} c_{\mathrm{dd}}(\ell_2, \ell_2', m_2; q) A_{\ell_1 m_1}(\mathbf{q}, t) a_{\ell_2 m_2}(\mathbf{k} - \mathbf{q}, t) \quad (\mathrm{A.1})$$

where $Y_{\ell m}(\mathbf{\Omega})$ is the spherical harmonics of projection m and rank ℓ. Here, $a_{\ell m}(\mathbf{k}, \mathbf{t})$ is defined as follows

$$a_{\ell m}(\mathbf{k}, \mathbf{t}) = \int d\mathbf{\Omega}\ Y^{\star}_{\ell m}(\mathbf{\Omega})\ \delta\rho(\mathbf{k}, \mathbf{\Omega}, t) \tag{A.2}$$

where $Y^{\star}_{\ell m}(\mathbf{\Omega})$ is the complex conjugate to $Y_{\ell m}(\mathbf{\Omega})$. $A_{\ell m}(\mathbf{q}, \mathbf{t})$ is the ℓmth component of the wavenumber- and time-dependent fluctuations in solute density. Consequently,

$$C_{EE}(t) = 2(k_{\mathrm{B}}T)^2\pi^2\rho_0 \sum_{\ell_1, m_1} \delta_{\ell_1,1}[\delta_{m_1,0} + \delta_{m_1,1} + \delta_{m_1,1}]$$
$$\times \int_0^{\infty} dq\ q^2 S^{\ell_1 m_1}_{\mathrm{self}}(q, t)|c_{\mathrm{dd}}(\ell_1, \ell_1, m_1; q)|^2 S^{\ell_1 m_1}_{\mathrm{solv}}(k - q, t) \tag{A.3}$$

where $S^{\ell m}_{\mathrm{self}}(\mathbf{q}, t)$ is the ℓmth component of the wavenumber-dependent self-orientational dynamic structure factor, which is defined as follows:

$$S^{\ell m}_{\mathrm{self}}(\mathbf{q}, t) = \langle A_{\ell m}(-\mathbf{q}, 0) A_{\ell m}(\mathbf{q}, t)\rangle \tag{A.4}$$

$S^{\ell m}_{\mathrm{solv}}(\mathbf{k} - \mathbf{q}, t)$ is the ℓmth component of the wavenumber-dependent solvent dynamic structure factor, which is defined as follows:

$$S^{\ell m}_{\mathrm{solv}}(\mathbf{k} - \mathbf{q}, t) = \frac{\langle a_{\ell m}(\mathbf{q} - \mathbf{k}, 0) a_{\ell m}(\mathbf{k} - \mathbf{q}, t)\rangle}{N} \tag{A.5}$$

where $\langle\ \rangle$ denotes the equilibrium ensemble average over all orientations (i.e., over all probable $\mathbf{\Omega}_0$ and $\mathbf{\Omega}$ values) and the δ-functions stands for the Kronecker δ-functions.

In Eq. (4.20) the effects of the self-motion of the dipolar solute was included through the self-dynamic structure factor $S^{\ell m}_{\mathrm{self}}(\mathbf{q}, t)$. The term $\ell = 0$, $m = 0$ corresponds to the translational mode of the solute, and other terms include the contribution of the rotational motion. As in the case of ion solvation, the coupling between the solute dipole and the solvent dipolar molecules enter through the direct correlation function term $c_{\mathrm{dd}}(q)$ which couples the solvent dynamics $S^{\ell m}_{\mathrm{solv}}(q, t)$ with the solvation process. Here again, the $\mathbf{k} = 0$ limit of Eq. (2.62) allows one to study the progress of solvation of a mobile dipolar solute probe with time.

In our present calculation, we calculated these correlation functions from MSA (properly corrected both at the $q \to 0$ and the $q \to \infty$ limits), because the analytical solutions for these functions for a binary mixture are available. The nonvanishing components of these direct correlation functions are the $c(000; q)$, $c(110; q)$, and $c(111; q)$ components only. Consequently, the expression for SETCF takes up the following form

$$C_{EE}(t) = 2\left[k_B T\pi\right]^2 \rho_0\left[F_{10}(t) + 2F_{11}(t)\right] \tag{A.6}$$

So, the expression for normalized SEETCF of mobile dipolar solute becomes

$$S_E(t) = \frac{\left[F_{10}(t) + 2F_{11}(t)\right]}{\left[F_{10}(t=0) + 2F_{11}(t=0)\right]} \tag{A.7}$$

where

$$F_{10}(t) = \int_0^\infty dq\, q^2 S^{10}_{\text{self}}(q, t)|c_{\text{dd}}(110; q)|^2 S^{10}_{\text{solv}}(q, t) \tag{A.8}$$

and

$$F_{11}(t) = \int_0^\infty dq\, q^2 S^{11}_{\text{self}}(q, t)|c_{\text{dd}}(111; q)|^2 S^{11}_{\text{solv}}(q, t) \tag{A.9}$$

APPENDIX B. Dynamic Structure Factor for Calculating $S(q, t)$

To calculate $S(q, t)$, we use the following dynamic structure factor obtained from viscoelastic approximation given by the Mori continued fraction [174,175]:

$$S(q, z) = \frac{S(q)}{z + \langle\omega_q^2\rangle[z + \Delta_q(z + \tau_q^{-1})^{-1}]^{-1}} \tag{B.1}$$

where $S(q)$ is the static structure factor obtained using the Weeks–Chandler–Andersen scheme [213]. The details regarding the calculation of the other static quantities are discussed elsewhere [249–252]; here we give only the necessary expressions.

$$\langle\omega_q^2\rangle = \frac{kTq^2}{mS(q)} \qquad \text{and} \qquad \Delta_q = \omega_l^2(q) - \langle\omega_q^2\rangle,$$

where $\omega_l^2(q)$ is the second moment of the longitudinal current correlation function, given by [26b,249–252]

$$\omega_l^2(q) = 3m^{-1}q^2 k_B T + \omega_0^2 + \gamma_d^l(q) \tag{B.2}$$

where, the longitudinal component of the vertex function $\gamma_d^l(q)$ and the Einstein frequency of the solvent ω_0 can be calculated from the interacting potential $v(r)$ and the radial distribution function $g(r)$ from the following expressions [26b,249–252]:

$$\gamma_d^l(q) = -m^{-1}\rho \int d\mathbf{r}\ \exp(-i\mathbf{q}.\mathbf{r})g(r)\frac{\delta^2}{\delta z^2}v(r) \tag{B.3}$$

and

$$\omega_0^2 = \frac{\rho}{3m}\int d\mathbf{r}\ g(r)\nabla^2 v(r) \tag{B.4}$$

Here we assume that the interacting potential is that given by the Lennard–Jones potential; then τ_q can be obtained from the relation $\tau_q = 2(\Delta_q/\pi)^{\frac{1}{2}}$. Thus once Δ_q and ω_0 are calculated, we can put these quantities in Eq. (6.11) and Laplace invert it numerically [64] to obtain $S(q, t)$.

APPENDIX C. Calculation of $R_{tt}(t)$

Here we give the expression for $R_{tt}(t)$, which is shown to be given by [175,249–251]

$$R_{tt}(t) = \frac{1}{\rho}\int [d\mathbf{q}'/(2\pi)^3][1 - (\hat{q}.\hat{q}')^2][\gamma^t{}_{d12}(q')]^2\Omega_o^{-4}[F^s(q', t) - F^o(q', t)]C_{tt}(q', t) \tag{C.1}$$

One input parameter necessary for the calculation of $R_{tt}(t)$ is the vertex function of the solute–solvent mixture $\gamma_{d12}^t(q)$, which actually takes care of the interaction of the solute motion with the current mode of the solvent. The other parameters required are the Einstein frequency of the solute in presence of the solvent molecules Ω_0, the dynamic structure factor of the solute, and the transverse current autocorrelation function of the solvent $C_{tt}(q, t)$. $C_{tt}(q, t) = \mathcal{L}^{-1}[C_{tt}(q, z)]$, where $C_{tt}(q, z)$ has the following expression [174,175]:

$$C_{tt}(q, z) = \frac{1}{z + \frac{\omega_t^2(q)}{z + \tau_t^{-1}(q)}} \tag{C.2}$$

where $\omega_t^2(q)$ is the second moment of the transverse current correlation function which is given by [174,175,249–251]

$$\omega_t^2(q) = q^2\frac{k_B T}{m} + \Omega_o^2 + \gamma_d^t(q) \tag{C.3}$$

where $\gamma_d^t(q)$ is the transverse component of the vertex function, which has the same expression as $\gamma_{d12}^t(q)$, only replacing $g_{12}(r)$ by $g(r)$ and $v_{12}(r)$ by $v(r)$.

For $\tau_t(q)$ we used the following expression [250–252]:

$$\tau_t^{-2}(q) = 2\omega_t^2(q) + \frac{\tau_t^{-2}(0) - 2\omega_t^2(q) + 2q^2\frac{k_B T}{m}}{1 + (q/q_o)^2} \tag{C.4}$$

where q_o is an adjustable parameter that actually determines the transition of the behavior of $C_{tt}(q, z)$ from small q to large q. For argon $q_o = 1.5\,\text{Å}^{-1}$ and $\tau_t^{-1}(0) = \lim_{q\to 0}[m\rho\omega_t^2(q)]/q^2\eta$.

Here η is the zero frequency shear viscosity that is calculated from the mode-coupling expression given by [174,175],

$$\eta = \eta_E + k_B T/60\pi^2 \int_0^\infty dq\, q^4[S'(q)/S(q)]^2 \int_0^\infty dt\, [F(q,t)/S(q)]^2 \tag{C.5}$$

where $S'(q)$ is the first derivative of the static structure factor and η_E is the Enskog shear viscosity, given by [174,175]

$$\eta_E = \eta_B \frac{[1 + 3.2\phi g_{12}(\sigma_{12}) + 12.18\phi^2 g_{12}{}^2(\sigma_{12})]}{g_{12}(\sigma_{12})} \tag{C.6}$$

where $\phi = \pi\rho^*/6$, $\eta_B = 0.179(mk_B T)^{\frac{1}{2}}/\sigma^2$ and $\sigma_{12} = \frac{\sigma + \sigma_2}{2}$.

We can thus straightforwardly calculate the shear viscosity by using the intermediate scattering function. We compared the value of the viscosity calculated from the Eq. (C.5) with simulation results [249–252]. The agreement is excellent in the normal liquid regime. For example, at $T^*(= k_B T/\varepsilon) = 0.728$ and $\rho^*(= \rho\sigma^3) = 0.844$, $\eta(\text{simulation}) = 2.53$ and $\eta(\text{calculated}) = 2.51$. Here σ and ε are the usual Lennard–Jones parameters for the diameter and the well depth, respectively. The values of η are scaled by $(mk_B T)^{\frac{1}{2}}/\sigma^2$.

Acknowledgments

It is a pleasure to thank our collaborators, Dr. Srabani Roy, Dr. Nilashis Nandi, Ms Sneha Sudha Komath, Ms N. Gayathri, and Ms Sarika Bhattacharyya. We are much indebted to Professor Graham Fleming for his continued encouragement, help and many suggestions throughout the course of the work reported here. We gratefully acknowledge our enlightening correspondence with Professor Mark Maroncelli, who has helped improve many aspects of our work. One of us (BB) has received continued support and encouragement from Professor CNR Rao, FRS, which is thankfully acknowledged. We thank Professor Amalendu Chandra for help in writing Section XVIII and for discussions. We thank Professors Paul Barbara, Mark Berg, Kankan Bhattacharyya, Nikolaus Ernstring, Branka Ladanyi, Abraham Nitzan, Iwao Ohmine, David Oxtoby, Peter Rossky, Shinji Saito, James Skinner, Richard Stratt, Keisuke Tominaga, S. Umapathy, and Keitaro Yoshihara for sending preprints of their work and for much useful correspondence. RB acknowledges the help received from Dr. Sanjay Bandyopadhyay and Dr. Swapan Kumar Pati. The work reported here was supported in part by grants from Department of Science and Technology, India and the Council of Scientific and Industrial Research (CSIR), India. RB thanks the CSIR for a research fellowship.

References

1. G. R. Fleming and P. G. Wolynes, *Phys. Today* **43**, 36 (1990).
2. G. A.Voth and R. M. Hochstrasser, *J. Phys. Chem.* **100**, 13034 (1996), and references therein.
3. S. Mukamel and Y. J. Yan, *Acc. Chem. Res.* **22**, 301 (1989) and references therein; S. Mukamel, *Principles of Non-linear Optical Spectroscopy.* Oxford University Press, New York, 1995.
4. H. A. Kramers, *Physica* **7**, 284 (1940); E. Montroll and K. E. Schuler, *Adv. Chem. Phys.* **1**, 361 (1958).

4a. D. Chandler, *J. Chem. Phys.* **68**, 2959 (1978); J. L. Skinner and P. G. Wolynes, *ibid.* **69**, 2143 (1978).

5. R. F. Grote and G. T. Hynes, *J. Chem. Phys.* **73**, 2715 (1980); J. T. Hynes, in *The Theory of Chemical Reactions* (M, Baer, ed.),Vol. 4. Chem. Rubber Publ. Co., Boca Raton, FL, 1985.
6. B. Bagchi, *Int. Rev. Phys. Chem.* **6**, 1 (1987); B. Bagchi and D. Oxtoby, *J. Chem. Phys.* **78**, 2735 (1983).
7. M.W. Makinen, S. A. Schichman, S. C. Gill, and H. B. Gray, *Science* **222**, 929 (1983); G. L. Closs and J. R. Miller, *ibid.* **240**, 440 (1988); G. Mclendon, *Acc. Chem. Res.* **21**, 160 (1988).
8. A. Nitzan, *Adv. Chem. Phys.* **70**, 489 (1988); B. Carmeli and A. Nitzan, *Phys. Rev. Lett.* **51**, 233 (1983); *Chem. Phys. Lett.* **106**, 329 (1984); B. J. Matkowsky, A. Nitzan, and Z. Schuss, *J. Chem. Phys.* **88** 4765 (1988).
9. E. Pollak, *J. Chem. Phys.* **86**, 3944 (1987); P. Hangii, T. Talkner, and M. Berkovec, Rev. Mod. Phys. **62**, 250 (1990).
10. E. M. Kosower and D. Hupert, *Annu. Rev. Phys. Chem.* **37**, 127 (1986).
11. R. A. Marcus and N. Sutin, *Biochim. Biophys. Acta* **811**, 275 (1985); H. Sumi and R. A. Marcus, *J. Chem. Phys.* **84**, 4894 (1986).
12. P. F. Barbara and W. Jarzeba, *Adv. Photochem.* **15**, 1 (1990); *Acc. Chem. Res.* **21**, 195 (1988).
13. G. R. Fleming, *Chemical Applications of Ultrafast Spectroscopy.* Oxford University Press, Oxford, 1986.

13a. D. A. Wiersma, ed., *Femtosecond Reaction Dynamics.* North-Holland, Amsterdam, 1994; A. H. Zewail, ed., *Femtochemistry: Ultrafast Dynamics of the Chemical Bond.* World Scientific, Singapore, 1994.

14. J. T. Hynes, *Annu. Rev. Phys. Chem.* **36**, 573 (1985); J. D. Simon, *Acc. Chem. Res.* **21**, 128 (1988); P. J. Rossky and J. D. Simon, *Nature (London)* **370**, 263, (1994).
15. B. Bagchi, *Annu. Rev. Phys. Chem.* **40**, 115 (1989).

15a. B. Bagchi and A. Chandra, *Adv. Chem. Phys.* **80**, 1 (1991).

16. S. Roy and B. Bagchi, *J. Chem. Phys.* **102**, 6719, 7937 (1995).
17. B. Bagchi and N. Gayathri, *Adv. Chem. Phys.* (in press); J. Jortner and M. Bixon, eds., *Electron Transfer Reactions—From Isolated Molecules to Biomolecules*, Spec. Issue, Part II, Chap. 1; N. Gayathri and B. Bagchi, *J. Phys. Chem.* **100**, 3056 (1996); *J. Chim. Phys.* **93**, 1652 (1996).
18. S. Mukamel, *Adv. Chem. Phys.* **70**, (Part I), 165 (1988); S. Mukamel and R. F. Loring, *J. Opt. Soc. Am. B* 3, 595 (1986); Y. J. Yan and S. Mukamel, *J. Chem. Phys.* **87**, 5840 (1987); M.

Sparpaglione and S. Mukamel, *ibid.* **88**, 3263, 4300 (1988); Y. J. Yan, M. Sparpaglione, and S. Mukamel, *J. Phys. Chem.* **92**, 4842 (1988).
19. R. M. Hochstrasser, *Pure Appl. Chem.* **52**, 2683 (1980).
20. D. H. Waldeck, *Chem. Rev.* **91**, 415 (1991), and references therein.
21. See the papers in *Activated Barrier Crossing*. G. R. Fleming and P. Hangii, eds., World Scientific, Singapore, 1993.
22. H. Heitele, *Angew. Chem., Int. Ed. Engl.* **32**, 359 (1993); M. J. Weaver, *Chem. Rev.* **92**, 463 (1992); M. J. Weaver and G. E. McMannis, III, *Acc. Chem. Res.* **23**, 294 (1990).
23. K. Tominaga and K. Yoshihara, *Phys. Rev. Lett.* **74**, 3061 (1995); *J. Chem. Phys.* **104**, 1159, 4419 (1996).
23a. N. Gayathri, S. Bhattacharyya, and B. Bagchi, *J. Chem. Phys.* **107**, 10381 (1997).
24. L. Landau and E. Teller, *Z. Sowjetunion* **10**, 34 (1936).
25. J. E. Straub, M. Berkovec, and B. J. Berne, *J. Chem. Phys.* **89**, 4833 (1988); B. J. Berne, M. E. Tuckerman, J. E. Straub, and A. L. R. Bug, *ibid.* **93**, 5084 (1990).
26. T. R. Kirkpatrick, *J. Non-Cryst. Solids* **75**, 437 (1985); *Phys. Rev. Lett.* **53**, 1735 (1984); *Phys. Rev.* **A 32**, 3130 (1985); T. R. Kirkpatrick and J.C. Nieuwoudt, *ibid.* **33**, 2651 (1986).
26a. L. Sjögren and A. Sjölander, *J. Phys.* **C 12**, 4369 (1979).
26b. W. Götze, in *Liquids, Freezing and Glass Transition* p. 292 (D. Levesque, J. P. Hansen, and J. Zinn-Justin, eds.). Elsevier North Holland, Amsterdam, 1991; J. Bosse, W. Gotze, and M. Lueke, *Phys. Rev. A* **17**, 434 (1978); M. C. Marchetti, *ibid.* **33**, 3363 (1986).
27. S. J. Rosenthal, X. Xie, M. Du, and G. R. Fleming, *J. Chem. Phys.* **94** 4715 (1991).
27a. M. Maroncelli, P.V. Kumar, A. Papazyan, M. L. Horng, S. J. Rosenthal, and G. R. Fleming, *AIP Conf. Proc.* **298**, 310 (1993).
28. M. Maroncelli, J. McInnis, and G. R. Fleming, *Science* **243**, 1674 (1989).
29. R. Jimenez, G. R. Fleming, P. V. Kumar, and M. Maroncelli, *Nature (London)* **369**, 471 (1994).
30. G. R. Fleming and M. Cho, *Annu. Rev. Phys. Chem.* **47**, 109 (1996).
31. M. Maroncelli, *J. Mol. Liq.* **57**, 1 (1993).
32. S. Passino, Y. Nagasawa and G. R. Fleming, *J. Chem. Phys.* **107**, 6094 (1997).
33. M. Cho, J. Y. Yu, T. Joo, Y. Nagasawa, S. A. Passino, and G. R. Fleming, *J. Phys. Chem.* **100**, 11944 (1996).
34. M. Maroncelli and G. R. Fleming, *J. Chem. Phys.* **86**, 6221 (1987).
35. M. Maroncelli and G. R. Fleming, *J. Chem. Phys.* **89**, 5044 (1988).
36. M. Maroncelli, P.V. Kumar, and A. Papazyan, *J. Phys. Chem.* **97**, 13 (1993); P.V. Kumar and M. Maroncelli, *J. Chem. Phys.* **103**, 3038 (1995).
37. E. A. Carter and J. T. Hynes, *J. Chem. Phys.* **94**, 5961 (1991).
38. Y. J. Chang and E. W. Castner, Jr., *J. Phys. Chem.* **98**, 9712 (1994).
38a. C. F. Chapman, R. S. Fee, and M. Maroncelli, *J. Phys. Chem.* **99**, 4811 (1995); R. S. Fee and M. Maroncelli, *Chem. Phys. Lett.* **183**, 235 (1994); M. Maroncelli, *J. Chem. Phys.* **106**, 1545 (1997).
39. Y. T. Mazurenko and N. G. Bakshiev, *Opt. Spectrosc.* **28**, 490, (1970).
40. G. van der Zwan and J. T. Hynes, *J. Phys. Chem.* **89**, 4181 (1985).
41. B. Bagchi, D. W. Oxtoby, and G. R. Fleming, *Chem. Phys.* **86**, 257 (1984).
42. B. Bagchi, E. W. Castner, Jr., and G. R. Fleming, *J. Mol. Struct. Theochem.* **194**, 171 (1989).

43. E. W. Castner, Jr., G. R. Fleming, and B. Bagchi, *J. Chem. Phys.* **89**, 3519 (1988).
44. M. Born, *Z. Phys.* **1**, 45 (1920).
45. L. Onsager, *J. Am. Chem. Soc.* **58**, 1485 (1935).
46. A. Declemy, C. Rullierie, and P. H. Kottish, *Chem. Phys. Lett.* **133**, 448 (1987).
47. L. A. Halliday and M. R. Topp, *Chem. Phys. Lett.* **40**, 45 (1977); *J. Phys. Chem.* **82**, 2273 (1978).
48. H. Lessing and M. Richert, *Chem. Phys. Lett.* **46**, 111 (1977).
49. T. Okamura, M. Sumitani, and K. Yoshihara, *Chem. Phys. Lett.* **94**, 339 (1983).
50. M. A. Kahlow, T. J. Kang, and P. F. Barbara, *J. Chem. Phys.* **88**, 2372 (1988).
51. E. W. Castner, B. Bagchi, M. Maroncelli, S. P. Webbs, A. J. Ruggiero, and G. R. Fleming, *Ber. Bunsenges. Phys. Chem.* **92**, 363 (1988).
52. S. W. Yeh, L. A. Philips, S. P. Webb, L. F. Buhse, and J. H. Clark, in *Ultrafast Phenomena* (D. H. Oston and K. B. Eisenthal, eds.), Vol. 4, p. 359. Springer, Berlin and New York, 1987.
53. V. Nagarajan, A. M. Brearly, T. J. Kang, and P. F. Barbara, *J. Chem. Phys.* **86**, 3183 (1987).
54. E. W. Castner, M. Maroncelli, and G. R. Fleming, *J. Chem. Phys.* **86**, 1090 (1987).
55. S. G. Su and J. D. Simon, *J. Phys. Chem.* **91**, 2693 (1987).
56. J. D. Simon and S. G. Su, *J. Phys. Chem.* **87**, 7016 (1987).
57. E. W. Castner, Jr., Ph.D. Thesis, University of Chicago, Chicago, 1988.
58. C. Rullierie, A. Declemy and P. H. Kottish, in *Ultrafast Phenomena* (G. R. Fleming and A. E. Siegman, eds.), Vol. 5, p. 312. Springer, Berlin and New York, 1987.
59. A. Declemy and C. Rullierie, *Chem. Phys. Lett.* **146**, 1 (1988)
60. R. Kohlrausch, *Pogg's Ann.* (Liepzig) **12**, 393 (1847).
61. G. Williams and D. C. Watts, *Trans. Faraday Soc.* **66**, 80 (1970).
62. M. Maroncelli and G. R. Fleming, *J. Chem. Phys.* **89**, 875 (1988).
63. N. Nandi and B. Bagchi, *J. Phys. Chem.* **100** 13914 (1996).
64. A. M. Jonkman, P. van der Meulen, H. Zhang, and M. Glasbeek, *Chem. Phys. Lett.* **256**, 21 (1996); P. van der Meulen, H. Zhang, A. M. Jonkman, and M. Glasbeek, *J. Phys. Chem.* **100**, 5367 (1996).
65. D. Bingemann and N. P. Ernsting, *J. Chem. Phys.* **102**, 2691 (1995); N. P. Ernsting, N. E. Konig, K. Kemeter, S. Kovalenko, and J. Ruthmann, *AIP Conf. Proc.* **441** (1995).
66. M. L. Horng, J. A. Gardecki, A. Papazyan, and M. Maroncelli, *J. Phys. Chem.* **99**, 17311 (1995).
67. C. A. Eckert, B. L. F. Knutson and P. G. Debenedetti, *Nature* (*London*) **383**, 313 (1996); J. S. Sewald, *ibid.* **370**, 285 (1994); M. M. Hoffman and M. S. Conradi, *J. Am. Chem. Soc.* **119**, 3811 (1997); R. F. Prini and M. L. Japas, *Chem. Soc. Rev.* **23**, 155 (1994).
68. S. Kim and K. P. Johnston, *ACS Symp. Ser.* **329** (1987).
69. D. G. Peck, A. Mehta, and K. P. Johnston, *J. Phys. Chem.* **93**, 4297 (1989).
70. K. P. Johnston and C. Haynes, *AIChE* J. **33**, 2017 (1987).
71. M. Re and D. Laria, *J. Phys. Chem.* **B 101**, 10494 (1997).
72. A. Chandra and B. Bagchi, *J. Phys. Chem.* **95**, 2529 (1991); A. Chandra, *Chem. Phys.* **195**, 93 (1995).
73. T. Fonseca and B. M. Ladanyi, *J. Chem. Phys.* **93**, 8148 (1990).
74. M. S. Skaf and B. M. Ladanyi, *J. Chem. Phys.* **102**, 6542 (1995).
75. B. M. Ladanyi and M. S. Skaf, *J. Phys. Chem.* **100**, 1368 (1996).

76. T. J. E. Dey and G. N. Patey, *J. Chem. Phys.* **106**, 2782 (1997); A. Chandra and B. Bagchi, *J. Mol. Liq.* **57**, 39 (1993); A. Chandra, *Chem. Phys. Lett.* **235**, 133 (1995); Y. P. Puhovski and B. M. Rode, *J. Chem. Phys.* **107**, 6908 (1997).
77. R. A. Averbach, J. A. Synowiec and G. W. Robinson, in *Picosecond Phenomena II* (R. M. Hochstrasser, ed.), p. 215. Springer, Berlin and New York, 1980.
78. G. W. Robinson, R. J. Robbins, G. R. Fleming, J. M. Morris, A. E. W. Knight, and R. J. S. Morrison, *J. Am. Chem. Soc.* **100**, 7145 (1978).
79. I. Rips, *Chem. Phys. Lett.* **245**, 79 (1995).
80. A. Migus, Y. Gauduel, J. Martin, and A. Antonetti, *Phys. Rev. Lett.* **58**, 1559 (1987).
81. F. Long, H. Lu, and K. Eisenthal, *Phys. Rev. Lett.* **64**, 1469 (1990).
82. Y. Gauduel, S. Pommeret, A. Migus, and A. Antonetti, *J. Phys. Chem.* **95**, 533 (1991).
83. Y. Gauduel, S. Pommeret, and A. Antonetti, *J. Phys. Chem.* **97**, 134 (1993).
84. J. Alfano, P. Walhout, Y. Kimura, and P. F. Barbara, *J. Chem. Phys.* **98**, 5996 (1993).
85. C. Silva, P. K. Walhout, K. Yokoyama, and P. F. Barbara, *Phys. Rev. Lett.* **80**, 1086 (1998).
86. B. J. Schwartz and P. J. Rossky, *Phys. Rev. Lett.* **72**, 3282 (1994).
87. B. J. Schwartz and P. J. Rossky, *J. Chem. Phys.* **101**, 6917 (1994).
88. B. J. Schwartz and P. J. Rossky, *J. Chem. Phys.* **101**, 6902 (1994).
89. B. J. Schwartz and P. J. Rossky, *J. Chem. Phys.* **105**, 6997 (1996).
90. B. J. Schwartz, E. R. Bittner, O.V. Prezdo, and P. J. Rossky, *J. Chem. Phys.* **104**, 5942 (1996).
91. A. Staib and D. Borgis, *J. Chem. Phys.* **103**, 2642 (1995).
92. S. Bratos and J. C. Leicknam, *Chem. Phys. Lett.* **261**, 117 (1996).
93. S. Bratos, J. C. Leicknam, D. Borgis, and A. Staib, *Phys. Rev.* **E 55**, 7217 (1997).
94. R. Biswas and B. Bagchi, *Proc.—Indian Acad. Sci., Sect. A* **109**, 347 (1997).
95. J. T. Fourkas and M. Berg, *J. Chem. Phys.* **98**, 7773 (1993).
96. M. Berg, *Chem. Phys. Lett.* **228**, 317 (1994).
97. J. Yu and M. Berg, *J. Chem. Phys.* **96**, 8741 (1992).
98. J. Yu, P. Earvolino, and M. Berg, *J. Chem. Phys.* **96**, 8750 (1992).
99. J. Yu and M. Berg, *J. Phys. Chem.* **97**, 1758 (1993).
100. J. T. Fourkas, A. Benigno and M. Berg, *J. Chem. Phys.* **99**, 8552 (1993).
101. J. Ma, D. V. Bout, and M. Berg, *Phys. Rev. E* **54**, 2786 (1996).
102. J. Ma, D. V. Bout, and M. Berg, *J. Chem. Phys.* **103**, 9146 (1995).
103. J. Ma, J. T. Fourkas, D. V. Bout, and M. Berg, *ACS Symp. Ser.* **676** (1997).
104. L. Reynolds, J. A. Gardecki, S. J.V. Frankland, M. L. Horng, and M. Maroncelli, *J. Phys. Chem.* **100**, 10337 (1996).
105. B. Bagchi, *J. Chem. Phys.* **100**, 6658 (1994).
106. A. M. Walsh and R. F. Loring, *Chem. Phys. Lett.* **186**, 77 (1991).
107. J. G. Saven and J. L. Skinner, *J. Chem. Phys.* **99**, 4391 (1993).
108. M. D. Stephens, J. G. Saven, and J. L. Skinner, *J. Chem. Phys.* **106**, 2129 (1997).
109. G. T. Evans, *J. Chem. Phys.* **103**, 8980 (1995).
110. M. Berg, *J. Phys. Chem.* **A 102**, 17 (1998); M. A. Berg and H. W. Hubble, *Chem. Phys.* (in press).
111. R. M. Stratt and M. Cho, *J. Chem. Phys.* **100**, 6700 (1994).
112. R. M. Stratt., *Acc. Chem. Res.* **28**, 201 (1995).

113. G. Goodyear, R. E. Larsen, and R. M. Stratt, *Phys. Rev. Lett.* **76**, 243 (1996); R. E. Larsen, E. F. David, G. Goodyear, and R. M. Stratt, *J. Chem. Phys.* **107**, 524 (1997).

114. B. M. Ladanyi and R. M. Stratt, *J. Phys. Chem.* **99**, 2502 (1995); **100**, 1266 (1996).

115. M. Buchner, B. M. Ladanyi, and R. M. Stratt, *J. Chem. Phys.* **97**, 8522 (1992); B. M. Ladanyi, in *Electron and Ion Transfer in Condensed Media* (A. A. Kornyshev, M. Toshi, J. Ulstrup and M. P. Tosi, eds.), p. 110. World Scientific, Singapore, 1997.

116. R. M. Stratt amd M. Maroncelli, *J. Phys. Chem.* **100**, 12981 (1996).

117. T. S. Kalbfleisch, D. L. Ziegler, and T. Keyes, *J. Chem. Phys.* **105**, 7034 (1996).

118. T. S. Kalbfleisch and D. L. Ziegler, *J. Chem. Phys.* **107**, 9878 (1997).

119. D. W. Oxtoby, *Adv. Chem. Phys.* **47**, 487 (1981).

120. D. W. Oxtoby, *Adv. Chem. Phys.* **40**, 1 (1979).

121. D. W. Oxtoby, *Annu. Rev. Phys. Chem.* **32**, 77 (1981).

122. D. W. Oxtoby, *J. Chem. Phys.* **70**, 2605 (1979).

123. A. Laubereau and W. Kaiser, *Annu. Rev. Phys. Chem.* **26**, 83 (1975); *Rev. Mod. Phys.* **50**, 607 (1978).

124. K. Eisenthal, *Annu. Rev. Phys. Chem.* **28**, 207 (1977); W. A. Steele, *Adv. Chem. Phys.* **34**, 1 (1976).

125. B. Faltermeier, R. Protz, and M. Maier, *Chem. Phys.* **62**, 377 (1981).

126. R. Rey and J. T. Hynes, *J. Chem. Phys.* **104**, 2356 (1996).

127. S. Glasstone, *An Introduction to Electrochemistry.* Litton Education Publishing, New York, 1942; J. O'M. Bockris and A. K. N. Reddy, *Modern Electrochemistry.* Plenum, New York, 1973; P. W. Atkins, *Physical Chemistry*, 5th ed., Part III, Chap. 24. Oxford University Press, Oxford, 1994.

128. G.W. Castellan, *Physical Chemistry*, 3rd ed., Chap. 31. Addison-Wesley, Reading, MA, 1971.

129. H. S. Harned and B. B. Owen, *The Physical Chemistry of Electrolyte Solutions*, 3rd ed. Reinhold, New York, 1958.

130. H. S. Frank, *Chemical Physics of Ionic Solutions.* Wiley, New York, 1956; R. A. Robinson and R. H. Stokes, *Electrolyte Solutions*, 2nd ed. Butterworth, London, 1959.

131. R. L. Kay and D. F. Evans, *J. Phys. Chem.* **70**, 2325 (1966).

131a. R. D. Shannon and C. T. Prewitt, *Acta Crystallogr. Sect. B* **25**, 925–946 (1969).

132. L. Onsager, *Can. J. Chem.* **55**, 1819 (1977).

133. D. F. Calef and P. G. Wolynes, *J. Chem. Phys.* **78**, 4145 (1983).

134. P. G. Wolynes, *J. Chem. Phys.* **86**, 5133 (1987).

135. M. S. Wertheim, *J. Chem. Phys.* **55**, 4291 (1971).

136. P. J. Rossky, *Annu. Rev. Phys. Chem.* **36**, 321 (1985).

137. G. Stell, in *Statistical Mechanics. Part A. Equilibrium Techniques* (B. J. Berne, ed.). Plenum, New York, 1977.

138. I. Rips, J. Klafter, and J. Jortner, *J. Chem. Phys.* **88**, 3246 (1988).

139. I. Rips, J. Klafter, and J. Jortner, *J. Chem. Phys.* **89**, 4288 (1988).

140. B. Bagchi and A. Chandra, *J. Chem. Phys.* **90**, 7338 (1989); *Chem. Phys. Lett.* **155**, 533 (1989).

141. A. Chandra and B. Bagchi, *J. Phys. Chem.* **93**, 6996 (1989); *Chem. Phys. Lett.* **151**, 47 (1988).

142. A. Chandra and B. Bagchi, *J. Chem. Phys.* **91**, 2954 (1989); *J. Phys. Chem.* **94**, 3152 (1990).

143. P. C. Hohenberg and B. I. Halperin, *Rev. Mod. Phys.* **49**, 435 (1977).

144. L. E. Fried and S. Mukamel, *J. Chem. Phys.* **93**, 932 (1990).

145. D. Wei and G. N. Patey, *J. Chem. Phys.* **93**, 1399 (1990).

146. N. G. Bakshiev, *Opt. Spectrosc.* **16**, 446, (1964); N. G. Bakshiev, Y. T. Mazurenko, and I. Piterskaya, *ibid.* **21**, 307 (1966).

147. W. Jarzeba, A. E. Johnson, M. A. Kahalow, and P. F. Barbara, *J. Phys. Chem.* **92**, 7039 (1989).

148. M. A. Kahalow, W. Jarzeba, T. J. Kang, and P. F. Barbara, *J. Chem. Phys.* **90**, 151 (1989); M. A. Kahalow, T. J. Kang, and P. F. Barbara, *ibid.* **88**, 2372 (1988).

149. G. C. Walker, W. Jarzeba, T. J. Kang, A. E. Johnson and P. F. Barbara, *J. Opt. Soc. Am.* **B 7**, 1521 (1990); T. Tominaga, G. C. Walker, T. J. Kang, and P. F. Barbara, *J. Phys. Chem.* **95**, 10485 (1991).

150. E. W. Castner, Jr., M. Maroncelli, and G. R. Fleming, *J. Chem. Phys.* **86**, 1090 (1987); M. Maroncelli and G. R. Fleming, *ibid.* **92**, 3251 (1990).

151. J. S. Bader and D. Chandler, *Chem. Phys. Lett.* **157**, 501 (1989).

152. M. Maroncelli, *J. Chem. Phys.* **94**, 2084 (1991).

153. F. O. Raineri and H. L. Friedman, *J. Chem. Phys.* **98**, 8910 (1993).

154. F. O. Raineri, H. Resat, B. C. Perng, F. Hirata, and H. L. Friedman, *J. Chem. Phys.* **100**, 1477 (1994).

155. Y. Zhou, H.L. Friedman, and G. Stell, *J. Chem. Phys.* **91**, 4885 (1989); F. O. Raineri, Y. Zhou, H. L. Friedman, and G. Stell, *Chem. Phys.* **152**, 201 (1991).

156. H. L. Friedman, F. O. Raineri, and H. Resat, in *Molecular Liquids* (J. Teixeira-Diaz, ed.), NATO-ASI Ser. p. 183, Kluwer Academic Publishers,, Amsterdam, 1992.

157. F. O. Raineri, H. Resat, and H. L. Friedman, *J. Chem. Phys.* **96**, 3068 (1992).

158. H. Resat, F. O. Raineri, and H. L. Friedman, *J. Chem. Phys.* **97**, 2618 (1992); **98**, 7277 (1993); F. O. Raineri, B.-C. Perng, and H. L. Friedman, *Chem. Phys.* **183**, 187 (1994).

159. S. Roy, S. Komath, and B. Bagchi, *J. Chem. Phys.* **99**, 3139 (1993).

160. S. Roy and B. Bagchi, *J. Chem. Phys.* **99**, 9938 (1993).

161. S. Roy and B. Bagchi, *J. Chem. Phys.*. **98**, 1310 (1993).

162. S. Roy and B. Bagchi, *J. Chem. Phys.* **101**, 4150 (1994).

163. N. Nandi, S. Roy, and B. Bagchi, *J. Chem. Phys.* **102**, 1390 (1995).

163a. N. Nandi, R. Biswas, and B. Bagchi, *J. Chem. Phys.* (submitted for publication).

164. S. S. Komath and B. Bagchi, *J. Chem. Phys.* **98**, 8987 (1993).

165. S. Ravichandran, S. Roy, and B. Bagchi, *J. Phys. Chem.* **99**, 2489 (1995).

166. S. Ravichandran and B. Bagchi, *Int. Rev. Phys. Chem.* **14**, 271 (1995).

167. E. Neria and A. Nitzan, *J. Chem. Phys.* **96**, 5433 (1992).

168. M. Cho, S. J. Rosenthal, N. F. Scherer, L. D. Ziegler, and G. R. Fleming, *J. Chem. Phys.* **96**, 5033 (1992).

169. M. Cho, M. Du, N. F. Scherer, G. R. Fleming, and S. Mukamel, *J. Chem. Phys.* **99**, 2410 (1993).

170. C. Kittel, *Introduction to Solid State Physics*, 7th ed. Wiley, Singapore and New York, 1996.

171. F. H. Stillinger and T. A. Weber, *Science* **225**, 983 (1984).

172. R. W. Zwanzig, *Phys. Rev.* **144**, 170 (1966).

173. R. W. Zwanzig, in *Statistical Mechanics: New Concepts, New Problems, New Applications* (S. A. Rice, K. F. Freed, and J. C. Light, eds.), University of Chicago Press, Chicago, 1972.

174. J. P. Hansen and I. R. McDonald, *Theory of Simple Liquids*, 2nd ed. Academic Press, London, 1986.
175. U. Balucani and M. Zoppi, *Dynamics of the Liquid State*. Clarendon Press, Oxford, 1994.
176. P. G. deGennes, *Physica* **25**, 825 (1959).
177. J. P. Boon and S. Yip, *Molecular Hydrodynamics*. McGraw-Hill, New York, 1980.
178. R. Kubo, J. Math. Phys. **4**, 174 (1963).
179. K. Ding, D. Chandler, S. J. Smithline, and A. D. J. Haymet, *Phys. Rev. Lett.* **59**, 1698 (1987).
180. B. Bagchi, A. Chandra, and S. A. Rice, *J. Chem. Phys.* **93**, 8991 (1990).
181. I. de Schepper and E. G. D. Cohen, *J. Stat. Phys.* **27**, 225 (1982).
182. J. L. Colot, X-G. Wu, H. Xu, and M. Baus, Phys. Rev. **A 38**, 2022 (1988).
183. J. L. Lebowitz and J. K. Percus, J. Math. Phys. **4**, 116 (1964).
184. M. S. Wertheim, *Phys. Rev. Lett.* **10**, 321 (1963).
185. E. Thiele, *J. Chem. Phys.* **39**, 474 (1963).
186. J. K. Percus and G. J. Yevick, *Phys. Rev.* **110**, 1 (1958).
187. D. Isbister and R. J. Bearman, *Mol. Phys.* **28**, 1297 (1974); S. A. Adelman and J. M. Deutch, *J. Chem. Phys.* **59**, 3971 (1973).
188. C. G. Gray and K. E. Gubbins, *Theory of Molecular Fluids*, Vol. 1. Clarendon Press, Oxford, 1984.
189. B. Bagchi and A. Chandra, *J. Chem. Phys.* **97**, 5126 (1992).
190. B. J. Berne and R. Pecora, *Dynamic Light Scattering*. Wiley, New York, 1976.
191. B. J. Berne, *J. Chem. Phys.* **62**, 1154 (1975).
192. See the papers in Y. Gauduel and P. J. Rossky, eds., *Ultrafast Reaction Dynamics and Solvent Effects*. AIP Press, New York, 1994.
193. See the articles in J. D. Simon, ed., *Ultrafast Dynamics of Chemical Systems*. Kluwer Academic Publishers, Dordrecht, the Netherlands, 1994.
194. C. F. Chapman, R. S. Fee, and M. Maroncelli, *J. Phys. Chem.* **94**, 4929 (1990).
195. R. Biswas and B. Bagchi, *J. Phys. Chem.* **100**, 1238 (1996).
196. H. X. Zhou, B. Bagchi, A. Papazyan, and M. Maroncelli, *J. Chem. Phys.* **97**, 9311 (1992).
197. S. Roy and B. Bagchi, *Chem. Phys.* **183**, 207 (1994).
198. A. Chandra and B. Bagchi, *J. Chem. Phys.* **91**, 7181 (1989).
199. J. C. Rasaiah, D. J. Isbister, and G. Stell, *Chem. Phys.* **79**, 189 (1981); *J. Chem. Phys.* **75**, 4704 (1981).
200. A. Yoshimori, *J. Chem. Phys.* **105**, 5971 (1996).
201. C.-M. Hu and R. Zwangig, *J. Chem. Phys.* **60**, 4353 (1974); G. K. Youngren and A. Acrivos, *ibid.* **63**, 3846 (1975).
202. R. Biswas and B. Bagchi, *J. Chem. Phys.* **100**, 4261 (1996).
203. S. Chandrasekhar, *Rev. Mod. Phys.* **15**, 1 (1943).
204. R. Biswas, S. Roy, and B. Bagchi, *Phys. Rev. Lett.* **75**, 1098 (1995).
205. J. G. Kirkwood, *J. Chem. Phys.* **14**, 180 (1946).
206. D. Y. C. Chan, D. J. Mitchell, and B. W. Ninham, *J. Chem. Phys.* **70**, 2946 (1979).
207. B. Bagchi and S. Roy, in *Ultrafast Chemical Reactions and Solvent Effects* (Y. Gauduel and P. J. Rossky, eds.), p. 296. AIP, New York, 1994.
208. R. Biswas, N. Nandi, and B. Bagchi, *J. Phys. Chem.* **B 101**, 2968 (1997).

209. G. V. Vijaydamodar, A. Chandra, and B. Bagchi, *Chem. Phys. Lett.* **161**, 413 (1989).
210. R. S. Moog, L. D. Blanket, and M. Maroncelli, *J. Phys. Chem.* **97**, 1496 (1993).
211. A. Papazyan and M. Maroncelli, *J. Chem. Phys.* **102**, 2888 (1995).
212. L. Verlet and J. J. Weiss, *Phys. Rev.* **A 5**, 939 (1972).
213. J. D. Weeks, D. Chandler, and H. C. Andersen, *J. Chem. Phys.* **54**, 5237 (1971).
214. H. Stehfest, *Commun. ACM* **13**, 624 (1970).
215. I. Ohmine, H. Tanaka, and P. G. Wolynes, *J. Chem. Phys.* **89**, 5852 (1988).
216. I. Ohmine and H. Tanaka, *J. Chem. Phys.* **93**, 8138 (1990); *ibid.* **91**, 6318 (1989).
217. F. H. Stillinger and A. Rahaman, *J. Chem. Phys.* **60**, 1545 (1974).
218. J. T. Kindt and C. A. Schmuttenmaer, *J. Phys. Chem.* **100**, 10373 (1996).
219. P. A. Madden and R. W. Impey, *Chem. Phys. Lett.* **123**, 502 (1986); R. W. Impey, P. A. Madden, and I. R. McDonald, *Mol. Phys.* **40**, 515 (1982).
220. P. A. Madden, in *Molecular Liquids: Dynamics and Interactions* (A. J. Barnes, W. J. Orville-Thomas, and J. Yarwood, eds.), p. 413. Reidel Publ., Dordrecht, the Netherlands, 1984.
221. J. B. Hasted, S. K. Hussain, F. A. M. Frescura, and R. Birch, *Chem. Phys. Lett.* **118**, 622 (1985).
222. D. Bertolini and A. Tani, *Mol. Phys.* **75**, 1065 (1992).
223. See the articles in F. Franks, ed., *Water: A Comprehensive Treatise*, Vol. 1. Plenum, New York, 1972.
224. See the articles in D. Eisenberg and W. Kauzman, eds., *The Structure and Properties of Water.* Oxford University Press, London, 1969.
225. U. Kaatze, *Chem. Phys. Lett.* **203**, 1 (1993).
226. G. Nemethy and H. Scheraga, *J. Chem. Phys.* **41**, 680 (1964).
227. F. Y. Joo and G. R. Freeman, *J. Phys. Chem.* **83**, 2383 (1979).
228. F. H. Long, H. Lu, and K. B. Eisenthal, *Chem. Phys. Lett.* **160**, 464 (1989).
229. Y. Nagasawa, Ph.D. Thesis, Chap. 6. Graduate University for Advanced Studies, Okazaki, Japan, 1993; H. Pal, Y. Nagasawa, K. Tominaga, S. Kumazaki, and K. Yoshihara, *J. Chem. Phys.* **102**, 7758 (1995).
230. J. Barthel, K. Bachhuber, R. Buchner, J. B. Gill, and H. Hetzenauer, *Chem. Phys. Lett.* **165**, 369 (1990); J. Barthel, K. Bachhuber, R. Buchner, J. B. Gill, and M. Kleebauer, *ibid.* **167**, 62 (1990).
231. J. Barthel and R. Buchner, *Pure Appl. Chem.* **63**, 1473 (1991).
232. S. Roy, Ph.D. Thesis, Indian Institute of Science, India, 1995.
233. A. L. McCellan, *Tables of Experimental Dipole Moments.* Freeman, San Fransisco, 1963. The experimental dipole moments reported here are measured in dilute solutions of benzene.
234. T. R. Dyke and J. S. Muenter, *J. Chem. Phys.* **59**, 3125 (1973).
235. D. A. Draegert, N. W. B. Stone, B. Curnutte, and D. Williams, *J. Opt. Soc. Am.* **56**, 64 (1966).
236. G. E. Walrafen, *J. Chem. Phys.* **36**, 1035 (1962); **44**, 1546 (1966); **47**, 114 (1967).
237. C. P. Hsu, X. Song, and R. A. Marcus, *J. Phys. Chem.* **B 101**, 2546 (1997).
238. B. Guillot, *J. Chem. Phys.* **95**, 1543 (1991).
239. I. M. Svishchev and P. G. Kusalik, *J. Chem. Soc., Faraday Trans.* **90**, 1405 (1994).
239a. M. J. Lang, X. J. Jordanides, X. Song, and G. R. Fleming, *J. Chem. Phys.* (submitted for publication).
240. R. N. Barnett, U. Landman, and A. Nitzan, *J. Chem. Phys.* **90**, 4413 (1990).

241. N. A. Smith, S. Lin, S. R. Meech, H. Shirota, and K. Yoshihara, *J. Phys. Chem. A* **101**, 9578 (1997).
242. R. Olender and A. Nitzan, *J. Chem. Phys.* **102**, 7180 (1995).
243. T. Joo, Y. Jia, J.-Y. Yu, M. J. Lang, and G. R. Fleming, *J. Chem. Phys.* **104**, 6089 (1996).
244. D. S. Alavi and D. H. Waldeck, *J. Chem. Phys.* **94**, 6196, (1991); also in *Understanding Chemical Reactivity*, p. 249. Kluwer Academic Publishers, Amsterdam, 1994; M. G. Kurnikova, D. H. Waldeck, and R. D. Coalson, *J. Chem. Phys.* **105**, 628 (1996).
245. J. T. Edwards, *J. Chem. Educ.* **47**, 261 (1970).
246. S. Mashimo, S. Kuwabara, S. Yagihara, and K. Higasi, *J. Chem. Phys.* **90**, 3292 (1989).
247. S. K. Garg and C. P. Smyth, *J. Phys. Chem.* **69** 1294 (1965).
248. D. McMorrow and W. T. Lotshaw, *J. Phys. Chem.* **95**, 10395 (1991).
249. S. Bhattacharyya and B. Bagchi, *J. Chem. Phys.* **106**, 1757 (1996).
250. S. Bhattacharyya and B. Bagchi, *J. Chem. Phys.* **106**, 7262 (1997).
251. S. Bhattacharyya and B. Bagchi, *J. Chem. Phys.* **107**, 5852 (1997).
252. S. Bhattacharyya and B. Bagchi, *J. Chem. Phys.* (submitted for publication).
253. M. Cho, G. R. Fleming, S. Saito, I. Ohmine, and R. M. Stratt, *J. Chem. Phys.* **100**, 6672 (1994).
254. R. A. Marcus, *J. Chem. Phys.* **24**, 966, (1956).
255. R. A. Marcus, *Annu. Rev. Phys. Chem.* **15**, 155 (1964).
256. C. A. Angell, *Chem. Rev.* **90**, 523 (1990); C. A. Angell *et al.*, *AIP Conf. Proc.* **256**, 3 (1992).
256a. U. Mohanty, *Adv. Chem. Phys.* **89**, 89 (1995).
257. R. Richert, F. Stickel, R. S. Fee, and M. Maroncelli, *Chem. Phys. Lett.* **229**, 302 (1994); A. Wagner and R. Richert, *ibid.* **176**, 329 (1991); R. Richert and A. Wagner, *J. Phys. Chem.* **95**, 10115, (1991); R. Richert, in *Disorder Effects on Relaxation Processes* (R. Richert and A. Blumen, eds.), p. 333. Springer-Verlag, Berlin, 1994.
258. P. Debye, *Polar Molecules.* Dover, New York, 1929.
259. H. Frohlich, *Theory of Dielectrics.* Clarendon Press, Oxford, 1958.
260. S. Brawer, *Relaxation in Viscous Liquids and Glasses.* American Society, Columbus, OH, 1985.
261. C. F. Chapman, R. S. Fee, and M. Maroncelli, *J. Phys. Chem.* **94**, 4929 (1990).
262. Y. J. Chang and E. W. Castner, Jr., *J. Phys. Chem.* **98**, 9712 (1994).
263. C. J. F. Bottcher and P. Bordewijk, *Theory of Electric Polarization*, Vol. 2. Elsevier, Amsterdam, 1978.
264. C. P. Hsu, Y. Georgievskii, and R. A. Marcus, *J. Phys. Chem. A* **102**, 2658 (1998).
265. R. Biswas, Ph.D. Thesis, Indian Institute of Science, India, 1998.
266. Y. J. Chang and E. W. Castner, Jr., *J. Chem. Phys.* **99**, 113 (1993).
267. Y. C. Guillaume and C. Guinchard, *Anal. Chem.* **69**, 183 (1997).
268. D. S. Venables and C. A. Schmuttenmaer, *J. Chem. Phys.* **108**, 4935 (1998).
269. N. Ernsting, private communication.
270. A. Z. Gadzhiev, *Russ. J. Phys. Chem.* (*Engl. Transl.*) **56**, 1660 (1982).
271. E. W. Castner, Jr. and M. Maroncelli, *J. Mol. Liq.* **77**, 1 (1998).
271a. M. Madhusoodanan and B. L. Tembe, *J. Phys. Chem.* **98**, 7090 (1994); **99**, 45 (1995); A. K. Das, M. Madhusoodanan, and B. L. Tembe, *ibid.* **101**, 2862 (1997); A. K. Das and B. L. Tembe, *J. Chem. Phys.* **108**, 2930 (1998).

272. R. Biswas, S. Bhattacharyya, and B. Bagchi, *J. Chem. Phys.* **108**, 4963 (1998).

273. R. Biswas and B. Bagchi, *Indian J. Chem.* **A 36**, 635 (1997).

274. N. Matubayasi, C. Wakai, and M. Nakahara, *J. Chem. Phys.* **107**, 9133 (1997).

275. M. Kalinichev and J. D. Bass, *J. Phys. Chem.* **A 101**, 9720 (1997).

276. K. Okada, Y. Imashuku and M. Yao, *J. Chem. Phys.* **107**, 9302 (1997).

277. D. D. Becker, D. D. Doring, and D. D. Zeldovich, in *Nucleation* (A. C. Zettlemoyer, ed.). MD Inc., New York, 1969.

278. S. L. Walen, B. J. Palmer, D. M. Pfund, J. L. Fulton, M. Newville, Y. Ma, and E. A. Stern, *J. Phys. Chem.* **A 101**, 9632 (1997).

279. M. Berg, private communication.

280. C. B. Harris, D. E. Smith, and D. J. Russel, *Chem. Rev.* **90**, 481 (1990); J. C. Owrutsky, D. Raftery, and R. M. Hochstrasser, *Annu. Rev. Phys. Chem.* **45**, 519 (1994); S. A. Adelman and R. H. Stote, *J. Chem. Phys* **88**, 4397 (1988); R. H. Stote and S. A. Adelman, *ibid.* p. 4415.

281. M. Fixman, *J. Chem. Phys.* **34**, 369 (1961).

282. R. Zwanzig, *J. Chem. Phys.* **34**, 1931 (1961).

283. K. F. Herzfeld, *J. Chem. Phys.* **36**, 3305 (1962).

284. S. A. Kovalenko, N. P. Ernsting, and J. Ruthmann, *J. Chem. Phys.* **106**, 3504 (1997).

285. D. A. McQuarrie, *Statistical Mechanics*. Harper-Row, New York, 1976.

286. I. Ohmine, *J. Chem. Phys.* **85**, 3342 (1986).

287. K. F. Herzfeld and T. A. Litovitz, *Absorption and Dispersion of Ultrasonic Waves* Academic Press, New York, 1959.

288. B. J. Berne, M. E. Tuckerman, J. E. Straub, and A. L. R. Bug, *J. Chem. Phys.* **93**, 5084 (1990).

288a. J. E. Straub, M. Borkovec, and B. J. Berne, *J. Chem. Phys.* **89**, 4833 (1988).

289. H. Haken, *Synergetics: An Introduction*. Springer-Verlag, Berlin, 1978.

290. B. P. Hills, *Mol. Phys.* **35**, 1471 (1978).

291. D. E. Edelstein, P. Graf, and A. Nitzan, *J. Chem. Phys.* **107**, 10470 (1997).

292. J. S. Bader and B. J. Berne, *J. Chem. Phys.* **100**, 8359 (1994); S. A. Egorov and B. J. Berne, *ibid.* **107**, 6050 (1997).

293. K. F. Evritt, S. A. Egorov and J. L. Skinner, *Chem. Phys.* **235**, 115 (1998).

294. R. Kubo, in *Fluctuation, Relaxation and Resonance in Magnetic Systems* (D. Ter. Harr, ed.), pp. 23–68. Oliver & Boyd, London, 1962.

295. A. Myers and F. Markel, *Chem. Phys.* **149**, 21 (1990).

296. M. J. Clouter and H. Keifte, *J. Chem. Phys.* **66**, 1736 (1977).

297. P. Debye and E. Huckel, *Phys. Z.* **24**, 185 (1923); L. Onsager, *ibid.* **27**, 388 (1926).

298. M. Born, *Z. Phys.* **1**, 221 (1920).

299. R. H. Boyd, *J. Chem. Phys.* **35**, 1281 (1961); **39**, 2376 (1963).

300. R. Zwanzig, *J. Chem. Phys.* **38**, 1603 (1963); **52**, 3625 (1970).

301. J. B. Hubbard and L. Onsager, *J. Chem. Phys.* **67**, 4850 (1977); J. Hubbard, *ibid.* **68**, 1649, (1978); T. Tominaga, D. F. Evans, J. B. Hubbard, and P. G. Wolynes, *J. Phys. Chem.* **83**, 2669 (1979).

302. J. B. Hubbard and R. F. Kayser, *J. Chem. Phys.* **74**, 3535 (1981); *Chem. Phys.* **66**, 377 (1982).

303. B. Bagchi and R. Biswas, *Acc. Chem. Res.* **31**, 181 (1998), and references therein.

304. P. G. Wolynes, *Annu. Rev. Phys. Chem.* **31**, 345 (1980).
305. P. G. Wolynes, *J. Chem. Phys.* **68**, 473 (1978).
306. P. Colonomos and P. G. Wolynes, *J. Chem. Phys.* **71**, 2644 (1979).
307. B. Bagchi, *J. Chem. Phys.* **95**, 467 (1991).
308. G. van der Zwan and J. T. Hynes, *J. Chem. Phys.* **76**, 2993 (1982); *Chem. Phys. Lett.* **101** 367 (1983); *J. Chem. Phys.* **78**, 4174 (1983).
309. M. Berkowitz and W. Wan, *J. Chem. Phys.* **86**, 376 (1987); L. Perera, U. Essman and M. L. Berkowitz, *ibid.* **102**, 450 (1995); M. R. Reddy and M. J. Berkowitz, *Chem. Phys.* **88**, 7104 (1988).
310. R. Biswas, S. Roy, and B. Bagchi, *Phys. Rev. Lett.* **75**, 1098 (1995).
311. R. Biswas and B. Bagchi, *J. Chem. Phys.* **106**, 5587 (1997).
312. R. Biswas and B. Bagchi, *J. Am. Chem. Soc.* **119**, 5946 (1997).
313. A. V. Indrani and S. Ramaswamy, *Phys. Rev. Lett.* **73**, 360 (1994).
314. See, for example, in *CRC Handbook of Chemistry and Physics*, 57th ed. WEAST, CRC Press, Boca Raton, FL, 1976.
315. T. Gover and P. G. Sears, *J. Am. Chem. Soc.* **60**, 330 (1956)
316. R. D. Shannon and C. T. Prewitt, *Acta Crystallogr., Sect. B* **25**, 925 (1969).
317. S. H. Lee and J. C. Rasaiah, *J. Chem. Phys.* **101**, 6964 (1994).
318. S. H. Lee and J. C. Rasaiah, *J. Phys. Chem.* **100**, 1420 (1996).
319. B. Bagchi and G.V. Vijayadamodar, *J. Chem. Phys.* **98**, 3351 (1993).
320. B. Bagchi, *J. Chem. Phys.* published Sept. 7, 1998 issue.
321. T. W. Nee and R. Zwanzig, *J. Chem. Phys.* **52**, 6353 (1970); J.B. Hubbard and P.G. Wolynes, *ibid.* **69**, 998 (1978).
322. T. Gaskell and S. Miller, *J. Phys. C* **11**, 3749 (1978); U. Balucani, R. Vallauri, and T. Gaskell, *Ber. Bunsenges. Phys. Chem.* **94**, 261 (1990).
323. Y. Pomeau and P. Resibois, *Phys. Rep.* **19**, 63 (1975)
324. J.T. Hynes, R. Kapral, and M. Weinberg, *J. Chem. Phys.* **70**, 1456 (1979).
325. T. Keyes, in *Statistical Mechanics. Part B: Time Dependent Processes* (B.J. Berne, ed.). Plenum, New York, 1977.
326. D. Huppert, V. Ittah, and E.M. Kosower, *Chem. Phys. Lett.* **159**, 267 (1989).
327. C. Chapman and M. Maroncelli, *J. Phys. Chem.* **95**, 9095 (1991).
328. E. Neria and A. Nitzan, *J. Chem. Phys.* **100**, 3855 (1994).
329. G. van der Zwan and J. T. Hynes, *Chem. Phys.* **152**, 169 (1991).
330. P. Debye and H. Falkenhagen, *Phys. Z.* **29**, 121, 401 (1928).
331. A. Chandra and G.N. Patey, *J. Chem. Phys.* **100**, 1552 (1994).
332. A. Chandra, *Chem. Phys. Lett.* **244**, 314 (1995).
333. A. Chandra, D. Jana, and S. Bhattacharjee, *J. Chem. Phys.* **104**, 8662 (1996).
334. K. Mahajan and A. Chandra, *J. Chem. Phys.* **106**, 2360 (1997).
335. S. Vajda, R. Jimenez, S. J. Rosenthal, V. Fidler, G. R. Fleming, and E. W. Castner, Jr., *J. Chem. Soc., Faraday. Trans.* **91**, 867 (1995).
336. N. Sarkar, K. Das, A. Datta, S. Das, and K. Bhattacharyya, *J. Phys. Chem.* **100**, 10523 (1996).
337. N. Sarkar, A. Datta, S. Das, and K. Bhattacharyya, *J. Phys. Chem.* (in press).

338. N. Nandi and B. Bagchi, *J. Phys. Chem.* **B 101**, 10954 (1997).
339. M. M. Teeter, *Annu. Rev. Biophys. Biophys. Chem.* **20**, 577 (1991); J. A. Rupley and G. Careri, *Adv. Protein Chem.* **41**, 37 (1991); R. Pethig, *Annu. Rev. Phys. Chem.* **43**, 177 (1992).
340. R. Pethig, in *Protein Solvent Interactions* (R. B. Gregory, ed.). Dekker, New York, 1995.
341. E. H. Grant, *Nature (London)* **196**, 1194 (1962); *J. Mol. Biol.* **19**, 133 (1966); G. P. South and E. H. Grant, *Proc. R. Soc. London, Ser. A* **328**, 371 (1972); E. H. Grant, B. G. R. Mitton, G. P. South, and R. J. Sheppard, *Biochem. J.* **139**, 375 (1974); E. H. Grant, V. E. R. McClean, N. R. V. Nightingale, R. J. Sheppard, and M. J. Chapman, *Bioelectromagnetics* **7**, 151 (1986).
342. N. Nandi and B. Bagchi, letters to the Editor, *J. Phys. Chem. B* (in press).
343. J. L. Oncley, in *Proteins, Amino Acids and Peptides* (E. J. John and J. T. Edsall, eds.), p. 543. Reinhold, New York, 1950.
344. A. Chandra, R. Biswas, and B. Bagchi, unpublished work.
345. R. Broadwater and R. L. Kay, *J. Phys. Chem.* **74**, 3802 (1970).
346. Poiseuille, *Ann. Chim. Phys.* **21**, 76 (1847).
347. G. Jones and M. Dole, *J. Am. Chem. Soc.* **51**, 2950 (1929).
348. M. P. Applebey, *J. Chem. Soc.* **97**, 2000 (1910).
349. R. Biswas and B. Bagchi, *J. Phys. Chem.* (submitted for publication).
350. P. B. Balbuena, K. P. Johnston, P. J. Rossky, and J. K. Hyun, *J. Phys. Chem. B* **102**, 3806 (1998).
351. G. R. Fleming, T. Joo, and M. Cho, *Adv. Chem. Phys.* **101**, 141 (1997).
352. P. van der Meulen, A. M. Jonkman and M. Glasbeek, *J. Phys. Chem.* **A 102**, 1906 (1998).

SPATIAL PATTERNS AND SPATIOTEMPORAL DYNAMICS IN CHEMICAL SYSTEMS

A. DE WIT

Service de Chimie Physique, Centre for Nonlinear Phenomena and Complex Systems CP 231, Université Libre de Bruxelles, Campus Plaine, 1050 Brussels, Belgium

CONTENTS

Advances in Chemical Physics, Volume 109, Edited by I. Prigogine and Stuart A. Rice
ISBN 0-471-32920-7

I. INTRODUCTION

In chemical systems, a spatial differentiation of concentrations is valuable for all applications that rely on a selective reactivity organized in space. Spatially varying chemical activity can, of course, be manufactured by building up systems in which different chemical species are distributed at desired locations through externally imposed separations. Nevertheless, chemical systems are able to spontaneously self-organize in space if they are maintained out of equilibrium, and if their kinetic and diffusional characteristics allow for local activation processes balanced by long-range inhibition. The concentrations of the different chemical species then form stationary spatial patterns that periodically span the space. These spontaneous spatial organizations emerge out of a base state when this latter one becomes unstable as the result of the change of parameters, such as the temperature or the concentration of some species. In two-dimensional systems, the spatial patterns resulting from such an instability take the form of higher concentration stripes or hexagons in a lower concentration background (Fig. 1). Such rolls and honeycombs are similar to striped or hexagonal convection cells arising in a fluid layer sandwiched between two plates and heated from below when it undergoes a Rayleigh–Bénard instability. In chemical systems, such patterns arise through a so-called Turing instability resulting from the sole coupling between nonlinear chemical kinetics and diffusion processes. This instability, first described by the mathematician Turing in 1952 [1] has for a long time been a paradigm of pattern-forming instabilities in chemical [2–11] and biological [12–15] systems. Sustained steady periodic Turing structures were observed experimentally for the first time in 1989. Since this experimental discovery, the study of Turing structures has gained increased attention.

The aim of this chapter is to review a variety of theoretical and numerical results that allow us to better understand the characteristics of the Turing patterns and to discuss some related spatiotemporal dynamics. We will focus principally on the advances made since 1989. Some recent reviews on patterns in chemical systems can be found in Refs. [4,5,7,9,16–18]. A com-

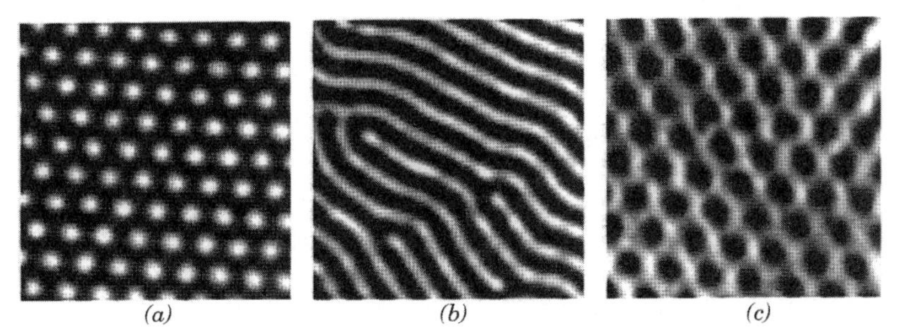

Figure 1. Experimental stationary Turing structures obtained using the chlorite–iodide–malonic acid (CIMA) reaction in a continuously fed unstirred disk gel reactor. The *black* and *white* regions correspond to regions with high and low concentrations in iodide, respectively, made visible to the eye by a color indicator (starch). The wavelength is on the order of 0.2 mm. These photographs show only a small part of the patterned zone of the reactor that encompasses several hundreds of wavelengths. (*a*) Triangular array of clear spots on an hexagonal dark background, (*b*) stripes, (*c*) transient array of dark triangles on a clear background. Courtesy of P. De Kepper (CRPP/CNRS).

prehensive review of pattern formation outside of equilibrium (including examples in hydrodynamic systems, solidification fronts, nonlinear optics, heterogeneous catalysis, semiconductors, and excitable biological media) was provided by Cross and Hohenberg [19], who discuss the theory used to study pattern formation, emphasizing on the universal characteristics of spatial structures in the framework of amplitude equations. Other reviews on pattern formation can be found in Refs. [20–26].

Although Turing patterns fit into that general framework, we will here rather focus on the peculiarities of reaction–diffusion systems that are not often encountered in other physical systems featuring spatial and spatiotemporal patterns. Chemical systems indeed exhibit several characteristic properties that we want to stress here.

One of the main originalities of the Turing instability lies in the fact that it leads to patterns with an intrinsic wavelength function only of the kinetic constants and diffusion coefficients. In most of other spatial structures, such as the Rayleigh–Bénard patterns, the wavelength is set by geometric factors and patterns are one-dimensional (1D) or two-dimensional (2D). In chemical systems on the contrary, three-dimensional (3D) patterns are obtained as soon as the length, width, and depth of the pattern-forming zone are on the order of or larger than the Turing wavelength. Chemical systems also allow for the study of patterns in the presence of ramps of parameter values as the chemical reactors used to study Turing patterns exhibit genuine gradients of concentration as they are fed from the sides. Another important characteristic of reaction–diffusion systems is that their nonlinearities stem from local kinetics, contrary to hydrodynamic systems, for instance, in which, for normal fluids, nonlinearities emerge in general from inertial or advective terms in the evolution equation for the velocity field. Hence, even when the effects of transport processes are quenched by turbulent mixing, the nonlinear chemical kinetics are responsable for numerous dynamic behaviors, such as temporal oscillations of the concentrations arising through a so-called Hopf instability, excitability, bistability, and chaos [7]. Chemical systems are thus privilegied systems in which one can observe the wealth of spatiotemporal dynamics that exists when a pattern-forming instability competes with other instabilities [27]. This is the case for the coupling between a Turing instability and temporal oscillations or for pattern formation in bistable systems, as will be detailed later.

This review is mainly restricted to spatial structures arising through a diffusive Turing instability. Little will be said about other pattern-forming mechanisms (such as front instabilities, global control, or mechanisms related to pulses) that can be important in chemical systems. In addition, the major part of this chapter will focus on the Turing structures observed in the chlorite-iodide-malonic acid (CIMA) system and its variants, in which lots of results on sustained Turing patterns have been obtained, and not on chemical patterns observed in Liesegang rings [28,29] or heterogeneous catalytic systems [17,30,31], for instance. Because introductions to nonlinear theory can be found in numerous books and reviews [3,5,6,9,12,13,16,19,21,32–34], we will limit the description of theoretical tools to an overall introduction, refering the reader to these more detailed sources. Eventually, let us note that several works have been devoted to the study of the Turing instability in a biological framework [12–14] but we do not intend to review this aspect.

This chapter is organized as follows. We first review in Section II what is meant by a Turing instability and what are the conditions for its occurence in reaction–diffusion systems. Section III focuses on the experimental observations of Turing structures and related spatiotemporal dynamics. After having described the general basis of pattern selection theory in Section IV, we then show how the experimental findings can be understood theoretically by the analysis of the 2D and 3D pattern-selection problems in monostable systems. We review in Section VI what theory tells us about the possible spatiotemporal dynamics that can occur because of a Turing–Hopf interaction. Specificities of bistable systems are addressed in Section VII.

II. THE TURING INSTABILITY

In 1952 [1], Turing developed the original idea that the coupling between reactions and diffusion of chemical species might play a role in morphogenesis, i.e., in the creation in living organisms of differentiated structures out of initially identical elementary cells. Turing showed that a uniform state may in some circumstances evolve because of a diffusive instability toward a new state in which the concentrations are stationary and periodically organized in space. The spatial symmetry of the initial state of the system is thus broken during the transition. The fact that this symmetry breaking results from the sole coupling between chemistry obeying mass action laws and diffusion ruled by Fick's law is *a priori* counterintuitive, as diffusion on its own is usually a stabilizing process, smoothing out any concentration heterogeneities. In fact, detailed studies have shown that this spontaneous pattern-forming instability can occur only in chemical systems maintained out of equilibrium and in which autoactivation processes are present [1–3,6,8,12,35,36]. This last criterion can be expressed in different ways, depending on the number of variables in the system. For the sake of simplicity, we will restrict ourselves to two-variable systems. In that case, three ingredients must be gathered for a stable steady state to become unstable because of a Turing instability:

1. An activator X implied in an autocatalytic reaction enhances its own production (or consumption).
2. An inhibitor Y slows down the preceding activation step.
3. The inhibitor diffuses quicker than the activator ($D_y > D_x$ where D_x and D_y are the diffusion coefficients of the activator and the inhibitor, respectively).

A spatial pattern settles down because of a balance between the local activation processes and the long-range inhibition provided by molecular diffusion. This mechanism is quite general and hence the principle of a Turing instability can be recovered in other fields, such as heterogeneous catalysis [17,30,31], nonlinear optics [24], gas discharges [37], semiconductor devices [20,26,38], and materials irradiated by energetic particles [9,39,40] or light [40,41]. The common denominator of these various systems is that they can be modeled by reaction–diffusion-type equations, such as those that naturally describe chemical systems. In all cases, the wavelength of the Turing-type spatial pattern accounts for the balance between the reaction–type mechanisms and the diffusion-like transport processes and is, therefore, intrinsic to the system.

Let us now look, from a more quantitative point of view, at which conditions a reaction–diffusion system can go through a diffusive instability. Let us consider a concentration field $\underline{C}$ the components of which are the concentrations of the various variables of the system. The spatiotemporal evolution of $\underline{C}$ is described by the following reaction–diffusion equations:

$$\partial_t \underline{C} = \underline{F}(\underline{C}, \gamma) + \underline{\underline{D}} \nabla^2 \underline{C} \tag{2.1}$$

where ∂_t and ∇ are the partial derivatives with regard to time and space, respectively; $\underline{F}(\underline{C}, \gamma)$ represents the nonlinear reaction speed; and γ stands for the tunable parameters in the system. For given boundary conditions, this system usually admits a homogeneous steady state $\underline{C}_s$ such that $\underline{F}(\underline{C}_s) = 0$. Perturbing this homogeneous steady state by small local inhomogeneous perturbations, we take $\underline{C} = \underline{C}_s + \underline{u}$, where $\underline{u}$ can be written as

$$\underline{u}(\mathbf{r}, t) = \sum_n \underline{c}_n e^{\omega_n t} \phi_n(\mathbf{r}) \tag{2.2}$$

where the $\phi_n(\mathbf{r})$ satisfy

$$\nabla^2 \phi_n(\mathbf{r}) = -k_n^2 \phi_n(\mathbf{r}) \tag{2.3}$$

for the given boundary conditions. Inserting this into Eq. (2.1) and linearizing around the homogeneous steady state, we get down to the following eigenvalue problem

$$\|\underline{\underline{L}} - \underline{\underline{D}} k_n^2 - \underline{\underline{I}} \omega_n\| = 0 \tag{2.4}$$

where $\underline{\underline{L}}$ is the Jacobian matrix and $\underline{\underline{I}}$ is the identity matrix. The sign of the real part of the eigenvalues $\omega_n = \omega_n(k_n^2)$ controls the stability of the system. If the real part ω of all eigenvalues is negative for any $\mathbf{k}_n$, then the perturbations $\underline{u}$ decay exponentially in time, and the system is defined as asymptotically stable. The system is said to be marginally stable if one eigenvalue has a real part vanishing for $|\mathbf{k}_n| = k_c$ and is negative otherwise, while all the other eigenvalues have negative real parts. The system is unstable as soon as one eigenvalue has a positive real part for all wave vectors $\mathbf{k}_n$ of length k_c, because then perturbations grow exponentially in time. The change of the value of one parameter can lead to a switch from a stable state toward an unstable state. In that case, the solutions of the nonlinear system modify their qualitative character, and a bifurcation takes place. The parameter ruling this transition is dubbed the bifurcation parameter or the control parameter. The bifurcation point is the value γ_c of the control parameter for which the system becomes unstable.

Let us consider two possible symmetry breaking instabilities. If the critical eigenvalue is real and positive for $|\mathbf{k}_n| = k_c$ (Fig. 2), the system evolves toward a new state, breaking the spatial symmetry, and a Turing bifurcation occurs. The concentrations are then modulated spatially with a periodicity given by the intrinsic critical wavelength $\lambda_c = 2\pi/k_c$.

If the critical eigenvalues correspond to a pair of complex conjugated roots with a nonvanishing imaginary part $i\omega^i$ and if the real part is zero at $|\mathbf{k}_n| = k_c$, the system evolves toward a new state in which the concentrations oscillate in

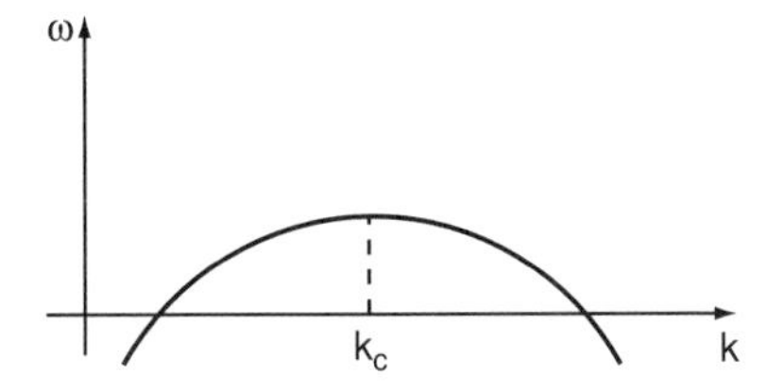

Figure 2. Dispersion relation displaying the growth rate ω of perturbations versus their wavenumber in the case of a Turing instability. The most unstable mode is that for which $|\mathbf{k}| = k_c$.

time with frequency ω^i. This is the Hopf instability. For two variable reaction–diffusion systems, the first mode to become unstable always has $k_c = 0$, and the temporal oscillations are homogeneous in space. For three or more variable systems, the Hopf instability can occur for a finite k_c, which then gives rise to propagating and standing waves.

III. EXPERIMENTAL BACKGROUND

The study of spatial patterns in reaction–diffusion media has recently boomed because of the first experimental observations of stationary Turing patterns in a chemical system. These structures, as such, can be sustained only if the system is maintained far from equilibrium, which implies to continuously feed the reactor with fresh reactants and to eliminate the products. This principle has been applied since the 1970s in open continuously stirred tank reactors, devoted to the study of stationary and temporal behaviors of chemical reactions out of equilibrium [7,42,43]. Incorporation of the spatial component was achieved in the 1980s in a series of unstirred open reactors, developed to produce and sustain chemical dissipative structures [44–46].

In 1989, De Kepper and co-workers used such a single-phase open reactor to obtain the first sustained standing Turing patterns [47]. Their reactor consisted of a thin, flat piece of gel, the sides of which were in contact with nonreacting chemical reservoirs containing subsets of reactants of the oscillating CIMA reaction. The overall redox CIMA reaction consists of the oxidation of iodide by chlorite complicated by the iodination of malonic acid [48,49]. The mechanism of this reaction was obtained by Epstein and co-workers [50]. In De Kepper's experiment, the gel was used to avoid any perturbing hydrodynamical current. The chemicals leaked on to the gel, where they were solely transported through diffusion and where the reactions took place. To make the concentration changes visible to the eye, the gel was loaded with starch, a specific color indicator that turns blue in the presence of polyiodide and is colorless in the absence of iodide [51]. At the beginning of the experiment, several clear and dark lines parallel to the feeding edges developed in the central region of the reactor inside the front between the reduced and oxydized states present at the opposite boundaries. Beyond a given value of malonic acid concentration, some of these lines split up into periodic spots that broke the symmetry of the imposed gradients of concentration [52] (Fig. 3). The typical wavelength of this array of spots was 0.2 mm, a length much smaller than any geometric size of the gel slab [Fig. 4(a) and 4(b)]. Hence the first observed Turing structure was a three-dimensional object with an intrinsic wavelength appearing solely from the interaction between chemical reactions and molecular diffusion. At that time, it was still unclear whether the difference of diffusion coefficients between the activator and inhibitor species of the reaction was inherent to that specific reaction or if the gel was playing an active role in the process.

Thereafter, Ouyang and Swinney built an open reactor using an analogous geometry but a different direction of visualization [Fig. 4(c)]. Their setup allowed one to visualize quasi-2D Turing structures in the same CIMA reaction [53]. The patterns observed were hexagons and stripes, analogous to those shown in Fig. 1, and developing in a plane perpendicular to the feeding direction.

These two experiments set the stage for a complete renewal of the study of chemical patterns. Indeed, they are at the start of different streams of works devoted to unraveling the characteristics of these experimentally observed patterns and answering the newly raised questions. The first challenge is to understand the origin of the difference in diffusivity necessary for a Turing instability to occur. In parallel, several authors set out to answer questions related to the possible symmetries of the structures in two and three dimensions. Are the hexagons and stripes observed the only stable patterns in two dimensions? What are the bifurcation scenarios that can be obtained? Do they match the theoretical predictions available at that time? What is the influence of the gradients of concentration owing to the feeding on the selection, spatial localization, and orientation of the patterns? Let us review some of the works that have focused on these problems.

A. Role of the Gel and the Color Indicator

The simple chlorine dioxide–iodine–malonic acid (CDIMA) reaction is known to be at the core of the temporal oscillations of the CIMA system [49,50,54,55], and Turing patterns have been obtained experimentally in the CDIMA reaction [56]. Epstein and co-workers [50] extracted from their kinetic studies a five-variable reaction–diffusion model of the CDIMA reaction. Lengyel and Epstein noted that a two-variable version of this model can sustain Turing structures if ClO_2^- (inhibitor) diffuses

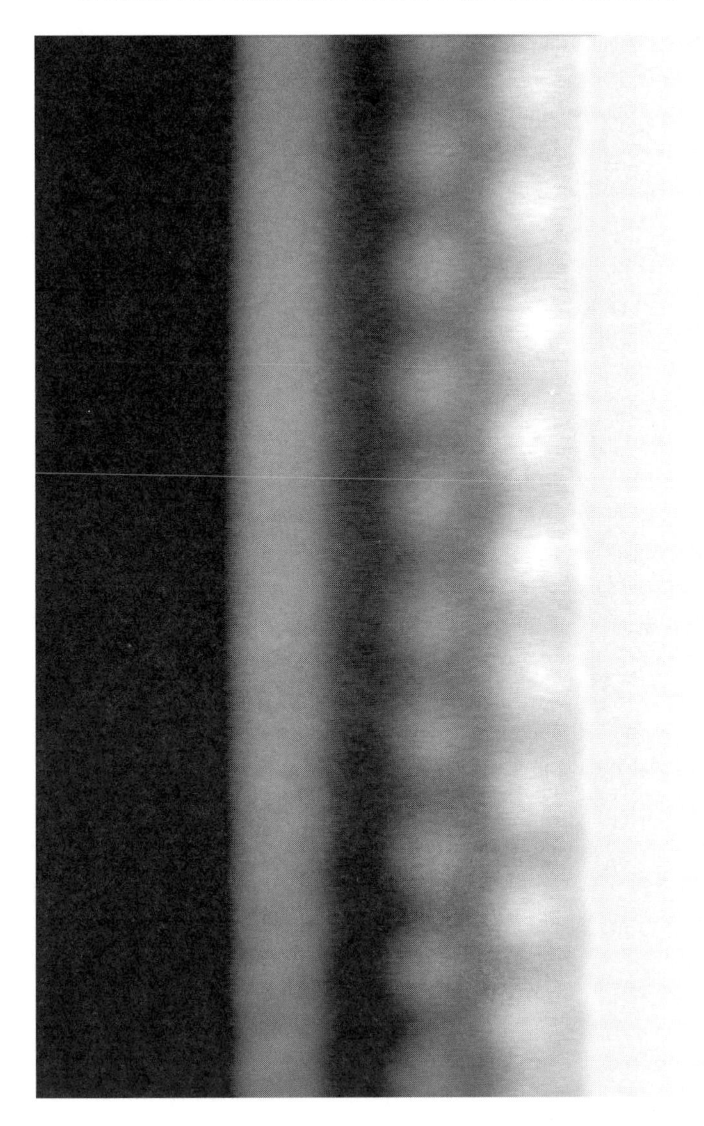

Figure 3. Experimental stationary Turing structure obtained using the CIMA reaction in a continuously fed, unstirred, thick strip gel reactor fed from the lateral boundaries. Iodide and malonic acid are injected at the left and the chlorite and iodide enter from the right. The system is hence in a reduced state with high concentrations of iodide (*black*) to the left; the oxydized (*white*) state dominates at the right. Turing spots breaking the feed symmetry develop in the central region of the gel where the reactants meet. The wavelength is on the order of 0.2 mm. Courtesy of P. De Kepper (CRPP/CNRS).

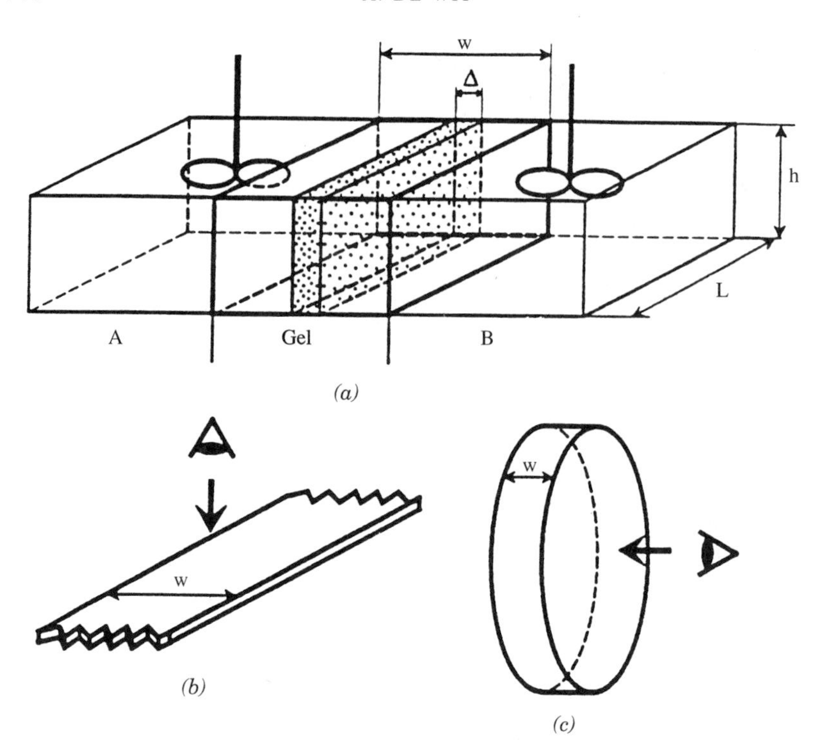

Figure 4. Open spatial reactors. (*a*) The basic principle. A block of hydrogel of length L, height h, and width w is in contact with the contents of two separated reservoirs (A and B). The reservoirs are vigorously stirred and fed with fresh solutions of reactants. The reactants diffuse into the gel from opposite sides, and Turing structures with a characteristic wavelength λ develop in the central zone of the gel where the reactants meet. The patternforming region is of width Δ. Two main types of reactors have been used in the experiments. (*b*) The thin strip reactor for which $\lambda \sim h \leq w \leq L$. Patterns are looked at perpendicularly to the feeding direction. (*c*) The disk reactor, in which $\lambda \sim w \leq h = L$. Patterns are looked at along the feeding direction. Courtesy of P. De Kepper (CRPP/CNRS).

more rapidly than I^- (activator) [57]. At that time, the origin of the low diffusivity of iodide in the CIMA and CDIMA reactions was not clearly understood. Since then, it has been shown that a reversible complexation of the activator into an immobile unreactive complex slows down its effective diffusivity, thereby facilitating the development of Turing patterns [55,58]. This effect renormalizes the evolution equations by a factor proportional to the complexation constant. The complexation mechanism, alluded to by Hunding and Sœrensen [59] in a biological framework, was proposed as a systematic way to design chemical systems able to produce stationary spatial structures [58,60,61].

Several authors have enlightened the role played in such complexation events by gels and the color indicators of iodide [62]. Agladze and co-workers [63,64] showed that Turing patterns can be obtained in gels of different natures and even in gel-free solutions. In absence of gels, the concentration of the color indicator is critical for the spontaneous development of the spatial structures. In that case, the low diffusivity of iodide originates from its binding to this indicator, a large molecule that diffuses more slowly than the ions involved in the CIMA reaction, and its variants [62]. Therefore, if the concentration of the color indicator is decreased, the standing Turing spots can be changed into a region of pulsatile waves [63,64]. If gels are not necessary to act on the diffusivities of the species in the CIMA family of reactions, they are, however, not always inert to the chemistry involved. Stationary patterns have indeed been observed by light diffraction to remain printed in the matrix of polyacrylamide gels in absence of starch [65]. In that case, polar groups of the gel matrix play an essential role in the process. Agarose, polyvinyl alcohols, and silica gels are more inert [48,63]; therefore, they have been mostly used in later experiments related to the CIMA system. Today, the dependence of the Turing wavelength on the diffusivity of the species and the concentration of starch has been studied in detail [66].

B. Two-Dimensional Patterns

The observation of quasi 2D Turing patterns [53,67–71], such as those seen in Fig. 1, and the drawing of experimental phase diagrams [48,69,71,72] that gather the domain of existence of the various structures have launched the comparison of theoretical predictions of bifurcation diagrams and experimental data. Several works [73–75] predicted that in 2D reaction–diffusion systems, hexagons should generally be the first spatial structures to appear subcritically followed by supercritical stripes, as in most pattern-forming systems [19,76–79]. It was also predicted that hexagons and stripes should coexist in some regions of parameters.

Experimental findings are supporting these predictions, as a direct transition from a uniform state to an hexagonal planform owing to a variation of temperature has been recorded [53]. Bistability between hexagons and stripes has also been obtained [68,72]. The predicted subcriticality of hexagons could not be unambiguously found within the available experimental resolution, but localized hexagons embedded into a homogeneous background, a signature of possible subcriticality, have been recently put forth in the CDIMA system [80]. In parallel, hexagons and stripes, analogous to those obtained experimentally, were obtained in numerical integrations of reaction–diffusion models [81–88].

The traditional hexagon and stripe competition has been recovered in chemical patterns; but several experimental observations have, however, called for additional theoretical work. First, the experimentally obtained Turing structures mostly develop in large aspect ratio reactors, i.e., in reactors with a characteristic length much larger than the Turing wavelength. Hence the recorded Turing patterns typically exhibit several hundreds of concentration cells in which defects unavoidably appear [53,81]. This fact underlines the need to distinguish the characteristics of patterns in small- and large-aspect ratios systems. Next, unexpected bifurcations were also observed, such as a transition from a stationary 2D Turing pattern to chemical turbulence [67], a direct transition from a uniform base state to stripes [68], and re-entrance of other phases far from onset [68,70]. In addition, planforms such as 2D rhomb-like [53,89–91], triangular [69], and even more intricate structures [69,70,91] did not fit into the list of theoretically predicted stable patterns. Eventually, several growth mechanisms of Turing structures out of a homogeneous background [80,92,93] have attracted attention to the nucleation mechanisms of Turing patterns. Most of these findings have triggered new theoretical and numerical work.

C. Ramps and Dimensionality of Patterns

The geometry of the open reactors used in the study of Turing patterns unavoidably introduces gradients of concentration as the gel is fed from its boundaries (Fig. 4). Hence it is important to understand the influence of these ramps on the dimensionality of the structures and on their selection. Two main types of reactors have been used in the experimental approach. The first one is the thin strip reactor of dimensions $h \leq w \leq L$, developed by De Kepper and co-workers [48], in which observations are made perpendicularly to the feed direction [Fig. 4(b)]. This geometry provides a direct view of the width Δ of the area of the pattern-forming region. Patterns develop in rows of spots parallel to the feed boundaries (Fig. 3), which are orthogonal to the direction of the concentration gradients. If the gel strip is thin enough, i.e., if its height h is on the order of the Turing wavelength λ, only one layer of structure can develop. These patterns will then be 1D or 2D, depending on the width Δ of the region in which the gradients localize [70]. On the other hand, if all three sides of the Turing zone are wider than λ, the pattern is 3D [64].

The second geometry developed by Ouyang and Swinney [91] is a disk reactor that is fed perpendicularly to the faces [Fig. 4(c)]. Observation is made along the direction of feeding. This geometry gives a view of planes parallel to the faces of feeding and hence of uniform values of input concentration. Depending on the thickness Δ of the pattern-forming region, the structures are then 2D or 3D. Ouyang and co-workers [68] modified such a reactor to demonstrate that the hexagonal and stripe-type patterns they had obtained before had a thickness Δ on the order of λ and not more, proving that they are actually quasi-2D patterns.

The dimensionality of the patterns has been investigated in detail by Dulos and co-workers [69,70], who used beveled gel reactors specially designed to make possible the unfolding of a pattern sequence in one direction of the plane of observation. Later, this geometry was adapted to yield a reactor fed only on one side [71,94].

From the theoretical point of view, conditions on the position along the gradient and the possible three-dimensionality of the structures were obtained by Lengyel, Kádár and Epstein [95] in a linear stability analysis of their model of the CDIMA reaction. Several theoretical studies examined the influence of gradients of concentrations on the selection and localization of 1D [96,97], 2D [52,81,82,85,98–102,102a] and 3D [102,103] patterns in the framework of reaction–diffusion models. More recently, conditions for which structures develop in monolayers or bilayers were studied by Dufiet and Boissonade [104] and Bestehorn [105]. Chemical systems thus genuinely present the opportunity to test the general theoretical works that were devoted to the analysis of the effect of ramps in pattern-forming systems (see also [106–116] and references therein).

D. Three-Dimensional Patterns

The fact that chemical patterns can be true 3D structures when their intrinsic wavelength is smaller than any dimension of the pattern-forming zone in the gel gives to chemical systems a specific role in pattern-forming media. It is indeed one of the few systems that can generate a true 3D symmetry-breaking instability. This fact was clearly evidenced in the first experimental finding of a Turing pattern [47]. Further observations made under different angles [64] show beady structures that could be consistent with a body-centered cubic symmetry. Because of the presence of gradients of concentration, various modes can sometimes develop in different depths of the gel [64,69,70,92] and multilayer spatial organizations are obtained. In that specific case, the actual 3D structures are made of a juxtaposition of 2D patterns and resolution of the involved symmetries can become much more complicated [69,104]. This resolution is also impaired by the presence of defects that in 3D can become quite involved [117].

E. Turing–Hopf Interaction

One of the most interesting aspects of studying pattern formation in chemical systems lies in the fact that reaction–diffusion media genuinely sustain different types of instabilities, such as a Hopf instability, bistability, or excitability. Chemical systems provide possibilities of studying interactions between different instabilities. In particular, several studies of the interaction between

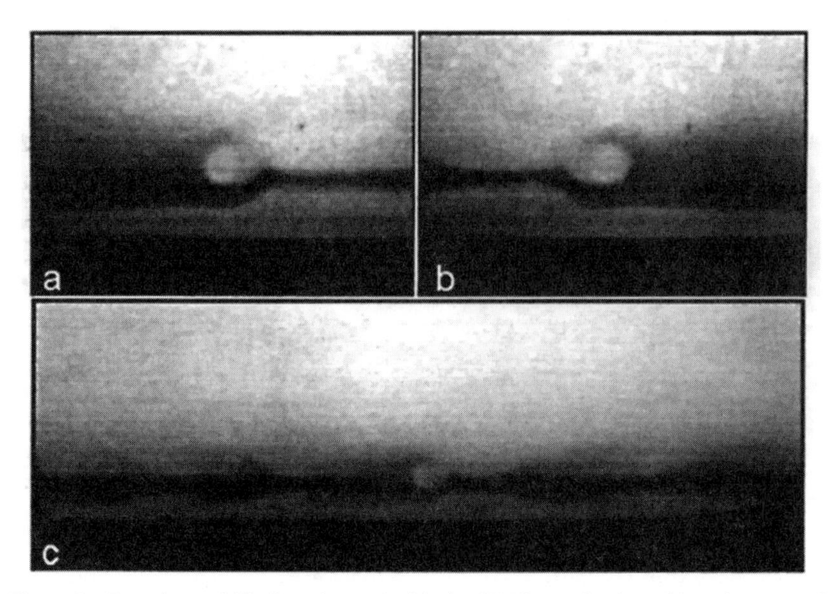

Figure 5. Experimental flip-flop observed with the CIMA reaction in a thin strip reactor for parameter values that allow interactions between Turing patterns and temporal oscillations. A stationary central Turing dot emits waves alternatively to each side (*a*, *b*), giving rise to a train of plane waves traveling along a line parallel to the feeding edges, seen here at the top and bottom (*c*). Courtesy of P. De Kepper (CRPP/CNRS).

spatial and temporal symmetry-breaking instabilities have been conducted in the CIMA reaction, because the thresholds of the Turing and Hopf instabilities can be brought to coincide in this system.

Today, it is clearly understood that in the CIMA system, the color indicator (for instance starch or polyvinyl alcohol) can play a key role in obtaining Turing patterns by slowing down the diffusivity of the activator of the reaction through a specific complexation with it. In agarose gels and in gel-free media, a transition from standing Turing patterns toward traveling waves is observed when the concentration of starch is decreased [63,64,92,118]. In the vicinity of this transition point, complex spatiotemporal dynamics resulting from the interaction between the Turing and Hopf modes, i.e., between a steady spatial mode and a homogeneously temporally oscillating mode, are obtained. In 1992, Perraud and co-workers [64,118] reported the first of such spatiotemporal dynamics owing to a Turing-Hopf interaction. It consists in an unusual wave source corresponding to an isolated Turing spot that emits wave trains along a thin band parallel to the feed surfaces (Fig. 5). The experiment was conducted in a thin strip gel reactor, and the thickness of the gel was small enough to ensure that the dynamics was one-dimensional.

A peculiarity of these waves is that they are not emitted synchronously by the source but alternatively to each side. This dynamics has, therefore, been coined chemical flip-flop [64,92,98,118,119]. Theoretical approaches have unraveled the different bifurcation scenarios that can arise thanks to the interactions between a Turing and a Hopf instability, as will be detailed later. In particular, they have shown the chemical flip-flop to be a Turing structure localized in a Hopf oscillating medium and existing in a region of bistability between the two solutions. In 2D systems, the equivalent of the flip-flop is a spiral, the core of which is a Turing dot, as observed experimentally [48,92,98] and numerically [120,121]. De Kepper and co-workers reported that, in 2D experiments, the Turing–Hopf interaction can also lead to spatiotemporal intermittency [69,92] and an interaction between standing Turing structures and spiral waves in geometries in which different modes develop into adjacent layers of the reactor [92].

F. New Systems

The first experiments performed on Turing patterns dealt with the CIMA reaction. Meanwhile, progress has been made in obtaining chemical patterns in other systems. First, Lengyel and Epstein proposed a methodology to design new Turing systems, exploiting the complexation step between the activator of the reaction and a slowly diffusing species [56,58,122]. This mechanism is also at play in the CDIMA reaction, at the core of the CIMA chemical scheme [56,95], which also exhibits Turing structures [56]. Because the CDIMA reaction is described quantitatively by the two variable Lengyel–Epstein model, it provides a good system to compare analytically predicted behaviors and experimental findings [95,123]. Moreover, transient Turing patterns were obtained in a closed reactor using this CDIMA system [122,123], making the phenomenon accessible to lecture demonstrations.

In 1995, Watzl and Münster obtained Turing-like patterns in the polyacrylamide–methylene blue–sulfide–oxygen (PA–MBO) system [124,125]. This oscillating reaction, discovered by Burger and Field [126], is essentially a redox relationship between the colorless reduced form MBH and the blue MB^+ form of the methylene blue monomer. The mechanism for the MBO temporal oscillations is explicitly known [127] and can be cast into a five-variable model [128]. In this system, the Turing structures are transient because experiments are performed in a semiclosed Petri dish. Nevertheless, the system is rich and allows the observation of hexagons, stripes, and zigzags [124,125]. An advantage of these patterns is that their wavelengths are on the order of 2 mm. They can thus be visualized straigthforwardly. In this system, the effect of an externally applied electrical field [93,125] and light [93] has been shown to affect the selection and orientation of the obtained structures. In the PA–MBO system, the polyacrylamide gel plays a role in the pattern formation, since no Turing structures could be obtained in experiments with the MBO reaction in agarose [93,124] or methylcellulose [93] gels. As a structuring of the gel's surface accompanies the formation of chemical patterns [124], it remains to be checked whether these patterns appear through a pure Turing instability of the PA–MBO system or if a possible mechanical response of the gel also plays a role in the pattern-forming process.

In addition, new highly irregular labyrinthine patterns have been found by Swinney and co-workers [129–131] in the ferrocyanide-iodate-sulfite (FIS) reaction. This reaction is also sometimes called the EOE reaction after Edblom, Orbán, and Epstein [132], who discovered it. The FIS reaction is bistable and can sustain large oscillations of pH in continuously stirred reactors. Models of the FIS kinetics are available [133,134], and a four-variable model [135] provides good insight into the experiments. One of the main differences of the FIS labyrinths compared to regular Turing stripes is that they are initiated only by large amplitude perturbations [129,130]. The same system also exhibits self-replicating spots [130,131,136,137], breathing spots [138], and other phenomena arising through front instabilities [130]. The bistable character of the FIS system is important for understanding these new aspects, as shown by numerous theoretical works.

Let us now develop the theoretical framework in which characteristics of pattern formation in monostable reaction–diffusion systems can be understood.

IV. PATTERN SELECTION THEORY

When a physicochemical system develops spatial patterns, different types of symmetries, such as stripes or hexagons, are typically observed. The pattern selection theory is devoted to determine which pattern will be observed among all possible ones for a given set of parameters and what their characteristics will be, such as orientation and wavelength [19,21,34]. Therefore, it is first necessary to fix the existence and stability conditions of each of the possible solutions of the reaction–diffusion equations and next to study their relative stability to account for the competition among patterns with different symmetries.

To treat this problem, the starting point consists in choosing a model that, even if it does not describe in detail the physical, chemical, or biological mechanisms of the system, summarizes at least its essential characteristics. For many problems, for instance in hydrodynamics, the starting equations are known but difficult to treat analytically. Simpler models that synthetize the relevant properties and cast the symmetries of the problem are then most useful. An example is the well-studied Swift–Hohenberg model

[139]. In chemistry and biology, evolution equations are often simply not known, and the use of reaction–diffusion models can then be justified *a fortiori*.

A linear stability analysis of the stationary steady states of these models determines at which values of the parameters different instabilities occur. In particular, it gives the critical value γ_c of the control parameter, above which the steady state becomes unstable because of a Turing instability and a new spatially organized solution appears. This Turing instability occurs when the growth rates of perturbations around the steady state are real and when one of them becomes positive for a wavenumber $|\underline{\mathbf{k}}| = k_c$ (see Section II). Beyond this critical point, a certain number of spatial modes grow exponentially in time (Fig. 2). This linear exponential growth is saturated when the nonlinear terms in the evolution equations come into play. The nonlinear competitions between modes then select the preferred spatial planform and lead to a new spatial structure of finite amplitude, constructed with N-dominating modes. The description of the asymptotic behavior of the system beyond the instability threshold hence calls for a nonlinear approach of the problem.

A. Weakly Nonlinear Analysis

The new spatially organized states existing beyond the bifurcation can be characterized if the temporal evolution of the amplitude T of each of the N modes underlying it is known. Indeed, the variables of the model can then be approached by a linear combination of these N modes as

$$\underline{C}(\mathbf{r}, t) = \underline{C}_0 + \sum_{j=1}^{N}[T_j(\tau)e^{i\mathbf{k}_j\cdot\mathbf{r}} + T_j^*(\tau)e^{-i\mathbf{k}_j\cdot\mathbf{r}}]\underline{w} + O(\cdots) \qquad |\mathbf{k}_j| = k_c \quad (4.1)$$

where $\underline{w}$ is the critical eigenvector of the linear evolution matrix of the problem; and τ, the slow-time scale on which the amplitude evolves, is inversely proportional to $\mu = \gamma - \gamma_c$, the distance from threshold (i.e., $\tau = \mu t$). Because the concentrations are real, the active modes involve pairs of opposite wavevectors $\pm\mathbf{k}_j$. If the linear combination [Eq. (4.1)] is spatially regular, it must be a solution of the evolution equation for the perturbations:

$$\frac{\partial \underline{u}}{\partial t} = \underline{\underline{L}}\,\underline{u} + M(\underline{u}) \qquad (4.2)$$

where $\underline{u} = \underline{C} - \underline{C}_o$ and is the concentration change around the reference steady state $\underline{C}_o$, $\underline{\underline{L}}$ is the linear evolution matrix, and M is the nonlinear

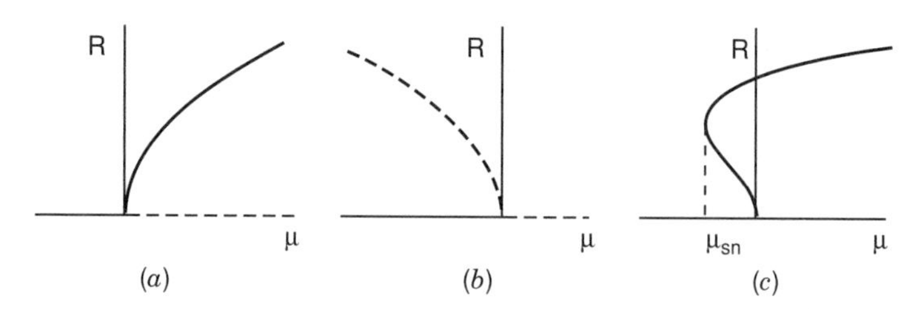

Figure 6. Bifurcation diagrams displaying the modulus R versus the bifurcation parameter μ. The *plain line* and *dashed line* represent stable and unstable branches of solutions, respectively. (*a*) Supercritical case; (*b*) subcritical case; (*c*) saturated subcritical case, in which the stable non-linear branch of solutions appears with a finite amplitude at a secondary saddle-node bifurcation when $\mu = \mu_{SN}$.

part in $\underline{u}$. In the vicinity of the bifurcation, one seeks to determine $\underline{u}$ through an asymptotic expansion:

$$\underline{u} = \varepsilon\, \underline{u}_1 + \varepsilon^2\, \underline{u}_2 + \varepsilon^3\, \underline{u}_3 + \dots \tag{4.3}$$

where the small expansion parameter ε is related to the distance μ from threshold as

$$\mu = \gamma - \gamma_c = \varepsilon\, \gamma_1 + \varepsilon^2\, \gamma_2 + \dots \tag{4.4}$$

Solving the system of equations at each order in ε allows us to obtain $\underline{u}_1, \underline{u}_2, \dots,$ which define the structure of the nonlinear solution through the expansion [Eq. (4.3)]. If $\gamma_1 \neq 0$, the new solution reduces at lowest order to $\underline{u} \sim \underline{u}_1$, with $\varepsilon \sim (\gamma - \gamma_c)/\gamma_1$. The solution $\underline{u}$ exists for both positive and negative ε, and the bifurcation is said to be transcritical. If $\gamma_1 = 0$, we still have $\underline{u} \sim \underline{u}_1$, but then $\varepsilon \sim \sqrt{(\gamma - \gamma_c)/\gamma_2}$. The bifurcation occurs for $\gamma > \gamma_c$ if $\gamma_2 > 0$ or $\gamma < \gamma_c$ if $\gamma_2 < 0$. The transition is then called supercritical or subcritical, respectively (Fig. 6). The dynamics of the system in the vicinity of the instability occur on different time scales than those of the reference steady state. Hence the partial differential operator in time is also developed in ε as

$$\partial_t = \varepsilon\, \partial_{\tau_1} + \varepsilon^2\, \partial_{\tau_2} + \dots \tag{4.5}$$

As the control parameter γ usually comes into play in $\underline{\underline{L}}$, we have

$$\underline{\underline{L}} = \underline{\underline{L}}_c + \varepsilon \underline{\underline{L}}_1 + \varepsilon^2 \underline{\underline{L}}_2 + \dots \tag{4.6}$$

where $\underline{\underline{L}}_c$ is the linear operator computed at B_c. Substituting expressions (4.3)–(4.6) into (4.2) and isolating the different orders in ε, the nonlinear initial problem gets down to solving successions of linear equations:

$$\underline{\underline{L}}_0 \, \underline{u}_1 = 0 \tag{4.7}$$

$$\underline{\underline{L}}_0 \, \underline{u}_i = \underline{I}_i \tag{4.8}$$

where $i = 2, 3, \ldots$ and $\underline{\underline{L}}_0 = \partial_{\tau_o} - \underline{\underline{L}}_c$. Eq. (4.7), at first-order, defines the critical wave vector $\underline{u}_1$. This equation is homogeneous in $\underline{u}_1$ so that the amplitudes T_j of the modes constructing this first-order solution remain unknown. The higher-order equations are nonhomogeneous and have a nontrivial solution only if their right-hand side is orthogonal to the kernel of $\underline{\underline{L}}_0^+$, the adjoint operator of $\underline{\underline{L}}_0$. This solvability condition, also called the Fredholm alternative, determines at successive orders the different coefficients of the perturbation expansion (4.3) expliciting hence the new solution appearing beyond the instability. In particular, the Fredholm alternative makes explicit the amplitude T of the first-order solution by providing its temporal evolution equation, the so-called amplitude equation, which takes the general form:

$$\frac{dT_j}{dt} = \mu T_j + G_j(\{T_i\}) \tag{4.9}$$

where $G_j(\{T_i\})$ are nonlinear polynomials in the active amplitudes. Details on the standard bifurcation techniques used to derive amplitude evolution equations can be found in [6,19,21,32,34].

The main advantage of the description in terms of amplitude equations is that close to the bifurcation point, the amplitude evolves on the slow-time scale of the critical modes that is inversely proportional to the distance from threshold. On this scale, the dynamics of the amplitude depends only on the type of instability. Indeed, the terms appearing in the amplitude equations are functions only of the broken symmetries and not of the details of the system that appear only in the value of the coefficients of these equations [140]. Note that the amplitude equations were obtained by a perturbation expansion and are, therefore, valid only in the vicinity of the bifurcation point. Other descriptions of the system will be necessary farther away from threshold.

B. Degeneracies

The nonlinear analysis of the problem must take into account the degeneracies of the system. In small systems (the size of which is on the order of the critical wavelength), the spectrum of the linearized operator is discrete and at most finitely degenerate [3,12,21,141]. Only a small number of modes are excited and the variables of the model can be constructed as a finite linear combination of these unstable modes. In that case, it can be shown that the reaction–diffusion system is correctly described by the reduced dynamics of the amplitude equations derived by standard bifurcation theory. This is not true, however, for large systems, the boundaries of which are either at infinity or at such a far distance that they do not constrain the spectrum of spatial modes. The majority of experimental Turing patterns belong to the class of these large systems because several hundred of wavelengths are commonly obtained in the experiments.

The spectrum of unstable modes in large systems is then degenerated for two reasons. The first degeneracy is an orientational degeneracy. In reaction–diffusion systems, the linear stability analysis shows that the growth rate of the unstable modes depends only on the modulus of the critical wave vectors. This means that the structures break the translational symmetry but not the rotational symmetry. All wave vectors lying on the sphere (or on the circle in 2D) of modulus $|\mathbf{k}| = k_c$ are equally excited and must be included in the nonlinear treatment of the model. In other words, the number of linear combinations such as Eq. (4.1) is infinite. In practice, one chooses from among all possible combinations those that are linked to regular pavements of space, unless the focus is on more complex structures. This number N of modes used fixes the geometrical aspect of the pattern. For reaction–diffusion systems, the patterns considered in 2D for instance, are typically stripes ($N = 1$), squares ($N = 2$), or hexagons ($N = 3$). The linear combinations for $N > 3$ give rise to more complex multiperiodic structures [142], such as those observed in experiments with parametric excitation [143,144] or in nonlinear optics [145]. These platforms have not been obtained in chemical systems and will thus not be considered here. A temporal evolution equation for the amplitude of the modes can be derived for each linear combination. To study the nonlinear competition between modes, one must first find solutions to the amplitude equations and then study their relative stability. This procedure (discussed below) shows which structure will be observed based on the values of the parameters of the system and is thus the basis of the pattern selection theory [19,21,78,79,146].

The second degeneracy that we must deal with is continuous band quasi-degeneracy. When the control parameter's value is above criticality, there is a finite but continuous band of modes that become unstable in addition to the critical wave vectors (Fig. 2). In large systems, the number of such modes is so large that they form a quasi-continuous ensemble of modes of various lengths close to k_c, spanning degenerate irregular spatial structures. This degeneracy can be treated by defining the amplitude as a slowly variable function not only of time but also of space. In other words, we write

$$\underline{C}(\mathbf{r}, t) = \underline{C}_0 + \sum_{j=1}^{N} [T_j(\tau, \mathcal{X})e^{i\mathbf{k}_j\cdot\mathbf{r}} + T_j^*(\tau, \mathcal{X})e^{-i\mathbf{k}_j\cdot\mathbf{r}}]\underline{w} \qquad |\mathbf{k}_j| = k_c \quad (4.10)$$

where the amplitude is now also a function of a slow space scale $\mathcal{X}$. This space scale is proportional to $\xi_o/(\gamma - \gamma_c)$, where ξ_o is the coherence length on the order of k_c^{-1}. The amplitude equation, in that case, is a partial differential equation in space and time of the following form:

$$\frac{\partial T_j}{\partial t} = \mu T_j + G_j(\{T_i\}) + \xi_o^2 \Box^2 T_j \qquad (4.11)$$

where $\Box^2$ is a spatial operator describing the modulation of the patterns on the long length scale. Amplitude Eq. (4.11) describes the nonlinear interactive behavior of the wave packets that account for the dynamics of all the modes included in the unstable band. Such envelope equations have become dynamic models on their own, because they reproduce numerous properties of nonequilibrium systems [19,21]. They allow the study of defect dynamics, of localized structures in weakly nonlinear regimes, and of spatiotemporal chaotic dynamics.

The amplitude of a spatial pattern is a complex variable. Its modulus corresponds to the intensity of the spatial modulation of the model's variables. Its phase is related to the breaking of translational symmetry. If the system is perturbed, variations of intensity relax on a slow characteristic time scale and are inversely proportional to the distance from the bifurcation treshold. The system remains neutral, however, in regard to a uniform phase change $T_j \to T_j e^{i\Theta}$ that corresponds to a global translation of the pattern. The phase Θ thus evolves on an even slower time scale. Far from the bifurcation point, the phase evolution on this time scale can then suffice to totally describe some properties of the system, such as typically long-wave instabilities. In that respect, phase equations have become a subject of research on their own [19].

C. Reaction–Diffusion Models vs Amplitude Equations

We have just seen that to tackle the pattern selection problem, two complementary points of view are available for theoreticians. The first one is to look at reaction–diffusion models. These models are the basis for an understanding of the effect of changing parameters. The advantage of their numerical integration is that the active modes are selected by the internal nonlinearities of the problem and not imposed *a priori*. Nevertheless, from an analytical point of view, the thresholds of instability are often the only relevant quantities that can be obtained. The nonlinear regime must then be studied numerically.

The second tool available is that of amplitude equations that allow analytical insight into the selection and possible transitions between different spatial planforms. Because the amplitude equations have a universal form and are a function only of the symmetries broken at the bifurcation point, the advantage of studying them is that all bifurcation scenarios predicted on their basis are applicable to any physicochemical system that presents the related breaking of symmetries. The disadvantages are that amplitude equations are valid only in the vicinity of the bifurcation point and that they depend on the modes considered. They are hence of no help if one does not know which modes are involved *a priori*.

V. TURING PATTERNS

The theoretical approach devoted to understand which types of patterns can be observed in reaction–diffusion systems and what their succession will be when a control parameter is varied is now outlined for 2D and 3D systems.

A. Reaction–Diffusion Models

To study pattern formation in chemical systems owing to the coupling of chemical reactions and diffusion processes, it is natural to turn to reaction–diffusion models [12,147]. The best model is, of course, the one that is the closest to the experimental system, as the ultimate goal of any theory is quantitative predictions. Unfortunately, chemical kinetics are often complicated; thus it is useful to study typical bifurcation scenarios via simpler models that are more easily mathematically handled.

Several works on Turing patterns have focused on quantitative comparison with experimental results. The Lengyel–Epstein model is the most realistic model available for quantitative comparisons with the CDIMA reaction [50,54,57]. This model provides structures with wavelengths that are in good agreement with those observed experimentally [58,122,123] and has been used to obtain conditions on possible three-dimensionality of the patterns in parameter space [58]. Jensen and co-workers thoroughly investigated the 1D and 2D pattern selection problem of the Lengyel–Epstein model [88,100,120,121,148,149]. They show that, in one-dimension, it exhibits a strong subcriticality of stripes and bistability between the stripes and a Hopf state. In the subcritical regime, a study of the wavenumber selection and propagation speed was performed in the case of a moving front between the stripes and the homogeneous steady state [120]. Pinning effects resulting in a vanishing front velocity because of interaction of the front with the underlying Turing pattern were evidenced [100,120,121].

In two dimensions, both stripes and hexagons appear subcritically. As a consequence, many localized structures exist in the domain of parameters for which bistability between two different states occur. In particular, stable spatial coexistence of stripes and hexagons [88], patches of hexagons embedded into the homogeneous steady state [88,120], and growing mechanisms of hexagons into a homogeneous background [120] were obtained in the Lengyel–Epstein model and enlighten the recent experimental observations of some of these phenomena [80]. Localized Turing–Hopf structures were also observed in the Turing–Hopf bistability regime of the Lengyel–Epstein [100,120,149].

Another model that has been extensively studied in the framework of pattern formation in chemical systems is the two-variable Schnackenberg model [150]. Dufiet and Boissonade showed that this model reproduces 2D patterns seen in the experiments and clarifies long-wave instabilities, such as zigzag or Eckhaus instabilities of patterns [86,87]. Quantitative comparison between analytical predictions and numerical simulations made with the Schnackenberg model have greatly helped test pattern selection theories [48,86,87,151].

Recently, a good insight into such comparisons was provided by the *ad hoc* construction of reaction–diffusion models in which the coefficients in front of the variables in the model are simply related to those of the amplitude equations [48,104]. Note that prototype models for pattern formation, such as the Swift–Hohenberg model [139], have also been studied in relation to chemical problems [90,99,152].

In this chapter, we mainly focus on the Brusselator model [3,153]. This two-variable model can exhibit both a Turing and a Hopf instability. Its advantage is that the base state, the thresholds of both instabilities, and the coefficients of the related amplitude equations for pattern formation or temporal oscillations are straightforwardly obtained analytically [9,97,154,155]. Moreover the Brusselator model has been the subject of many studies, so we have a good knowledge of the possible spatiotemporal dynamics it can exhibit [3,75,79,119]. It is, therefore, the model we will focus on in this review, because it has been used to analyze most of the topics discussed here. The reaction–diffusion equations of the irreversible Brusselator are as follows:

$$\begin{aligned} \partial_t X &= A - (B+1)X + X^2 Y + D_x \nabla^2 X \\ \partial_t Y &= BX - X^2 Y + D_y \nabla^2 Y \end{aligned} \tag{5.1}$$

The concentration of species B is usually chosen as the bifurcation parameter. The homogeneous steady state $(X_s, Y_s) = (A, B/A)$ of system Eqs.

(5.1) undergoes a Turing instability when $B > B_c^{\mathrm{T}} = (1 + A\sqrt{D_x/D_y})^2$. A stationary spatial pattern then emerges, characterized by an intrinsic critical wave vector $k_c^2 = A/\sqrt{D_x D_y}$. The steady state may also go through a Hopf instability if $B > B_c^{\mathrm{H}} = 1 + A^2$, evolving then into an homogeneous limit cycle characterized by a critical frequency $\omega_c = A$. The thresholds of these two instabilities coincide at a codimension-two Turing–Hopf point, defined as the point at which $B_c^{\mathrm{H}} = B_c^{\mathrm{T}}$. This condition is achieved when the ratio of the diffusion coefficients $\sigma = D_x/D_y$ reaches its critical value $\sigma_c = [(\sqrt{1 + A^2} - 1)/A]^2$.

Note that these models do not exhibit some of the fundamental characteristics of the experimental reaction–diffusion systems, such as bistability, excitability, and formation of traveling waves owing to a Hopf bifurcation with a nonzero wavenumber. This latter instability occurs only in three-variable models. Because bistability is an important characteristic of the FIS reaction, some studies devoted to understanding spatial patterns formed in that system have used the bistable FitzHugh–Nagumo model [156,157], the Gáspár–Showalter model of the FIS reaction [134,135] and the Gray and Scott model [158].

B. Two-Dimensional Pattern Selection

In this section, we will briefly illustrate the nonlinear analysis techniques sketched in Section IV to study the 2D pattern selection problem in reaction–diffusion systems in the weakly nonlinear regime. To do so, we will first neglect any possible spatial variation of the mode's amplitude and obtain standard bifurcation diagrams that are valid in the vicinity of the bifurcation point. We will then discuss specificities of chemical systems, such as re-entrance of hexagons and localized structures in subcritical regimes before commenting on effects induced by spatial modulations of the amplitudes.

1. *Standard Bifurcation Diagram*

Let us successively consider amplitude equations for the regular 2D spatial structures constructed with one (stripes), two (squares), or three (hexagons) pairs of wave vectors.

A perfect periodic structure in only one direction of space, such as stripes, is built on one pair ($N = 1$) of wave vectors; and in that case, the concentration field [Eq. (4.1)] can be constructed as

$$\underline{C}(\mathbf{r}, t) = \underline{C}_0 + [T(\tau)e^{i\mathbf{k}_1 \cdot \mathbf{r}} + T^*(\tau)e^{-i\mathbf{k}_1 \cdot \mathbf{r}}]\underline{w} \qquad |\mathbf{k}_1| = k_c \tag{5.2}$$

At the lowest order, the amplitude equation for T, derived using techniques discussed in Section IV.A, reads

$$\frac{dT}{dt} = \mu T - g|T|^2 T \tag{5.3}$$

The amplitude T is imaginary, and if we separate its modulus and phase, writing $T = Re^{i\theta}$, we get

$$\frac{dR}{dt} = \mu R - gR^3 \tag{5.4}$$

$$\frac{d\theta}{dt} = 0 \tag{5.5}$$

The phase can take any constant value, a signature of translation invariance. The phase may thus be set to zero by a suitable choice of coordinates. The modulus equation [Eq. (5.4)] has two stationary solutions: either $R_s = 0$, characterizing the homogeneous steady state, or $R_s = \sqrt{\mu/g}$, corresponding to stripes. Two subcases can be distinguished: if $g > 0$, the bifurcation is supercritical, and the new solution exists for $\mu > 0$; whereas if $g < 0$, the new solution arises for $\mu < 0$, and the bifurcation is subcritical. In both cases, we have a pitchfork bifurcation [19,21].

To investigate the stability of these solutions, a standard linear stability analysis of each of them must be performed. The difficulty in the analysis arises from the fact that for each state one must discuss the stability with respect to all possible types of perturbations: phase, modulus, orientation of the wavevector, and resonant perturbations to other structures with different symmetries. Let us first consider perturbations of the modulus (the other perturbations will be considered later). Writing $R = R_s + \delta R$, and inserting it into Eq. (5.4), we see that the trivial state $R_s = 0$ looses stability when $\mu > 0$, whereas the stripes are stable for $\mu > 0$ if $g > 0$. If $g < 0$, the stripes that arise subcritically are unstable for $\mu < 0$, as the instability is not saturated by nonlinear terms. The bifurcation calculation must then be carried out to higher orders, where we get the following amplitude equation:

$$\frac{dT}{dt} = \mu T - g|T|^2 T - g'|T|^4 T \tag{5.6}$$

If g' is positive, the bifurcation saturates, leading to stripes that appear subcritically at the point $\mu_{\rm SN} = -g^2/4g'$, where a secondary saddle-node bifurcation occurs [Fig. 6(c)]. The stability of the various branches are then calculated in a standard fashion. For $\mu_{\rm SN} < \mu < 0$, there is bistability between the stripes and the trivial homogeneous state, with the possibility of observing localized structures (see Section V.B.3). If g' is negative, the bifurcation is not yet saturated, and we must proceed at still higher orders.

Let us now consider 2D structures built on more than one pair of wave vectors that are regular pavements, i.e., pavements for which the N pairs of modes make π/N angles between them. When $N = 2$, this corresponds to squares for which the concentration field becomes

$$\underline{C}(\mathbf{r}, t) = \underline{C}_0 + [T_1(\tau)e^{i\mathbf{k}_1\cdot\mathbf{r}} + T_2(\tau)e^{i\mathbf{k}_2\cdot\mathbf{r}} + c.c]\underline{w} \qquad |\mathbf{k}_1| = |\mathbf{k}_2| = k_c \quad (5.7)$$

where $c.c$ stands for complex conjugate. The corresponding amplitude equations for squares are

$$\frac{dT_j}{dt} = \mu T_j - g|T_j|^2 T_j - g_{\text{ND}}(\theta)|T_l|^2 T_j \qquad (5.8)$$

with $j, l = 1, 2$ and $\theta = \pi/2$. We see that a nonlinear coupling term between the two sets of modes must now be taken into account. Let us consider here the cases in which the instability is saturated at this order, i.e., $g > 0, g_{\text{ND}} > 0$. Writing $T_j = R_j e^{i\theta_j}$, we get

$$\frac{dR_j}{dt} = \mu R_j - gR_j^3 - g_{\text{ND}}(\theta)R_l^2 R_j \qquad (5.9)$$

$$\frac{d\theta_1}{dt} = \frac{d\theta_2}{dt} = 0 \qquad (5.10)$$

In the case of stripes, the constant factor phases correspond to a simple translation of the pattern. The solutions to the evolution equation for the modulus R_j are the trivial homogeneous steady state ($R_1 = R_2 = 0$) and stripes corresponding to $R_1 \neq 0, R_2 = 0$ (stripes perpendicular to $\mathbf{k}_1$) or $R_1 = 0, R_2 \neq 0$ (stripes perpendicular to $\mathbf{k}_2$). The squares correspond to the case $R_1 = R_2 = R_R$, with

$$R_R = \sqrt{\frac{\mu}{g + g_{\text{ND}}}} \qquad (5.11)$$

A linear stability analysis of these supercritical branches shows that squares and stripes are mutually exclusive, i.e., stripes are the stable pattern if $g < g_{\text{ND}}$, and squares are stable if $g > g_{\text{ND}}$. Note that, if the system is even slightly anisotropic, θ might be different from $\pi/2$, and the squares then become rhombs. As the coefficient g_{ND} takes values that vary with θ, the stability domain of these rhombs may differ with that of the squares; but nevertheless, they remain unstable in regard to stripes as long as $g < g_{\text{ND}}(\theta)$. When $\theta = \pi/3$, the Eq. (5.8) for rhombs is no longer valid, because in such a case, the vector $\mathbf{k}_1 + \mathbf{k}_2$ falls on the circle of critical wave vectors and

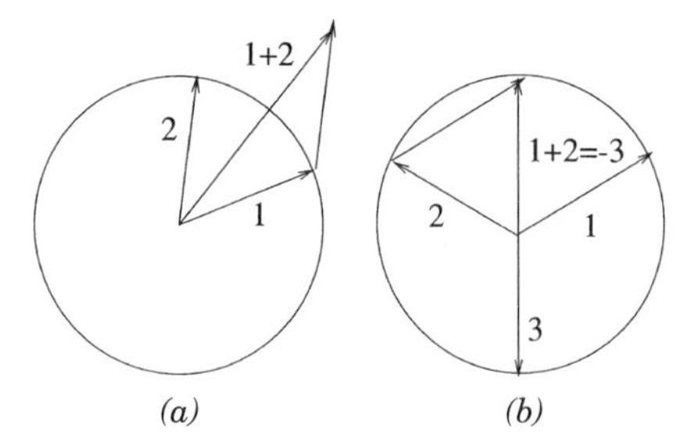

Figure 7. The resonance condition on the circle of radius k_c. (*a*) Nonresonant combination of two wave vectors $\mathbf{k}_1$ and $\mathbf{k}_2$: Their sum is wave vector $\mathbf{k}$ with $|\mathbf{k}| \neq k_c$. (*b*) Resonant combination of two wave vectors that make an angle of $2\pi/3$ between them and, therefore, excite a third critical wave vector.

is excited as well (Fig. 7). Its dynamics must then be taken into account. The resulting pattern is a regular structure corresponding to hexagons and constructed with three pairs of wavenumbers ($N = 3$) such that $\mathbf{k}_1 + \mathbf{k}_2 + \mathbf{k}_3 = 0$. The amplitude equations for these three modes are

$$\frac{dT_1}{dt} = \mu T_1 + v T_2^* T_3^* - g|T_1|^2 T_1 - h(|T_2|^2 + |T_3|^2)T_1 \tag{5.12}$$

The amplitude for the two other modes are obtained by cyclic permutations of the indices. These amplitude equations and in particular the quadratic term present in Eq. (5.12) can be obtained only in systems for which the $T \to -T$ symmetry is broken. This is, for instance, the case in non-Boussinesq Rayleigh–Bénard convection [19] or typically in chemical systems. As in the case of rhombs, the coefficients g and h of the coupling term are functions of the angles between the three wave vectors. For the sake of simplicity, let us assume that g and h are positive and that the hexagonal pavement is regular, i.e., the angle between modes is strictly equal to $\pi/3$. The evolution equation for the sum of the phases $\Theta = \theta_1 + \theta_2 + \theta_3$ reads

$$\frac{d\Theta}{dt} = -v\left[\frac{R_1^2 R_2^2 + R_1^2 R_3^2 + R_2^2 R_3^2}{R_1 R_2 R_3}\right] \sin\Theta. \tag{5.13}$$

Contrary to the cases of stripes and rhombs, the phases of hexagons are not free to translate independently. Two phases are free, because we have two degrees of translation freedom on a plane and the third phase is fixed by the dynamics of the system. The evolution Eq. (5.13) has two stationary sol-

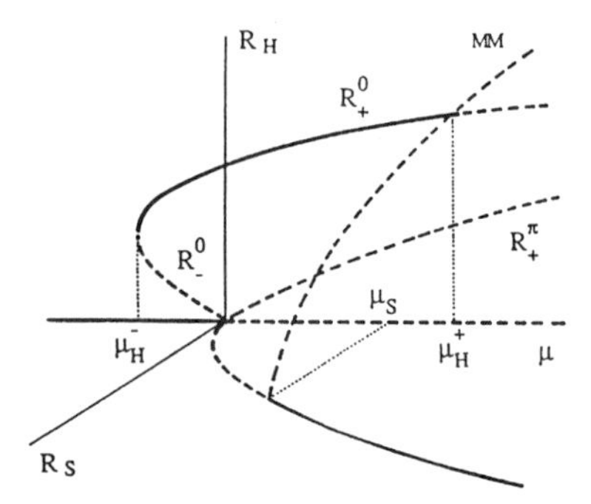

Figure 8. 2D bifurcation diagram displaying the modulus R versus the bifurcation parameter μ when $v > 0$. The *plain line* and *dashed line* represent stable and unstable branches of solutions, respectively. The H0 hexagons with amplitude R_+^0 appear subcritically and are stable for $\mu_{\rm H}^- < \mu < \mu_{\rm H}^+$, whereas the stripes with amplitude R_s appear supercritically and are stable for $\mu > \mu_s$. Bistability between hexagons and stripes is observed for $\mu_s < \mu < \mu_{\rm H}^+$.

utions: $\Theta_s = 0$ and $\Theta_s = \pi$. In regard to perturbations of the phase (i.e., writing $\Theta = \Theta_s + \delta\Theta$), we see that the stable phases are [84]

$$\Theta_s = \pi \text{ when } v < 0 \tag{5.14}$$

$$\Theta_s = 0 \text{ when } v > 0 \tag{5.15}$$

For the moduli, we get the following equation:

$$\frac{dR_1}{dt} = \mu R_1 + vR_2R_3\cos\Theta - gR_1^3 - h(R_2^2 + R_3^2)R_1 \tag{5.16}$$

and two other equations with cyclic permutations. This set of equations features as solutions the homogeneous steady state, the stripes $R_1 = R_s = \sqrt{\mu/g}$, $R_2 = 0$, $R_3 = 0$ (and permutations), and mixed modes

$$R_1 = \frac{v}{h-g} \qquad R_2 = R_3 = \sqrt{\frac{\mu - gR_1^2}{g+h}} \tag{5.17}$$

(and permutations) that are always unstable. There also exists a regular hexagonal solution for which $R_1 = R_2 = R_3 = R_\pm^\Theta$ with $\Theta = 0$ or π. For $v > 0$, R_+^π exists only for $\mu > 0$ and is always unstable to total phase perturbations, as seen above (Fig. 8). If $\Theta = 0$, solutions exist for $\mu > \mu_{\rm H}^- = -v^2/4(g+2h)$. The upper branch R_+^0 is stable up to $\mu = \mu_{\rm H}^+ = v^2(2g+h)/(h-g)^2$. The lower branch R^0 is unstable. The reverse conditions with respect to the total phase hold for $v < 0$.

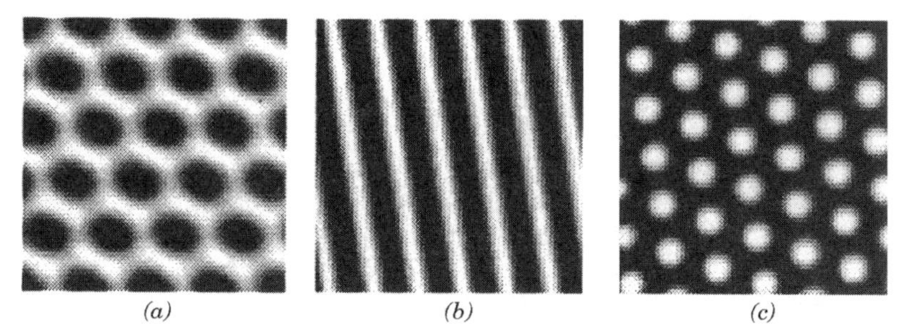

Figure 9. Stationary 2D Turing structures. The concentrations vary between their absolute minimum (*black*) and maximum (*white*) values. (*a*) Hπ hexagons for which the maxima are on an hexagonal honeycomb lattice, (*b*) stripes, (*c*) H0 hexagons for which the maxima are on a triangular lattice.

Depending on the sign of the quadratic coupling v, the first hexagonal phases to appear subcritically are thus either H0 hexagons ($v > 0$ and $\Theta = 0$), for which the maxima of concentrations are organized as a triangular lattice [Fig. 9(c)], or Hπ hexagons ($v < 0$ and $\Theta = \pi$), for which the maxima of concentrations span a honeycomb lattice [Fig. 9(a)]. In chemical systems, one type of hexagon is commonly observed in experiments, whereas the other type of hexagon appears only transiently (Fig. 1). Two types of hexagons have also been observed in Faraday instability [159], in oscillated granular layers [160,161], in nonlinear optics [162–164], and in hydrodynamics [19].

To complete the bifurcation diagram, it is also necessary to study the stability of stripes in regard to perturbations favoring hexagons. Writing $R_1 = R_s + \delta R_1$, $R_2 = \delta R_2$, $R_3 = \delta R_3$ and inserting this into Eq. (5.16), we see that stripes are unstable with respect to the formation of hexagons, if $\mu < \mu_s = v^2 g/(h - g)^2$.

To summarize the pattern selection in 2D, a supercritical branch of stripes is stable if $g < g_{\mathrm{ND}}$, whereas a supercritical branch of rhombs is obtained when the reverse is true. In all reaction–diffusion models studied to date, one usually has $g < g_{\mathrm{ND}}$, and stripes are observed. In addition, a branch of hexagons can appear subcritically with a finite amplitude (Fig. 8). Depending on the sign of the quadratic term v, these hexagons are H0- or Hπ hexagons and become unstable when $\mu > \mu_{\mathrm{H}}$. The stripes are unstable for $\mu < \mu_S$. Stripes and hexagons thus coexist for $\mu_S < \mu < \mu_{\mathrm{H}}$. This bifurcation scenario corresponds to the standard hexagon–stripe competition, widely described in the literature [19] and observed in hydrodynamics [73,77], nonlinear optics [162,164], and gas discharges [165,166] among others. This standard roll-hexagon competition is recovered in the experimental Turing patterns [48,91] and in reaction–diffusion models [84,86,120]. In that respect, chemical systems join the group of pattern-forming systems that present a generic behavior. Modification of this scenario in the vicinity of the primary bifurcation point $\mu = 0$ arises if g and/or h are not positive. If $g < 0$ but $g + h > 0$, stripes appear subcritically, as seen in the Lengyel–Epstein model [120], whereas hexagons are still well described by the third-order amplitude Eq. (5.12). If both g and h are negative, the amplitude equations for the stripes and hexagons are both saturated only at higher orders, and the pattern selection is consequently different.

Let us now focus on the peculiarities of chemical systems that bring some complexity into this standard 2D picture of pattern selection.

2. *Re-entrant Hexagons*

We have seen that the sign of the quadratic term v controls the type of hexagons (H0 or Hπ) observed. Chemical systems are characterized by the fact that the sign of v may change within a given experiment following an increase of the control parameter. This arises because the control parameter often multiplies one variable in the kinetic terms of the reaction-diffusion equations. In the Brusselator model for instance, the control parameter B appears in terms proportional to BX in the evolution equations for the two variables X and Y. This results in the fact that, sufficiently far away from the bifurcation point, the coefficients ξ of the amplitude equations are renormalized by the distance μ from the bifurcation threshold, i.e., we typically have $\xi = \bar{\xi} + \mu\xi_1$. In the Brusselator model, for example, the quadratic term v is equal to [84]

$$v = \bar{v} + \frac{2}{A}\frac{(B - B_c)}{B_c} \tag{5.18}$$

where A is a parameter of the model [Eqs. (5.1)]. This affects the stability of hexagons and the stability of stripes in regard to hexagons. If $\bar{v}$ is positive, then the overall quadratic term v remains positive when B is increased beyond B_c, and H0 hexagons are stable toward stripes up to $\mu_{\mathrm{H}}^{+} = v^2(2g + h)/(h - g)^2$. If on the contrary, $\bar{v}$ is negative, which means that Hπ hexagons are the first stable structure to appear subcritically, then an increase of B can change the sign of v and lead to the switch from one type of hexagon to the other. We thus have the succession Hπ, Hπ/S, S, S/H0, H0 (where A/B indicates bistability of structures A and B). This sequence was first observed numerically in reaction–diffusion models [84,86] (Fig. 10) and then confirmed experimentally in the CIMA reaction [66]. A complete analysis of the effect of renormalizations carried out for the Schnackenberg model [150] showed that, depending on the parameter values, other scenarios such as S, S/H0, H0; H0, H0/S, S, S/H0, H0; H0, H0/S, H0; and H0 are

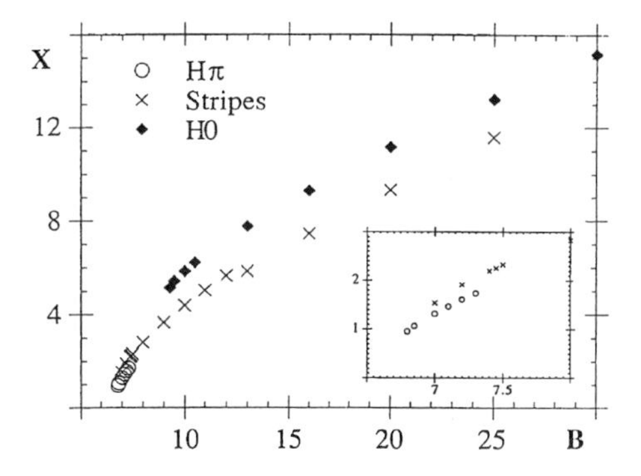

Figure 10. Numerical bifurcation diagram for the variable X of the Brusselator model as a function of parameter B. Here the amplitude is defined as $X_{\max} - X_s$. The parameters are $A = 4.5$, $D_x = 7$, and $D_y = 56$. Near the bifurcation threshold (B_c =6.71), we recover the standard hexagon–stripe competition with an hysterisis loop, and hexagons with the reverse total phase become stable for higher values of B.

also possible and are sometimes seen in experiments [48]. Note that a renormalization of the quadratic term v can also result from a coupling with a bistable regime (see Section VII.A). Re-entrance of various planforms can also be the result of the presence of higher-order terms in the amplitude equation because in that case the standard bifurcation scenario is also modified [99].

3. *Localized Structures in Subcritical Regimes*

The standard bifurcation theory predicts that the hexagons should appear subcritically, leading to a bistability regime between the hexagons and the homogeneous stable steady state. In addition, stripes may appear subcritically, as seen before. This situation is encountered in the Lengyel–Epstein model, which features a strong subcritical regime of stripes in 1D and of both hexagons and stripes in 2D [88]. In this subcritical domain, different steady states coexist; and the system usually evolves toward one or the other solution, depending on the initial condition. A common way to know which state is dominant is to look at the propagation of wavefronts connecting the two states, because in this case the prefered state invades the other one.

In 1D, Jensen and co-workers [120] studied such propagating fronts on the Lengyel–Epstein model in the subcritical regime. They observed that the wavenumber selected when a stable striped structure invades the homogeneous steady state is different from that obtained from spontaneous growth of the pattern out of noise added on the homogeneous steady state [12,167]. Moreover, there exists a band of values of the control parameter for which the velocity of the front vanishes, giving rise to a stable stationary front between the homogeneous steady state and the Turing structure. The stability of such a front is related to the interaction of the front with the periodicity of the spatial organization [79,168–170], a so-called nonadiabatic effect common in solid state physics. This effect, which is not contained in the amplitude equation formalism, can occur for fronts between two states, one of which is periodic in space [168]. It appears, for instance, in the growth of crystals, in which the interaction between the interface and the periodic structure gives rise to a periodic potential. If the difference in free energy between the two phases is smaller than the energy required to move the front by one wavelength, the front remains pinned. The Lengyel–Epstein model is a nonpotential model; thus one cannot define a function to minimize. The picture of an interaction between the front and the Turing structure, however, remains qualitatively correct and gives rise to an intrinsic pinning of the front for a large set of values of the control parameter. Calculation of the front velocity via the usual techniques [171–173] shows a change in behavior at the crossover between the subcritical and the supercritical regimes. In particular, in the subcritical domain, the front no longer moves uniformly but jumps one wavelength at a time, the interval between two successive jumps increases as the pinning band is approached. Such interactions are also the result of the interaction between the front and the Turing pattern. The interaction between two fronts can lead to the formation of stable pulses [174–178].

In 2D, the Lengyel–Epstein model exhibits subcriticality of both hexagons and stripes. There thus exists a range of parameters for which tristability among the hexagons, stripes, and homogeneous steady state occurs. As in 1D, localized structures of one of these states into another stable one can occur; and indeed, stable patches of hexagons inside a homogeneous background are obtained (Fig. 11). Note that such localized hexagons have been recently observed experimentally in the CDIMA reaction and that this observation could point toward a subcritical regime [80,91]. Such localized hexagons can also appear in the strong resonant forcing of oscillators [179,180] or because of localized heating in thermocapillary convection [181]. Another possible localized structure consists of stripes coexisting with hexagons [88,120]. In 2D, the growth of fronts between different types of structures [170,182] is, nevertheless, not as simple as in 1D, because pinning occurs only when the front is perpendicular to the wave vectors of the pattern [120]. Hence the growth of subcritical localized hexagons outside the pinning zone is qualitatively different from that of supercritical hexagons inside an unstable background. In the former case, hexagons grow by adding new points in the directions in which the pinning is the weakest, as observed in the subcritical region of the Lengyel–Epstein model. In

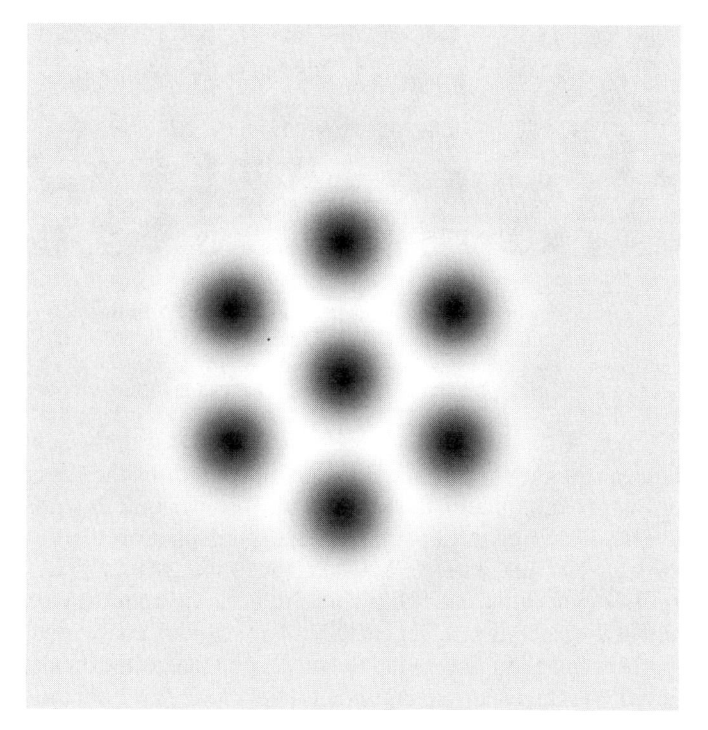

Figure 11. Subcritical hexagons localized in a homogeneous background obtained by numerical integration of a generalized Swift–Hohenberg model. From Ref. [101].

the growth of supercritical hexagons, stripes form along the sides of the hexagons and successively break up into dots, as seen in experiments with the PA-MBO system [93], in convection cells [170], and in the Brusselator [120] model, for instance.

4. Boundaries

Let us now examine how the perfect stripes and hexagons can be affected by boundaries. In reaction–diffusion systems, boundaries have an important effect on the characteristics of patterns, such as selection and orientation of patterns and the relaxation time necessary to obtain a stationary spatial structure or to relax a defect. These effects are particularly important in small systems in which only a few wavelengths develop [3,21,85,183–188].

Figure 12 compares hexagons obtained at a same time for the same values of parameters and starting from the same random initial condition in a small system with periodic, no flux, or fixed boundary conditions. The periodic boundary conditions lead to regular planforms, and the periodicity for-

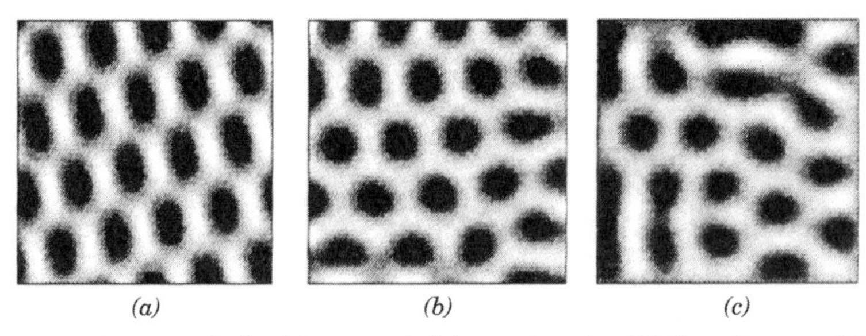

(a) (b) (c)

Figure 12. Hexagons obtained by numerical integration of a 2D Brusselator model with $A = 4.5, D_x = 7, D_y = 56$, and $B = 7$, starting from the same random noise initial condition in a system of size 64×64. (*a*) Periodic boundary conditions; (*b*) no flux boundary conditions; (*c*) conditions fixing $X = X_s = A$ and $Y = Y_s = B/A$ at the boundaries.

ces the pattern's alignment. Nevertheless, if the length of the system is not exactly an integer multiple of the wavelength, the periodic boundary conditions give rise to a distortion of the hexagonal planform, as shown in Figure 12(a), in which the angle between the wave vectors is not exactly $\pi/3$ and the hexagons look rather like rhombs. Periodic boundary conditions partially lift the orientational degeneracy of the initial condition, because the wave vectors can align only along the directions that allow the periodicity to be achieved. Symmetry group arguments [189] have shown that in a square domain with periodic boundary conditions, rhombic patterns and even more complex structures have the potential of being stable [141]. Rhombs can also have other origins: Temporal forcing of hexagons can select rhombic arrays in some cases [190]. Nonequilateral patterns based on two or three wave vectors of different lengths also occur when rotational invariance is broken [191]. It is, therefore, necessary to be cautious when determining whether rhomblike structures observed experimentally in small aspect ratio systems result from a distortion of hexagons because of the boundaries or from a genuine stable solution of the pattern selection problem. A Fourier transform of the pattern can be of some help in that regard, as pure rhombs are characterized by two Fourier modes of same amplitude, whereas deformed hexagons correspond to three peaks of different amplitudes.

No flux boundary conditions also lead to the rapid development of the structures but often result in a distortion of the pattern close to the boundaries [Fig. 12(b)]. Dufiet and Boissonade showed for instance, that striped patterns develop preferentially orthogonally to the walls when no flux boundary conditions are applied [87]. In the middle of the system, the intrinsic wavelength is not constrained, and hexagons develop according to the directions privileged by the initial condition. The fixed boundary conditions, on the other hand, strongly constrain the structure and easily lead to the formation of defects [Fig. 12(c)]. Note that several studies have used mesoscopic lattice-gas cellular automaton models to examine the effects of fluctuations and small system size on Turing structures [192–196].

5. *Long-Wavelength Instabilities and Phase Equations*

In systems large enough that several dozens of wavelengths develop, boundary conditions usually play a role limited to a little layer close to the boundaries. In such a case, starting from random initial conditions, the orientational degeneracy leads to the formation of domains with different orientations. The compatibility between these domains is ensured by the presence of defects. The defects that evolve on a long time scale are typically dislocations, disclinations, or grain boundaries in stripe structures or penta-hepta defects in hexagons. Numerous works characterized such defects in detail [9,19,21,34,197–200] and clarified in particular the interactions among them [201,202]. A particular type of defect [170,182,203] is the boundary between two different types of 2D spatial structures. Defects also play a role in the transition between patterns of different symmetries [204]. Some studies have indeed shown that the unstable planforms are present in the heart of the stable patterns defects [76,202].

Beyond the influence of boundaries and defects, patterns can also be deformed because of modulational instabilities owing to a spatial modulation of the amplitude of their underlying modes. For example, Figure 13 shows zigzag stripes obtained in the Brusselator model. These stripes coexist with straight rolls for the same parameter values. One or the other structure is obtained, depending on the initial condition. In this context, Dufiet and Boissonade studied the generical instabilities of the stripes in great detail on the Schnackenberg model [86]. The zigzags are stable until the angle of the deformation reaches $\pi/3$, when they become unstable toward hexagons [85]. The zigzag branch exists because of the competition between wave vectors belonging to the same band of unstable modes. This effect can be described only if we take a spatial variation of the amplitude into account. Straight stripes with their wave vector aligned along x with $|\mathbf{k}| = k_c$ can be described as

$$\underline{C} = \underline{C}_o + (Te^{ik_c x} + c.c.)\ \underline{w} \tag{5.19}$$

On the other hand, zigzags imply a transverse modulation along direction y. They can thus be described as a first approximation as

$$\underline{C} = \underline{C}_o + (Te^{i(k_c x + \alpha \cos \beta y)} + c.c.)\ \underline{w} \tag{5.20}$$

Figure 13. Zigzag stripes observed in the Brusselator model for the parameter values given Figure 10, and $B = 8$.

This modulation leads to a change in the local value of the wave vector or equivalently of the stripes phase. The amplitude equation taking this spatial effect into account is the Newell–Whitehead–Segel (NWS) equation [205,206]

$$\frac{\partial T}{\partial \tau} = \mu T - |T|^2 T + \left(\frac{\partial}{\partial \mathcal{X}} - i \frac{\partial^2}{\partial \mathcal{Y}^2} \right)^2 T \tag{5.21}$$

where $\mathcal{X}$ and $\mathcal{Y}$ are the length scales on which the spatial modulation of the amplitude occurs. This equation can be derived [21] by the standard perturbation expansion techniques if the slow time and space scales are developed as $\tau = \varepsilon^2 t$, $\mathcal{X} = \varepsilon x$, $\mathcal{Y} = \varepsilon^{\frac{1}{2}} y$. The NWS equation features as solution the straight stripes with wavevector Q:

$$T_Q = \sqrt{\mu - Q^2}\, e^{iQ\mathcal{X}} + c.c. \qquad Q^2 < \mu \tag{5.22}$$

where Q belongs to the band of unstable wave vectors, i.e., is proportional to $k - k_c$. To study the stability of these straight stripes in regard to modulus and phase perturbations, let us take

$$T = \left(\sqrt{\mu - Q^2} + u\right)e^{i[Q\mathcal{X} + \theta(\tau, \mathcal{X}, \mathcal{Y})]} \tag{5.23}$$

In the absence of spatial derivatives in the amplitude equations, we recover the stability analysis described previously in the construction of the bifurcation diagrams. If this expression is now inserted into the NWS equation we get at the lowest order the evolution equation for the phase $\theta(\tau, \mathcal{X}, \mathcal{Y})$:

$$\partial_\tau \theta = D_\mathcal{X} \nabla_\mathcal{X}^2 \theta + D_\mathcal{Y} \nabla_\mathcal{Y}^2 \theta \tag{5.24}$$

with

$$D_\mathcal{X} = \frac{\mu - 3Q^2}{\mu - Q^2}, \qquad D_\mathcal{Y} = Q \tag{5.25}$$

When the phase diffusion coefficients $D_\mathcal{X}$ and $D_\mathcal{Y}$ become positive, the longitudinal and transverse perturbations relax through a diffusive process, and the straight rolls are stable. On the other hand, the stripes are unstable toward long-wavelength instabilities if these diffusion coefficients are negative. When $Q < 0$, $D_\mathcal{Y}$ becomes negative, and the stripes are transversally modulated through a zigzag instability [207–209]; but if $Q^2 < \mu < 3Q^2$, $D_\mathcal{X}$ becomes negative, and the bands are longitudinally modulated (Eckhaus instability) [210,211]. Higher-order terms in Eq. (5.24) could saturate this instability, which explains why stable zigzags can be obtained in experiments [48,66,91,124] and numerical simulations [86].

Analogous modulational instabilities occur in hexagonal planforms. For example, an Eckhaus instability of one of the three modes forming a hexagonal planform is shown in Figure 14. Phase equations for hexagonal planforms have recently been derived [182,212–214], and their long-wavelength instabilities have been analyzed [212–214].

Here it is important to note that the spatial derivatives appearing in the NWS Eq. (5.21) for stripes take this form only because we assumed at the beginning that the stripes were aligned along x. This *a priori* choice supposes that one direction is privileged. Recently, a debate appeared in the literature arguing that amplitude equations should satisfy rotational invariance [34,90,215–219]. In that case, the spatial derivatives present in the amplitude equations are different from those in the NWS. The main consequence of the new nonlinear gradient terms [220,221] that appear in the amplitude equations satisfying such rotational invariance is that they affect the stability of the different planforms and, in particular, allow the stabilization of rhombs

Figure 14. Eckhaus instability of one of the three wave vectors that underly an H0 hexagonal planform. The succession in time must be read from top to bottom and left to right. The starting H0 hexagons are obtained in the Brusselator model for $A = 2, D_x = 4, D_y = 20$, and $B = 4.5$. When B is increased to 5.5, the original wavelength is too large and an Eckhaus instability takes place.

[89,90,215,221], such as those observed in the CIMA reaction, which are otherwise unstable in the standard bifurcation theory.

C. Three-Dimensional Pattern Selection

One of the main characteristics of the experimental Turing structures is that they are generally three-dimensional objects [47,48,64,69,70,72,98] as soon as the wavelength of the pattern is smaller than any side of the region of the gel in which the symmetry-breaking instability takes place. Hence the pattern selection problem must take into account the possible new structures that arise in 3D.

1. *Bifurcation Diagrams*

As in 2D, the 3D pattern selection approach starts by writing the concentration in the system as a linear combination of N modes, such as Eq. (4.1). Each 3D structure is also characterized by the number N of modes that underlie its construction. The general temporal evolution equations for the amplitudes T_i with $i = 1, \ldots, N$ become in absence of spatial modulations [79,117,222]:

$$\frac{dT_i}{dt} = \mu T_i + v \sum_j \sum_k T_j^* T_k^* \delta(\mathbf{k}_i + \mathbf{k}_j + \mathbf{k}_k) - \sum_{j \neq i} g_{\mathrm{ND}}(ij) |T_j|^2 T_i$$
$$- g_{\mathrm{D}} |T_i|^2 T_i - \sum_j \sum_k \sum_l \beta(ijkl) T_j^* T_k^* T_l^* \delta(\mathbf{k}_i + \mathbf{k}_j + \mathbf{k}_k + \mathbf{k}_l) \quad (5.26)$$

The coefficients $g_{\mathrm{ND}}(ij)$ and $\beta(ijkl)$ are functions of the angles between the wave vectors considered. Among the solutions to these amplitude equations, we recover the $N = 1, 2, 3$ solutions already studied in 2D. If $N = 1$, the stripes become 3D parallel isoconcentration planes, also called lamellae. The 3D extension of the rhombs with $N = 2$ are prisms, with a rhombic base; whereas the hexagons ($N = 3$) correspond to hexagonally packed cylinders (HPC). The relative stability of these patterns extend into 3D the conclusions drawn in 2D, i.e., in the simplest case, lamellae appear supercritically, the rhombic prisms are unstable with respect to the lamellae if $g_{\mathrm{D}} < g_{\mathrm{ND}}$ and HPC arise through a subcritical bifurcation. The total phase of the HPC's hexagonal basis relaxes either to 0 or π, giving rise to two possible solutions: the HPC0 or the HPCπ.

The interest of the 3D pattern selection problem lies in the possibility of getting new structures that have no 2D equivalent. If $N = 4$, for instance, the basic wave vectors can form noncoplanar quadrilaterals ($\mathbf{k}_1 + \mathbf{k}_2 + \mathbf{k}_3 + \mathbf{k}_4 = 0$) in Fourier space, a combination corresponding to a face-centered cubic (FCC) structure in real space [223]. Again, the condition $g_{\mathrm{D}} < g_{\mathrm{ND}}$ ensures that the FCC structure is unstable. If $N = 6$, a maximum number of resonances [224] is obtained if the six wave vectors are aligned along the edges of a regular octahedron (Fig. 15) which corresponds

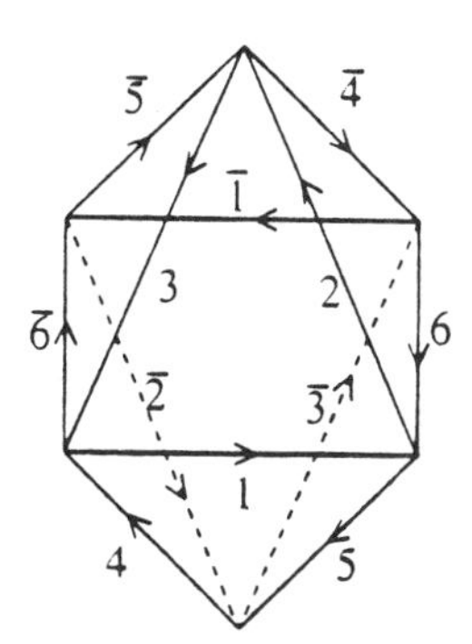

Figure 15. 3D body-centered cubic symmetry. In Fourier space, the six pairs of wave vectors are parallel to the edges of a regular octahedron. Each pair is involved into two nonplanar resonant triads. From Ref. [151].

in real space to a body-centered cubic (bcc) structure. In this case, each wave vector belongs to two equilateral triangles, and the following three independent resonance conditions must be satisfied [151]:

$$\begin{aligned} \mathbf{k}_1 + \mathbf{k}_2 + \mathbf{k}_3 &= 0 \\ \mathbf{k}_1 + \mathbf{k}_4 + \mathbf{k}_5 &= 0 \\ \mathbf{k}_2 - \mathbf{k}_4 + \mathbf{k}_6 &= 0 \end{aligned} \tag{5.27}$$

Recalling that the amplitudes $T_j = R_j e^{i\theta_j}$, the resonance conditions on the wave vectors in Fourier space come down to conditions on the phases θ_j. It can be shown that three independant phases $\Theta_1 = \theta_1 + \theta_2 + \theta_3$, $\Theta_2 = \theta_1 + \theta_4 + \theta_5$, and $\Theta_3 = \theta_2 - \theta_4 + \theta_6$ must be known to completely characterize the BCC pattern. A stability analysis of the BCC pattern indicates that only two possibilities exist:

$$\Theta_1 = \Theta_2 = \Theta_3 = 0 \tag{5.28}$$

$$\Theta_1 = \Theta_2 = \Theta_3 = \pi \tag{5.29}$$

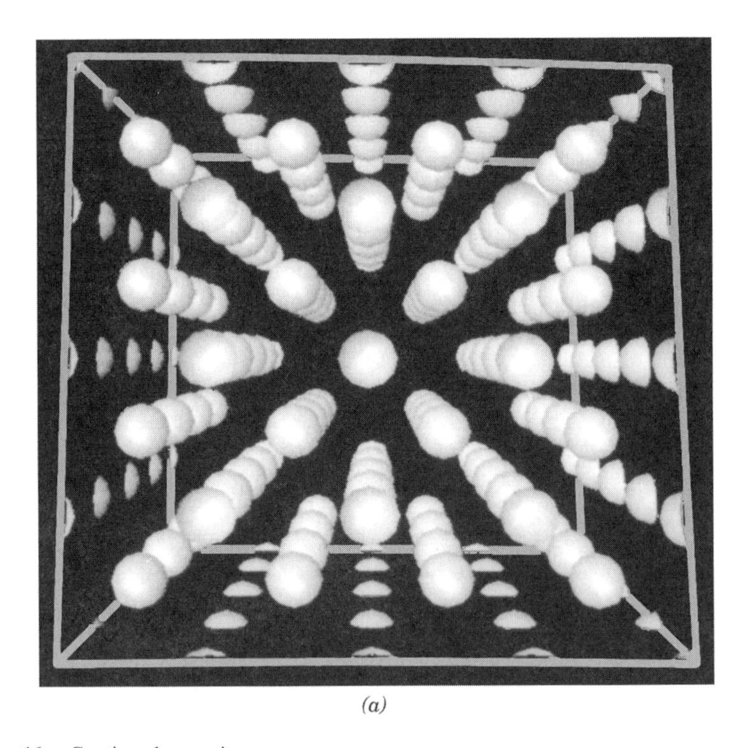

(a)

Figure 16. Continued opposite.

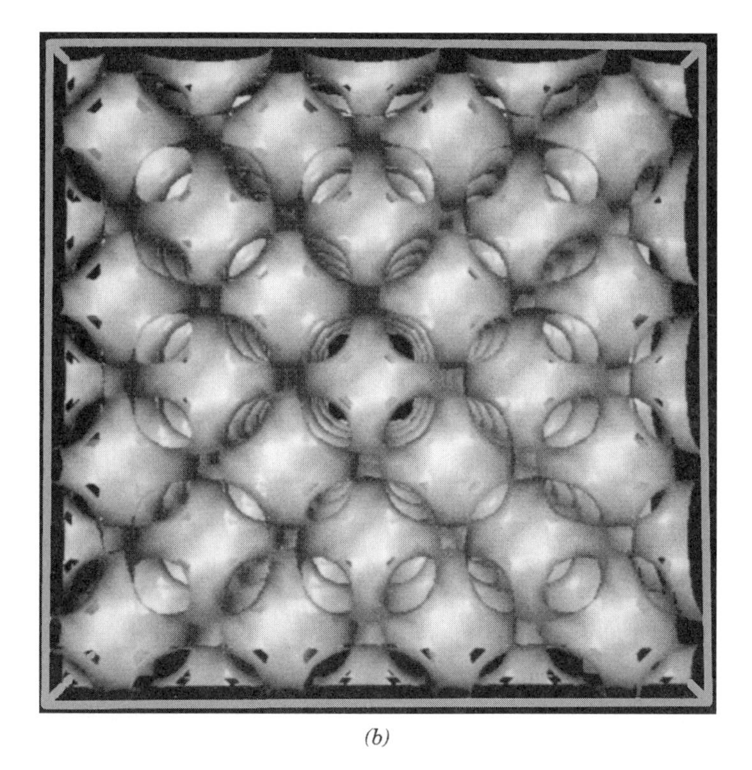

(b)

Figure 16. 3D BCCπ Turing pattern obtained by numerical integration of the Brusselator model in a cube with sides equal to 40 and periodic boundary conditions along the three axes. The parameters are $A = 4.5$, $D_x = 2$, $D_y = 16$, and $B = 6.9$. Isoconcentration surfaces are looked at perpendicularly to one face of the cube. (*a*) Spheres of lower isoconcentrations ($X = 2.737$) organized with the BCC symmetry; (*b*) higher isoconcentrations ($X = 5.153$) filling in the interspace between the lower isoconcentrations.

The corresponding structures are thus called BCC0 and BCCπ. In real space, BCC0 has its maxima of concentrations organized as a body-centered cubic lattice, whereas the minima fill in the interspace between the maxima as filamental structures with cubic symmetry. The reverse situation is obtained in the BCCπ case (Fig. 16). The relative stability analysis of the BCC structure with the other 3D patterns shows that the BCC pattern arises subcritically and becomes unstable toward the HPC at higher values of the control parameter. A succession of BCC–HPC–lamellae with regions of bistability is, therefore, predicted analytically; the total phase of the BCC depends on the sign of the quadratic term v [75,117,222,225]. This bifurcation scenario was confirmed

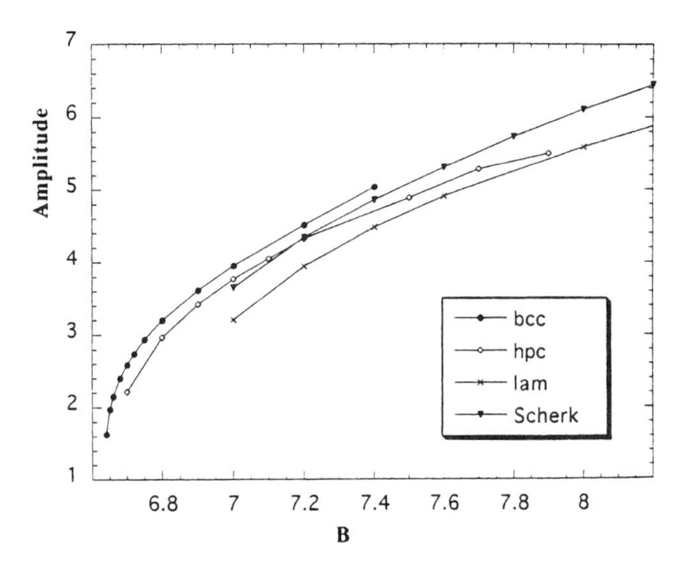

Figure 17. 3D numerical bifurcation diagram obtained with the Brusselator model for the set of parameters given in Figure 16 and varying the control parameter B. The amplitude here is defined as $X_{\max} - X_{\min}$.

by numerical integration of the Brusselator model (Fig. 17) [117] and bears analogies with those occuring in equilibrium systems [226]. It was suggested that other 3D patterns, for instance a double-diamond structure in which the maxima and minima form two interlocked diamond lattices, could be stable as well [79,227]. In fact, a classification of the primary solution branches of different lattices with periodic boundary conditions was performed in 3D by Dionne and co-workers [228,229]. Callahan and Knobloch checked the stability conditions of several of these 3D patterns (including FCC and double-diamond structures) on three cubic lattices [227,230,231]. In particular, they specified parameter regions in the Brusselator model [227,231] and the Lengyel–Epstein model [231] for which rolls, FCC, or the double-diamond should be stable with respect to perturbations on these lattices. Unfortunately, these values are outside those already scanned in 3D numerical simulations [117,222]. It remains to be checked how these conclusions generalize to arbitrary boundary conditions and spatially extended systems. The problem is even more complicated as evidence exists that outside the weakly nonlinear regime, the traditional pattern selection competition is drastically changed when higher harmonics come into play [232,233]. Such effects should be kept in mind when experimental data are analyzed.

2. *Minimal Surfaces*

The experimentally observed 3D Turing patterns could be compatible with the BCC, HPC, or lamellae symmetries [63,64,72] predicted by the theoretical works. Nevertheless, a clear interpretation of the experiments is often made cumbersome by the difficulty in resolving the changes in concentration in the depth of the gel and by the presence of gradients and defects [72]. Ramps localize the 3D structures in subregions of the reactor [102], and different patterns can develop and coexist spatially at different depths in the gel reactor [48,68–70,92,91,104]. This localization favors the appearance of defects in the transition zone between two different symmetries. It is, therefore, important to have insight into the possible defects existing in 3D. The defects of 3D patterns were the subject of extensive studies in solid state [223], liquid crystal [234–237], and macromolecular [238] physics. The 3D chemical Turing structures contain the traditional defects of 3D crystals, such as dislocations and disclinations (Fig. 18). The description of these defects can be made as in 2D in the framework of phase diffusion equations [19,239,240]. In particular, Pismen [239] used phase equations to show that resonance conditions can cause confinement of dislocations in 3D.

In addition to point and line defects, a class of defects organized along a minimal surface was recently noted in the Brusselator model [117]. This defect consists in a twist-grain boundary continuously joining two orthogonal sets of perfect lamellae (Fig. 19). The same kind of twist-grain boundary was evidenced in block copolymers [238,241], amphiphilic systems [242] and liquid crystals [237]. It can be shown that such a twist-grain boundary embodies a whole family of constant mean curvature surfaces spanned by the isoconcentration surfaces. Among them, the zero mean curvature surface corresponding to the unstable reference state's isoconcentration surface is of particular interest, as it corresponds to the first Scherk minimal surface [238]. In this minimal surface, the connection between the two orthogonal sets of lamellae consists in a doubly periodic array of saddle surface regions (Fig. 19). Such minimal surfaces are frequently encountered in numerical simulations of the Brusselator model in the region of parameters where the lamellae are stable. It can be understood theoretically on the basis of phase equations that both the lamellae and the Scherk surface have the same domain of stability [117] (Fig. 17).

The observation of a Scherk minimal surface in simulations of a reaction–diffusion model points toward a new vision of the 3D pattern selection problem that should now also take into account all possible structures built on continuous sheets and surfaces rather than on only discrete points or centers on a lattice. This particularity comes from the fact that Turing structures are based on the spatial structuration of concentrations that are by nature

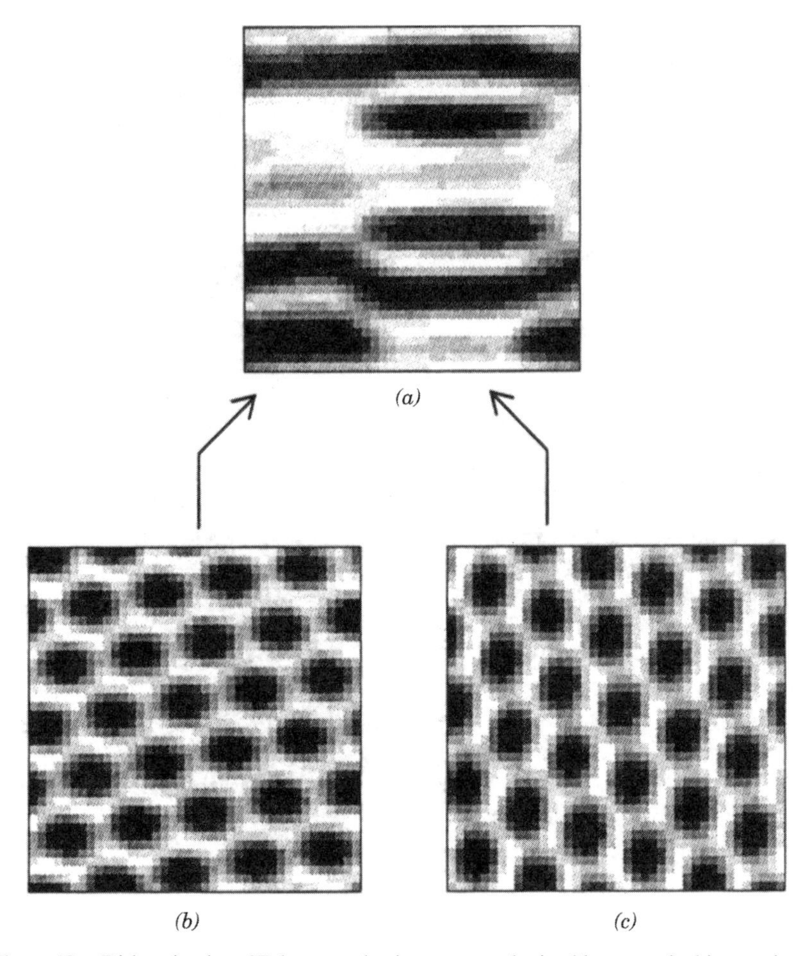

Figure 18. Dislocation in a 3D hexagonal prism pattern obtained by numerical integration of the Brusselator model in a cube with sides equal to 40 and with periodic boundary conditions along the three axes. The parameters are those given in Figure 16, with $B = 7.0$. (*a*) One upper face of the cube displays two zones of prisms with different orientations; (*b*, *c*) planes cutting the previous plane perpendicularly at the locations indicated to by the *arrows*.

continuous variables rather than on discrete elements, such as atoms in the case of crystals. Chemical Turing structures in that respect join the domain of flexocrystals [243,244] observed in soft matter, i.e., in polymers, microemulsions, and biological vesicles in which complicated geometries (e.g., hexagonally packed hollow loops [245]) have been observed. It is, therefore, probable that other organizations known to embody various kinds of minimal surfaces such as the Schwarz surface, the gyroid structure, and the lamellar

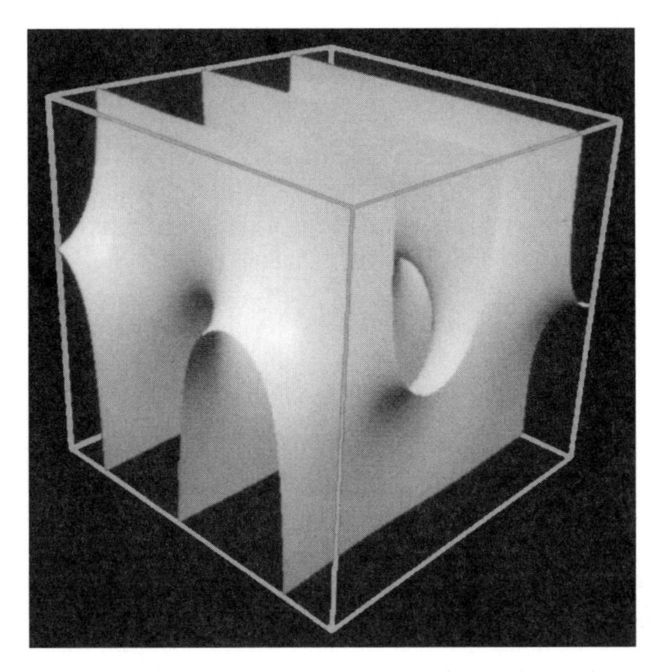

Figure 19. Scherk surface observed in the numerical integration of the Brusselator model for the set of parameters given in Figure 16 and $B = 7.2$. Periodic boundary conditions are applied along the x- and y-axes, and no flux boundary conditions are applied along the vertical direction. To have a better visualization of the saddle region, typical of the Scherk surface, the figure zooms in on half the system. The isoconcentration surface corresponds to that of the uniform unstable reference state.

catenoid [242,246], could exist in 3D nonlinear chemical systems because of a Turing instability. If not stable *per se*, these minimal surfaces could appear as transient states between two stable steady states. In that respect, perforated lamellar states have been evidenced in the transition from lamellae to HPC using a time-dependent Ginzburg–Landau equation [247,248]. It can be reasonably expected that similar effects should exist in reaction–diffusion models as well. In the same spirit for bistable systems, sponge phases [117,249] can be expected to provide the 3D equivalent of the 2D labyrinthine patterns observed in the FIS reaction.

If it may seem trivial at first sight that the pattern selection problem is more involved in 3D systems than in 2D systems, today we are beginning to understand that this complexity is even greater because the nature of the spatial organization of 3D and 2D systems differ. In this context, it is important to bear in mind that in the experiments, the presence of concentration gradients sometimes leads to distinguish a true 3D situation from a succession of coupled 2D layers [68–70,92]. Bestehorn [105] indeed showed that the coupling of two pattern-forming layers may lead to the stabilization in one layer of mixed states and triangles such as those observed experimentally [68,90,92], which are otherwise unstable in a single 2D system. Dufiet and Boissonade [104] analyzed how the 2D pattern selection is affected when the Turing structures are confined to a monolayer by a gradient of parameters in a 3D system. Note that many studies have also been devoted to the study of 3D oscillatory, bistable, and excitable regimes [5,151,250].

VI. TURING–HOPF INTERACTION

The Turing patterns obtained experimentally exist thanks to the reversible formation of a complex of low mobility between iodide, the activator of the CIMA reaction and the color indicator of the system. Lengyel and Epstein [58] and Pearson [60,61] showed that this complex-forming step does not affect the threshold of the Turing instability but moves the threshold of the Hopf instability away. Therefore, if the concentration of the color indicator is progressively decreased, a transition from standing stationary spatial structures toward temporal oscillations occurs [63,64]. In the transition region, several complex spatiotemporal dynamics have been observed for parameters suggesting the presence of a codimension-two Turing–Hopf point (CTHP) [48,63,64,92,118]. Such a codimension-two point is defined as the point in parameter space for which the thresholds of the Turing and Hopf instabilities are equal.

Degenerated bifurcation points were the subject of several detailed mathematical studies, which mainly considered the temporal dynamics of systems for which a real eigenvalue and a pair of purely imaginary eigenvalues simultaneously cross the imaginary axis [32,251–253]. Near such a degeneracy point, the system can present complex or even chaotic temporal dynamics. Simplified models have also allowed the study of the competition between instabilities breaking the temporal and spatial symmetries, respectively [254–258]. Studies of codimension-two bifurcation have been achieved in hydrodynamics [259–264], electronic networks [265], semiconductors [266], and nonlinear optics [267,268]. In some of these examples, the oscillating regime occurs through a Hopf bifurcation with a finite wavenumber k. Here we will study the interaction between steady and oscillating instabilities for a Hopf bifurcation with a zero wavenumber, *i.e.*, when the temporal oscillations are homogeneous. This situation was treated theoretically in several works that focused mainly on small systems [147,155,269,270]. Here we enlarge the description of the spatiotemporal dynamics arising in the vicinity of such a CTHP in connection with the recent experimental observations. In particular, experiments performed in large systems call for additional insight into long-wavelength instabilities, leading to complex dynamics not studied before.

Let us first review the various 1D dynamics that can be obtained close to a CTHP as predicted on the basis of amplitude equations that account for the coupling between a steady Turing-type mode and a Hopf mode. Owing to resonance possibilities between the Turing and Hopf modes and their harmonics, new solutions arise in addition to the pure modes. These predicted solutions are recovered in numerical integrations of reaction–diffusion models. We will see, however, that several dynamics observed in these models (including some that are observed experimentally) cannot be casted into the amplitude equation formalism. We will also review some of the recent works that focused on the Turing–Hopf interaction in 2D.

A. Interaction between Steady and Hopf Modes

In the vicinity of a CTHP, the concentration field $\underline{C}$ of a 1D reaction–diffusion system may be expressed in terms of two complex amplitudes T and H:

$$\underline{C}(x, t) = \underline{C}_o + T(\tau, \mathcal{X})e^{ik_c x}\underline{w}_{\mathrm{T}} + H(\tau, \mathcal{X})e^{i\omega_c t}\underline{w}_{\mathrm{H}} + c.c. \tag{6.1}$$

where $\underline{w}_{\mathrm{T}}$ and $\underline{w}_{\mathrm{H}}$ are the critical Turing and Hopf eigenvectors of the linearized reaction–diffusion operator, respectively; ω_c is the critical frequency of the limit cycle; and k_c is the critical Turing wave vector. The competition between these modes can be described by two coupled amplitude equations for the Turing and Hopf modes [32,155]:

$$\frac{\partial T}{\partial \tau} = \mu_{\mathrm{T}} T - g|T|^2 T - \lambda |H|^2 T + D^T \frac{\partial^2 T}{\partial \mathcal{X}^2} \tag{6.2}$$

$$\frac{\partial H}{\partial \tau} = \mu_{\mathrm{H}} H - (\beta_r + i\beta_i)|H|^2 H - (\delta_r + i\delta_i)|T|^2 H + (D_r^H + iD_i^H)\frac{\partial^2 H}{\partial \mathcal{X}^2} \tag{6.3}$$

where μ_{H} and $\mu_{\mathrm{T}} = \mu_{\mathrm{H}} + v$ are the distances from the Turing and Hopf thresholds, respectively. Let us here mention that in the CIMA experiments, the transition between Turing patterns and temporal oscillations can be obtained by varying the malonic acid concentration for a fixed starch concentration or by varying the starch concentration for a given malonic acid concentration. These two concentrations fix the CTHP. The starch concentration determines the ratio between the diffusion coefficients of activator and inhibitor and hence the distance v between the thresholds of the Turing and Hopf instabilities. Varying the concentration of malonic acid would here correspond to vary μ_{H}. The two thresholds μ_{H} and μ_{T} can thus be varied independently.

We assume in the following that g, β_r, D^{T}, and D_r^{H} are positive so that both bifurcations are supercritical. Eqs. (6.2) and (6.3) have nonvariational forms because of their asymmetry and the presence of imaginary terms. This set of equations has three nontrivial global solutions:

1. A family of Turing structures:

$$T = \left\{ \frac{\mu_{\mathrm{T}} - D^{\mathrm{T}} Q^2}{g} \right\}^{\frac{1}{2}} e^{iQ\mathcal{X}} \qquad H = 0 \tag{6.4}$$

2. A one-parameter family of plane waves:

$$T = 0 \qquad H = \left\{ \frac{\mu_{\mathrm{H}} - D_r^{\mathrm{H}} K^2}{\beta_r} \right\}^{\frac{1}{2}} e^{i(\Omega_K \tau - K\mathcal{X})} \tag{6.5}$$

   with the frequency renormalization: $\Omega_K = -\beta_i |H_K|^2 - D_i^{\mathrm{H}} K^2$, where H_K is the pre-exponential factor in H.

3. A two-parameter family of mixed modes:

$$T = \left\{ \frac{\beta_r(\mu_{\mathrm{T}} - D^{\mathrm{T}} Q^2) - \lambda(\mu_{\mathrm{H}} - D_r^{\mathrm{H}} K^2)}{\Delta} \right\}^{\frac{1}{2}} e^{iQ\mathcal{X}}$$

$$H = \left\{ \frac{g(\mu_{\mathrm{H}} - D_r^{\mathrm{H}} K^2) - \delta_r(\mu_{\mathrm{T}} - D^{\mathrm{T}} Q^2)}{\Delta} \right\}^{\frac{1}{2}} \times e^{i(\Omega_{KQ}\tau - K\mathcal{X})} \tag{6.6}$$

with $\Delta = \beta_r g - \lambda \delta_r$ and $\Omega_{KQ} = -\beta_i |H_{KQ}|^2 - \delta_i |T_{KQ}|^2 - D_i^{\mathrm{H}} K^2$, where H_{KQ} and T_{KQ} are the pre-exponential factors of H and T.

Various bifurcation scenarios may be obtained by studying, along the lines of the procedure detailed previously, the stability of each of these solutions in regard to homogeneous perturbations and perturbations favoring the other solutions. When $\Delta < 0$, the mixed mode is always unstable and bistability between the limit cycle and the Turing mode occurs [Fig. 20(a)]. When $\Delta > 0$, the mixed mode exists and is stable in the domain in which both pure Turing and Hopf modes are unstable [155,269] [Fig. 20(b)].

The coupling between the Turing and Hopf instabilities allows us to observe Turing–Hopf bistability or a Turing–Hopf mixed mode, depending on the values of parameters. In the bistability regime, the system is expected to give rise, for the same set of parameters, either to a steady spatial pattern or to a homogeneously oscillating concentration, depending on the initial condition. In the mixed mode stability domain, a spatial structure

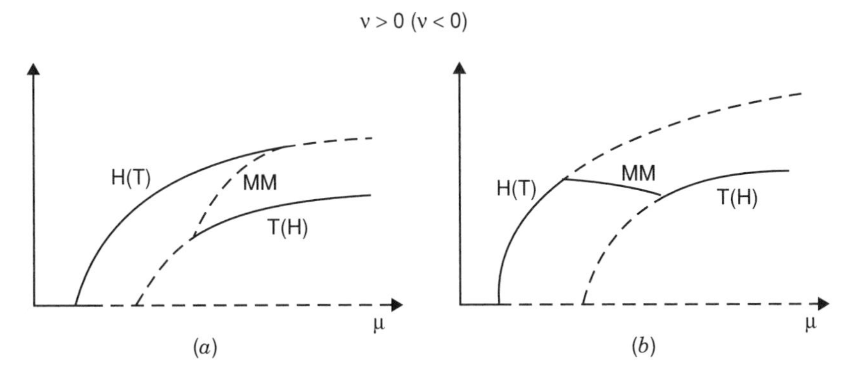

Figure 20. Bifurcation diagrams that are valid close to a CTHP. The *plain line* and *dashed line* represent stable and unstable branches of solutions, respectively. When the distance ν between the Turing and Hopf thresholds is fixed, we have either (*a*) Turing–Hopf bistability when $\Delta < 0$ or (*b*) a stable mixed mode when $\Delta > 0$. If ν is positive, the first bifurcation occurs toward the Hopf temporal oscillations (H), whereas the first instability is toward Turing patterns (T) when $\nu < 0$.

characterized by the Turing wavelength is expected, each point of the structure oscillating in phase with the same Hopf frequency.

These predictions are altered in large systems, in which spatial modulations of the amplitude can lead to long-wavelength instabilities. In the absence of spatial modulations, Eqs. (6.2) and (6.3) are invariant under the transformations $T \to Te^{i\theta}$ and $H \to He^{i\phi}$. As a result, the corresponding matrix linearized around the mixed mode has two zero eigenvalues. When spatially inhomogeneous perturbations are taken into account, these marginal modes may induce diffusive instabilities of the phases. The marginal mode associated to the amplitude equation for T is related to the long-wavelength instabilities of stationary structures, such as the zigzag or Eckhaus instabilities [19] (see Section V.B.5). The marginal mode associated with H can undergo the Benjamin–Feir instability of the limit cycle. This instability leads to a desynchronization of the temporal oscillations at various locations in the system, resulting in spatiotemporal chaos. The Benjamin–Feir instability occurs when the inequality

$$D_i^{\mathrm{H}} \beta_i + D_r^{\mathrm{H}} \beta_r < 0 \tag{6.7}$$

is true. An analogous criterion of instability can be obtained for the Turing–Hopf mixed mode [271]. In particular, the most stable mixed mode ($Q = 0, K = 0$) undergoes such an instability when

$$\mathcal{D} = \frac{D_i^{\mathrm{H}}(\beta_i g - \lambda \delta_i) + D_r^{\mathrm{H}}(\beta_r g - \lambda \delta_r)}{\Delta} < 0 \tag{6.8}$$

Note that the Benjamin–Feir instability criterion [Eq. (6.7)] of the Hopf mode is recovered when all the parameters related to the coupling between the two modes are set equal to zero in Eq. (6.8). It is nevertheless important to note that Eq. (6.8) may be satisfied even when Eq. (6.7) is not fulfilled, i.e., when the limit cycle is stable with respect to the modulational instability. Thus the destabilization of the Turing–Hopf mixed mode exists genuinely because of the coupling between these two modes.
Numerical integration of Eqs. (6.2) and (6.3) shows that when $\mathcal{D} < 0$ the mixed mode is indeed unstable [271]. According to the values of the parameters, the system then enters either a phase-turbulent regime, similar to that of the Kuramoto–Sivashinsky equation [272,273], or a defect chaos regime, characterized by phase defects and large amplitude fluctuations on both T and H [274–276]. The domain of existence of these two types of chaos bears analogies with the dynamics obtained in the 1D complex Ginzburg–Landau equation in its Benjamin–Feir unstable regime [277].

Let us now examine how these predictions based on the amplitude equation formalism are recovered in the Brusselator reaction–diffusion model. We recall that this model (introduced in Section V.A) exhibits for a given value of parameter A a CTHP when the ratio $\sigma = D_x/D_y$ reaches its critical value

$$\sigma_c = \left[\frac{\sqrt{1+A^2}-1}{A^2}\right]^2 \tag{6.9}$$

where D_x and D_y are the diffusion coefficients of the two variables of the model.

1. *Mixed Modes*

For several values of parameters close to a CTHP, the Brusselator model [119] and other reaction–diffusion models [268,278] exhibit a stable mixed mode, corresponding to a spatial pattern with the Turing wave number oscillating in time with the Hopf frequency [119,279] (Fig. 21). A space–time map of these dynamics shows the polygonal space–time structure characteristic of mixed modes [Fig. 22(a)]. This invalidates the conclusions of Rovinsky and Menzinger [280], who stated that the mixed mode is always unstable in the Brusselator model, and the statement of Tlidi and Haelterman [268], who noted that the mixed mode is not generic in reaction–diffusion models. Using the size L of the system as a bifurcation parameter, the mixed mode becomes phase unstable when L is increased and the system enters a regime of spatiotemporal chaos [119]. The fact that this chaos appears when using the length of the system as a control parameter confirms that we

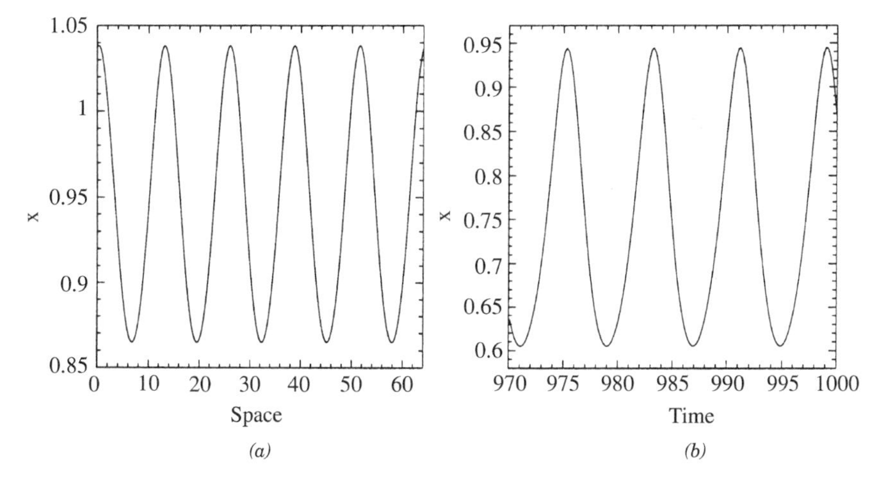

Figure 21. A Turing–Hopf mixed mode observed in a 1D Brusselator model in a system of length 64 with periodic boundary conditions: $A = 0.8, D_x = 1.11, D_y = 10$, and $B = 1.675$. (*a*) The spatial profile taken at one given time. The spatial structure corresponds to a Turing pattern with wavenumber k_c. (*b*) Periodic temporal dynamics at one given fixed point of the system, showing oscillations with the Hopf frequency ω_c.

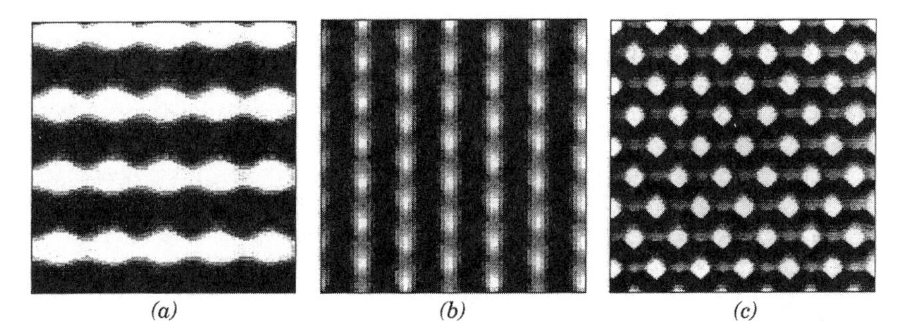

Figure 22. A space–time map of the three types of mixed modes existing near a CTHP, displaying the 1D spatial dynamics in the Brusselator model versus time running upward. (*a*) Turing–Hopf mixed mode with $A = 0.8, D_x = 1.11, D_y = 10$, and $B = 1.675$; (*b*) subharmonic Turing mode with $A = 3, D_x = 5.1949, D_y = 10$, and $B = 10.45$; (*c*) subharmonic Turing–Hopf mode with $A = 3, D_x = 5.71, D_y = 10$, and $B = 11$.

are dealing with the predicted long-wavelength instability and not with a homoclinic type of chaos. Such spatiotemporal chaos obtained close to a CTHP may be relevant to experimental observations in gas discharges [281] or to those observed in the CIMA reaction [64,92]. It might also be the mechanism leading to chemical turbulence numerically obtained in an enzymatic reaction–diffusion system [282]. Complex mixing of different spa-

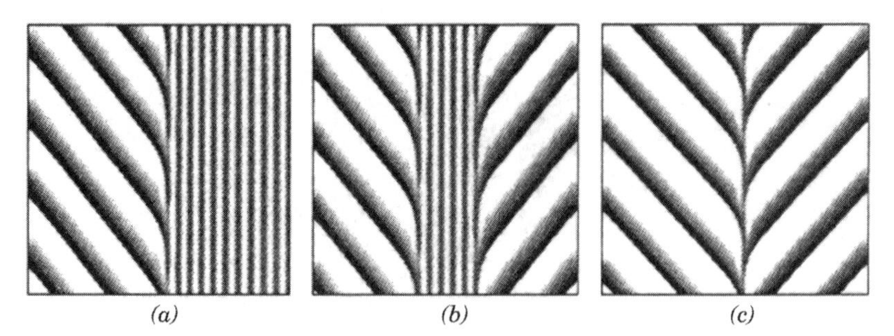

Figure 23. Space–time map of various localized structures that are stable in the Turing–Hopf bistability regime occuring in a 1D Brusselator model with no flux boundary conditions for $A = 2.5$, $D_x = 4.11$, $D_y = 9.73$, and $B = 10$. Time is running upward. (*a*) Turing–Hopf front; (*b*) Turing structure imbedded into an oscillating background; (*c*) flip-flop.

tiotemporal dynamics are also obtained: The stable mixed mode can appear as a localized structure in a Turing pattern when the size of the system is increased [119]. This mixed mode, generic of the CTHP, is characterized by one wavenumber k_c and one frequency ω_c. Other types of mixed modes are also observed close to the CTHP, but in such cases they occur as a result of subharmonic instabilities, as will be described later.

2. *Bistability and Localized Structures*

In the Turing–Hopf bistability domain, the system evolves for a given set of parameters either to homogeneous temporal oscillations or to a stationary spatial pattern, depending on the initial conditions. For several values of parameters near the CTHP, bistability is observed numerically in the Brusselator model. In this regime, a stable front can exist between a Turing domain and a train of plane waves [Fig. 23(a)]. The stability of such a front is related to a nonadiabatic effect owing to the interaction of the front with the periodicity of the spatial organization [79,88,98,118–120,168–170], such as that leading to stable fronts between Turing patterns and a homogeneous steady state in subcritical regimes (see Section V.B.3). The intrinsic pinning of the Turing–Hopf front occurs for a large set of values of the control parameter B (Fig. 24). The nonadiabatic effect also accounts for a stepwise progression of the Turing–Hopf front outside the pinning domain [79,119,120]. In this process, the mode-locking phenomenon shows up as a tendency of the average velocity to lock into rational multiples of the Hopf frequency [283]. The simplest mode locking is one wavelength for one frequency, but other ratios are possible, as long as there is an integer number of wavelengths per period of oscillation or vice versa. In these situations, the front may progress faster or slower; and to satisfy the nonadiabatic

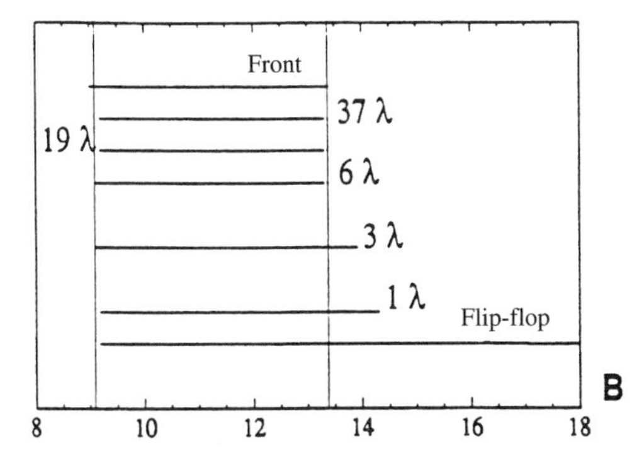

Figure 24. Stability domain of various localized Turing structures imbedded inside a Hopf background versus the control parameter B. The parameters are those given in Figure 23. If the localized Turing pattern contains more than four wavelengths, the nonadiabatic effect is dominant, and the pinning zone limited by the two *vertical lines* is the same as that for the front. Beyond this pinning band, the only stable localized Turing patterns are those that have a small enough core (less or equal to four wavelengths) for nonvariational interactions to come into play. The localized structure stable in the widest domain is the flip-flop.

constraint, the system then sometimes creates temporary localized subzones [119]. Recently, the velocity of such Turing–Hopf fronts has been studied in detail by Or-Guil and Bode [283a].

Two interacting fronts may then build up droplets of one global state imbedded into another [174–178]. The juxtaposition of two Turing–Hopf fronts leads to stable localized droplets of a Turing (Hopf) state imbedded into a Hopf (Turing) domain [Fig. 23(b)]. We observe that, if the Turing core contains several wavelengths, the stability domain of such localized structures is the same as that of the front (Fig. 24); and their stabilization can correspondingly be ascribed to nonadiabatic pinning effects, as in the case of the simple front. If the localized Turing domain contains few wavelengths, this stabilizing nonadiabatic effect can no longer be invoked alone. Stable localized Turing patterns with few wavelengths are nevertheless observed in the Brusselator model; and the fewer wavelengths they contain, the larger their stability domain. Thus their stability results from nonvariational effects that in other systems [284–287] stabilize localized structures if they provide a repulsive interaction between two fronts that otherwise attract each other [19,174,175,288]. They can, therefore, account for the existence of localized droplets of one state imbedded into the other state. This effect is strongest for the so-called flip-flop localized pattern with the smallest core [Fig. 23(c)] and, therefore, the widest stability domain (Fig. 24). This could explain why the flip-flop is the only localized pattern to be observed experimentally in the CIMA reaction for concentration values near the CTHP [64] (Fig. 5).

Turing–Hopf localized structures have also been observed experimentally in 1D arrays of resistively coupled nonlinear LC oscillators [265] and in binary fluid convection [262].

Bistability between the Turing and Hopf modes near a CTHP had long been predicted in the amplitude equation formalism [155]. Recent studies have shown that in this bistability regime, localized structures of one state embedded into the other can be stabilized by a combination of nonadiabatic and nonvariational effects [118–120].

B. Subharmonic Instabilities

When numerical integration of reaction–diffusion models are carried out near a CTHP, scenarios [119,278] occur that do not fit in the bifurcation diagrams of Figure 20 predicted by the amplitude equations Eqs. (6.2) and (6.3) for the coupled Turing and Hopf modes. In particular, two new mixed modes are observed that contain more than one wavenumber. Fourier transforming these spatiotemporal dynamics point out the presence of subharmonic modes of both Turing and Hopf states. These subharmonic modes can be excited because of subharmonic instabilities of the pure Turing and Hopf modes. Indeed, if the bifurcation parameter near the

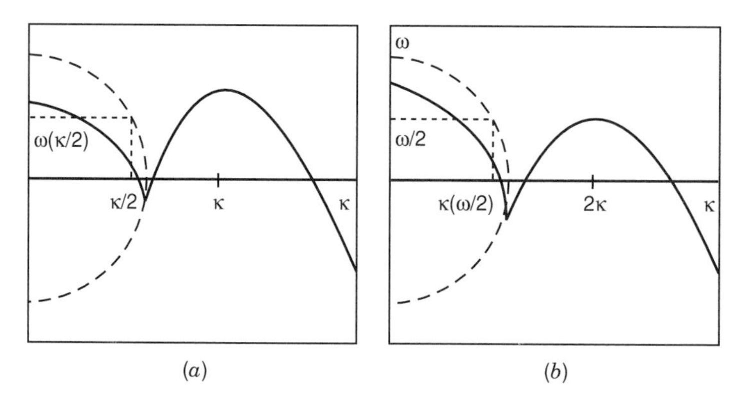

Figure 25. Dispersion relations explaining the resonances between the Turing and Hopf modes, with their subharmonic in the vicinity of a CTHP. The *plain line* and *dashed line* represent the real and imaginary part, respectively, of the eigenvalues of the linear stability analysis. (*a*) The subharmonic $k/2$ of the Turing wavenumber k corresponds to an eigenvalue with a nonvanishing frequency $\omega(k/2)$. The coupling between the Turing mode and its subharmonic leads to a subharmonic Turing mode with two wavenumbers k, $k/2$ and one frequency $\omega(k/2)$. (*b*) The subharmonic $\omega/2$ of the Hopf mode is a wave with wavenumber $k(\omega/2)$. If this wavenumber is on the order of the subharmonic of the Turing wavenumber $2k$, then resonance between the Hopf mode, its subharmonic, and the Turing mode leads to a subharmonic Turing–Hopf mode with two wavenumbers $k(\omega/2)$, $2k$ and two frequencies ω, $\omega/2$.

CTHP is increased, conditions are often such that the root of the characteristic equation corresponding to the subharmonic of the Turing mode, with wavenumber $k_c/2$, crosses the imaginary axis (i.e., this mode becomes active), and its imaginary part $i\omega$ is different from zero, i.e., $\omega(k_c/2) \neq 0$ [Fig. 25(a)]. Therefore, the subharmonic mode lies in the Hopf part of the linear dispersion relation. A resonance can then occur between the Turing mode with a wavenumber–frequency couple $(k_c, 0)$ of amplitude T and its subharmonic mode $(k_c/2, \omega[k_c/2])$ corresponding to left and right traveling waves of amplitude A_{R} and A_{L} [289]. The concentration field is then described as

$$\begin{aligned}\underline{C}(x, t) = \underline{C}_0 + Te^{ik_c x}\underline{w}_T + A_{\mathrm{L}}e^{i[\omega(\frac{k_c}{2})t + \frac{k_c}{2}x]}\underline{w}_{\mathrm{L}} \\ + A_{\mathrm{R}}e^{i[\omega(\frac{k_c}{2})t - \frac{k_c}{2}x]}\underline{w}_{\mathrm{R}} + c.c.\end{aligned} \tag{6.10}$$

where $\underline{w}_{\mathrm{L}}$ and $\underline{w}_{\mathrm{R}}$ are the critical eigenvectors corresponding to the left- and right-going waves of wave number $k_c/2$ and frequency $\omega(k_c/2)$. The coupled amplitude equations for the modes read [119,290–292]

$$\frac{\partial T}{\partial \tau} = \mu T - g|T|^2 T - \lambda(|A_{\mathrm{R}}|^2 + |A_L|^2)T + \nu A_{\mathrm{R}}^* A_L + D^{\mathrm{T}}\frac{\partial^2 T}{\partial \mathcal{X}^2} \tag{6.11}$$

$$\begin{aligned}\frac{\partial A_{\mathrm{R}}}{\partial \tau} &= \mu_{\mathrm{H}} A_{\mathrm{R}} - g'|A_{\mathrm{R}}|^2 A_{\mathrm{R}} - h'|A_{\mathrm{L}}|^2 A_{\mathrm{R}} \\ &\quad - \lambda'|T|^2 A_{\mathrm{R}} + \nu' T^* A_{\mathrm{L}} + D'\frac{\partial^2 A_R}{\partial \mathcal{X}^2}\end{aligned} \tag{6.12}$$

$$\frac{\partial A_{\mathrm{L}}}{\partial \tau} = \mu_{\mathrm{H}} A_{\mathrm{L}} - g'|A_{\mathrm{L}}|^2 A_{\mathrm{L}} - h'|A_{\mathrm{R}}|^2 A_{\mathrm{L}} - \lambda'|T|^2 A_{\mathrm{L}} + \nu' T A_{\mathrm{R}} + D'\frac{\partial^2 A_{\mathrm{L}}}{\partial \mathcal{X}^2} \tag{6.13}$$

A linear stability analysis of the solutions to this set of equations shows that for some values of coefficients a mixed standing wave mode $(T \neq 0, A_{\mathrm{R}} = A_{\mathrm{L}} \neq 0)$, characterized by two wavenumbers k_c, $k_c/2$ and one frequency $\omega(k_c/2)$, can bifurcate from the pure Turing mode $(T \neq 0, A_{\mathrm{R}} = A_{\mathrm{L}} = 0)$ [291]. This situation corresponds to one of the mixed modes observed in the Brusselator [119] [Fig. 22(b)] and in the Gray and Scott model [293]. It also describes oscillating subharmonic patterns observed experimentally in the flow of a viscous fluid inside a partially filled rotating horizontal cylinder [263,264]. This subharmonic Turing mode can furthermore become phase unstable [291], giving rise to spatiotemporal chaos, as seen in the hydrodynamics experiments.

A subharmonic instability can also affect the Hopf mode $(0, \omega_c)$ when it interacts with its subharmonic mode $k(\omega_c/2), \omega_c/2$. A theoretical analysis in the same spirit as that described above shows that a pure Hopf mode may become unstable toward a mixed mode composed of this Hopf state and its subharmonic. This new mixed mode is characterized by one wavenumber $k(\omega_c/2)$ and two frequencies ω_c, $\omega_c/2$. Such a mode is not seen in the simulations with the Brusselator model, because near the CTHP it may itself easily resonate with a pure Turing mode of wavenumber $k_c \approx 2k(\omega_c/2)$ [Fig. 25(b)]. The resulting dynamics is another mixed state, dubbed a subharmonic Turing–Hopf mode, characterized by two wavenumbers $k(\omega_c/2)$, $2k(\omega_c/2)$ and two frequencies ω_c, $\omega_c/2$. At a given time, the concentration exhibits alternation of high- and low-amplitude peaks in the spatial structure [Fig. 26(a)]. Each spatial location oscillates in time, alterning high- and low-amplitude temporal oscillations resulting from the superposition of the Hopf mode and its subharmonic [Fig. 26(b)]. These spatiotemporal dynamics were observed in numerical simulations of the Brusselator [119] model [Figs. 22c and 26], the Gray–Scott model [294], and a reaction–diffusion model of a semiconductor device [266,278,295]. In large systems, this subharmonic Turing–Hopf mode can also become phase unstable, leading the system to spatiotemporal chaos [117,278].

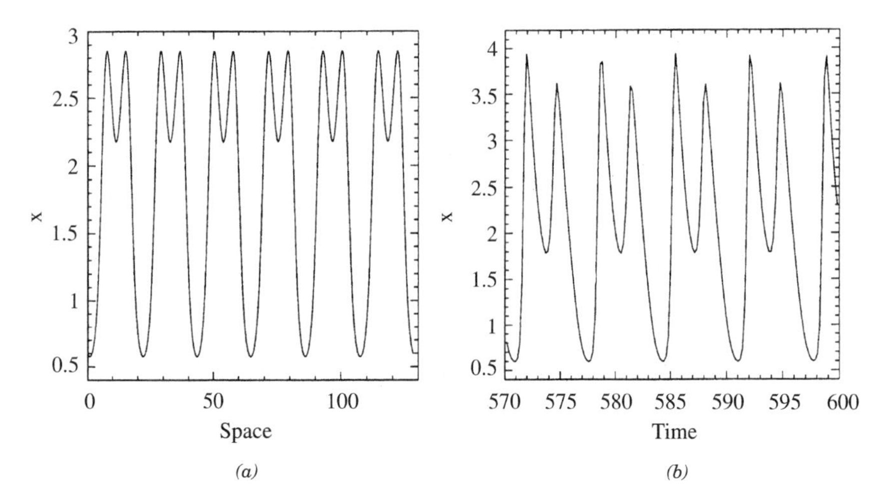

Figure 26. Subharmonic Turing–Hopf mixed mode observed in a 1D Brusselator model in a system of length 128 with periodic boundary conditions: $A = 3, D_x = 5.71, D_y = 10$, and $B = 11$. (*a*) Spatial profile taken at a given time. Two wavelengths — the Turing one and the double of it (corresponding to the subharmonic of the Turing wave vector) — are underlying the spatial structure. Later, the minima become maxima and vice versa. (*b*) Periodic temporal dynamics at a given fixed point of the system, showing oscillations with two frequencies: the Hopf one and its subharmonic.

C. Genericity

To summarize, four main types of 1D spatiotemporal dynamics are commonly encountered near a CTHP:

1. Turing–Hopf bistability, which leads to various localized structures, such as the flip-flop.
2. A Turing–Hopf mixed mode, a pure Turing structure oscillating in time with the Hopf frequency.
3. A subharmonic Turing mode, a spatial structure with two wavenumbers and one frequency.
4. A subharmonic Turing–Hopf mode, which leads in the vicinity of a Turing bifurcation, to structures with two wavenumbers and two frequencies.

In the Brusselator model, these dynamics were classified in an A versus σ/σ_c parameter space (Fig. 27). Here it is important to recall that in the Brusselator model, the frequency of Hopf temporal oscillations is equal to the value of the parameter A, which thus fixes the time scale of the system. Similarly, σ relates to the diffusion coefficients that fix the space scale. Figure 27 displays a time scale versus a space scale parameter space. An analogous classification has been performed for the spatiotemporal dynamics occurring in a reaction– diffusion model that describes semiconductor transport near a CTHP [278]. In that system, Turing–Hopf bistability (and related localized structures) and the simple and subharmonic mixed modes have been observed as well. In that case the spatiotemporal self-organization is that of the current density rather than that of chemical concentrations. In the semiconductor model, the classification in the time scale versus space scale of the various spatiotemporal dynamics near the CTHP presents strong similarities with the one shown for the Brusselator model.

We propose, therefore, that a time-scale versus length-scale diagram might be an appropriate way for organizing the comparison with other models that present CTHP, because it separates the different scales that reflect the spatial character of the Turing mode and temporal character of the Hopf mode. In this sense, this type of diagram might be useful for looking for the predicted spatiotemporal dynamics characteristic of an interaction between a steady instability (not necessarily the Turing one) and a Hopf instability. In particular, in nonlinear optics, Turing–Hopf bistability and related localized structures [267] as well as the Turing–Hopf mixed mode [267,268] were identified close to codimension-two Turing–Hopf points. We believe that the subharmonic mixed modes could also be found in the parameter space of these nonlinear optical systems using the above classification.

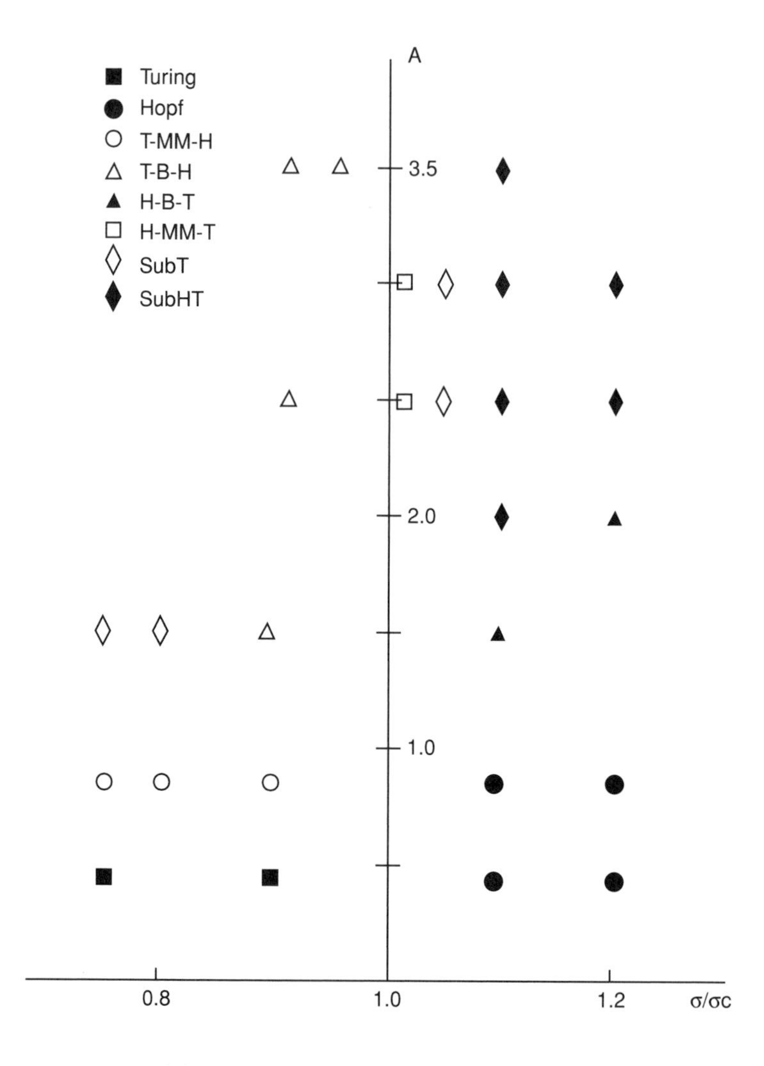

Figure 27. Summary of the various bifurcation scenarios observed for the Brusselator model and classified in the parameter space A versus σ/σ_c. B stands for Turing–Hopf bistability (with the corresponding localized structures), MM relates to the Turing–Hopf mixed mode (characterized by the wavevector k_c and the frequency ω_c). The subharmonic Turing mode with two wavevectors k_c, $k_c/2$ and one frequency ω_c is designated by sub T; the subharmonic Turing–Hopf mode with two wavevectors k_c, $k_c/2$ and two frequencies ω_c, $\omega_c/2$ is designated by sub HT. The succession T-MM-H, for example, means that, when increasing the bifurcation parameter for the given $A, \sigma/\sigma_c$ couple, successive transitions between pure Turing patterns, a mixed mode, and Hopf oscillations are observed. When $\sigma/\sigma_c > 1$, the first solution to appear beyond the critical point is a Hopf mode, whereas the Turing mode is dominant at criticality when $\sigma/\sigma_c < 1$.

D. Two-Dimensional Spatiotemporal Dynamics

All the Turing–Hopf spatiotemporal dynamics predicted in 1D systems have equivalents in 2D systems. Beyond the pure Turing and pure Hopf solutions, the Turing–Hopf interaction can lead to 2D mixed modes, such as stripes and hexagons, oscillating homogeneously in time with one frequency or to bistability between Turing patterns and a Hopf homogeneous temporal oscillation [9,119,296]. In the bistability regimes, Jensen and co-workers [120,121] showed that a Turing dot at the tip of a 2D spiral obtained numerically in the Lengyel–Epstein model is the equivalent of the 1D well-studied flip-flop [64,100]. Such a Turing–Hopf spiral was observed experimentally [48,92,98]. Localized structures with more than one Turing dot are difficult to obtain, because in 2D, pinning of a Turing–Hopf front owing to the underlying interaction of the front with the pattern occurs only in some privileged directions. This explains why fronts between 2D Turing structures with several wavelengths and an oscillating zone have been obtained only as transients up to now (Fig. 28). Pure 2D Turing–Hopf mixed modes involving only one wavenumber and one frequency have, to our knowledge, never been observed experimentally or numerically, although they should probably be stable for some values of parameters.

On the contrary, the other main types of spatiotemporal dynamic characteristics of the Turing–Hopf interaction, i.e., subharmonic instabilities, arise in 2D simulations of the Brusselator model for parameter values close to a CTHP (Fig. 29). In that case, hexagons or stripes oscillate in time with two frequencies. The hexagonal dots have a little bump in their center, which correspond to the intermediate minima observed in the case of the 1D subharmonic Turing mode [Fig. 26(a)] and are the signature of subhar-

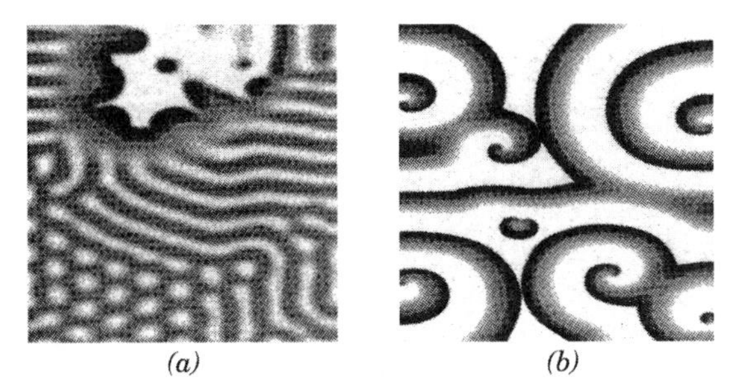

Figure 28. (*a*) Transient front between 2D Turing structures and a Hopf oscillating zone observed in the Brusselator model for $A = 2.5$, $D_x = 4.49$, $D_y = 8.91$, and $B = 8.6$ in a system of size 256×256 with no flux boundary conditions. The asymptotic state of the system was highly irregular spatiotemporal dynamics. (*b*) Here $B = 11$. The 2D spirals are obtained for the same parameter values that give rise to 1D flip-flops.

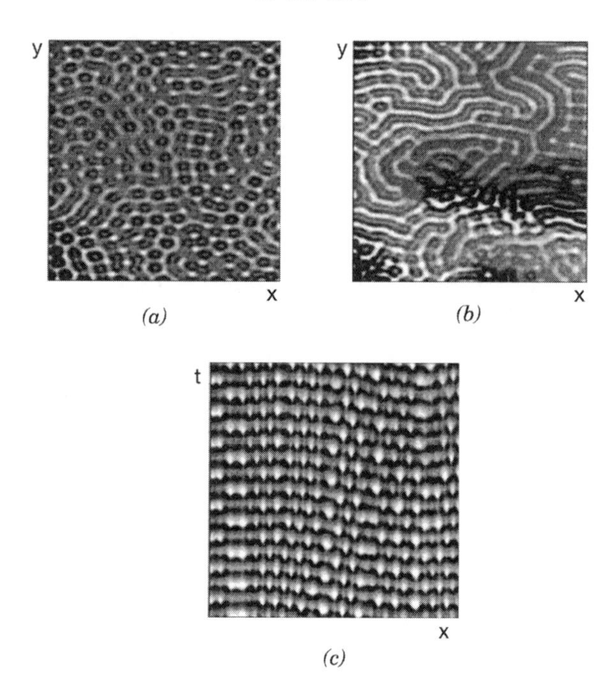

Figure 29. The 2D subharmonic Turing–Hopf modes observed in the Brusselator model for $A = 2.5, D_x = 4.49$, and $D_y = 8.91$ in a system of size 256×256 with no flux boundary conditions. (*a*) Here $B = 7.5$. The highly 2D irregular hexagonal-type patterns shown at a given time. Each spot features a little local minimum in its center, indicating the presence of subharmonics. (*b*) Here $B = 7.8$. The 2D stripes shown at a given time. Each band also exhibits a little local minimum in its middle. In both (*a*) and (*b*), each point of the system oscillates in time with two frequencies, as can be seen in the space–time map, showing the dynamics along one line parallel to the y-axis in part (*b*) versus time running upward. This space–time map is reminiscent of Figure 22(*c*), evidencing the spatiotemporal dynamics characteristic of a 1D subharmonic Turing–Hopf mixed mode.

monics. The stripes also present analogous little wiggles in their maxima. This behavior can be clearly seen in Figure 29(c), which displays a space–time map of the dynamics along one line of the 2D spatial system. This space–time map is reminiscent of the one characterizing the 1D subharmonic Turing–Hopf mixed mode with two wavenumbers and two frequencies [119] [Fig. 22(c)]. It is however not clear yet which spatial and temporal subharmonics are involved when more than one basic wave vector is present, as in the case of hexagons [297].

Earlier we saw that phase instabilities of the various Turing–Hopf mixed modes may lead to spatiotemporal chaos in 1D [271,291]. Chaotic 2D dynamics have often been observed experimentally [48,69,92] and numerically

[119,121,295] in the vicinity of a codimension-two Turing–Hopf point, but no characterization of the destabilizing instabilities have been provided. The study of the 2D Turing–Hopf spatiotemporal dynamics is thus still in its infancy.

In several systems, the Turing–Hopf interaction implies a Hopf mode with a nonzero wavenumber [256,258,298]. In that case, the number of possible resonances between the Turing and Hopf wavenumbers become much more important. These resonances provide interesting new 2D dynamics, such as winking hexagons [299], drifting rhombs [300], or hexagons [301], and even quasicrystalline patterns [302,303].

We saw that characteristics of the patterns emerging in a monostable reaction–diffusion system are greatly affected by the vicinity of another competing bifurcation, such as the Hopf one. The main advantage of chemical systems is that they allow competition with a bifurcation between different homogeneous steady states. Let us now examine which type of spatiotemporal dynamics such an interplay may induce.

VII. BISTABLE SYSTEMS

The labyrinthine patterns observed in the FIS reaction [129–131] call for insight into pattern formation in bistable systems. Indeed, bistability between two homogeneous steady states (HSS) characterizes the parameter space in which the FIS spatial structures are obtained. Several other pattern-forming systems, such as gas discharges [37,304], liquid crystals [305], and nonlinear optics [24,164,306], also exhibit bistable regimes. In bistable systems, patterns can result from Turing bifurcations of different homogeneous branches or from morphological instabilities of fronts connecting different HSS. Let us successively examine these two possibilities.

A. Zero Mode

In bistable systems, patterns can emerge through Turing instabilities on the various HSS branches. Recently, some studies focused on the pattern selection problem using the bistable FitzHugh–Nagumo model [27,307,308] or the Gray and Scott model [293,309]. Two situations leading to bistability are a pitchfork bifurcation (generally imperfect) from one HSS toward two different HSS or an hysterisis loop formed by two back-to-back saddle-node bifurcations, as encountered in the FIS system. The study of the pattern selection can be carried out in the vicinity of the cusp point of these transitions. This analysis relies on the fact that the homogeneous perturbations are quasi neutral. As a consequence, a zero mode becomes active and must be taken into account in the dynamics. If some of the HSS undergo diffusive instabilities, it is expected that for some values of parameters the Turing and homogeneous bifurcations can interact. In that case, the concentration field $\underline{C}$ is constructed as

$$\underline{C}(\mathbf{r}, t) = \underline{C}_0 + T_0 \underline{w}_0 + \sum_{j=1}^{N} [T_j e^{i\mathbf{k}_j \cdot \mathbf{r}} + T_j^* e^{-i\mathbf{k}_j \cdot \mathbf{r}}] \underline{w} + O(\cdots) \tag{7.1}$$

where $\underline{w}_0$ and $\underline{w}$ are the eigenvectors of the Jacobian matrix of the homogeneous and distributed systems, respectively. The pattern selection problem is then ruled by coupled amplitude equations for T_0 and the T_j. The coupling with the zero mode has a profound influence on the possible bifurcation scenarios. In 1D, it can give rise to a rich variety of subharmonic and superharmonic patterns [310] and even to spatial chaos [311]. In 2D, the corresponding amplitude equations are [32,225]

$$\frac{dT_0}{dt} = f(T_0) + \alpha \sum_{j=1}^{3} |T_j|^2 - \beta_1 \left[\sum_{j=1}^{3} |T_j|^2 \right] T_0 - \beta_2 (T_1^* T_2^* T_3^* + T_1 T_2 T_3) \tag{7.2}$$

$$\begin{aligned} \frac{dT_1}{dt} &= \mu T_1 + \alpha [T_0 T_1 + T_2^* T_3^*] - g_1 |T_1|^2 T_1 \\ &\quad - g_2 [|T_2|^2 + |T_3|^2] T_1 - g_3 T_1 T_0^2 - g_4 T_0 T_2^* T_3^* \end{aligned} \tag{7.3}$$

where β_1, β_2, g_3, and g_4 are taken to be positive and $\mathbf{k}_1 + \mathbf{k}_2 + \mathbf{k}_3 = 0$. The equations for T_2 and T_3 are obtained by cyclic permutations of the subscripts in Eq. (7.3). These equations admit the following classes of solutions:

1. Homogeneous solutions T_0^o given by $f(T_0^o) = 0$ with $T_j = 0$ corresponding to the reference HSS of the bistable regime.
2. Mixed spatial pattern solutions with smectic ($T_0^s \neq 0, |T_1| \neq 0$, $T_2 = T_3 = 0$) or hexagonal symmetry ($T_0^h \neq 0, |T_1| = |T_2| = |T_3| \neq 0$).

It is important to note here that pure stripes or hexagons are generally not solutions to the problem. Moreover, the presence of the amplitude of the homogeneous mode in the pseudo quadratic term $v = g_4 T_0^h - \alpha$ of the amplitude Eq. (7.3) for the Turing modes has several consequences for pattern selection. First of all, the sum of the phases $\Theta = \theta_1 + \theta_2 + \theta_3$ of the spatial component of the hexagons depends on the sign of T_0^h, the amplitude of the homogeneous component of the spatial pattern, because it now obeys an equation of the type

$$\frac{d\Theta}{dt} \sim (g_4 T_0^h - \alpha) \sin \Theta \tag{7.4}$$

The phase Θ relaxes to zero or π, depending on whether $(g_4 T_0^h - \alpha)$ is < 0 or > 0, respectively, forming H0 or Hπ mixed hexagons that may even coexist.

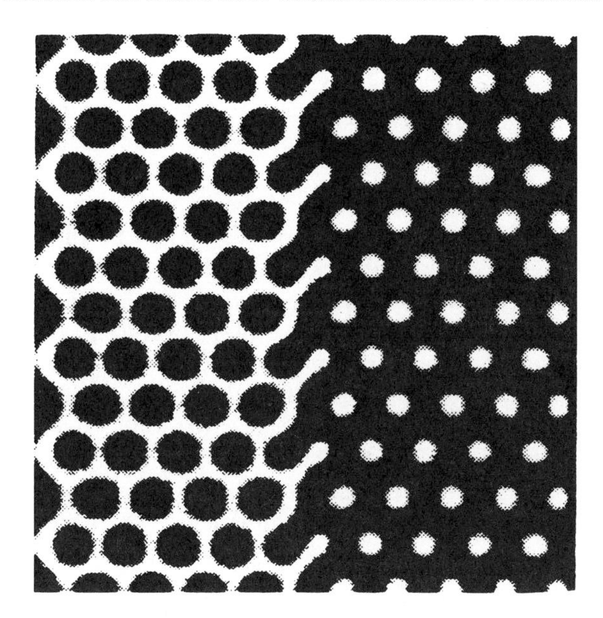

Figure 30. Front between H0 and Hπ hexagons observed in a numerical integration of the FitzHugh–Nagumo model for conditions such that a Turing instability occurs in the vicinity of a bifurcation between different homogeneous steady states. From Ref. [101].

In this latter case, fronts between H0 and Hπ hexagons can be sustained (Fig. 30). Such fronts have been observed experimentally in convection with SF_6 near its critical point [312] and in oscillated granular layers [161]. As T_0^h is modified when varying the bifurcation parameter, it may bring about a change of sign of the overall quadratic term v, leading to possible re-entrance of hexagons, as observed in liquid crystals [305] or in a gas discharge experiment [304]. In 3D systems, the same process applies, and re-entrant BCC or HPC phases are also obtained [225], leading to bifurcation diagrams similar to those obtained for block copolymers [313]. Note that this re-entrance should not be confused with the model–dependent re-entrance that occurs when the nonlinear terms vary with the bifurcation parameter [84,86] (see Section V.B.2). The re-entrance from a zero mode is more generic and occurs as soon as the Turing instability arises in the vicinity of bifurcations among different HSS. The fact that the quadratic term is a function of T_0 also leads to the large subcritical appearance of the various planforms and, as a consequence, to possible localized structures, such as 2D hexagonal domains of H0 or Hπ embedded into the uniform state [99]. More strikingly, if $\alpha = 0$, the vicinity of a bifurcation toward HSS can give rise to the stabilization beyond the stripes of both hexagonal phases, even in systems exhibiting inversion symmetry [27,225,314].

The possible bifurcation scenarios in bistable systems can thus be different from those occuring in the absence of the zero mode. Let us also note that, in bistable systems, the amplitude of the mixed Turing-homogeneous mode patterns is on the order of the difference in amplitudes between the two HSS and is thus usually quite large. In some cases, isolated branches of such patterns can form [307,308], and the corresponding high amplitude structures can then be reached only by local finite perturbations applied to the HSS as observed in the FIS experiment [129]. Such perturbations generate fronts connecting the two HSS. If such a front now undergoes a morphologic instability, the system may evolve toward the isolated Turing branch and settle down as a high-amplitude stationary labyrinthine or hexagonal pattern, as observed in numerical simulations of the FitzHugh–Nagumo model [151]. In addition to the labyrinthine structures oberved by Lee and co-workers [129], this could also explain experimental observations with the CDIMA system conducted near a cusp point in the monostable regime in the vicinity of two different HSS. There, a flowerlike structure develops when a circular front between the two HSS is unstable because of a morphologic instability. The flower petals successively break into a hexagonal pattern, which is the asymptotic state of the system [80].

B. Morphologic Instabilities

Many studies have been devoted to the properties of fronts linking two HSS in bistable systems [26,274,315–317]. If the system is composed of at least two species with sufficiently different diffusion coefficients, planar fronts can become unstable, because of morphologic instabilities that lead to a deformation of the interface between the two HSS [274,315]. When two fronts come close together, repulsion at short distances occurs, excluding fusion or breakup of the fronts and leading to labyrinthine patterns [129–131]. Numerous theoretical works focused on such instabilities that give rise to fronts and patterns of this type [26,138,174,318–330].

In such systems, a wealth of different spatiotemporal dynamics is observed, depending on the ratios between the characteristic time and space scales of the activator and inhibitor of the system [26,318,319]. For example, an investigation of the front dynamics in bistable systems was performed in both the fast inhibitor limit [26,318,320] and the slow inhibitor limit [321–324]. The models commonly used in these investigations on reaction–diffusion systems are the bistable FitzHugh–Nagumo-type models [26,138,156,157,319,324,325], the Gáspár–Showalter model of the FIS reaction [130,135,137], and the Gray–Scott model. This last model has allowed, among other things, detailed study of the spot replication process [130,131,136–138,293,309,319,326,331–333].

To account for the properties of labyrinthine structures observed in the FIS reaction, several complementary approaches have been used, such as a general asymptotic theory of instabilities for patterns with sharp interfaces [328], interfacial dynamic approaches [321,328,329], and parity-breaking bifurcations of planar fronts [138,324,330].

VIII. CONCLUSIONS AND PERSPECTIVES

The study of chemical temporal oscillations started with Belousov's experimental observations of the now well-known Belousov–Zhabotinskii reaction [42], which was skeptically received by the scientific community [334]. It was only in the 1970s that the number of works devoted to these temporal oscillations took off [7,42,43]. Today, oscillating reactions are a classical domain of nonlinear science and are recognized as a key element in the understanding of several biorythms [335]. They are now being considered in the engineering domain for applications such as temporally modulated drug release [336,337] and possible digital logic [338].

Spatial chemical patterns may now be on their way to an analogous history. Turing structures were predicted theoretically quite a long time ago; and the recently experimentally observed structures have led to many studies devoted to spatial chemical patterns and related spatiotemporal dynamics. Nevertheless, more work is needed to unravel the richness of these spatial chemical structures.

First, several aspects of the experimentally obtained Turing patterns call for additional studies, such as the 3D pattern selection problem. The fact that the Turing wavelength is intrinsic places reaction–diffusion systems in a unique position among pattern-forming systems for studying all possible 3D self-organizations, which are much richer than in 2D systems [117]. Unavoidably related to that is the problem of gradients and how spatial variation of parameters affects the selection and orientation of spatial structures. Another area of chemical systems is the interaction with other instabilities. The Turing–Hopf coupling provides very rich spatiotemporal dynamics, which are now being discovered in other systems, such as nonlinear optics [267]. It is expected that several dynamics, such as the subharmonic mixed modes and phase instabilities of the Turing–Hopf mixed modes that lead to spatiotemporal chaos, will soon be discovered in those systems as well. Note that the different mechanisms that give rise to chaotic space–time behaviors near a CTHP provide tools for testing to what extent measures that quantify chaotic dynamics are independent from the instability mechanism in the same system. In addition, the spatiotemporal chaos owing to a phase instability of the various mixed modes is much richer than the chaos in the complex Ginzburg–Landau equation, for instance, because the underlying unstable base state contains different Turing and/or Hopf modes. This fact was exploited by Petrov and co-workers [294], who stabilized a spatial pattern and a temporal oscillation out of a chaotic Turing–Hopf mixed mode using chaos-control techniques. In the same spirit, the observation of spatial patterns in bistable reaction–diffusion systems has shown that the pattern selection problem can be affected in the bistability regime by the presence of a homogeneous mode.

The analysis of the coupling between Turing modes with Hopf or homogeneous modes participate in a stream of studies that enlarge the number of modes considered to play a role in the buildup of spatial planforms and spatiotemporal dynamics. In the first classical papers devoted to the Rayleigh–Bénard instability [77,207] and in many papers that focused on spatial patterns, the only structures considered were stripes, squares, and hexagons. Experiments have shown, however, that more spatial modes sometimes need to be considered already in 2D systems, as in the case of quasicrystalline structures observed in the Faraday instability [143,144] and in nonlinear optical systems [145]. Other structures built on the whole circle of unstable modes, such as turbulent crystals [339], have also been described. In addition, modes of different lengths may sometimes coexist [340] or resonate to yield nonequilateral patterns, black-eyes, decagons, etc. [191,341]. In the same spirit, the coupling between a nonoscillatory short-wavelength instability and a long-wavelength neutrally stable mode was recently shown to produce new spatiotemporal dynamics intermediate between those of the traditional long- and short-wavelength ones [342,343]. Oscillatory modes with a finite wavenumber are also able to lead to pattern-forming dynamics in reaction–diffusion systems, such as traveling and standing waves and target patterns in 1D systems [344–346]. In 2D systems, resonance interactions among several pure Hopf waves produce a multiplicity of patterns, including traveling and blinking rolls, rhombs, and hexagons [301,347,348]. All these possibilities are potentially viable in chemical systems and remain topics of future research.

Beyond the complexity related to the increasing number of modes considered, one must also deal with the complexity arising from heterogeneities and anisotropies. Voroney and co-workers [349] analyzed the effect of spatial inhomogeneities in the distribution of complexing agents on Turing patterns and their switch toward oscillating behaviors. Such a distribution arises if one assumes that diffusion coefficients are space dependent owing to the geometry of the medium [350] or to an imposed spatial distribution of the color indicator in the CIMA system, as was done for the study of waves in patterned systems [351]. This could allow for an understanding of the interplay between the wavelength of the pattern and the typical length scale of the inhomogeneities that sometimes leads to time-dependent solutions [349,352,353]. Such possible coupling is connected with the interaction of different spatial modes alluded to above.

In the same spirit, Turing patterns can be affected by anisotropies, such as those arising from gradients [85], electrical fields [93,125,354–356], and anisotropic diffusion coefficients [357]. An important source of anisotropies is hydrodynamic flows. In the past, hydrodynamic effects prevented the observation of Turing patterns [358,359]. Now that the conditions for obtaining Turing structures in gels to avoid any perturbing hydrodynamic current are better understood, possible competition of Turing structures with flows can be studied. Two different situations occur, depending on whether hydrodynamic currents are of internal or external origins. Internal currents may arise owing to chemically driven convection. In gel-free media [63], the density gradients inherent to the spatial variation of concentrations in the Turing pattern can generate local convective motions [360–362]. External currents refer to global flows imposed along a given direction as the result of gradients of pressure or shearing, which lead to advected Turing patterns [196,363–368]. In that case, an absolute and a convective instability [369] of the Turing modes can arise, depending on the flow rate; and the linear stability analysis of the patterns is modified [48,368]. In such systems, one needs to be cautious with the origin of patterns. It is indeed known that noise-sustained spatial structures may emerge in convectively unstable regimes [370], which should not be confused with Turing patterns. Moreover, systems with reactive flows exhibit differential flow-induced chemical instabilities (DIFICI) [371] that are typically convectively unstable [79]. This traveling wave-forming instability occurs when one reactive is binded to a support and another one flows onto it. Owing to the richness of different pattern-forming mechanisms in advection–reaction–diffusion systems and the possible interactions among them [193,366,368,372,373], more work must be done to understand the properties of spatial structures in the presence of flows.

Although the Turing instability and the Rayleigh–Bénard instabilities have been the paradigms of pattern formation instabilities in chemical and hydrodynamic systems, respectively, other pattern-forming conditions related to chemical systems have also been studied. For example, Petrov and co-workers [374] recently confirmed experimentally that the external periodic forcing by light of a photosensitive variant of the Belousov–Zhabotinsky reaction can lead to a resonant pattern formation, which was predicted theoretically by Coullet and Emilsson [179,180]. Another source of pattern formation is global coupling, which plays an important role in some reaction–diffusion systems. Global coupling arises as a limiting case of infinite diffusivity of one of the variables. The temporal evolution equation of this variable is then given by an integral equation of the type $x_t = \langle f(x, y) \rangle$, where $\langle ... \rangle$ denotes averaging over the entire reactive domain. Such a global control is encountered, e.g., in experiments with catalytic wires [375,376], in which global coupling results from a constant mean temperature. Similarly, conservation of the total current leads to a global control in electrochemical [377] and glow discharge devices [286]. In such systems, propagating waves owing to oscillatory local kinetics are transformed into standing waves by global coupling, which can also lead to complex dynamics of pulses and kinks [285]. Standing waves owing to a global control occur in catalytic reactions as well [17,378–382]. There global control is achieved through the gas phase, which mixes the substances and thus imposes a rapid communication among different locations of the catalytic surface and even among different surfaces [383]. Recently, Middya and Luss [384] suggested that the same effect could apply when a reaction–diffusion system operates in a thin slab in contact with a surrounding vessel in which rapid homogeneitization of the concentrations take place. They showed that, if global interaction between the gel and the vessel is taken into account, then spatial patterns can result from the global control, even when reaction–diffusion mechanisms cannot create patterns on their own. In that case, the patterns bear analogies with those observed in the FIS experiments [129,384,385], and transitions between striped and spotted patterns with changing levels of global control can be obtained [384,386].

To conclude, let us say that a movement is under way to complement the analysis of temporal complexity in chemical reactions by their spatial aspect. This evolution toward an increase in complexity is also parallel to an increase of diversity and richness of the possible spatial and related spatiotemporal behaviors to be discovered in chemical systems. Chances are, therefore, great that in the near future spatial patterns will provide numerous potential applications wherever a spatial distribution of chemical activity is necessary.

Acknowledgments

I wish to thank Professor G. Nicolis for his continuous support and interest in this work. I am much indebted to Pierre Borckmans and Guy Dewel, in collaboration with whom all my contributions to the field of Turing patterns have been obtained, for numerous enlightening discussions. I thank them for suggesting various modifications of this manuscript. I also acknowledge discussions with P. De Kepper, E. Dulos, J. Boissonade, D. Lima, M'F. Hilali, J. Verdasca, S. Métens, J. Pontes, J. Lauzeral, and D. Walgraef. I thank in particular P. De Kepper for his comments on the manuscript and for providing me with the figures of experimental results. I acknowledge the financial support from the Belgian Federal Office for Scientific, Technical, and Cultural Affairs under the Pôles d'Attraction Interuniversitaire program.

References

1. A. Turing, *Philos. Trans. R. Soc. London, Ser. B* **327**, 37 (1952).
2. P. Glansdorff and I. Prigogine, *Thermodynamic Theory of Structure, Stability and Fluctuations*. Wiley, New York, 1971.
3. G. Nicolis and I. Prigogine, *Self-Organization in Nonequilibrium Systems*. Wiley, New York, 1977.
4. F. Baras and D. Walgraef, eds., *Physica A* **188** Spec. Issue (1992).
5. R. Kapral and K. Showalter, eds., *Chemical Waves and Patterns*. Kluwer Academic Publishers, Dordrecht, The Netherlands, 1995.
6. G. Nicolis, *Introduction to Nonlinear Science*. Cambridge University Press, Cambridge, UK, 1995.
7. I.R. Epstein and K. Showalter, *J. Phys. Chem.* **100**, 13132 (1996).
8. C. Vidal, G. Dewel, and P. Borckmans, *Au-delà de l'équilibre*. Hermann, Paris, 1996.
9. D. Walgraef, *Spatiotemporal Pattern Formation*. Springer-Verlag, New York, 1997.
10. O. Vafek, P. Pospisil, and M. Marek, *Sci. Pap. Prague Inst. Chem. Technol., Sect. K* **14**, 179 (1979).
11. J.I. Gmitro and L.E. Scriven, in *Intracellular Transport* (K.B. Warren, ed.), p. 179. Academic Press, New York, 1966.
12. J.D. Murray, *Mathematical Biology*, 2nd ed. Springer-Verlag, Berlin, 1993.
13. H. Meinhardt, *Models of Biological Pattern Formation*. Academic Press, New York, 1982.
14. A.J. Koch and H. Meinhardt, *Rev. Mod. Phys.* **66**, 1481 (1994).
15. A. Babloyantz, *Molecules, Dynamics and Life*. Wiley, New York, 1986.
16. E. Meron, *Phys. Rep.* **218**, 1 (1992).
17. G. Ertl, *Science* **254**, 1750 (1991).
18. E. Mosekilde and O.G. Mouritsen, eds., *Modelling the Dynamics of Biological Systems*. Springer, Berlin, 1995.
19. M.C. Cross and P.C. Hohenberg, *Rev. Mod. Phys.* **65**, 851 (1993).
20. E. Schöll, *Nonequilibrium Phase Transitions in Semiconductors*. Springer, Berlin, 1987.
21. P. Manneville, *Dissipative Structures and Weak Turbulence*. Academic Press, Boston, 1990.
22. S. Kai, *Physics of Pattern Formation in Complex Dissipative Systems*. World Scientific, Singapore, 1992.
23. A.C. Newell and J.V. Moloney, *Nonlinear Optics*. Addison-Wesley, Redwood City, CA, 1992.
24. L. Lugiato, ed., *Chaos, Solitons Fractals* **4** (8–9), Spec. Issue (1994).
25. M. Seul and D. Andelman, *Science* **267**, 476 (1995).
26. B.S. Kerner and V.V. Osipov, *Autosolitons*. Kluwer Academic Publishers, Amsterdam, 1994.
27. G. Dewel, A. De Wit, S. Métens, J. Verdasca, and P. Borckmans, *Phys. Scr.* **T67**, 51 (1996).
28. M. Flicker and J. Ross, *J. Chem. Phys.* **60**, 3458 (1974).
29. J. Ross, A.P. Arkin, and S.C. Müller, *J. Phys. Chem.* **99**, 10417 (1995).
30. J. Falta, R. Imbihl, and M. Henzler, *Phys. Rev. Lett.* **64**, 1409 (1990).
31. R. Imbihl, A.E. Reynolds, and D. Kaletta, *Phys. Rev. Lett.* **67**, 275 (1991).
32. J. Guckenheimer and P. Holmes, *Nonlinear Oscillations, Dynamical Systems, and Bifurcations of Vector Fields*. Springer-Verlag, New York, 1983.

33. S.K. Scott, *Chemical Chaos.* Clarendon Press, Oxford, 1991.
34. A.C. Newell, T. Passot, and J. Lega, *Annu. Rev. Fluid Mech.* **25**, 399 (1993).
35. L. Szili and J. Tóth, *Phys. Rev. E* **48**, 183 (1993).
36. J.A. Vastano, J.E. Pearson, W. Horsthemke, and H.L. Swinney, *Phys. Lett. A* **124**, 320 (1987).
37. C. Radehaus, H. Willebrand, R. Dohmen, F.-J. Niedernostheide, G. Bengel, and H.-G. Purwins, *Phys. Rev. A* **45**, 2546 (1992).
38. F.-J. Niedernostheide, M. Ardes, M. Or-Guil, and H.-G. Purwins, *Phys. Rev. B* **49**, 7370 (1994).
39. K. Krishan, *Nature (London)* **287**, 420 (1980).
40. D. Walgraef and N.M. Ghoniem, eds., *Patterns, Defects and Material Instabilities.* Kluwer Academic Publishers, Dordrecht, the Netherlands, 1990.
41. V.J. Emelyanov, *Laser Phys.* **2**, 389 (1992).
42. R.J. Field and M. Burger, *Oscillations and Traveling Waves in Chemical Systems.* Wiley, New York, 1985.
43. J. Ross, S.C. Müller, and C. Vidal, *Science* **240**, 460 (1988).
44. Z. Noszticzius, W. Horsthemke, W.D. McCormick, H.L. Swinney, and W.Y. Tam, *Nature (London)* **329**, 619 (1987).
45. W.Y. Tam, J.A. Vastano, H.L. Swinney, and W. Horsthemke, *Phys. Rev. Lett.* **61**, 2163 (1988).
46. Q. Ouyang, J. Boissonade, J.C. Roux, and P. De Kepper, *Phys. Lett. A* **134**, 282 (1989).
47. V. Castets, E. Dulos, J. Boissonade, and P. De Kepper, *Phys. Rev. Lett.* **64**, 2953 (1990).
48. J. Boissonade, E. Dulos, and P. De Kepper, in *Chemical Waves and Patterns* (R. Kapral and K. Showalter, eds.), p. 221. Kluwer Academic Publishers, Dordrecht, the Netherlands, 1995.
49. P. De Kepper, J. Boissonade, and I.R. Epstein, *J. Phys. Chem.* **94**, 6525 (1990).
50. I. Lengyel, G. Rábai, and I.R. Epstein, *J. Am. Chem. Soc.* **112**, 4606 (1990).
51. A. Cesáro, J.C. Benegas, and D. R. Ripoll, *J. Phys. Chem.* **90**, 2787 (1986).
52. J. Boissonade, *J. Phys. (Paris)* **49**, 541 (1988).
53. Q. Ouyang and H.L. Swinney, *Nature (London)* **352**, 610 (1991).
54. I. Lengyel, G. Rábai, and I.R. Epstein, *J. Am. Chem. Soc.* **112**, 9104 (1990).
55. I. Lengyel and I.R. Epstein, in *Chemical Waves and Patterns* (R. Kapral and K. Showalter, eds.), p. 297. Kluwer Academic Publishers, Dordrecht, the Netherlands, 1995.
56. I.R. Epstein, I. Lengyel, S. Kádár, M. Kagan, and M. Yokoyama, *Physica A* **188**, 26 (1992).
57. I. Lengyel and I.R. Epstein, *Science* **251**, 650 (1991).
58. I. Lengyel and I.R. Epstein, *Proc. Natl. Acad. Sci. U.S.A.* **89**, 3977 (1992).
59. A. Hunding and P.G. Sœrensen, *J. Math. Biol.* **26**, 27 (1988).
60. J.E. Pearson, *Physica A* **188**, 178 (1992).
61. J.E. Pearson and W.J. Bruno, *Chaos* **2**, 513 (1992).
62. Z. Noszticzius, Q. Ouyang, W.D. McCormick, and H.L. Swinney, *J. Phys. Chem.* **96**, 6302 (1992).
63. K. Agladze, E. Dulos, and P. De Kepper, *J. Phys. Chem.* **96**, 2400 (1992).
64. J.-J. Perraud, K. Agladze, E. Dulos, and P. De Kepper, *Physica A* **188**, 1 (1992).
65. K.J. Lee, W.D. McCormick, H.L. Swinney, and Z. Noszticzius, *J. Chem. Phys.* **96**, 4048 (1992).
66. Q. Ouyang, R. Li, G. Li, and H.L. Swinney, *J. Chem. Phys.* **102**, 2551 (1995).

67. Q. Ouyang and H.L. Swinney, *Chaos* **1**, 411 (1991).

68. Q. Ouyang, Z. Noszticzius, and H.L. Swinney, *J. Phys. Chem.* **96**, 6773 (1992).

69. B. Rudovics, E. Dulos, and P. De Kepper, *Phys. Scr.* **T67**, 43 (1996).

70. E. Dulos, P. Davies, B. Rudovics, and P. De Kepper, *Physica D* **98**, 53 (1996).

71. B. Rudovics, Ph.D. Thesis, Bordeaux University, 1995.

72. P. De Kepper, V. Castets, E. Dulos, and J. Boissonade, *Physica D* **49**, 161 (1991).

73. L.M. Pismen, *J. Chem. Phys.* **72**, 1900 (1980).

74. H. Haken and H. Olbrich, *J. Math. Biol.* **6**, 317 (1978).

75. D. Walgraef, G. Dewel, and P. Borckmans, *Adv. Chem. Phys.* **49**, 311 (1982).

76. S. Ciliberto, P. Coullet, J. Lega, E. Pampaloni, and C. Perez-Garcia, *Phys. Rev. Lett.* **65**, 2370 (1990).

77. F.H. Busse, *J. Fluid Mech.* **30**, 625 (1967).

78. L.M. Pismen, in *Dynamics of Nonlinear Systems* (V. Hlavacek, ed.), p. 47. Gordon & Breach, New York, 1986.

79. P. Borckmans, G. Dewel, A. De Wit, and D. Walgraef, in *Chemical Waves and Patterns* (R. Kapral and K. Showalter, eds.), p. 323. Kluwer Academic Publishers, Dordrecht, the Netherlands, 1995.

80. P.W. Davies, P. Blanchedeau, E. Dulos, and P. De Kepper, *J. Phys. Chem. A* **102**, 8236 (1998).

81. J. Boissonade, V. Castets, E. Dulos, and P. De Kepper, *Int. J. Numer. Math.* **97**, 67 (1991).

82. T.C. Lacalli, D.A. Wilkinson, and L.C. Harrison, *Development (Cambridge, UK)* **104**, 105 (1988).

83. P. Borckmans, G. Dewel, and A. De Wit, *Rev. Cytol. Biol. Vég.—Bot.* **14**, 209 (1991).

84. J. Verdasca, A. De Wit, G. Dewel, and P. Borckmans, *Phys. Lett. A* **168**, 194 (1992).

85. P. Borckmans, A. De Wit, and G. Dewel, *Physica A* **188**, 137 (1992).

86. V. Dufiet and J. Boissonade, *J. Chem. Phys.* **96**, 664 (1992).

87. V. Dufiet and J. Boissonade, *Physica A* **188**, 158 (1992).

88. O. Jensen, V.O. Pannbacker, G. Dewel, and P. Borckmans, *Phys. Lett. A* **179**, 91 (1993).

89. Q. Ouyang, G.H. Gunaratne, and H.L. Swinney, *Chaos* **3**, 707 (1993).

90. G.H. Gunaratne, Q. Ouyang, and H.L. Swinney *Phys. Rev. E* **50**, 2802 (1994).

91. Q. Ouyang and H.L. Swinney, in *Chemical Waves and Patterns* (R. Kapral and K. Showalter, eds.), p. 269. Kluwer Academic Publishers, Dordrecht, the Netherlands, 1995.

92. P. De Kepper, J.-J. Perraud, B. Rudovics, and E. Dulos, *Int. J. Bifurcation Chaos* **4**, 1215 (1994).

93. M. Watzl and A.F. Münster, *J. Phys. Chem. A* **102**, 2540 (1998).

94. R.D. Vigil, Q. Ouyang, and H.L. Swinney, *Physica A* **188**, 17 (1992).

95. I. Lengyel, S. Kádár, and I.R. Epstein, *Phys. Rev. Lett.* **69**, 2729 (1992).

96. M. Herschkowitz-Kaufman and G. Nicolis, *J. Chem. Phys.* **56**, 1890 (1972).

97. J.F. Auchmuty and G. Nicolis, *Bull. Math. Biol.* **37**, 323 (1975).

98. G. Dewel, P. Borckmans, A. De Wit, B. Rudovics, J.-J. Perraud, E. Dulos, J. Boissonade, and P. De Kepper, *Physica A* **213**, 181 (1995).

99. M'F. Hilali, S. Métens, G. Dewel, and P. Borckmans, *Phys. Rev. E* **51**, 2046 (1995).

100. O. Jensen, E. Mosekilde, P. Borckmans, and G. Dewel, *Phys. Scr.* **53**, 243 (1996).

101. M'F. Hilali, Ph.D. Thesis, Brussels University, 1997.

102. A. De Wit, P. Borckmans, and G. Dewel, in *Instabilities and Nonequilibrium Structures IV* (E. Tirapegui and W. Zeller, eds.), p. 247. Kluwer Academic Publishers, Dordrecht, the Netherlands, 1993.

102a. E. Barillot, PhD Thesis (Bordeaux University, 1996).

103. A. Hunding and M Brøns, *Physica D* **44**, 285 (1990).

104. V. Dufiet and J. Boissonade, *Phys. Rev. E* **53**, 4883 (1996).

105. M. Bestehorn, *Phys. Rev. E* **53**, 4842 (1996).

106. P.M. Eagles, *Proc. R. Soc. London, Ser. A* **371**, 359 (1980).

107. L. Kramer, E. Ben-Jacob, H. Brand, and M.C. Cross, *Phys. Rev. Lett.* **49**, 1891 (1982).

108. Y. Pomeau and S. Zaleski, *J. Phys. Lett. (Orsay, Fr.)* **44**, L135 (1983).

109. I.C. Walton, *J. Fluid Mech.* **131**, 455 (1983).

110. M.A. Dominguez-Lerma, D.S. Cannell, and G. Ahlers, *Phys. Rev. A* **34**, 4956 (1986).

111. H. Riecke and H.-G. Paap, *Phys. Rev. Lett.* **59**, 2570 (1987).

112. G. Dewel and P. Borckmans, *Phys. Lett. A* **138**, 189 (1989).

113. B.A. Malomed and A.A. Nepomnyashchy, *Europhys. Lett.* **21**, 195 (1993).

114. J. da Rocha Miranda Pontes, Ph.D. Thesis, Brussels University, 1994.

115. R.B. Hoyle, *Phys. Rev. E* **51**, 310 (1995).

116. R.J. Wiener, G.L. Snyder, M.P. Prange, D. Frediani, and P.R. Diaz, *Phys. Rev. E* **55**, 5489 (1997).

117. A. De Wit, P. Borckmans, and G. Dewel, *Proc. Natl. Acad. Sci. U.S.A.* **94**, 12765 (1997).

118. J.-J. Perraud, A. De Wit, E. Dulos, P. De Kepper, G. Dewel, and P. Borckmans, *Phys. Rev. Lett.* **71**, 1272 (1993).

119. A. De Wit, D. Lima, G. Dewel, and P. Borckmans, *Phys. Rev. E* **54**, 261 (1996).

120. O. Jensen, V.O. Pannbacker, E. Mosekilde, G. Dewel, and P. Borckmans, *Phys. Rev. E* **50**, 736 (1994).

121. O. Jensen, V.O. Pannbacker, E. Mosekilde, G. Dewel, and P. Borckmans, *Open Syst. Inf. Dyn.* **3**, 215 (1995).

122. I.R. Epstein and I. Lengyel, *Physica D* **84**, 1 (1995).

123. I. Lengyel, S. Kádár, and I.R. Epstein, *Science* **259**, 493 (1993).

124. M. Watzl and A.F. Münster, *Chem. Phys. Lett.* **242**, 273 (1995).

125. A.F. Münster, M. Watzl, and F.W. Schneider, *Phys. Scr.* **T67**, 58 (1996).

126. M. Burger and R.J. Field, *Nature (London)* **307**, 720 (1984).

127. P. Resch, A.F. Münster, and F.W. Schneider, *J. Phys. Chem.* **95**, 723 (1991).

128. Y.X. Zhang and R.J. Field, *J. Phys. Chem.* **95**, 6270 (1991).

129. K.J. Lee, W.D. McCormick, Q. Ouyang, and H.L. Swinney, *Science* **261**, 192 (1993).

130. K.J. Lee and H.L. Swinney, *Phys. Rev. E* **51**, 1899 (1995).

131. G. Li, Q. Ouyang and H.L. Swinney, *J. Chem. Phys.* **105**, 10830 (1996).

132. E.C. Edblom, M. Orbán, and I.R. Epstein, *J. Am. Chem. Soc.* **108**, 2826 (1986).

133. E.C. Edblom, L. Gyorgyi, and I.R. Epstein, *J. Am. Chem. Soc.* **109**, 4876 (1987).

134. V. Gáspár and K. Showalter, *J. Am. Chem. Soc.* **109**, 4869 (1987).

135. V. Gáspár and K. Showalter, *J. Phys. Chem.* **94**, 4973 (1990).

136. K.J. Lee, W.D. McCormick, J. Pearson and H.L. Swinney, *Nature (London)* **369**, 215 (1994).

137. K.J. Lee and H.L. Swinney, *Int. J. Bifurcation Chaos* **7**, 1149 (1997).

138. D. Haim, G. Li, Q. Ouyang, W.D. McCormick, H.L. Swinney, A. Hagberg, and E. Meron, *Phys. Rev. Lett.* **77**, 190 (1996).
139. J. Swift and P.C. Hohenberg, *Phys. Rev. A* **15**, 319 (1977).
140. F. Hynne and P.G. Sœrensen, *Phys. Rev. E* **40**, 4106 (1993).
141. B. Dionne, M. Silber, and A.C. Skeldon, *Nonlinearity* **10**, 321 (1997).
142. B.A. Malomed, A.A. Nepomnyashchy, and M.I. Tribelsky, *Sov. Phys.—JETP (Engl. Transl.)* **69**, 388 (1989).
143. B. Christiansen, P. Alstrøm, and M.T. Levinsen, *Phys. Rev. Lett.* **68**, 2157 (1992).
144. W.S. Edwards and S. Fauve, *Phys. Rev. E* **47**, R788 (1993).
145. E. Pampaloni, P.L. Ramazza, S. Residori, and F.T. Arecchi, *Phys. Rev. Lett.* **74**, 258 (1995).
146. B.A. Malomed and M.I. Tribelsky, *Sov. Phys.—JETP (Engl. Transl.)* **65**, 305 (1987).
147. G. Nicolis, T. Erneux, and M. Herschkowitz-Kaufman, *Adv. Chem. Phys.* **38**, 263 (1978).
148. E. Mosekilde, O. Jensen, G. Dewel, and P. Borckmans, *Syst. Anal. Model. Simul.* **18–19**, 45 (1995).
149. E. Mosekilde, F. Larsen, G. Dewel, and P. Borckmans, *Int J. Bifurcation Chaos* **8**, 1003 (1998).
150. J. Schnackenberg, *J. Theor. Biol.* **81**, 389 (1979).
151. S. Métens, Ph.D. Thesis, Brussels University, 1998.
152. M'F. Hilali, S. Métens, G. Dewel, and P. Borckmans, in *Instabilities and Nonequilibrium Structures V* (E. Tirapegui and W. Zeller, eds.). Kluwer Academic Publishers, Dordrecht, the Netherlands, 1996.
153. I. Prigogine and R. Lefever, *J. Chem. Phys.* **48**, 1695 (1968).
154. Y. Kuramoto and T. Tsuzuki, *Prog. Theor. Phys.* **52**, 1399 (1974).
155. H. Kidachi, *Prog. Theor. Phys.* **63**, 1152 (1980).
156. R. FitzHugh, *Biophys. J.* **1**, 445 (1961).
157. J. Nagumo, S. Arimoto, and S. Yoshizawa, *Proc. IRE* **50**, 2061 (1962).
158. P. Gray and S.K. Scott, *Chem. Eng. Sci.* **38**, 29 (1983).
159. K. Kumar and K.M.S. Bajaj, *Phys. Rev. E* **52**, R4606 (1995).
160. C. Bizon, M.D. Shattuck, J.B. Swift, W.D. McCormick, and H.L. Swinney, *Phys. Rev. Lett.* **80**, 57 (1998).
161. F. Melo, P.B. Umbanhowar, and H.L. Swinney, *Phys. Rev. Lett.* **75**, 3838 (1995).
162. W.J. Firth and A. J. Scroggie, *Europhys. Lett.* **26**, 521 (1994).
163. G. D'Alessandro and W.J. Firth, *Phys. Rev. A* **46**, 537 (1992).
164. T. Ackemann, Y.A. Logvin, A. Heuer, and W. Lange, *Phys. Rev. Lett.* **75**, 3450 (1995).
165. Y. Astrov, E. Ammelt, S. Teperick, and H.-G. Purwins, *Phys. Lett. A* **211**, 184 (1996).
166. E. Ammelt, Y. Astrov, and H.-G. Purwins, *Phys. Rev. E* **55**, 6731 (1997).
167. G. Dee and J.S. Langer, *Phys. Rev. Lett.* **50**, 383 (1983).
168. Y. Pomeau, *Physica D* **23**, 3 (1986).
169. D. Bensimon, B.I. Shraiman, and V. Croquette, *Phys. Rev. A* **38**, 5461 (1988).
170. B.A. Malomed, A.A. Nepomnyashchy, and M.I. Tribelsky, *Phys. Rev. A* **42**, 7244 (1990).
171. E. Ben-Jacob, H. Brand, G Dee, L. Kramer, and J.S. Langer, *Physica D* **14**, 348 (1985).
172. W. van Saarloos, *Phys. Rev. A* **37**, 211 (1988).
173. W. van Saarloos, *Phys. Rev. A* **39**, 6367 (1989).

174. S. Koga and Y. Kuramoto, *Prog. Theor. Phys.* **63**, 106 (1980).

175. O. Thual and S. Fauve, *J. Phys.* (*Paris*) **49**, 182 (1988).

176. W. van Saarloos and P.C. Hohenberg, *Physica D* **56**, 303 (1992).

177. V. Hakim, P. Jakobsen, and Y. Pomeau, *Europhys. Lett.* **11**, 19 (1990).

178. G. Dewel and P. Borckmans, *Europhys. Lett.* **17**, 523 (1992).

179. P. Coullet and K. Emilsson, *Physica A* **188**, 190 (1992).

180. P. Coullet and K. Emilsson, *Physica D* **61**, 119 (1992).

181. A.V. Gaponov-Grekhov, A.S. Lomov, G.V. Osipov, and M.I Rabinovich, in *Nonlinear Waves 1: Dynamics and Evolution* (A.V. Gaponov-Grekhov, M.I. Rabinovich, and J. Engelbrecht, eds.), *Res. Rep. Phys.*, p. 65. Springer-Verlag, Berlin, 1989.

182. H. Herrero, C. Pérez-Garcia, and M. Bestehorn, *Chaos* **4**, 15 (1994).

183. M.D. Graham, I.G. Kevrekidis, K. Asakura, J. Lauterbach, K. Krischer, H.-H. Rotermund, and G. Ertl, *Science* **264**, 80 (1994).

184. C. Varea, J.L. Aragón, and R.A. Barrio, *Phys. Rev. E* **56**, 1250 (1997).

185. B. von Haeften, G. Izús, R. Deza, and C. Borzi, *Physica A* **236**, 403 (1997).

186. B. Hasslacher, R. Kapral, and A. Lawniczak, *Chaos* **3**, 7 (1993).

187. S. Kondo and R. Asai, *Nature* (*London*) **376**, 765 (1995).

188. G. Izús, R. Deza, C. Borzi, and H. Wio, *Physica A* **237**, 135 (1997).

189. J.D. Crawford and E. Knobloch, *Annu. Rev. Fluid Mech.* **23**, 341 (1991).

190. V. Pérez-Munuzuri, M. Gómez-Gesteira, A.P. Munuzuri, L.O. Chua, and V. Pérez-Villar, *Physica D* **82**, 195 (1995).

191. B.A. Malomed, A.A. Nepomnyashchy, and A.E. Nuz, *Physica D* **70**, 357 (1994).

192. R. Kapral, A.T. Lawniczak, and P. Masiar, *Phys. Rev. Lett.* **66**, 2539 (1991).

193. R. Kapral, *Physica D* **86**, 149 (1995).

194. D. Dab, J.-P. Boon and Y.-X. Li, *Phys. Rev. Lett.* **66**, 2535 (1991).

195. S. Ponce Dawson, S. Chen, and G.D. Doolen, *J. Chem. Phys.* **98**, 1514 (1993).

196. J.R. Weimar and J.-P. Boon, *Physica A* **224**, 207 (1996).

197. A.C. Newell, T. Passot, C. Bowman, N. Ercolani, and R. Indik, *Physica D* **97**, 185 (1996).

198. C. Bowman and A.C. Newell, *Rev. Mod. Phys.* **70**, 289 (1998).

199. L.M. Pismen and A.A. Nepomnyashchy, *Europhys. Lett.* **24**, 461 (1993).

200. M.I. Rabinovich and L.S. Tsimring, *Phys. Rev. E* **49**, R35 (1993).

201. S. Rica and E. Tirapegui, *Phys. Rev. Lett.* **64**, 878 (1990).

202. L.S. Tsimring and M.I. Rabinovich, *Spatio-Temporal Pattern Nonequilib. Complex Syst., Proc. NATO-ARW*, Santa Fe, 1993.

203. L.M. Pismen and A.A. Nepomnyashchy, *Europhys. Lett.* **27**, 433 (1994).

204. S. Ciliberto, E. Pampaloni, and C. Pérez-Garcia, *J. Stat. Phys.* **64**, 1045 (1991).

205. A.C. Newell and J.A. Whitehead, *J. Fluid Mech.* **38**, 279 (1969).

206. L.A. Segel, *J. Fluid Mech.* **38**, 203 (1969).

207. F.H. Busse, *Rep. Prog. Phys.* **41**, 1929 (1978).

208. H. Sakaguchi, *Prog. Theor. Phys.* **86**, 759 (1991).

209. S. Sasa, *Prog. Theor. Phys.* **84**, 1009 (1990).

210. W. Eckhaus, *Studies in Nonlinear Stability Theory.* Springer, Berlin, 1965.

211. J.S. Stuart and R.C. Di Prima, *Proc. R. Soc. London, Ser. A* **362**, 27 (1978).

212. J. Lauzeral, S. Métens, and D. Walgraef, *Europhys. Lett.* **24**, 707 (1993).
213. M.M. Sushchik and L.S. Tsimring, *Physica D* **74**, 90 (1994).
214. M. Bestehorn, *Intern. J. Bifurcation Chaos* **4**, 1085 (1994).
215. G.H. Gunaratne, *Phys. Rev. Lett.* **71**, 1367 (1993).
216. R. Graham, *Phys. Rev. Lett.* **76**, 2185 (1996); **80**, 3887, 3888 (1998).
217. K. Matsuba and K. Nozaki, *Phys. Rev. Lett.* **80**, 3886 (1998).
218. M.C. Cross and A.C. Newell, *Physica D* **10**, 299 (1984).
219. M. Bestehorn and H. Haken, *Phys. Rev. A* **42**, 7195 (1990).
220. H.R. Brand, *Prog. Theor. Phys. Suppl.* **99**, 442 (1989).
221. H. Sakaguchi and H.R. Brand, *Phys. Lett. A* **227**, 209 (1997).
222. A. De Wit, G. Dewel, P. Borckmans, and D. Walgraef, *Physica D* **61**, 289 (1992).
223. C. Kittel, *Introduction to Solid State Physics*, 6th ed. Wiley, New York, 1986.
224. S. Alexander and J. McTague, *Phys. Rev. Lett.* **41**, 702 (1978).
225. G. Dewel, S. Métens, M'F. Hilali, P. Borckmans, and C.B. Price, *Phys. Rev. Lett.* **74**, 4647 (1995).
226. S. Sakurai, *Trends Polym. Sci.* **3**, 90 (1995).
227. T.K. Callahan and E. Knobloch, *Phys. Rev. E* **53**, 3559 (1996).
228. B. Dionne and M. Golubitsky, *Z. Angew. Math. Phys.* **43**, 36 (1992).
229. B. Dionne, *Z. Angew. Math. Phys.* **44**, 673 (1993).
230. T.K. Callahan and E. Knobloch, *Nonlinearity* **10**, 1179 (1997).
231. T.K. Callahan and E. Knobloch, *Physica D* (submitted for publication) (1998).
232. M. Olvera de la Cruz, *Phys. Rev. Lett.* **67**, 85 (1991).
233. C.M. Marques and M.E. Cates, *Europhys. Lett.* **13**, 267 (1990).
234. S. Chandrasekhar, *Liquid Crystals*. Cambridge University Press, Cambridge, UK, 1992.
235. J. Toner and D.R. Nelson, *Phys. Rev. B* **23**, 216 (1981).
236. E. Guazelli, E. Guyon, and J.E. Wesfreid, *Philos. Mag.* [8] **48**, 709 (1983).
237. S.R. Renn and T.C. Lubensky, *Phys. Rev. A* **38**, 2132 (1988).
238. E.L. Thomas, D.M. Anderson, C.S. Henkee, and D. Hoffman, *Nature* (*London*) **334**, 598 (1988).
239. L.M. Pismen, *Phys. Rev. E* **50**, 4896 (1994).
240. Y. Shiwa, *Phys. Lett. A* **228**, 279 (1997).
241. S.P. Gido, J. Gunther, E.L. Thomas, and D. Hoffman, *Macromolecules* **27**, 4506 (1994).
242. S. Andersson, S.T. Hyde, K. Larsson, and S. Lidin, *Chem. Rev.* **88**, 221 (1988).
243. A.L. Mackay, *Proc. R. Soc. London, Ser. B* **442**, 47 (1993).
244. A.L. Mackay, *Curr. Sci.* **69**, 151 (1995).
245. L. Zhang, C. Bartels, Y. Yu, H. Shen, and A. Eisenberg, *Phys. Rev. Lett.* **79**, 5034 (1997).
246. W.T. Góźdź and R. Holyst, *Phys. Rev. E* **54**, 5012 (1996).
247. S. Qi and Z.-G. Wang, *Phys. Rev. E* **55**, 1682 (1997).
248. M. Laradji, A.-C. Shi, R.C. Desai, and J. Noolandi, *Phys. Rev. Lett.* **78**, 2577 (1997).
249. D. Roux, *Physica A* **213**, 168 (1995).
250. A. Malevanets and R. Kapral, *Phys. Rev. E* **55**, 5657 (1997).

251. J. Guckenheimer, in *Dynamical Systems and Turbulence* (D.A. Rand and L.-S. Young, eds.). Springer-Verlag, Berlin, 1981.
252. P. Holmes, *Ann. N.Y. Acad. Sci.* **357**, 473 (1980).
253. M. Golubitsky, J.W. Swift, and E. Knobloch, *Physica D* **10**, 249 (1984).
254. H.R. Brand, P.C. Hohenberg, and V. Steinberg, *Phys. Rev. A* **30**, 2548 (1984).
255. W. Zimmermann, D. Armbruster, L. Kramer, and W. Kuang, *Europhys. Lett.* **6**, 505 (1988).
256. R. Becerril and J.B. Swift, *Phys. Rev. E* **55**, 6270 (1997).
257. A. Arnéodo and O. Thual, *Le chaos* Synthèses Commissariat à l'Energie Atomique - Eyrolles, France, 1988.
258. P. Colinet, P. Géoris, J.C. Legros, and G. Lebon, *Phys. Rev. E* **54**, 514 (1996).
259. I. Rehberg and G. Ahlers, *Phys. Rev. Lett.* **55**, 500 (1985).
260. T. Mullin and T.J. Price, *Nature (London)* **340**, 294 (1988).
261. B.J.A. Zielinska, D. Mukamel, and V. Steinberg, *Phys. Rev. A* **33**, 1454 (1986).
262. P. Kolodner, *Phys. Rev. E* **48**, R665 (1993).
263. D.P. Vallette, W.S. Edwards, and J.P. Gollub, *Phys. Rev. E* **49**, R4783 (1994).
264. D.P. Vallette, G. Jacobs, and J.P. Gollub, *Phys. Rev. E* **55**, 4274 (1997).
265. G. Heidemann, M. Bode, and H.-G. Purwins, *Phys. Lett. A* **177**, 225 (1993).
266. A. Wacker, S. Bose, and E. Schöll, *Europhys. Lett.* **31**, 257 (1995).
267. M. Tlidi, P. Mandel, and M. Haelterman, *Phys. Rev. E* **56**, 6524 (1997).
268. M. Tlidi and M. Haelterman, *Phys. Lett. A* **239**, 59 (1998).
269. J.P. Keener, *Stud. Appl. Math.* **55**, 187 (1976).
270. M. Herschkowitz-Kaufman and T. Erneux, *Ann. N.Y. Acad. Sci.* **316**, 296 (1979).
271. A. De Wit, G. Dewel, and P. Borckmans, *Phys. Rev. E* **48**, R4191 (1993).
272. Y. Kuramoto, *Prog. Theor. Phys., Suppl.* **64**, 346 (1978).
273. G.I. Sivashinsky, *Acta Astron.* **6**, 569 (1979).
274. Y. Kuramoto, *Chemical Oscillations, Waves and Turbulence*. Springer, Tokyo, 1984.
275. H. Sakaguchi, *Prog. Theor. Phys.* **84**, 792 (1990).
276. P. Coullet, L. Gil, and J. Lega, *Phys. Rev. Lett.* **62**, 1619 (1989).
277. B.I. Shraiman, A. Pumir, W. van Saarloos, P.C. Hohenberg, H. Chaté, and M. Holen, *Physica D* **57**, 241 (1992).
278. M. Meixner, A. De Wit, S. Bose, and E. Schöll, *Phys. Rev. E* **55**, 6690 (1997).
279. M. Sangalli and H.-C. Chang, *Phys. Rev. E* **49**, 5207 (1994).
280. A. Rovinsky and M. Menzinger, *Phys. Rev. A* **46**, 6315 (1992).
281. H. Willebrand, F.-J. Niedernostheide, R. Dohmen, and H.-G. Purwins, in *Oscillations and Morphogenesis* (L. Rensing, ed.), p. 81. Dekker, New York, 1993.
282. P. Strasser, O.E. Rössler, and G. Baier, *J. Chem. Phys.* **104**, 9974 (1996).
283. P. Bak, T. Bohr, and M.H. Jensen, *Phys. Scr.* **T9**, 50 (1985).
283a. M. Or-Guil and M. Bode, *Physica A* **249**, 174 (1998).
284. G. Ahlers, *Physica D* **51**, 421 (1991).
285. U. Middya, M. Sheintuch, M.D. Graham, and D. Luss, *Physica D* **63**, 393 (1993).
286. H. Willebrand, T. Hünteler, F.-J. Niedernostheide, R. Dohmen, and H.-G. Purwins, *Phys. Rev. A* **45**, 8766 (1992).
287. H.H. Rotermund, S. Jakubith, A. von Oertzen, and G. Ertl, *Phys. Rev. Lett.* **66**, 3083 (1991).

288. F. Barra, O. Descalzi, and E. Tirapegui, *Phys. Lett. A* **221**, 193 (1996).

289. A. Hill and I. Stewart, *Dyn. Stab. Syst.* **6**, 149 (1991).

290. K. Fujimura and Y. Renardy, *Physica D* **85**, 25 (1995).

291. D. Lima, A. De Wit, G. Dewel, and P. Borckmans, *Phys. Rev. E* **53**, R1305 (1996).

292. M. Cheng and H.-C. Chang, *Phys. Fluids A* **4**, 505 (1992).

293. K.E. Rasmussen, W. Mazin, E. Mosekilde, G. Dewel, and P. Borckmans, *Int. J. Bifurcation Chaos* **6**, 1077 (1996).

294. V. Petrov, S. Métens, P. Borckmans, G. Dewel, and K. Showalter, *Phys. Rev. Lett.* **75**, 2895 (1995).

295. M. Meixner, S. Bose, and E. Schöll, *Physica D* **109**, 128 (1997).

296. D. Walgraef, "Lecture Notes delivered at the Latin American School on Complex Systems," in *Centro Latinoamericano de Estudios*, I. Prigogine, ed., San Luis, Argentina, 1998.

297. Y.A. Logvin, T. Ackemann, and W. Lange, *Phys. Rev. A* **55**, 4538 (1997).

298. M. Hoyuelos, P. Colet, M. San Miguel, and D. Walgraef, *Phys. Rev. E.* **58**, 2992 (1998).

299. Y.A. Logvin, T. Ackemann, and W. Lange, *Europhys. Lett.* **38**, 583 (1997).

300. Y.A. Logvin, B.A. Samson, A.A. Afanas'ev, A.M. Samson, and N.A. Loiko, *Phys. Rev. E* **54**, R4548 (1996).

301. Y.A. Logvin and N.A. Loiko, *Phys. Rev. E* **56**, 3803 (1997).

302. B.Y. Rubinstein and L.M. Pismen, *Phys. Rev. A* **56**, 4264 (1997).

303. L.M. Pismen and B.Y. Rubinstein, preprint (1998).

304. W. Breazeal, K.M. Flynn, and E.G. Gwinn, *Phys. Rev. E* **52**, 1503 (1995).

305. G. Ahlers, L.I. Berge, and D.S. Cannell, *Phys. Rev. Lett.* **70**, 2399 (1993).

306. W.J. Firth, A.J. Scroggie, and G.S. McDonald, *Phys. Rev. A* **46**, 3609 (1992).

307. S. Métens, G. Dewel, P. Borckmans, and R. Engelhardt, *Europhys. Lett.* **37**, 109 (1997).

308. S. Métens, P. Borckmans, and G. Dewel, in *Instabilities and Non-equilibrium Structures VI* (E. Tirapegui and W. Zeller, eds.). Kluwer Academic Publishers, Dordrecht, the Netherlands, 1998.

309. W. Mazin, K.E. Rasmussen, E. Mosekilde, P. Borckmans, and G. Dewel, *Math. Comput. Simul.* **40**, 371 (1996).

310. M'F. Hilali, G. Dewel, and P. Borckmans, *Phys. Lett. A* **217**, 263 (1996).

311. K.A. Gorshkov, L.N. Korzinov, M.I. Rabinovich, and T.S. Tsimring, *J. Stat. Phys.* **74**, 1033 (1994).

312. M. Assenheimer and V. Steinberg, *Phys. Rev. Lett.* **70**, 3888 (1993).

313. F.S. Bates and G.H. Fredrickson, *Annu. Rev. Phys. Chem.* **41**, 525 (1990).

314. C.B. Price, *Phys. Lett. A* **194**, 385 (1994).

315. D. Horvath, V. Petrov, S.K. Scott, and K. Showalter, *J. Chem. Phys.* **98**, 6332 (1993).

316. L.M. Pismen, *J. Chem. Phys.* **71**, 462 (1979).

317. P. Ortoleva and J. Ross, *J. Chem. Phys.* **63**, 3398 (1975).

318. B.S. Kerner and V.V. Osipov, *Sov. Phys.—Usp.* (*Engl. Transl.*) **32**, 101 (1989).

319. V.V. Osipov and A.V. Severtsev, *Phys. Lett. A* **227**, 61 (1997).

320. D.M. Petrich and R.E. Goldstein, *Phys. Rev. Lett.* **72**, 1120 (1994).

321. T. Ohta, M. Mimura, and R. Kobayashi, *Physica D* **34**, 115 (1989).

322. T. Ohta, A. Ito, and A. Tetsuka, *Phys. Rev. A* **42**, 3225 (1990).

323. A. Hagberg and E. Meron, *Phys. Rev. E* **48**, 705 (1993).
324. C. Elphick, A. Hagberg, and E. Meron, *Phys. Rev. E* **51**, 3052 (1995).
325. C.B. Muratov and V.V. Osipov, *Phys. Rev. E* **53**, 3101 (1996).
326. J.E. Pearson, *Science* **261**, 189 (1993).
327. C.B. Muratov, *Phys. Rev. E* **55**, 1463 (1997).
328. C.B. Muratov and V.V. Osipov, *Phys. Rev. E* **54**, 4860 (1996).
329. R.E. Goldstein, D.J. Muraki, and D.M. Petrich, *Phys. Rev. E* **53**, 3933 (1996).
330. P. Coullet, J. Lega, B. Houchmanzadeh, and J. Lajzerowich, *Phys. Rev. Lett.* **65**, 1352 (1990).
331. W.N. Reynolds, S. Ponce Dawson, and J.E. Pearson, *Phys. Rev. E* **56**, 185 (1997).
332. A. Doelman, T.J. Kaper, and P.A. Zegeling, *Nonlinearity* **10**, 523 (1997).
333. N. Parekh, V. Ravi Kumar, and B.D. Kulkarni, *Phys. Rev. E* **52**, 5100 (1995).
334. A. Pacault and J.-J. Perraud, *Rythmes et formes en chimie*, (Que sais-je, Presses Universitaires de France, 1997).
335. A. Goldbeter, *Biochemical Oscillations and Cellular Rythms*. Cambridge University Press, New York, 1997.
336. S.A. Giannos and S.M. Dinh, *Polym. News* **21**, 118 (1996).
337. R. Yoshida, H. Ichijo, T. Hakuta, and T. Yamaguchi, *Macromol. Rapid Commun.* **16**, 305 (1995).
338. A. Hjelmfelt, F.W. Schneider, and J. Ross, *Science* **260**, 335 (1993).
339. A.C. Newell and Y. Pomeau, *J. Phys. A: Math. Gen.* **26**, L429 (1993).
340. S. Residori, P.L. Ramazza, E. Pampaloni, S. Boccaletti, and F.T. Arecchi, *Phys. Rev. Lett.* **76**, 1063 (1996).
341. M.A. Vorontsov and B.A. Samson, *Phys. Rev. A* **57**, 3040 (1998).
342. M.I. Tribelsky and M. Velarde, *Phys. Rev. E* **54**, 4973 (1996).
343. I.L. Kliakhandler and B.A. Malomed, *Phys. Lett. A* **231**, 191 (1997).
344. H. Levine and X. Zou, *Phys. Rev. E* **48**, 50 (1993).
345. A.M. Zhabotinsky, M. Dolnik, and I.R. Epstein, *J. Chem. Phys.* **103**, 10306 (1995).
346. A.B. Rovinsky, A.M. Zhabotinsky, and I.R. Epstein, *Phys. Rev. E* **56**, 2412 (1997).
347. M. Silber and E. Knobloch, *Nonlinearity* **4**, 1063 (1991).
348. H.R. Brand and R.J. Deissler, *Phys. Lett. A* **231**, 179 (1997).
349. J.-P. Voroney, A.T. Lawniczak, and R. Kapral, *Physica D* **99**, 303 (1996).
350. I. Bose and I. Chaudhuri, *Phys. Rev. E* **55**, 5291 (1997).
351. O. Steinbock, P. Kettunen, and K. Showalter, *Science* **269**, 1857 (1995).
352. G. Hartung, F.H. Busse, and I. Rehberg, *Phys. Rev. Lett.* **66**, 2741 (1991).
353. W. Zimmermann and R. Schmitz, *Phys. Rev. E* **53**, R1321 (1996).
354. H. Malchow, H. Rosé, and C. Sattler, *J. Non-Equilib. Thermodyn.* **17**, 41 (1992).
355. A.F. Münster, P. Hasal, D. Snita, and M. Marek, *Phys. Rev. E* **50**, 546 (1994).
356. J. Mosquera, M. Gomez-Gesteira, V. Perez-Munuzuri, A.P. Munuzuri, and V. Perez-Villar, *Int. J. Bifurcation Chaos* **5**, 797 (1995).
357. F. Mertens, N. Gottschalk, M. Bär, M. Eiswirth, and A. Mikhailov, *Phys. Rev. E* **51**, R5193 (1995).
358. P. Borckmans, G. Dewel, D. Walgraef, and Y. Katayama, *J. Stat. Phys.* **48**, 1031 (1987).
359. J.-C. Micheau, M. Gimenez, P. Borckmans, and G. Dewel, *Nature (London)* **305**, 43 (1983).

360. M. Diewald and H.R. Brand, *Phys. Rev. E* **51**, R5200 (1995).
361. D.A. Vasquez, J.W. Wilder, and B.F. Edwards, *Phys. Rev. Lett.* **71**, 1538 (1993).
362. M. Marlow, Y. Sasaki, and D.A. Vasquez, *J. Chem. Phys.* **107**, 5205 (1997).
363. E.A. Spiegel and S. Zaleski, *Phys. Lett. A* **106**, 335 (1984).
364. R. Friedrich, M. Bestehorn, and H. Haken, *Int. J. Mod. Phys.* **4**, 365 (1990).
365. C.R. Doering and W. Horsthemke, *Phys. Lett. A* **182**, 227 (1993).
366. S. Ponce Dawson, A. Lawniczak, and R. Kapral, *J. Chem. Phys.* **100**, 5211 (1994).
367. G.P. Bernasconi and J. Boissonade, *Phys. Lett. A* **232**, 224 (1997).
368. S.P. Kuznetsov, E. Mosekilde, G. Dewel, and P. Borckmans, *J. Chem. Phys.* **106**, 7609 (1997).
369. L. Landau and E.M. Lifshitz, *Fluid Mechanics*. Pergamon, London, 1959.
370. R.J. Deissler, *J. Stat. Phys.* **40**, 371 (1985).
371. A.B. Rovinsky and M Menzinger, *Phys. Rev. Lett.* **69**, 1193 (1992); **70**, 778 (1993).
372. A.B. Rovinsky, A. Malevanets, and M Menzinger, *Physica D* **95**, 306 (1996).
373. R. Satnoianu, J. Merkin, and S. Scott, *Phys. Rev. E* **57**, 3246 (1998).
374. V. Petrov, Q. Ouyang, and H.L. Swinney, *Nature (London)* **388**, 655 (1997).
375. G.A. Gordonier, F. Schuth, and L.D. Schmidt, *J. Chem. Phys.* **91**, 5374 (1989).
376. G. Philippou, F. Schulz, and D. Luss, *J. Chem. Phys.* **95**, 3224 (1991).
377. O. Lev, M. Sheintuch, L.M. Pismen, and C. Yarnitzky, *Nature (London)* **336**, 458 (1988).
378. S. Jakubith, H.H. Rotermund, W. Engel, A. von Oertzen, and G. Ertl, *Phys. Rev. Lett.* **65**, 3013 (1990).
379. H. Levine and X. Zou, *Phys. Rev. Lett.* **69**, 204 (1992).
380. M. Falcke and H. Engel, *J. Chem. Phys.* **101**, 6255 (1994).
381. F. Mertens, R. Imbihl, and A. Mikhailov, *J. Chem. Phys.* **101**, 9903 (1994).
382. K.C. Rose, D. Battogtokh, A. Mikhailov, R. Imbihl, W. Engel, and A.M. Bradshaw, *Phys. Rev. Lett.* **76**, 3582 (1996).
383. M. Ehsasi, O. Frank, J.H. Block, and K. Christmann, *Chem. Phys. Lett.* **165**, 115 (1990).
384. U. Middya and D. Luss, *J. Chem. Phys.* **100**, 6386 (1994).
385. K. Krischer and A. Mikhailov, *Phys. Rev. Lett.* **73**, 3165 (1994).
386. L.M. Pismen, *J. Chem. Phys.* **101**, 3135 (1994).

AUTHOR INDEX

Numbers in parentheses are reference numbers and indicate that the author's work is referred to although his name is not mentioned in the text. Numbers in *italic* show the pages on which the complete references are listed.

SUBJECT INDEX